Fundamentals of Forensic Biology

Avinash Puri •
Nithyanandam Mahalakshmi •
Tanya Chauhan • Alka Mishra •
Preeti Bhatnagar
Editors

Fundamentals of Forensic Biology

Springer

Editors
Avinash Puri
Department of DNA and Forensic Biology
Regional Forensic Science Laboratory
Indore, Madhya Pradesh, India

Nithyanandam Mahalakshmi
Department of DNA
Forensic Sciences
Chennai, Tamil Nadu, India

Tanya Chauhan
Department of Forensic Science
National Forensic Science University
Delhi Campus (LNJN NICFS)
New Delhi, Delhi, India

Alka Mishra
Department of Biology
Regional Forensic Science Laboratory
Gwalior, Madhya Pradesh, India

Preeti Bhatnagar
Department of DNA
Regional Forensic Science Laboratory
Bhopal, Madhya Pradesh, India

ISBN 978-981-99-3160-6 ISBN 978-981-99-3161-3 (eBook)
https://doi.org/10.1007/978-981-99-3161-3

This Springer imprint is published by the registered company Springer Nature Singapore Pte Ltd.
The registered company address is: 152 Beach Road, #21-01/04 Gateway East, Singapore 189721, Singapore

Paper in this product is recyclable.

Preface

Forensic science is the application of different disciplines of science in crime scene investigation. While criminal activities diversify and become complex in the modern world, the principles of forensic science remain instrumental in assisting enforcement agencies solve crimes. Biological evidences are an integral discipline of forensic science, since humans and living beings alike are often central to commonly occurring crimes. Homicides, assault, paternity/maternity disputes, accidents, identification, trafficking, and many other crimes depend on forensic biology to find answers and achieve justice.

Forensic biology uses different principles and applications of biological sciences to help identify and analyze biological evidences from the scene of crime and otherwise. DNA profiling, a prominent subfield of forensic biology, is now a fundamental part of crime scene investigation throughout the world. It has helped deliver justice to millions and solve many cold cases and exonerate innocent people. Other disciplines of the field, such as anthropology, serology, pathology, blood stain pattern analysis, botany, entomology, wildlife forensics, and odontology help analyze a diverse range of evidences found at crime scenes.

The book entitled *Fundamentals of Forensic Biology* aims to introduce the fields of Forensic Biology to students and forensic scientists alike. The book will cover the theory, principles, and application of the mentioned disciplines of forensic biology with a focus on easier understanding of the applicability of the topics. It discusses the subject with an aim to enhance the theoretical and practical knowledge of the subject and explore the potentials of the fields in modern-day crime scene investigation for researchers and practitioners of the field.

This book will be supported through infographics, visual cues, and illustrations to make the learning process engaging for readers. This book will be supplemented with real-life case studies from national and international cases, significant to the discipline or unique in their approach towards evidence analysis. At the last of each chapter of the book will be filled contain summaries and revision sections for enhanced learning experience.

This book aims to break the monotony of higher education learning resources and help students conceptualize complex topics through detailed deconstruction of the topic and using real-life examples and visual learning tools. The book will cover a range of important subdisciplines of the field with important emphasis on understanding the fundamentals.

Indore, Madhya Pradesh, India — Avinash Puri
Chennai, Tamil Nadu, India — Nithyanandam Mahalakshmi
New Delhi, Delhi, India — Tanya Chauhan
Gwalior, Madhya Pradesh, India — Alka Mishra
Bhopal, Madhya Pradesh, India — Preeti Bhatnagar

Contents

Editors and Contributors

About the Editors

Avinash Puri is working as a Scientific Officer in Forensic Biology and DNA division of Regional Forensic Science Laboratory, Home Ministry, Govt. of Madhya Pradesh. He has more than 7 years of experience in forensic biology, DNA, and serology. He has analyzed thousands of forensic exhibits of biological origin along with material objects of the deceased such as bone, muscle tissue, teeth, etc. paternity dispute cases using DNA typing and submitted reports and deposed evidence in the Courts of Law, M.P. India. His research interest includes forensic serology, forensic biology, population genetics, forensic DNA testing, and molecular biology. He has delivered and organized lectures at various institutes in forensic field for officers of the Criminal Justice System. He has published many research articles in peer-reviewed journals. He has developed and patented software regarding chain of custody (ACP). He has also published several books.

Tanya Chauhan is working as Assistant Professor in Forensic Biology Division of LNJN National Institute of Criminology and Forensic Science, Ministry of Home Affairs, Govt. of India, at Delhi. She has more than 11 years of experience in forensic biology, serology, and DNA profiling. Her research focuses on forensic DNA and molecular biology. She has organized short-term training programs in forensic field for senior officers of the Criminal Justice System. She has published 14 research articles in peer-reviewed journals. She has also contributed to the preparation and editing of several books. She is a member of Association for the promotion of DNA Fingerprinting and other DNA Technologies (ADNAT), Bio-informatics organization, Massachusetts, and Science Advisory Board, United States.

Nithyanandam Mahalakshmi is working as a deputy director, DNA Division, Forensic Sciences Department. She has also generated a database for forensically relevant DNA markers (such as autosomal and Y-STRs and mtDNA) and HLA-DRB1 system for endogamous population of Tamil Nadu. She has published many research articles along with books.

Alka Mishra is presently working as a Scientific Officer in Regional Forensic Science Laboratory, Gwalior. She has 7 years of experience in forensic biology and published various publications in national and international journals.

Preeti Bhatnagar is presently working as a Scientific Officer in DNA Fingerprinting Unit-Regional Forensic Science Laboratory Bhopal. She has 7 years of experience in forensic biology and DNA and also served as Assistant Professor of Microbiology in various educational institutions for more than 6 years. An eminent scholar and winner of the Young Scientist Award, she has various publications in national and international journals.

Contributors

Neelam Ahirwar Department of Microbiology, Kalinga University, Raipur, India

Shreya Arora CTM IRTE, Faridabad, Haryana, India

Neelam Arya FSL, Haryana, Madhuban, Haryana, India
DNA UNIT, Regional Forensic Science Laboratory, Indore, India

Astha LNJN National Institute of Criminology and Forensic Science, Delhi, India

Alok Atreya Department of Forensic Medicine, Lumbini Medical College, Tansen, Palpa, Nepal

Paromita Banerjee Department of Microbiology, Kalinga University, Raipur, India
Department of Bioinformatics, Kalinga University, Raipur, Chhattisgarh, India

Hansi Bansal Department of Forensic Science, Government Institute of Forensic Science, Nagpur, India

Anuj Bharadwaj CFSL Chandigarh, Chandigarh, India

Preeti Bhatnagar Department of DNA, Regional Forensic Science Laboratory, Bhopal, Madhya Pradesh, India

Ajay Biswas Department of Forensic Science, Medi-Caps University, Indore, Madhya Pradesh, India

Mehar Chadha Department of Life Sciences, Christ University, Bangalore, Karnataka, India

Chitrita Chakraborty Department of Forensic Science, Adamas University, Kolkata, India

Tilak Ram Chandrakar Faculty of Science, Department of Forensic Science, Medi-Caps University, Indore, Madhya Pradesh, India
Faculty of Forensic Science, Medi-Caps University, Indore, Madhya Pradesh, India

Rudresh Channe Department of Forensic Science, Institute for Excellence in Higher Education, Bhopal, Madhya Pradesh, India

Preetika M. Chatterjee Department of Forensic Science, MATS University, Raipur, Chhattisgarh, India

Tanya Chauhan Department of Forensic Science, National Forensic Science University Delhi Campus (LNJN NICFS), New Delhi, Delhi, India

Shivam Chourasiya Faculty of Science, Department of Forensic Science, Medi-Caps University, Indore, Madhya Pradesh, India

Rahul Ravindra Darunde Department of Forensic Science, Medi-Caps University, Indore, India

Ankit Debnath Department of Forensic Science, Adamas University, Kolkata, India

Sunanda Dhenge Forensic Science Education Society, Raipur, Chhattisgarh, India

Ulhas Gondhali Jindal Institute of Behavioural Sciences, O.P. Jindal Global University, Sonipat, India

Kam Salem Guite Department of Forensic Medicine and Toxicology, All India Institute of Medical Sciences, Jodhpur, Rajasthan, India

Arjun Rao Isukapatla Faculty of Forensic Science, Department of Life Sciences, Christ University, Bangalore, Karnataka, India
Department of Microbiology, Kalinga University, Raipur, India
Faculty of Forensic Science, School of Life Sciences, Christ (Deemed to be University), Bengaluru, Karnataka, India

Aditi Iyengar Impact Science, Cactus Communications, Mumbai, India

Tanuj Kanchan Department of Forensic Medicine and Toxicology, All India Institute of Medical Sciences, Jodhpur, Rajasthan, India

Aditya Kumar Kar Department of Forensic Science, Adamas University, Kolkata, India

Nisha Kaushik Forensic Science Education Society, Raipur, Chhattisgarh, India

Kewal Krishan Department of Anthropology, Panjab University, Chandigarh, India

Prashant Kumar Department of Microbiology, Kalinga University, Raipur, India
Department of Bioinformatics, Kalinga University, Raipur, Chhattisgarh, India

Nithyanandam Mahalakshmi Department of DNA, Forensic Sciences, Chennai, Tamil Nadu, India

H. Manderiya Department of Forensic Science, Institute for Excellence in Higher Education, Bhopal, Madhya Pradesh, India

Aditi Mishra Kristu Jayanti College, Bangalore, India

Alka Mishra Department of Biology, Regional Forensic Science Laboratory, Gwalior, Madhya Pradesh, India

Sarthak Misra Indira Gandhi Institute of Medical Sciences, Patna, India

Tanima Nandi Department of Forensic Science, Adamas University, Kolkata, India

Pratiksha H. Nimbkar School of Applied Sciences and Technology, Gujarat Technological University, Ahmedabad, Gujarat, India

Ragini Pandey Department of Forensic Science, Jain University, Bangalore, India

Radha Patel Department of Forensic Science, Adamas University, Kolkata, India

Varsha Rani Patel Faculty of Science, Department of Forensic Science, Medi-Caps University, Indore, Madhya Pradesh, India

Avinash Puri Department of DNA and Forensic Biology, Regional Forensic Science Laboratory, Indore, Madhya Pradesh, India

Monisha Samuel Department of Forensic Science, National Forensic Sciences University, Delhi Campus (LNJN NICFS), New Delhi, Delhi, India

Himani Sharma Faculty of Science, Department of Forensic Science, Medi-Caps University, Indore, Madhya Pradesh, India

Parth Sharma School of Applied Sciences and Technology, Gujarat Technological University, Ahmedabad, Gujarat, India

Rutwik Shedge School of Forensic Science, National Forensic Sciences University, Gandhinagar, Gujarat, India

Moumita Sinha Department of Forensic Science, Kalinga University, Raipur, Chhattisgarh, India
Department of Microbiology, Kalinga University, Raipur, India

Pooja Uppal Department of Forensic Science, Adamas University, Kolkata, India

Varsha Warrier Department of Forensic Medicine and Toxicology, All India Institute of Medical Sciences, Jodhpur, India

Prachi Yadav Department of Forensic Science, Guru Ghasidas Vishwavidyalaya, Bilaspur, Chhattisgarh, India

Introduction: Forensic Biology

1

Avinash Puri and Neelam Arya

Abstract

Forensic science is the accumulation and application of a wide field of combined knowledge borrowed from a large number of scientific disciplines to reveal facts about ambiguous situations pertaining to unexplained occurrence that holds relevance to the legal system of society. One of the most important and nowadays necessary aspects of investigation is scientific investigation implementing forensic science or the practice of scientifically examining physical evidence collected from the scene of a crime or a person of interest in a crime. Forensic biology is the scientific analysis of biological evidence to provide objective information on legal matters. Over the years, the application of scientific methods in the investigation of crime has developed into a full-fledged field of specialization such as forensic biology.

Every crime takes place at a certain time at a certain place or places and involves a victim and the person or persons committing the crime. The perpetrator may or may not use a weapon. Depending upon the case circumstances, the biological material is transferred between the scene, the victim, the perpetrator and the weapon. Using technology, forensic biologists collect and analyse biological evidence/material found on clothing, weapons and other surfaces to determine the time and cause of death and other crucial information. Forensic biologists examine blood and other bodily fluids, hair, bones, insects and plant and animal remains to help identify the perpetrator. For example, evidence/items found in a perpetrator's possession may be linked to a victim. This also applies to

A. Puri (✉)
Department of DNA and Forensic Biology, Regional Forensic Science Laboratory, Indore, Madhya Pradesh, India

N. Arya
Forensic science laboratory, Madhuban, Haryana, India

A. Puri et al. (eds.), *Fundamentals of Forensic Biology*,
https://doi.org/10.1007/978-981-99-3161-3_1

the transfer of evidence based on the principles of exchange material theory, also known as Locard's exchange principle, which theorizes that the cross-transfer of evidence occurs when a perpetrator came in contact with the victim.

In addition, forensic biology includes a number of disciplines and a number of principles to analyse different kinds of biological evidence taking advantage of DNA sequencing and other biomolecular techniques. As a result, DNA evidence has been used to help identify perpetrators of crimes and to exonerate innocent people before they become suspects. It is clear that forensic biological technology is here to stay and will continue to help forensic investigations, though it will keep advancing at a rapid pace, necessitating constant updating of relevant information on the part of investigative agencies, forensic professionals and biologists.

Keywords

Forensic biology · Scientific investigation · Biological evidence · DNA sequencing · Crime scene · Exchange of material

Forensic biology is the scientific analysis of biological evidence to provide objective information on legal matters. Over the years, the application of scientific methods in the investigation of crime has developed into a full-fledged field of specialization such as forensic biology. The field of forensic biology deals with the examination of the evidence pertaining to living beings, and their associated biological materials, commonly found at the scene of a crime. Forensic biologists analyse physiological fluids tissue and cellular samples that are relevant to crime investigation.

By comparison of the DNA profile from crime scene samples to known samples, the results can serve to link victims and suspects with the crime scene or can rule out a suspect from association with that particular crime. Moreover, scientific analysis of biological evidence may provide unbiased information to substantiate case circumstances, corroborate or refute an alibi or identify and establish a weapon used in a crime. Additionally, it is necessary because no human eyewitnesses are present to help solve the mysterious murder and rape, but forensic biology provides critical leads concerning the people who committed the crime. The society for its continued welfare and progress wishes to keep the crime in check by bringing into play all the forces at its command. It involves the dispensation of justice which is the role of the criminal justice system.

1.1 History and Development

The application of science and technology to the detection and investigation of crime and the administration of justice is not new to India. Although our ancestors did not know forensics in its present form but applied forensics in various fields, without knowing the science behind it. Now, forensics has turned into a wide branch, which is used to solve crimes and for other purposes also.

The recent technological and scientific development in the field of forensic biology is certainly stamping a huge impact in convicting criminals without any doubt. Forensic biology today has been expanded with numerous innovative technological advances and molecular discoveries. Following refinement and rigorous testing, many technologies/protocols have been standardized by forensic laboratories, including polymerase chain reaction, Genetic Analyzers, expert software systems and automated liquid handling.

In the ancient world, the king used to decide the culprits and punish them. These punishments were applied in favour of their existence, just to show how powerful they are. With the passage of time, everything changed; the man with knowledge replaced the king, and punishments were given with respect to the intensity of the crime the culprit had committed. Now when a crime occurs, the court of law decides the culprit, and punishment is based on the scientific evidence and statements produced by the investigating agency and prosecution before the court. In Roman times, the incident was narrated in front of the public. The culprit and the victim were given a chance to deliver their words based on their side of the story to prove their innocence and escape from the punishment. The person with the best argument would win the debate of proving the crime, and based on the points of argument, the public or the people with higher power determine the consequences.

Various branches have their own root of origin that gives us a simple outline of forensic biology itself. Forensics cannot be limited to a few branches; it is a wide stream of science that relates each and every branch to solve the mystery behind each crime. With each passing day, the technology develops, and its applications in forensic science also get increased. The advanced techniques help the investigating officers' prosecution and judiciary to solve a crime and come to a conclusion easily. As technology improves, the type and manner of crime also get improved. Still, many scientists are working on advanced techniques for the application of forensic biology.

1.2 Common Discipline

Forensic biology is a broad discipline that includes various areas of specialization such as DNA analysis, forensic anthropology, forensic entomology, forensic odontology, forensic pathology, forensic toxicology, forensic botany, forensic serology, forensic microbiology and diatom examination. This deals with the following analysis

1. Examination and identification of biological fluids like blood, urine, semen, sweat, saliva, milk, tears, juice and determining their origin either come from human, plant or animal sources (Fig. 1.1).
2. Examination and identification of materials like wood, fibre, faecal matter, nails, bones, teeth, leaves, seeds, pollen and other plant material.
3. Identification and examination of diatom in the water, human viscera and food items.

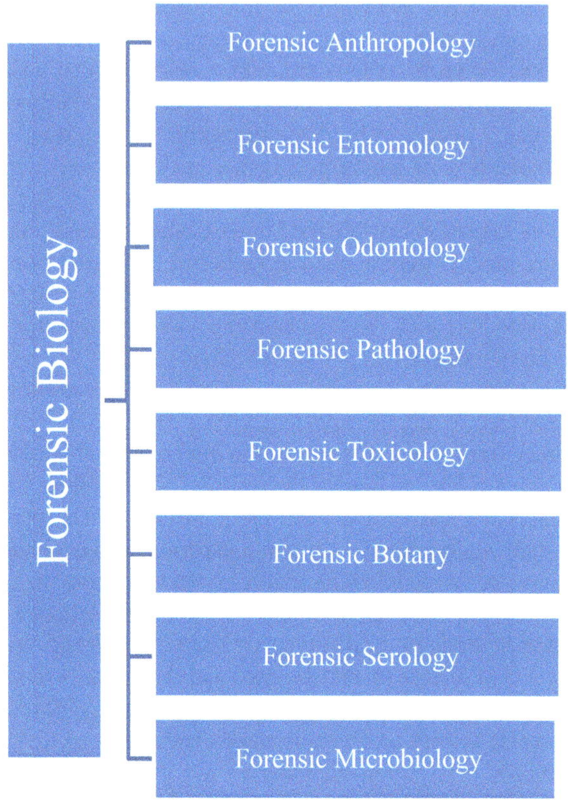

Fig. 1.1 Common discipline of forensic biology

4. Determination of the group of blood, semen, sweat and saliva if found to be of human origin.
5. Determination of paternity/maternity and identity of an unknown individual by DNA profiling.
6. Determining the time since death by examining the bones and teeth.
7. Determining the age and sex of human beings by examining the skeleton remains.
8. Determining the time since death and manner of death by examining the insects found on and near the dead body.

1.3 Basic Principles of Forensic Science

Science and technology have captured the mind of all agencies in the criminal justice delivery system. Forensic science is the scientific discipline which is engaged to the recognition, identification, individualization and evaluation of physical evidence by using the laws and principles of natural science for the purpose of administration to terminate doubtful questions in the court of law. All these principles of forensic science are essential in crime scene investigation to link a suspect/criminal with the scene of the crime as well as the victim.

1. Law of Individuality
   This law states that **'Every object whether natural or man-made has a distinctive quality or characteristic in it which is not duplicated in any other object'**, in other words, no two things in this universe are alike. The most common example is the human fingerprints, they are unique and permanent and prove the individuality of a person. Even twins do not have the same fingerprints. This principle is considered the most basic elementary unit of forensic science. Fingerprints, footprints, tool marks and marks on fired bullets obtained from the crime scene are studied and analysed on the principle of individuality.
2. Law of Progressive Change
   Change is inevitable. 'Longer the delay, greater the changes'. This principle emphasizes that **'Everything changes with the passage of time and nothing remains constant'**. The changing frequency varies from sample to sample and on different objects. The scene of occurrence undergoes rapid changes and it must be secured in time otherwise a change in weather like rain, heat, wind and presence of animals/humans affects the crime scene. Samples degrade with time, bodies decompose, tyre marks, footmarks and bite marks fade, the firearm barrel loosens, metal objects rust, etc. Without an appropriate preservative, tissue samples start degrading immediately and they need immediate analysis. The criminals undergo progressive changes with time. If he is not apprehended in time, he becomes unrecognizable except for his fingerprints or other characteristics of permanent nature.
3. Locard's Principle of Exchange (Law of exchange of material)
   This principle was stated by French scientist Edmond Locard (a pioneer in criminology and forensic science). Law of exchange states that, **'As soon as two things come in connection with each other, they mutually interchange the traces between them'**.

   Whenever a criminal or his weapon of crime comes in contact with the victim made a connection with the victim or the things surrounding him, they leave traces. Likewise, the criminal or his weapon picks up traces from the same contact. These traces are very helpful for investigation purposes as these traces are identified by the expert and linked to their original source resulting in the decisive linkage of the criminal with the crime scene and the victim. This law forms the basis of scientific crime investigation. The basic requirement of this law is the correct location of the physical evidence—(a) What are the areas and things with which the perpetrator or tool actually came in contact during the crime? (b) Investigating officer should establish the correct points of contact, it leads the investigation in the correct direction (Fig. 1.2).
4. Principle of Comparison
   The law state that **'Only the likes can be compared'**. For laboratory investigation, this law is very important. It highlights the requirement of providing similar

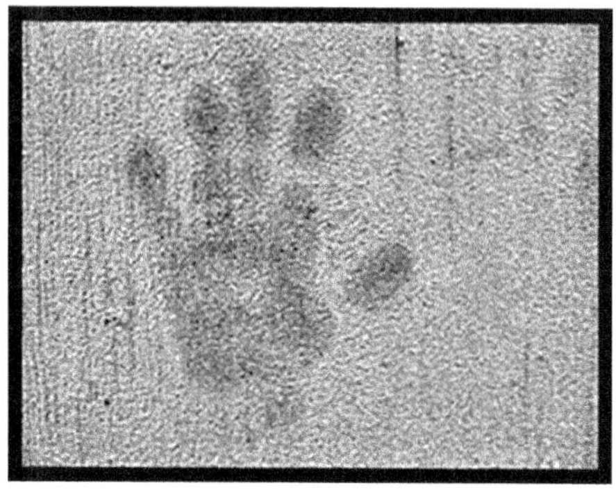

Fig. 1.2 Locard's exchange principle

samples and specimens for evaluation with the questioned items. For example, if the murder is done by a firearm weapon then it is useless to send a knife for comparison. So, the important condition of this principle is to supply specimens/ samples of similar nature for proper assessment with the questioned sample discovered from the crime scene.

5. Principle of Analysis
   This principle states that '**The quality of any analysis would be better by collection of correct sample and its correct preservation in the prescribed manner**'. Improper sampling and contamination render the best analysis useless. If you collect a hard disk in a paper bag, it can be damaged when it falls within the range of a strong electromagnetic field, resulting in poor results. Hence, always appropriate and effective collection and packaging techniques must be used.
6. Law of Probability
   This law states that '**All identifications (definite or indefinite), made consciously or unconsciously on the basis of probability**'. The probability is mostly misunderstood if we say that according to probability a particular fingerprint has come from the given source, but it is not a definite opinion. The perpetrator blood group is also the blood group of various people is high, but the probability of the same occurring in the case is low. Probability is a mathematical concept. It determines the chances of occurrence of a particular event in a particular way.
7. Law of Circumstantial Facts
   According to this law, **'Facts do not lie, cannot be wrong, men can and do'.** This law emphasizes the significance of circumstantial facts and supports that a statement given by eyewitnesses may or may not be accurate. In an investigation, identified and discovered facts are more accurate and reliable than any eyewitness.

Forensic science by these principles is used for recognition, identification; individualization of pieces of evidence collected from the scene of crime and guides the criminal proceedings from the discovery of a crime to the conviction of the accused, helping the process of investigation.

Multiple Choice Questions

1. Locard's principle of exchange relates to
   (a) Exchange of trace evidence material
   (b) Exchange of criminal information
   (c) Foreign exchange
   (d) Exchange of hearts
   Answer: (a)
2. Screening tests are performed for
   (a) Definite identification
   (b) Tentative identification
   (c) Quantitative analysis
   (d) Semiquantitative analysis
   Answer: (b)
3. Which of the following preservatives is used for preserving hair for DNA analysis
   (a) Dimethyl sulfoxide
   (b) Normal saline
   (c) Rectified spirit
   (d) No need of preservatives
   Answer: (d)
4. In case of poisoning of living persons, the following biological samples are preserved, except
   (a) Vomit
   (b) Excereta
   (c) Unsoiled clothing
   (d) Stomach wash
   Answer: (c)
5. Faecal matter as evidence is encountered in cases like
   (a) Rape
   (b) Murder
   (c) Hanging
   (d) Bestiality
   Answer: (d)
6. The colour of blood in deaths due to burns is
   (a) Pink
   (b) Blue
   (c) Cherry red
   (d) Brown
   Answer: (c)
7. Forensic palynology is the study of
   (a) Fossils
   (b) Spores
   (c) Pollens
   (d) Both (b) and (c)
   Answer: (d)

8. Following could be analysed from bite marks
   (a) Saliva
   (b) Denture
   (c) Caused by child or adult
   (d) All the above
   Answer: (d)

Further Reading

Arya N (2014) Advanced molecular biology: as a significant forensic tool. In: Proceedings-next generation science: vision 2020 and beyond. ISBN 978-81-920945-4-0

Arya N (2017) Microbial Forensics: A newly emerging crime fighting tool. In: Souvenir of international conference on microbes for health and wealth

Arya N, Jilowa CR (2013) Role and impact of advanced forensic technology in criminal justice system. Kurukshetra Law J III. ISSN-2231-5519

FBI (1999) Handbook of forensic services: evidence examinations, 1999. U.S. Department of Justice, FBI, Washington, DC

Federal Bureau of Investigation (2013) 'Clearances'. Accessed 16 April. http://www.fbi.gov/aboutus/cjis/ucr/crime-in-the-u.s/2011/crime-in-the-u.s.-2011/clearances

Frippiat C, Zorbo S, Leonard D, Marcotte A, Chaput M, Aelbrecht C, Noel F (2011) Evaluation of novel forensic DNA storage methodologies. Forensic Sci Int Genet 5(5):386–392. https://doi.org/10.1016/j.fsigen.2010.08.007

McClintock TJ (2014) Forensic Analysis of Biological Evidence: A Laboratory Guide for Serological and DNA Typing ISBN 9781466504561

Melton T, Holland C, Holland M (2012) Forensic mitochondrial DNA analysis: current practice and future potential. Forensic Sci Rev 2012(24):101–122

Ogden R (2010) Forensic Science, genetics and wildlife biology: getting the right mix for a wildlife DNA forensic lab. Forensic Sci Med Pathol 6:172. https://doi.org/10.1007/s12024-010-9178-5

Rana AK (2018) Crime investigation through DNA methylation analysis. Methods and applications in forensics. Egypt J Forensic Sci 8:7

Roberts KA, Johnson DJ (2012) Investigations on the use of sample matrix to stabilize crime scene biological samples for optimized analysis and room temperature storage. U.S. Department of Justice, Washington, DC, p 296

Schmedes SE, Sajantila A, Budowle B (2016) Expansion of microbial forensics. *J Clin Microbiol* 2016(54):1964–1974

Tewari RK, Ravikumar KV (2000) History and development of forensic science in India. J Postgrad Med 46(4):303–308

2 Investigation of Biological Evidence

Neelam Arya and Avinash Puri

Abstract

The investigation of crime is in fact a dynamic process of formulation and testing of the hypothesis of crime and there is no substitute for a careful and thoughtful approach. An investigator must not jump to hasty conclusions as to what happened based upon limited information but must generate several different theories of the crime, keeping the ones that are not eliminated by incoming information at the scene.

Physical evidence can range from massive objects to microscopic items, resulting from a crime and recovered at the scene or relevant locations. The purpose of the observation and documentation of the crime scene condition is to note the location of potential evidence and to mentally outline how the scene will be examined. Considering all sources of information available in investigations, physical evidence plays a vital role. Physical evidence, when recognized and properly handled, offers the best prospect for providing objective and reliable information regarding the incident under investigation. However, the value of even the most carefully recovered and preserved evidence can be lost if the chain of custody is not properly maintained. Chain of custody exists to ensure full transparency in the process of how evidence is collected, handled, and stored.

Once an item is recognized for its evidence value, the collection becomes a relatively simple matter of performing the following routine tasks; i.e., handling the evidence properly, marking the evidence for later recognition, packaging the evidence to prevent alteration or damage, tagging the evidence for package

N. Arya
FSL, Haryana, Madhuban, Haryana, India

A. Puri (✉)
Department of DNA and Forensic Biology, Regional Forensic Science Laboratory, Indore, Madhya Pradesh, India

A. Puri et al. (eds.), *Fundamentals of Forensic Biology*,
https://doi.org/10.1007/978-981-99-3161-3_2

recognition, and maintaining the chain of custody. The last phase of the crime scene investigation process aims at selecting the means of transportation and storage that are appropriate for the type of physical evidence to ensure the integrity of evidence to be forwarded to the laboratory.

In a broader sense, crime scene management— implementing new technologies and methodologies in forensic science—is the first and most important golden moment in solving any crime and strengthening the criminal justice system.

Keywords

Crime scene processing · Physical evidence · Biological evidence · Chain of custody · Collection of evidences · Recreation of crime scene

2.1 Crime Scene Management

One of the most important and nowadays necessary aspects of the criminal justice system is crime scene management which leads to scientific investigation implementing forensic biology or the practice of scientifically examining biological evidence collected from the scene of a crime or a person of interest in a crime. The investigation begins at the crime scene with the recognition and recovery of physical evidence. The investigation is basically the art of unearthing the truth for the purpose of successful detection and prosecution. The success and failure of criminal investigation depend on the thoroughness of the exercise at the scene of crime.

The examination of the scene of crime needs meticulous planning and care. The investigation may fail due to improper handling of the scene of the crime and may destroy and contaminate the evidence. There is no substitute for a careful and thoughtful approach. A thorough crime scene processing can lead to a sequential collection of evidence that constitutes a crucial part of the legal processing of the case. "Processing a crime scene" is a long, tedious process that involves purposeful documentation of the conditions at the scene and the collection of any physical evidence that could possibly illuminate what happened and point to who did it.

The scene of crime processing is the golden hour of investigation. Golden hour is the term used for the period immediately after an offence has been committed when the material is readily available in high volumes to the police which is of crucial importance in the identity of potential suspects and to obtain relevant facts and data. Positive action in the period immediately after the report of a crime minimizes the amount of material that could be lost to the investigation and maximizes the chance of securing the material that will be admissible in a court of law (Fig. 2.1).

There is no typical crime scene, there is no typical body of evidence, and there is no typical investigative approach. Each and every crime scene is unique and delicate in nature. The protocol for the crime scene investigation differs from the scenario associated with the crime. No specific standard is adopted because the delicacy and fragility of a crime scene differ with the type of the crime scene it is associated with.

Fig. 2.1 Scene of crime processing is the golden hour of investigation

For example, an indoor crime scene is considered more secure than an outdoor scene of crime. Likewise in a hit-and-run case, the safety of the crime scene is the primary concern because of the interference of the travelers and on-lookers around.

The crime scene investigator arrives on the scene and makes sure it is secure. The most important aspect of evidence collection and preservation is protecting the crime scene. This is to keep the potential evidence uncontaminated until it can be recorded and collected. Scene preservation starts soon after the incident is discovered and reported to the police station. The investigator does an initial walk-through to get an overall feel of the crime scene, finds out if anyone moved anything before he arrived, generates initial theories based on visual examination, and makes note of potential evidence. At this point, he touches nothing. Concerns for scene protection end only when the scene investigation process is completed and the scene is relieved.

Demarcation of the area to be protected is a complex activity, and the boundaries of the scene may change as the investigation unfolds. What appears to be obvious at the outset may change and need to be re-evaluated. Once demarcated, the area is clearly cordoned off using any kind of physical barriers. From the beginning to the end of the crime scene in the investigation, strict anti-contamination measures are important like wearing protective clothing and gloves., using a single path when entering the scene, and avoiding moving anything/anybody, unless it is of absolute necessity.

Investigators working at crime scenes may be exposed to various health and safety hazards. Not all hazards are immediately obvious and some may come up as the investigation unfolds. Potential health and safety hazards may arise from a number of sources, at a crime scene:

1. Chemicals, either present at the scene such as in the case of clandestine laboratories or used as part of the investigation.
2. Biological materials like blood and body fluids may be a source of risk for HIV and other infections.

3. Environmental factors, e.g., excessive heat or cold.
4. Unexploded explosives like booby traps and firearms.
5. Insecure environment, when the offender is still present at the scene.
6. Unsafe structure especially when collecting evidence at fire and bombing scenes.
7. Other risks include sharp objects, radiological, nuclear and electrical risks, gases, etc.

In a broader sense, crime scene management—implementing new technologies and methodologies in forensic science—is the first and most important golden moment in solving any crime and strengthening the criminal justice system.

2.2 Documentation and Legal Considerations at Crime Scene

2.2.1 Documentation of the Crime Scene Conditions

The goal of crime scene documentation is to create a visual record that will allow the prosecution and judiciary to easily recreate an accurate view of the scene. Documentation aims at producing a permanent, objective record of the scene, the physical evidence and any changes that take place. The purpose of the observation and documentation of crime scene condition is to note the location of potential evidence and to mentally outline how the scene will be examined. The crime scene conditions should be carefully observed and transient details, such as lighting on/off, weather, temperature, movement of furniture or other disturbances made, and conditions that would support or refute suicide/self-defense should be recorded. It is also important to be able to recognize what should be present at the scene of the crime but is not there, e.g., keys, watch, ornaments, vehicles, etc. Similarly, the objects which appear to be out of place and might have been left by the perpetrators should be taken note of. The most common documentation methods are note taking, drawing sketches, taking photographs, videography, and three-dimensional mapping of the scene of the crime (Fig. 2.2).

Note making: Note making at a crime scene is not as straightforward as it may seem. An expert or investigator training includes the art of scientific observation.

Fig. 2.2 Documentation of scene of crime by note making, photography, videography, and by three-dimensional (3D) scanner

Whereas a layperson may see a large, brownish-red stain on the carpet, spreading outward from the corpse, and write down "blood spreading outward from underside of corpse," an expert would write down "large, brownish-red fluid spreading outward from underside of corpse." This fluid might be blood or it might also be decomposition fluid, which resembles blood at a certain stage. In crime scene investigation, **opinions don't matter and assumptions are harmful.** When describing a crime scene, an expert makes factual observations without drawing any conclusions. An investigator first makes a **rough sketch** that is later turned into a finished sketch showing the positions of persons and objects with a scale.

Sketching: An investigator creates sketches to depict the entire scene, which is easier to do in a sketch than in notes because a sketch can span several rooms and particular aspects of the scene that will benefit from exact measurements. The sketch of the crime scene establishes the distance, size, and relationship between various objects in the scene. The sketch artist may indicate details like the height of a window frame, the exact size of the room, the distance from the window to the door, and the length of skid marks or tire marks on the road.

Photography: An investigator takes pictures of everything before touching or moving a single piece of evidence. The medical examiner will not touch the corpse until the investigator has taken photograph of it and of the surrounding area. The overview, mid-range, and close-up views of the scene should be photographed successively. Overview shots are the widest possible views of the entire scene. Mid-range photos come next. These shots show key pieces of evidence in context, so the photo includes not only the evidence but also its location in a room and its distance from other pieces of evidence. Close-ups of individual pieces of evidence show any serial numbers or other identifying marks that includes a ruler for scale.

Videography: Scene documentation may also include a video walk-through, especially in major cases involving large areas or multiple homicides. A video recording can offer a better feel for the layout of the crime scene like how long it takes to get from one room to another and how many turns are involved, for instance. Also, once the investigation is further along, it may reveal something that was overlooked at the scene because the investigator didn't know to look for it. During a video walk-through, the investigator captures the entire crime scene and surrounding areas from every angle and provides a constant audio narrative.

Three-dimensional (3D) scanning: This technology has been growing incredibly fast in just the past few years, changing forensic workflows worldwide. The documentation of crime scenes is a painstaking process. Once investigators arrive at the scene, there's a lot to swallow. First, the evidence can be completely unpredictable since all evidence is not created equally. Despite that, diagrams supported by various measurements are essential as the scene is documented. Traditionally, investigators used tools, such as paper sketchbooks, measuring tape, or rulers, and photography that was time consuming and often lacked accuracy and consistency. They may measure the length of a piece of evidence, but not how far away it was from another piece of evidence. In contrast, 3D scanning captures all the measurements in a room and all the objects, plus photo-realistic colors, textures, and geometry. It is far more accurate, data-rich, and faster, enabling investigators to explore the scene reconstructed digitally well after the actual scene is cleaned up. By capturing large amounts of data quickly, 3D scanners enable investigators to create a complete 360-degree image of a scene in minutes. Besides enabling investigators to clear a scene faster, this is also valuable if new evidence emerges or suspects change their narratives. The result is a more comprehensive investigation rather than relying on photographic imagery alone.

2.2.2 Legal Considerations

The legal considerations determine the admissibility of the evidence collected at the crime scene. Failure to comply with existing laws, rules, and regulations can result in a situation where the evidence cannot be used in court. Hence, it is a matter of importance for personnel working at the scene to be aware of and to ensure proper compliance with these rules. Local laws, rules, and regulations govern many activities of the crime scene investigation and forensic process such as how to obtain authority to enter the scene, to conduct the investigation, to handle evidence, and to submit physical evidence to the forensic science laboratory or other investigating agency. Regardless of local rules and regulations, codes of professional conduct outline the moral obligations of personnel working at crime scenes. Such codes typically stress on the importance of acting with care and professionalism, objectivity, open-mindedness, and impartiality. If there is a conflict between the preservation of evidence and the possibility of saving a human life, priority is always given to emergency medical care.

2.3 Collection, Packaging, Preservation, and Transportation of Biological Sample

Once the crime scene has been thoroughly documented, the **collection** process of physical evidence begins. Every piece of evidence of the slightest importance should be collected without any delay. Delay may result in the loosing of material evidence affecting the solving of the crime properly. Crime scene evidence could be found at any place where a crime has been committed, i.e., a crime scene. It may be found on anything worn or carried by the victim during the time of the offence or within or on the body of any person associated with the crime, i.e., offender/ perpetrator. Hence, the evidence could be collected not only from the scene of the crime rather than from the victim/deceased and from his environment. This collection process also involves the role of the medico-legal doctors or medical men along with the investigating agency, who deal with the accused or victim in any manner after the crime. It cannot be generalized as to what are the biological evidence to be collected during the crime investigation. Each kind of crime involves the collection of different types of evidence which can lead to the investigation of that particular crime. However, as general guidance following biological evidence and material may have evidentiary value:

- Blood, semen, saliva, vomit stains, tissue/fetal remains, bones/teeth, hair strands, nail clippings, sweat, tears, urine/fecal matter, etc.
- Clothing, caps, masks, eyeglasses, and gloves from any involved individual.
- Ligature material, i.e., cloth piece, rope, belt, wire/cable, etc.
- Bedding, i.e., pillow, mattresses, etc.
- Cigarette butts or other smoking devices.
- Licked items like envelopes, stamps, etc.

While **packaging** the material for dispatch to the laboratory, care should be taken to see that no article is inadvertently contaminated with extraneous material and for that purpose, the investigator must choose the proper packaging material and the same way the size of the packaging material is equally relevant and significant. Too small packing is likely to be inadequate and if too big it would be difficult to recover. Ideally, the size of the packing material should be according to the size of the sample. These are some general principles for the packaging of biological samples:

- Each sample should be separately packed and labeled indicating the serial number of an item.
- Never pack more than one type of sample together.
- All the packets belonging to one case should be enclosed in one box or an outer covering.
- Evidence belonging to different cases should not be forwarded under the same cover. All exhibits should be carefully sealed by the dispatching authority and packed in such a manner that they cannot be opened without destroying the seal and the seal should be the same throughout, along with the sample seal.

Certain environmental factors may break down biological evidence into smaller pieces, and environmental factors may be negated by maintaining the evidence in a climate-controlled environment, preventing direct exposure to light. If evidence containing DNA is packaged correctly and stored under proper conditions, it will be stable and therefore useful for forensic examination indefinitely. The forensic laboratory has successfully examined cases where the DNA evidence was stored for more than 25 years (Table 2.1).

- Liquid blood can be collected on cotton gauze cloth piece/thread, air dried in shade and packed in paper envelope.
- Blood can also be preserved on an FTA card.
- Blood stains can be scratched and placed in paper envelope. Blood-stained earth should be thoroughly dried and placed in a paper envelope, along with unstained earth collected as control.

Table 2.1 General guidelines for collection, packaging preservation, and transportation of biological evidence

Biological evidence/ sample	Source/location	Collection packaging and preservation	Transportation/ precautions
Blood	Living person	– Collect in an EDTA tube	Maintain cold chain with the coolant pack and transport to the lab at the earliest possible preferably within 24 h
		– On the FTA card	
Blood	Crime scene	– In an EDTA tube	Avoid placing in excessive heat and transport to the lab at the earliest possible
		– On the FTA card	
		– Lift on moist cotton thread and keep in an airy envelope	
Blood	Clothing of victim/suspect or from the scene of the crime	– Air dry in the shade	– Avoid blower or heater to dry blood-stained clothes transport to the lab at the earliest possible
		– Wrap in a paper sheet separately to avoid any cross contamination	
Blood/ vomit/other body fluid stains	Dried blood stains on any article/floor/weapon, etc. recovered during the investigation	– Scratch with sterile forceps/scalpel and collect in a paper envelope	– Dry in the shade and transport to the lab at the earliest possible
		– If in little amounts, lift with moist thread and keep in a paper envelope	– Avoid air-tight containers
		– Never mix blood scrapings	
Semen	Victim, suspect, crime scene	– Collect the sample with a sterile cotton swab and pack it in an airy envelope after properly drying it	– Transport to the lab without any delay
		– Medical officer also need to collect samples separately and keep them in an airy envelope separately	– Dry in the shade and transport to the lab at the earliest possible
			– Avoid air-tight containers
Vaginal/ anal stains / swab	Victim, suspect, crime scene	– Collect the sample with a sterile cotton swab and pack it in an airy envelope after properly drying it	– Transport to the lab without any delay
		– Medical officer also need to collect samples separately and keep them in an airy envelope separately	– Dry in the shade and transport to the lab at the earliest possible
			– Avoid air-tight containers

(continued)

Table 2.1 (continued)

Biological evidence/ sample	Source/location	Collection packaging and preservation	Transportation/ precautions
Teeth/bone pieces	Dead body, at crime site or any other suspected place	– Wash and clean to remove debris dry in the shade and wrap in a paper sheet or muslin cloth separately	– Transport to the lab without any delay – Dry in the shade and transport to the lab at the earliest possible – Bone sample should remain intact, so transport in a hard container or box – Charred/completely burnt bones are not suitable for DNA profiling
Body tissue/ organ	Dead body, fragments of the body at the crime site or any other suspected place	– Must be placed in a sterile container along with recommended saline for preservation	– Maintain cold chain with a coolant pack and transport to the lab without any delay
		– Medical officer need to take samples separately	– Never use formalin as a preservative
Small/long hair strands (pubic hair/ scalp hair)	– Transferred on clothing and body of victim/ suspect, at crime site or any other suspected place/ vehicle	– Collect with sterile forceps and keep in a paper envelope or zip-locking envelope – Can use cello tape to collect the samples	– Dry in the shade and transport to the lab at the earliest possible
		– If hair strands are found stuck on any small object, preferably don't remove them and pack the whole object in a paper sheet or airy envelope – Collect reference samples from the victim and suspect (50–60 hair strands with root)	– Never clean or wash the hair strands recovered during the investigation
Sample for DNA profiling	Blood/semen/blood-stained garments/bedding/ bones/teeth/smear slides/ vaginal swab/vehicle/ crime scene and vicinity	– On the FTA card	– Transport to the lab without any delay
		– Lift on moist cotton thread and keep in an airy envelope	– Bone sample should remain intact, so transport in a hard container or box
		– Air dry in the shade	– Charred/completely burnt bones are not

(continued)

Table 2.1 (continued)

Biological evidence/ sample	Source/location	Collection packaging and preservation	Transportation/ precautions
		– Wrap in a paper sheet separately to avoid any cross contamination	suitable for DNA profiling

- Blood/semen-stained garments should be air dried and packed/wrapped in paper sheets separately in such a manner that the stained portion (should be marked, if possible) should not contaminate the unstained area of the garment.
- Blood-stained area of any weapon should be wrapped with paper and packed in a paper bag.
- Soft tissue or organs taken during postmortem examination should not be preserved in formalin, as it affects the DNA typing results.
- Saliva can be lifted with a cotton swab from the bite marks on the victim.
- Hair strands of the victim/suspect should be collected with the help of sterile forceps and placed in a paper envelope along with control samples of the victim/ suspect.
- Bones/teeth should be washed and dried properly before wrapping them into a brown paper sheet or envelope. Remnants of burnt bones are very fragile and need to be handled carefully. They should be collected and placed in a wooden or hard cardboard box with a shock-absorbing material.
- Ashes from the cremation site should be placed in a wooden or hard cardboard box.
- Small fibers sticking to the ligature and ligature mark should be collected with the help of adhesive tape and secured on a glass slide.
- For the detection of diatom, water samples from the site of drowning and bone samples of dead bodies should be collected and forwarded to the laboratory. Any material present in the clenched hands should be properly dried and packed.

The investigator should try to collect and **transport** the exhibits as soon as possible since the evidentiary value of the collected material decreases with the passage of time. The proper time to collect the biological evidence is within 24–72 h of the occurrence; however, this is not applicable in all cases.

2.4 Chain of Custody

Biological evidence, when recognized and properly handled, offers the best prospect for providing objective and reliable information regarding the incident under investigation. However, the value of even the most carefully recovered and preserved evidence can be lost if the chain of custody is not properly maintained. From the first responder to the end user of the information, all personnel involved should have an adequate understanding of the scientific and legal processing of the scene of crime.

The chain of custody begins from the crime scene itself and continues till the evidence is presented in the court. The chain of custody exists to ensure full transparency in the process of how evidence is collected, handled, and stored. Chain of custody refers to the chronological and careful documentation of evidence to establish its connection to an alleged crime. It is of the utmost importance in a case to maintain each and every record of the evidence from the date of collection to the date of presentation before the court, because if even any one entry is missed, the evidence will become inadmissible in the court and non-presentation of even one evidence can assuredly affect the decision given by the court. The chain of custody helps in maintaining the integrity and authenticity of the evidence through careful and cautious handling of the evidence.

To prove someone guilty, a prosecutor must prove that the evidence presented in court is the same evidence that was recovered at the scene of an alleged crime. They must be able to show that the evidence was handled properly and was not contaminated or tampered with. If law enforcement does not properly handle evidence, the evidence can be challenged on the grounds that it was tampered with, that test results are faulty or inaccurate, or that evidence was planted at the scene of a crime. Because the prosecutions rely on evidence gathered by police officers, it is prosecutors who must establish the chain of custody.

Proving a chain of custody can be difficult. If law enforcement does not do it correctly, a chain of custody can be successfully challenged in a criminal case. Challenging the chain of custody can be a successful defense strategy to eliminate evidence that might be used to convict a defendant. If the judge finds that certain evidence is not admissible, the prosecutor may not have enough evidence to continue with the case. A typical evidence log will include the date and time the evidence was collected, the name of the investigator, the location where the evidence was collected, the reason the evidence was collected, relevant serial numbers, a description of the evidence, and the method that was used to collect the evidence. Any time evidence is examined by a forensic scientist, the examiner must list everyone who came in contact with the evidence and everything that was done to the evidence. A defendant can challenge the chain of custody of a piece of evidence by questioning whether the evidence presented at trial is the same evidence as what was collected at the scene of the crime. If there are any discrepancies in the chain of custody and law enforcement cannot prove who had the evidence at a particular time, the chain of custody is broken and the defendant can ask to have the evidence declared inadmissible. A broken chain of custody can result in failure to serve justice. When a proper chain of custody is maintained, it prevents the police officials and other lab or law officers to tamper with the evidence, as the record of who collected the evidence, who handled it, period of guardianship of the evidence, safeguarding conditions while storing the evidence, and how evidence is handed over to the subsequent custodians, each time a transfer takes place, and through this record, if any evidence is tampered it can be easily traced as to who played with the evidence and can thus be punished for the same.

2.5 Crime Scene Reconstruction

Reconstruction of the scene of the crime is the development of actions and circumstances based on the system of evidence discovered and examination in relation to a particular crime. Crime scene reconstruction is just like putting together a jigsaw puzzle but doing so without access to the box top. The analyst does not know what the picture is supposed to look like. Furthermore, not all of the pieces are likely to be present, so there will be holes in the picture.

The process of crime reconstruction is not an easy one because we are dealing with human action and are trying to tell details of what happened at a particular time in the past. In this philosophy, all elements of evidence that came to light in a given case are treated as interdependent, the significance of each piece, each action, and each event falls and rises on the backs of the others. Plan of reconstruction based on

- Available information and data
- Findings on the spot
- Sequence of events
- Missing link
- Hypothesis formation
- Human psychology
- Human behavior
- Determination of the significance of the evidence
- Theory formulation

The basic objective is to have an accurate accounting of the crime scene in order to determine the facts of the crime and to understand the events that might have occurred during that period. Frequently, the facts don't match the story, and solving a case can be in the smallest of details which are easily overlooked. Crime scene reconstruction or recreation is one of many profiling techniques used to build a hypothesis for an offender or to solve a crime. First, assumptions are made about how the crime was carried out. Then deductive and inductive reasoning is used to support the theory of how the crime occurred. Understanding the area of the crime scene, the position of physical evidence, etc. helps to get an idea of the events leading to crime, physical evidence at crime scenes is, in any case, amazingly important in the remaking of crimes and the action that has occurred at the crime scene.

Recreation not just includes logical crime scene investigation, understanding of the scene design evidence, and research center assessment of physical evidence, yet in addition, includes methodical investigation of related data and the coherent definition of a hypothesis and supplant the essential things or activities back at a crime scene through unique scene documentation. Criminal profiling is a procedure that depends on the mental and measurable investigation of the crime scene, which is utilized to decide the general attributes of the most probable suspect for the crime. Critical thinking is a purposeful and goal-directed activity that is based on the principles of science and the scientific method. Reconstruction not only involves the scientific scene analysis, interpretation of scene pattern evidence, and laboratory examination of physical evidence but also involves the systematic study of available information and data. Not all law enforcement personnel can do crime reconstruction. It must be someone who is a keen observer, understands science, recognizes evidence, and can apply logic.

Multiple Choice Questions

1. Dried seminal stains from clothing are preserved in
   (a) 40% Formalin
   (b) 10% Formalin
   (c) 20% Formalin
   (d) As such
   Answer: (d)
2. A wet bloody shirt should be
   (a) Packed as soon as possible
   (b) Air dried prior to packing
   (c) Air dried and not packed
   (d) Dried with heat
   Answer: (b).
3. In which of the following cases, the study of diatoms will be helpful
   (a) Immersion drowning
   (b) Secondary drowning
   (c) Filigree burns
   (d) Dry drowning
   Answer: (b)
4. The most prevalent plant fiber is
   (a) Polymer
   (b) Wool
   (c) Hemp
   (d) Cotton
   Answer: (d)
5. Nodes are seen in the following fiber
   (a) Cotton
   (b) Linen
   (c) Wool
   (d) Silk
   Answer: (b)
6. In searching the scene for evidentiary clues, one must first lift
   (a) Weapon of offense
   (b) Bloodstain
   (c) Fired bullet/cartridge case
   (d) Micro evidence like hair strand, fiber, etc.
   Answer: (d)
7. A crime committed in a vehicle falls in the category of an

   (a) Indoor crime scene
   (b) Outdoor crime scene
   (c) Both (a) and (b)
   (d) None of above
   Answer: (a)
8. Sex of the scalp hair can be determined by the examination of
   (a) Length of hair
   (b) Cross section of hair
   (c) Nucleated cells of the hair root
   (d) Medullary index ratio
   Answer: (c)
9. Comparison of soil to determine its origin can be detected through
   (a) Densitometer
   (b) Wet chemical analysis
   (c) Neutron activation analysis
   (d) Density gradient tube method
   Answer: (d)
10. Glass pieces broken from a window pane by the action of heat will generally be found.
   (a) Held together in small fragments
   (b) Scattered into fine pieces
   (c) In same direction as the source of heat
   (d) In opposite direction as the source of heat
   Answer: (c)

Further Reading

Amankwaa AO, McCartney C (2021) The effectiveness of the current use of forensic DNA in criminal investigations in England and Wales. https://doi.org/10.1002/wfs2.1414

Arya N (2011) High potential of plant DNA analysis in forensics. Souvneir on biodiversity: challenges & opportunities

Arya N (2017) Microbial forensics: a newly emerging crime fighting tool. Souvnenir of International Conference on Microbes for Health and Wealth

Arya N (2018) A hand book on forensic science and criminal justice system. Chandigarh Judicial Academy, Chandigarh

Arya N, Jilowa CR (2013) Role and impact of advanced forensic technology in criminal justice system. Kurukshetra Law J III. ISSN-2231-5519

Benecke M (2004) Forensic DNA samples—collection and handling. In: Fuchs J, Podda M (eds) Encyclopedia of medical genomics and proteomics. Dekker encyclopedias series, 1420, chap 103. CRC Press, New York

Caputo M, Bosio LA, Corach D (2011) Long-term room temperature preservation of corpse soft tissue: an approach for tissue sample storage. Investig Genet 2(17):6. https://doi.org/10.1186/2041-2223-2-17

Fujita Y, Kubo S (2006) Application of FTA technology to extraction of sperm DNA from mixed body fluids containing semen. Leg Med (Tokyo) 8(1):43

IvyPanda (2021) Forensic biology in crime scene investigations. https://ivypanda.com/essays/forensic-biology-in-crime-scene-investigations/

Li R (2008) A book on forensic biology. ISBN 13:978-1-4200-4344-0. eBook PDF
National Institute of Justice (2000) Crime scene investigation: a guide for law enforcement. NCJ 178280. U.S. Department of Justice, Office of Justice Programs, National Institute of Justice, Technical Working Group on Crime Scene Investigation, 58, Washington, DC
Pathak AG, Gadhari RK, Gavle PS (2018) Importance of visit to the scene of crime to determine manner of death-a case report. JKIMSU, vol 7-1. ISSN 2231-4261
Rana P, Ranjan AR (2020) Use of forensic science in investigation of crimes; a critical legal study. IJESC 10:7
Zhang ZL, Liang WB, Sun HB, Yang X, Ma LY (2021) Forensic application of plant. Evidence 37(1):87–90. https://doi.org/10.12116/j.issn.1004-5619.2019.490708

3 Investigation of the Dead Body

Alok Atreya, Rutwik Shedge, and Tanuj Kanchan

Abstract

One of the greatest sources of physical evidence is the dead body of the victim itself. A cadaver is a plethora of clues, that forensic scientists can investigate and reconstruct the circumstances surrounding the crime. Death can be defined as the permanent and irreversible cessation of functions of the three vital organs: the brain, the heart, and the lungs. Once an individual dies, the body undergoes certain orderly changes, that may be documented and used for the estimation of time passed since the individual's death, also known as the post-mortem interval (PMI) (Shedge et al., Postmortem changes, StatPearls Publishing, Treasure Island, 2020). The investigation of a dead body involves examination at the scene of crime, autopsy or post-mortem examination, and determination of miscellaneous information such as the mode, manner, and cause of death. This chapter is divided into three parts: First, it discusses the examination of a dead body at the scene of the crime, the post-mortem examination and its documentation, and then the determination of miscellaneous information.

Keywords

Death · Cadaver · Post-mortem examination · Post-mortem interval · Post-mortem changes

A. Atreya (✉)
Department of Forensic Medicine, Lumbini Medical College, Tansen, Palpa, Nepal

R. Shedge
School of Forensic Science, National Forensic Sciences University, Gandhinagar, Gujarat, India

T. Kanchan
Department of Forensic Medicine and Toxicology, All India Institute of Medical Sciences, Jodhpur, India

A. Puri et al. (eds.), *Fundamentals of Forensic Biology*,
https://doi.org/10.1007/978-981-99-3161-3_3

3.1 Definitions

- Autopsy: an examination of a dead body conducted primarily to ascertain the cause of death
- Death: the permanent and irreversible cessation of the functions of the brain, heart, and lungs
- Inquest: the legal inquiry into the cause of death
- Post-mortem changes: the changes that a cadaver undergoes after an individual's death
- Post-mortem interval: the time elapsed since an individual died

3.2 Introduction

Investigation of death starts immediately if there are evidence suggestive of death that has occurred with or without the presence of the dead body. For, e.g. there is a pool of human blood at the scene of the crime which is suggestive that the person would not survive with such exsanguination. The presence of a dead body is a prima facie that death has occurred. During homicidal trials in court, a dead body is considered 'corpus delicti' (Knight and Saukko 2016).

The practice of investigation of death and examination differs as per the jurisdiction and also upon who is conducting the inquest. In those countries where a medical examiner conducts the inquest, the investigation into the death and preliminary investigation are carried out in the scene of the crime itself. In countries where a police inquest or the coroner's inquest is followed, the body is sent to the mortuary where an examination of the dead body (autopsy) is carried out by a forensic pathologist. In the latter type, the forensic pathologist visits the scene of the crime as per the nature of the individual case. Irrespective of the type of inquest, the main objective of the examination of the dead however remains the same (De Matteis et al. 2020).

3.3 Examination of the Cadaver at the Scene of Crime

At any scene of the crime, if there is a cadaver present, the majority of the attention of forensic scientists goes towards the dead body. While the deceased has to be shifted to a nearby mortuary for a detailed post-mortem examination, some preliminary examination can be conducted at the scene itself. This preliminary examination of the cadaver includes making a detailed record of the following:

1. The position and location of the cadaver
2. The state of the cadaver (freshly dead, early stages of putrefaction, later stages of putrefaction, etc.)
3. Presence of any overt injuries, bleeding, etc.
4. The number, dimensions, and location of such injuries

5. Nature of injuries (abrasions, bruises, lacerations, stab wounds, incised wounds, etc.)
6. Presence of extrusion of fluids from such injuries
7. Complete photography of the cadaver (overview, mid-ranged, and close up photographs)
8. Sketching of the cadaver to denote injuries and their dimensions

3.4 Autopsy

An autopsy is a scientific examination of the dead body. Etymologically it is derived from the Greek word *autopsia* (*autos*—oneself, *opsis*—see). It is also popularly called post-mortem examination (Latin: *mors*-death). Autopsy of non-humans is generally referred to as necropsy (Greek: nekrós—dead). In all cases of unnatural deaths, an autopsy is done following the requisition from the authority. In cases of natural deaths where the cause of death is not known, a clinical (hospital) autopsy is conducted to find out the underlying pathology. Medicolegal autopsy in cases of unnatural deaths is part of death investigation globally. However, clinical (hospital) autopsies are not mandatory and are not done in all countries. In medicolegal autopsies, the law requires the examination of the body, so the consent of the relatives of the deceased is not required. However, in a clinical autopsy, the doctor wants to find out the underlying cause of death for a better understanding of the disease process therefore, the consent of the relatives is mandatory. In a broad sense, an autopsy is either conducted for medicolegal purposes or for education and research purposes, hence its classification:

- Medicolegal autopsy
- Clinical (hospital) autopsy
- Anatomical autopsy

Objectives of Autopsy

- To determine the cause of death
- To determine the mechanism of death
- To determine the manner of death
- To establish the time since death
- To ascertain whether the injuries sustained are antemortem or post-mortem in nature
- In the case of newborn
  - To distinguish live birth, dead born, or still birth
  - To estimate whether the child was viable or not
  - In the case of live birth—to find out the period of survival

(continued)

- To establish the identity of the deceased (in case of a mutilated body, decomposed body, and mass disasters)
- To determine the type of weapon/object used to cause death
- To establish the link between the crime and the criminal to aid in justice
- To establish the reason for death if possible.

A complete autopsy consists of external examination, internal examination, and ancillary procedures (Knight and Saukko 2016).

External examination refers to the examination and documentation of all the external features present on the body surface which includes clothing, the external genitalia (sex), any injuries, post-mortem feature, personal belongings like wristwatches, jewellery, identifying features like tattoos, scars, congenital deformity, acquired peculiarity, body length, body weight, hair color and eye color. All the injuries are to be measured and documented which are then photographed for later evidence. In cases of unidentified bodies, photographs should be obtained for identification purposes (Menezes and Monteiro 2022).

After the external examination is complete, internal examination is done which involves the opening of the body, and removal of the organs for dissection in order to examine and document the injury and/or underlying pathology (Cordner 1993). A complete autopsy involves the opening of four body cavities, i.e. cranial cavity, thoracic cavity, abdominal cavity, and pelvic cavity; and examination of their contents. Even if the lesion is evident in any one of the body cavities and the cause of death could be ascertained, for, e.g. a gunshot wound on the head, all the body cavities are to be opened. Further dissection of the body may be necessary apart from opening the four body cavities. However, there are instances where the internal examination is not conducted. In case of severe decomposition or in case of highly infectious disease, the autopsy may be limited to the examination of a particular organ/structure (Menezes and Monteiro 2022).

After a meticulous external and internal examination, the cause pertaining to death is not found. This leads to an obscure autopsy (Knight 1980). When the need arises, special procedures are conducted during, before, or after autopsy known as ancillary procedures. These procedures are either conducted in the mortuary itself in the presence of the forensic specialist or the body fluid sample, tissue sample, or trace evidence is subjected to laboratory testing. Even after the complete autopsy and ancillary procedures, the cause of death could not be identified which results in a negative autopsy (Cohle and Sampson 2001).

A forensic pathologist may at times need to interview the family members to know the circumstance and manner of death which is known as a follow-up investigation. At times, hospital treatment records need to be examined to fully understand the clinical picture and source of injury (venous catheter, injection marks, and fractured rib in case of cardiopulmonary resuscitation).

3.5 Cause of Death

After the completion of the autopsy, the forensic pathologist should document all the findings in detail that he/she encountered during the autopsy. After documentation, he/she is required to provide an opinion as to the cause of death and manner of death (Cordner 1993). It is therefore important and mandatory to mention the cause and manner of death in the autopsy report, which would aid the judiciary in solving the crime.

Death is an irreversible condition where there is a cessation of function of all three vital systems (respiratory, circulatory, and nervous system) of the body (Knight and Saukko 2016). If one system of the body ceases to function, the other two will ultimately fail. However, based upon which system failed first the mode of death is determined as asphyxia, syncope, or coma. Mode of death as such is not the cause or mechanism of death, therefore should not be used in death certification or autopsy reports.

3.6 Manner of Death

Manner of death hints at the circumstances surrounding the death and can be categorized into natural, unnatural, or undetermined (Advenier et al. 2016). Whenever a person dies due to old age or due to the ordinary course of a disease, the manner of death in such a case is considered to be natural. Unnatural manner of death could be further classified into homicide, suicide, or accidental. The manner of death when an individual is killed by someone else is a homicide, when the person kills themselves is a suicide, and when the person dies due to an accident such as a train derailment is accidental. Whenever there is no sufficient evidence that points to the mode being natural or unnatural, it is defined to be the undetermined manner of death.

3.7 Autopsy Proper

The autopsy starts with an external examination of the body. The body should be photographed before the removal of the clothes, jewellery, and other belongings.

Photographs

- Photographs should be preferably taken by a high-resolution camera.
- However, in case of unavailability any camera can be used provided the picture quality is good and the findings could be reviewed later on.
- Photographs should include a scale, autopsy number, and date.

(continued)

- Serial photographs should be taken at each and every step of the external examination.
- Photographs should be obtained before undressing, washing, or shaving and also thereafter.
- Wide-range photographs should supplement the close-up photographs for orientation and identification. All the injuries and special features present in the body should be depicted by close-range photographs.

3.8 External Examination

3.8.1 Clothing

The clothes should be examined for any stains and tears. The clothing should be noted in terms of colour, texture, material, type, design, size, manufacturer, and tears, stains and the sequence in which it was present in relation to other clothing. If any trace evidence is found it should be described and secured for further analysis. Similarly, all the belongings from the dead body are to be removed and documented with a description.

3.8.2 Identifying Features

Following the removal of the cloth, the body is examined for external identifying features like external genitalia, body length, body weight, nutritional status, hair distribution, colour of skin, colour of hair, and colour of eyes. In the case of foetuses and infants, apart from the body length, head circumference, crown-rump length, and crown heel length are also measured. In the case of an unidentified body, the fingerprints of all the fingers should be obtained. Furthermore, in the case of an unidentified body, the body should be examined for peculiar identifying features like congenital malformation (cleft lip, cleft palate, syndactyly, polydactyly, etc.), tattoo, scars, birthmarks, and acquired deformity (amputation of fingers/ limbs, club foot, kyphosis, scoliosis, etc.).

3.8.3 Post-Mortem *Lividity/Livor Mortis/*Post Mortem *Hypostasis*

After an individual dies, their heart stops circulating the blood throughout the entire body. As a result, the blood present in the vessels, start coalescing towards the dependent regions of the body (Shedge et al. 2020). For example, in the case of a person lying on their back, the dependent regions will be the back and the buttocks of the individual. Within 1–3 h after death, patches of reddish-blue discolourations can

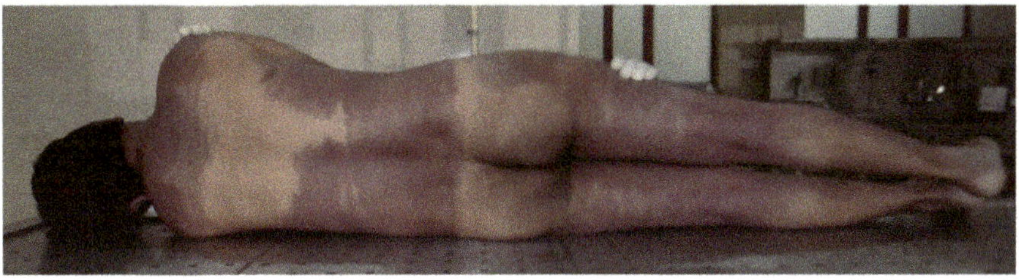

Fig. 3.1 Presence of livor mortis on a deceased's back

be seen in the dependent regions. Within 4–6 h, these patches coalesce and increase in size. Complete development of the livor mortis can be observed between 6 and 8 h after death. Before 6–8 h after death, if the cadaver's position is shifted, there is a shift in the post-mortem staining as well. This shifting may not be possible at 6–8 h after death, and the lividity is said to be 'fixed' in this case. This 'fixation of post-mortem staining' occurs as the capillaries of the dependent region that contain the blood rupture, and the blood that comes out of them stains the neighbouring tissues. This reddish-blue staining of the dead body discolourations after death is known as post-mortem lividity, livor mortis, hypostasis, or post-mortem staining. Post-mortem lividity is recorded in terms of colour, position, and whether it is fixed or not (Fig. 3.1).

3.8.4 Rigor Mortis

Immediately after death, the muscles of the cadaver get relaxed, a phenomenon which is known as primary relaxation. However, as time goes on, the muscles of the cadaver undergo stiffening, which is resolved only after the onset of putrefaction. This stiffening of muscles is known as rigor mortis or post-mortem rigidity/stiffening (Shedge et al. 2020). When an individual dies, there is a steady depletion of levels of adenosine triphosphate (ATP). ATP is necessary for the relaxation of muscles, and its lack causes the body to remain in the position it had acquired immediately after death. Rigor mortis is first observed in the eyelids, followed by the muscles of lower jaw, neck, chest, arms, abdomen, legs, and finally the fingers and the toes. Post-mortem rigidity appears within 1–2 h after death, is completely formed within 12 h of death, stays for another 12 h, and vanishes in the same order it appeared in the next 12 h. This order of appearance and disappearance is known as the 'march of rigor'.

The body is examined for the degree of rigor mortis whether it is developing or fully developed. The rigor is evaluated in all the joints starting from the face and downwards towards the limbs and is noted in terms of the presence or absence of stiffness (Fig. 3.2).

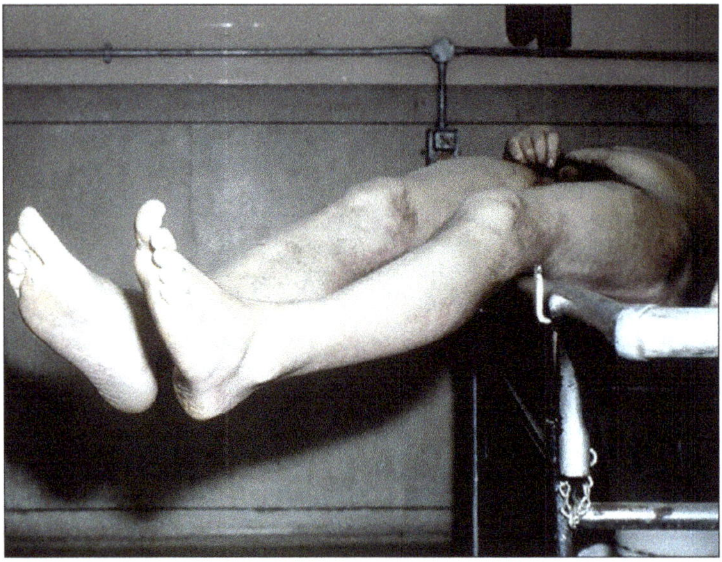

Fig. 3.2 Deceased showing rigor mortis in their lower limbs (Mesri M, Behzadnia M, Dorooshi G. Accelerated rigor mortis: A case letter. J Res Med Sci 2017; 22:126)

3.8.5 Algor Mortis

After the death of a person, the temperature of their body falls steadily with respect to the ambient environmental temperature. This fall in temperature is known as algor mortis (Shedge et al. 2020).

The body temperature is recorded at the mortuary. The temperature recorded is core body temperature, the measurement of which is obtained from a natural body orifice, most preferably the rectum. An electric thermocouple is used for recording the body temperature, the electrode of which is inserted into the rectum. However, in case of the absence of an electric thermocouple, a chemical thermometer can be used. If the body is refrigerated in the mortuary cold chamber, then recording the temperature is not necessary.

3.8.6 Injuries

All the external injuries are then to be examined and described. The injury is described as per their type, location in relation to anatomic landmark, shape, colour, and dimension. Photographs should be obtained for all the injuries. In cases where the body hair has masked the injury, shaving should be done prior to examination of such an injury. Sometimes the actual injury is masked by the blood stain which should be cleaned prior to the examination. In a suspected case of sexual-homicide, ano-genital swabs are obtained during the external examination. The swabs should be obtained before the measurement of body temperature.

3.8.7 Decomposition/Putrefaction

After the death of a person, on account of the natural flora within the body, the body's own enzymes, and the presence of external microorganisms, the body starts undergoing decomposition (Shedge et al. 2020). The action of the body's own enzymes is known as autolysis, while the rest is known as putrefaction. The different microorganisms act in a different manner, and the chief perpetrator amongst all the microbes is *Clostridium perfrigens* (previously known as *Clostridium welchii*) which causes putrefaction by secreting an enzyme known as lecithinase. Lecithinase causes haemolysis of the blood within the blood vessels and tissues.

The action of different microbes leads to the production of different putrefactive gases such as hydrogen sulphide, carbon dioxide, ammonia, and methane. The first external sign of putrefaction is the greenish discolouration of the right iliac fossa (Fig. 3.3). The greenish discolouration is due to the conversion of haemoglobin into sulphmethaemoglobin because of the action of hydrogen sulphide. Since the microorganisms use blood vessels as means of traversing throughout the body, the superficial vessels of the arms, thorax, and abdomen develop a deep green colour. These coloured vessels give an appearance similar to veins present over marble tiles, and hence the phenomenon is known as marbling (Fig. 3.4).

As the gases keep on developing in the body, they start accumulating in the different cavities of the body. The accumulation of gases causes bloating of the cadaver (Fig. 3.5). The gases lead to other findings such as distension of the abdomen, protrusion of eyes, lips, and tongue, swelling of face and external genitalia, and purging of putrefactive liquids from natural orifices of the body. The skin of the cadaver may slip, and there may be a formation of putrefactive bullae on the

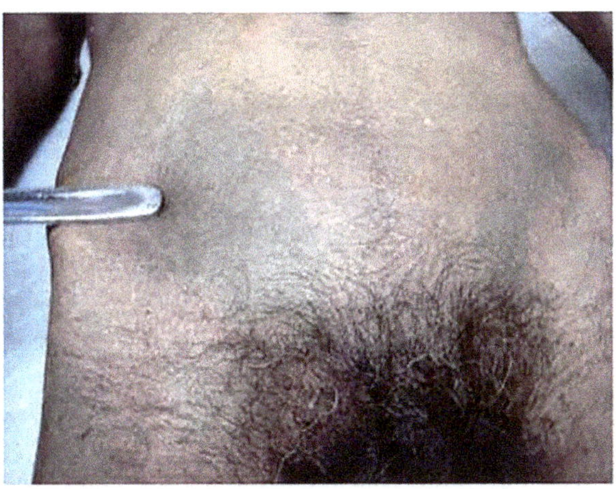

Fig. 3.3 Greenish discolouration of the right iliac fossa (Mesri M, Behzadnia M, Dorooshi G. Accelerated rigor mortis: A case letter. J Res Med Sci 2017; 22:126)

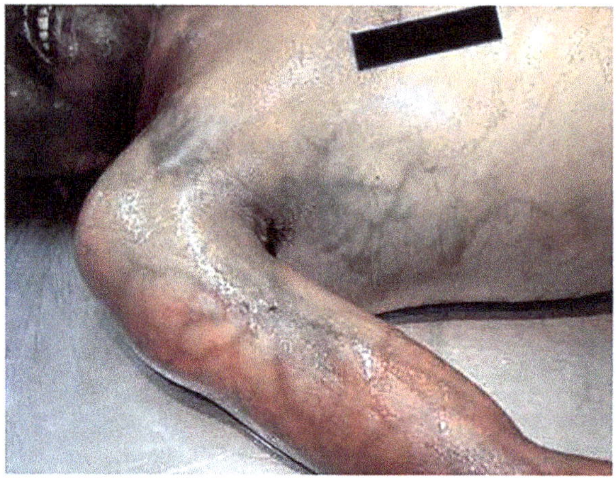

Fig. 3.4 Marbling (Mesri M, Behzadnia M, Dorooshi G. Accelerated rigor mortis: A case letter. J Res Med Sci 2017; 22:126)

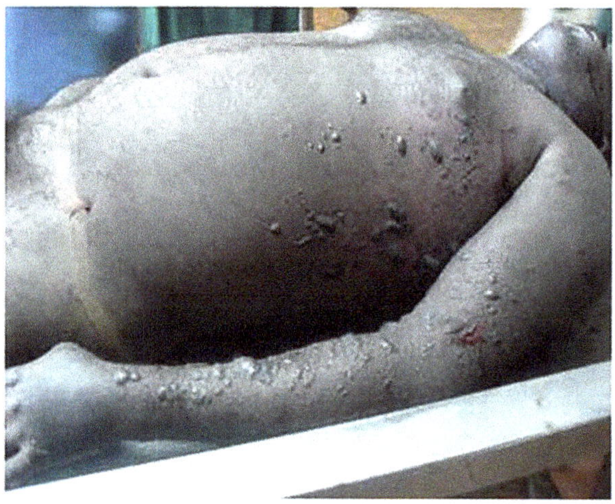

Fig. 3.5 Bloating (Mesri M, Behzadnia M, Dorooshi G. Accelerated rigor mortis: A case letter. J Res Med Sci 2017; 22:126)

cadaver (Fig. 3.6). Eventually, there is liquefaction of most of the organs, and the abdomen may rupture due to the continued gaseous pressure. During the post-mortem examination, the sign and degree of decomposition are to be examined.

3.8.8 Miscellaneous

Furthermore, the oral cavity is to be examined for any injuries (laceration of the frenulum, denture, tongue bite, and gagging). Nails are to be examined for disease and any peculiarity notes (break and heaping of tissue in nail bed). The nails in a suspected case of homicide are clipped and the contents of the nail bed are scrapped off with a toothpick and collected for further examination. In case of suspected firearm injury, a dermal nitrate test is done for the presence of gunshot residue.

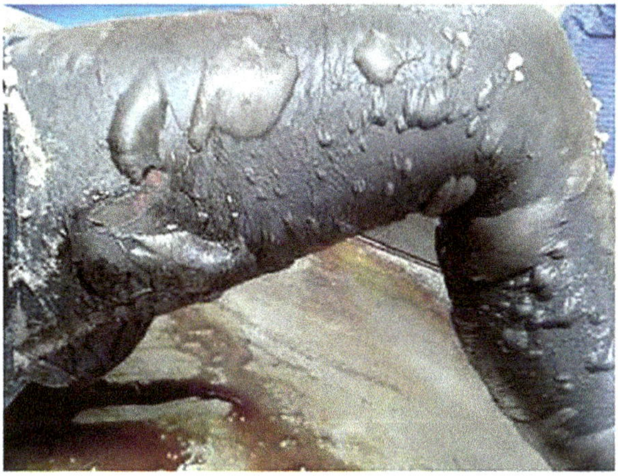

Fig. 3.6 Putrefactive blisters on the cadaver (Mesri M, Behzadnia M, Dorooshi G. Accelerated rigor mortis: A case letter. J Res Med Sci 2017; 22:126)

3.9 Internal Examination

After the external examination is complete, an internal examination is done which consists of a surgical opening of the body cavity, taking out the contents, and dissecting them for examination. Internal examination commences following the skin incision.

Types of Skin Incision

- Coronal incision (skull)
- I-incision (trunk)
- Y-incision (trunk)
- Modified Y incision (trunk)
- Fourth incision (trunk)
- Fifth incision (trunk and back) (Kapila et al. 2018)
- X-incision (back)

Following the skin incision, the body cavity is reached. In the head after reflection of the skin, the skull bone is sawed in order to facilitate the examination of the brain. For this purpose, an electric saw is used. In the absence of an electric saw, councilman's saw can also be used. Similarly, in the trunk the ribs are cut at the costochondral junction to remove the sternum so as to facilitate the examination of thoracic contents. There are various ways of removal of the thoraco-abdominal contents at autopsy. Once the organs are removed, they are individually studied with further dissection. All the organs are washed (except for the lungs which should not be washed before dissection), weighed, and dissected to facilitate the gross pathology or changes. The body cavity is then examined for any significant findings including injuries or pathology. The ribs are examined for any fractures.

Techniques of Organ Retrieval at Autopsy

- Technique of Ghon
- Technique of Letulle
- Technique of Virchow
- Technique of Rokitansky

Once all the organs are dissected and the changes noted, all the organs are put back into the body. Only those organs or tissues which show gross changes and are required for further examination (e.g. histopathology) are retrieved. Such retrieval of organs or tissue is always documented. The retrieved tissue is kept in formalin solution, sealed, labelled, and sent to the forensic science lab, maintaining the chain of custody. Once the internal examination is complete, the skin is sutured, following which the body is washed, cleaned, and wrapped. Gross deformity or mutilation of the body if present should be cosmetically restored in the best possible way during the suturing. The body should be handed over to the representative of the legal authority who requested the autopsy.

There are instances when a weapon or foreign body may be found lodged inside the body. Sometimes the police to bring a weapon of offence along with the body to the mortuary for examination. In such cases the forensic pathologist should further examine such objects and describe them and document with photographic evidence.

3.10 Ancillary (Special) Procedures

The forensic pathologist then takes the help of ancillary methods to determine the cause of death. These special procedures are conducted either to find out the definite cause of death or for the identification of the unknown deceased. These procedures are either conducted in the mortuary itself in the presence of the forensic specialist or the body fluid sample, tissue sample, or trace evidence is subjected to laboratory testing. Some examples of ancillary procedures are

1. Serology (blood testing) (Baleriola et al. 2012).
2. Histopathological (microscopic) examination (Knight and Saukko 2016).
3. Cytochemistry (Bohnert et al. 2019).
4. Radiological investigation: In case of foreign bodies and cases of firearm-related deaths, a radiological examination of the dead body is conducted. In case of unknown bodies, radiological assessment can be done to assess the fusion of epiphysis of the long bones.
5. Analytical toxicology: In suspected cases of poisoning or substance abuse, the quantitative and qualitative toxicological assay is conducted to identify the poison and the amount present in the blood.
6. Opening of back during autopsy in case of torture.

7. Opening of the vertebral column and examination of the spinal cord in case of suspected spinal injury.
8. Dissection of the epiphysis of a long bone to see the secondary ossification centre in case of fatal autopsy.
9. Microbiology culture.
10. Bite mark analysis.

3.11 Documentation and Furnishing Report

Proper documentation is an integral part of medicolegal investigation. Although slight variation can be seen in various autopsy reports, however, all the autopsy reports should include proper external and internal examination. Each and every autopsy is given a unique identification number which is integral for documentation and office records. An assistant should be present at the autopsy who will note all the external and internal findings during the autopsy. The forensic pathologist dictates the findings during the autopsy procedure where all the details are noted down. Body charts and diagrams are popularly used for the documentation of injuries.

Following are the integral sections of all autopsy reports

1. Unique identification number (autopsy number)
2. Authority requesting autopsy with letter reference number
3. Representative of the authority who identified and handed over the body for autopsy
4. Particulars of the deceased
5. Date and time of conduction of the autopsy
6. Date and time of completion of the autopsy
7. External examination findings including description of injuries
8. Internal examination findings
9. Conclusive opinion regarding cause and manner of death
10. Authorized signature and seal of pathologist conducting autopsy with names of mortuary technicians and assistants

After the completion of the autopsy, the reports are to be typed, printed, and copies of the signed reports are to be made for official records. The original is then released to the legal authority requesting the autopsy. In case ancillary procedures are done and the test results are awaited, the opinion pertaining to the cause of death is kept pending. Once all the reports arrive the opinion is updated, and the cause of death is then mentioned. In case of negative autopsy, after receipt of all the reports, the pathologist should state the opinion as negative autopsy rather than wrongly stating the cause of death.

The opinion of the autopsy report should include the following information:

- Identification of the deceased if not identified
- Cause of death

- Mechanism of death
- Manner of death (circumstance of death)
- Time since death

3.12 Identification

All the identifying features noted in the external examination would help in the identification of the deceased. Peculiar features like mole, scar, and tattoo would help to positively identify the person. Missing person record or complaint from police can be matched with the features present in the body which would also help in positive identification. Fingerprints obtained during the autopsy should be studied and matched with the database. Samples of DNA can also be obtained and analysed which would help to establish the identity of the person. In countries where there is a DNA database of the population, identification would not be difficult, however, in low- and middle-income countries that do not have a database, the DNA should be individually matched with the first-degree relative of the suspected missing person.

3.13 Cause and Mechanism of Death

No autopsy report is complete if the cause of death is not mentioned in the opinion section. Any disease or event (injury) which initiated an uninterrupted sequence of events that ultimately led to the death of the person is known as the cause of death. Any underlying physical, biochemical, or anatomical derangements that caused death are known as mechanisms of death.

Example: A 34-year-old male consumed over-the-counter NSAID for his long ongoing knee pain. He developed a duodenal ulcer which got perforated. He then developed peritonitis from the perforated duodenal ulcer and died. Here the cause of death is a duodenal ulcer, whereas the mechanism of death is peritonitis from a perforated duodenal ulcer.

If multiple causes of death are found in the same person, for example in a case of complex suicide, then all the causes of death are to be mentioned followed by the mechanism of death. Only mentioning the mechanism of death without the cause is not acceptable. For example, cardiac tamponade is the mechanism of death and cannot be written as a cause of death. There are various aetiologies for cardiac tamponade which could be trauma, myocardial infarction, or cardiovascular surgery. If post-myocardial infarction was a cause of tamponade, then the proper opinion to cause of death should be cardiac tamponade due to myocardial infarction.

If a lesion is found in any organ or tissue (example: caseous necrosis), then the sample should be sent for histopathological examination. The opinion in such cases should be made only after the confirmation of the lesion.

3.14 Time Since Death/Post Mortem Interval (PMI)

The interval between death and the conduction of post-mortem examination is the time since death and is also referred to post-mortem interval (Usumoto et al. 2019). Estimation of post-mortem interval is important in homicidal cases to narrow down the investigation period and which would further aid in narrowing down the suspects. There are various ways of estimating post-mortem interval.

- Physical method
  - Rigor mortis

    If the body is warm, and there is no appreciable stiffness, the PMI is said to be within a few hours.

    If the march of rigor is visible, but not complete (arms and legs show stiffening, but fingers and toes are still malleable), the PMI is said to be less than 12 h.

    If the rigor is completely developed, the PMI is said to be more than 24 h.
  - Algor mortis

    The rectal temperature of the cadaver is recorded.

    PMI can be estimated using the following formula:

    $$\text{PMI} = \frac{\text{Normal body temperature}(37\ {}^\circ\text{C}) - \text{Rectal temperature of the victim}}{\text{Rate of fall of temperature per hour}}$$

    The rate of fall of temperature per hour depends upon the geographical area, weather conditions, seasonal variations, etc. and is measured by recording the rectal temperature at regular intervals.
  - Livor mortis

    The dependent areas stained with reddish-blue colour are identified.

    If upon pressing that area, there is no whitish discolouration present, lividity is said to be fixed, and the PMI is said to be more than 6–8 h.

    If upon pressing, there is a whitish discolouration present (blanching), the PMI is said to be less than 8 h.

    If the lividity is present in terms of large patches which have not coalesced, the PMI is said to be less than 6 h.

    If only tiny patches over the dependent area portray the presence of post-mortem lividity, the PMI is said to be within 1–3 h.
  - Eye changes

    After death, the cornea starts getting hazy and the extent of haziness is directly proportional to PMI.

    The eyes also lose corneal and pupillary reflexes, there is a loss of intraocular tension, and the eyes become flaccid.

    If the eyelids are open after death, the dust gets deposited in the exposed areas, and there is a formation of the yellow triangular region that eventually becomes darker with time. This phenomenon is known as 'Tache noir de la sclerotique' (Fig. 3.7)
  - Gastric emptying time

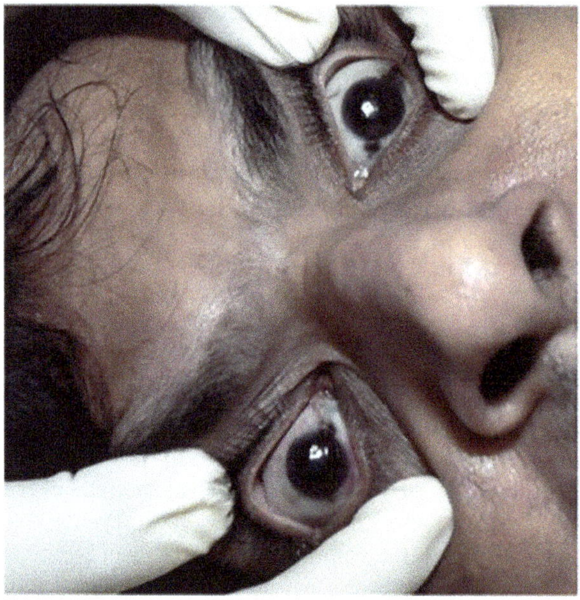

Fig. 3.7 Tache noir de la sclerotique

Under ordinary circumstances, it takes about 60–90 min for the stomach to become empty after ingestion of food. The amount of food in the stomach and its extent of digestion can comment upon PMI.

- Chemical method
  - Post-mortem cytochemistry

    Post-mortem cytochemistry deals with the study of the concentration/amount of different chemicals within the body and their relation to post-mortem interval.

    Body chemicals such as hypoxanthine, ammonia, NADH, and formic acid increase linearly after death and can thus be used to estimate PMI.

    Another biological fluid with great application in the estimation of PMI is the vitreous humour. After death, there is a steady increase in potassium levels within the vitreous humour.
- Biological method
  - Entomology of the cadaver

    As a cadaver starts undergoing putrefactive changes, certain foul-smelling gases are generated, that attract a hoard of different insects to the cadaver.

    Depending upon which insects are present over the body, as well as what stage of their life cycle these insects are in, PMI can be estimated.

Multiple Choice Questions

1. Which of the following is the first external sign of decomposition?
   (a) Formation of putrefactive blisters
   (b) Bloating
   (c) Greenish discolouration of the right iliac fossa
   (d) Protrusion of tongue
   Answer: (c)
2. What is the gradual drop in the temperature of an individual after death called?
   (a) Algor mortis
   (b) Livor mortis
   (c) Rigor mortis
   (d) Tache noir
   Answer: (a)
3. Post mortem staining or hypostasis is also known as?
   (a) Algor mortis
   (b) Livor mortis
   (c) Rigor mortis
   (d) Tache noir
   Answer: (b)
4. A post mortem report in which no cause of death can be provided, even after complete autopsy and ancillary procedures are done is known as __________
   (a) Clinical autopsy
   (b) Medical autopsy
   (c) Obscure autopsy
   (d) Negative autopsy
   Answer: (d)
5. Which of the following is considered to be a biological method of estimating PMI
   (a) Livor mortis
   (b) Rigor mortis
   (c) Cytochemistry
   (d) Entomology
   Answer: (d)

References

Advenier A-S, Guillard N, Alvarez J-C et al (2016) Undetermined manner of death: an autopsy series. J Forensic Sci 61(suppl 1):S154–S158. https://doi.org/10.1111/1556-4029.12924

Baleriola C, Johal H, Robertson P et al (2012) Infectious disease screening of blood specimens collected post-mortem provides comparable results to pre-mortem specimens. Cell Tissue Bank 13:251–258. https://doi.org/10.1007/s10561-011-9252-6

Bohnert S, Ondruschka B, Bohnert M et al (2019) Post-mortem cerebrospinal fluid diagnostics: cytology and immunocytochemistry method suitable for routine use to interpret pathological processes in the central nervous system. Int J Legal Med 133:1141–1146. https://doi.org/10.1007/s00414-019-02050-z

Cohle SD, Sampson BA (2001) The negative autopsy: sudden cardiac death or other? Cardiovasc Pathol 10:219–222. https://doi.org/10.1016/s1054-8807(01)00093-x

Cordner SM (1993) Deciding the cause of death after necropsy. Lancet 341:1458–1460. https://doi.org/10.1016/0140-6736(93)90892-k

De Matteis A, Del Fante Z, Santoro P (2020) Forensic pathology: past, present and future. Clin Ter 171:e302–e303. https://doi.org/10.7417/CT.2020.2232

Kapila P, Gupta R, Raina SK et al (2018) Development of a new skin incision for conduct of conventional autopsy. Egypt J Forensic Sci 8:54. https://doi.org/10.1186/s41935-018-0084-4

Knight B (1980) The obscure autopsy. Forensic Sci Int 16:237–240. https://doi.org/10.1016/0379-0738(80)90208-x

Knight B, Saukko PJ (2016) Knight's forensic pathology, 4th edn. CRC Press, Taylor & Francis, Boca Raton

Menezes RG, Monteiro FN (2022) Forensic autopsy. In: StatPearls. StatPearls Publishing, Treasure Island

Shedge R, Krishan K, Warrier V, Kanchan T (2020) Postmortem changes. In: StatPearls. StatPearls Publishing, Treasure Island

Usumoto Y, Kudo K, Tsuji A et al (2019) Predictive equation for post-mortem interval using spectrophotometric values of post-mortem lividity: a pilot study. Forensic Sci Int 297:47–55. https://doi.org/10.1016/j.forsciint.2019.01.014

Bloodstain Pattern Analysis

4

Aditi Mishra, Ragini Pandey, and Sarthak Misra

Abstract

Bloodstain pattern analysis plays a crucial role in determining the actions that have caused bloodshed. By combining the principle of math and physics, the bloodstain pattern could be interpreted. The detailed analysis of bloodstain shows different forces and activities influence the creation and appearance of these patterns so that they can be interpreted as part of the crime scene investigation. In this chapter, detailed information regarding the causes, types, factors, documentation, etc. has been discussed.

Keywords

Bloodstain evidence · Forensic science · Reconstruction · Passive pattern · Transfer pattern · Projected pattern

4.1 Definition of Terms

Bloodstain	Deposit of blood on a surface
Bloodstain pattern	Distribution of bloodstains in various shapes and sizes resulting from various factors such as surface, angle, distance, or type of weapon used

(continued)

A. Mishra
Kristu Jayanti College, Bangalore, India

R. Pandey
Department of Forensic Science, Jain University, Bangalore, India

S. Misra (✉)
Indira Gandhi Institute of Medical Sciences, Patna, India

A. Puri et al. (eds.), *Fundamentals of Forensic Biology*,
https://doi.org/10.1007/978-981-99-3161-3_4

Bloodstain	Deposit of blood on a surface
Bloodstain pattern	Distribution of bloodstains in various shapes and sizes resulting from various factors such as surface, angle, distance, or type of weapon used
Bloodstain pattern analysis (BPA)	Bloodstain pattern analysis is the technique used in forensic science to recreate the actions that caused the bloodshed
Passive bloodstain	Passive bloodstains patterns, such as drops, flows, and pools, are created by the force of gravity alone
Blood pool	A blood pool pattern is generally defined as an accumulation of blood found at the lowest point of flow patterns
Transfer stain	When any bloody object comes into contact with an unstained surface leaving a pattern known as transfer/contact stains
Projected patterns	Projected patterns are blood spatter, which is caused by a force other than impact
Splash patterns	A bloodstain pattern formed by a huge amount of liquid blood falling onto a surface
Splatter stain	A splatter stain is the result of blood being pressed against a liquid blood source
Expirated bloodstain	Expirated bloodstain patterns are generated by the blood that is evacuated through the mouth or nose as a result of an internal injury
Cast-off spatter	Cast-off spatter happens when blood is hurled from a blood-bearing object such as a hammer or knife
Void pattern	A void pattern is the absence of bloodstains in otherwise continuous staining patterns

4.2 Introduction

Bloodstain pattern analysis (BPA) is a technique used in forensic science to reconstruct the events that resulted in the bloodshed. In other words, it helps in the reconstruction of the crime by interpreting the bloodstains present at the scene of a crime. To form an opinion about the happenings of the crime scene, a forensic expert examines the size, distribution, shape, and location of the bloodstains. This technique is mostly used in the reconstruction of homicide cases such as murder or any other violent crimes where the presence of blood can be seen. Blood pattern analysis goes beyond mere speculation and instead utilizes evidence, reasoning skills, and expertise to determine what may have or not happened during a crime.

This technique is employed worldwide by forensic scientists, police officials, and medicos in an interdisciplinary manner. It is very common for the individual to be injured in case of any violent crimes and leave blood or bloodstain as evidence. This might leave some distinctive patterns and help in interpreting important investigative events that have happened during the commission of the crime. Bloodstain patterns occur because blood has biological (blood behavior), physical (cohesion, capillary action, and velocity), and mathematical (geometry, distance, and angle) property which reacts when acted upon by physical forces (Ablett 1983).

When liquids in flight are depicted in general, they are frequently depicted as having a teardrop shape. However, this is very different from reality. When a liquid falls from an object, it takes the shape of a drop. Once in the air, the drop takes on its smallest surface area and flies in a spherical shape. The flowing of blood patterns is affected by many physical forces such as air resistance and gravity. Although this form of stain is well known, there are other ways in which gravity affects blood like in a pool of blood. The effect of gravity would let down the sinking of blood cells and separate the serum above as a clear transparent liquid (Allen 1995). This process would create serious misunderstandings or doubts among the individual who is unfamiliar with this concept and may believe that liquid has been introduced to the blood. However, in actuality, this is a natural blood pool in which the blood was standing. This is where expert opinion would play a crucial role (Allery et al. 2001).

Another issue is that diverse sources can sometimes result in stains that seem the same. As a result, to make an appropriate assessment, a full understanding of physics and bloodstain pattern analysis, as well as differential diagnoses, is required. It is also worth noting that scenarios can emerge where onlookers are more bloodied than the perpetrator. This can happen when a weapon is flung overhead in front of a group of people, for example. Although such occurrences are uncommon, they are a compelling cause to visit a bloodstain pattern analysis professional. As a result, the bloodstain pattern aids investigators in addressing questions like:

- What was the source of the blood?
- What happened to cause the wounds?
- In what direction did the victim get hit?
- What was the position of the victim(s) and perpetrator(s)?
- What steps were taken following the bloodshed?
- How many criminals were involved there?
- Does the bloodstain evidence approve or disprove the witness's testimony?

Forensic serologist identify and study the bloodstain patterns with established scientific techniques and draw scientific conclusions about how the blood was shed. The analysis of these patterns gives some detailed information regarding the direction of travel of the blood, the location of the blood source which was acted upon to create the pattern, the level of force of injury, any movements during or after the bloodshed, and other activities. The detailed analysis of bloodstain shows how different forces and activities influence in creation and appearance of these patterns so that they can be interpreted as part of the crime scene investigation (Anoruo et al. 2007).

BPA gives information on not just what happened, but also what could have happened. This could help the investigator in the reconstruction of the crime, corroborating witness testimony, and including or excluding probable culprits from the investigation (Baxter 1973). In many of the world's most notorious criminal cases, bloodstain evidence has been a deciding factor. As a result, for individuals on both sides of the courtroom, substantiation of this evidence is much more critical.

4.3 Historical Perspective of BPA

Bloodstain pattern analysis has been frequently used in the investigation of homicide cases or any other violent crimes. The roots of bloodstain pattern analysis can be traced back more than 100 years. Some literature on bloodshed characteristics has been found by Herbert Leon MacDonnell which dated back to the 1500s.

However, the first person to scientifically investigate bloodstain patterns was Dr. Eduard Piotrowski. In 1895, an article was published that examined bloodstain patterns resulting from head wounds. Following Piotrowski's discovery, other articles were published that shed light on certain elements of bloodstains but did not lead to a systematic examination. Balthazard and colleagues (1939) published findings of reconstructing the angle of impact from the impact pattern by measuring the width and length of small bloodstains.

Another research conducted by Schmidtmann showed the possibilities of bloodstain patterns in the reconstruction of the crime scene. The beginning of the bloodstain pattern in the forensics as crucial evidence was marked in the homicide case of the State of Ohio v. Samuel Sheppard (1955). Dr. Sheppard was charged guilty in the death of his wife Marilyn Sheppard and later got convicted. Dr. Kirk examined the bloodstain patterns present on the walls of Marilyn Sheppard's bedroom and testified as an expert in the courtroom. Dr. Sheppard was acquitted and this shed some light on bloodstain pattern analysis for forensic purposes. Herbert Leon Macdonell's in 1971 published the book "Flight Characteristics of Human Blood and Stain Patterns."

Later in 1972, one training program was also conducted on bloodstain patterns by Macdonell. Many institutes have participated in this training program which gave more popularity to this technique. With more popularity, several kinds of research have been conducted and got published in reputed journals. Various professional forensic associations have been created such as "The International Association of Bloodstain Pattern Analysts (IABPA)." Several conferences and certification programs have been conducted with worldwide participants.

One Scientific working group for bloodstain pattern analysis (SWGStain) was also hosted by the Federal Bureau of Investigation. This group published recommendations for training bloodstain pattern analysts, recommendations for the content of a basic bloodstain pattern analysis workshop, guidelines for developing standard operating procedures, quality assurance guidelines, and a recommended terminology for bloodstain pattern analysis. Recently, software-based programs have been also made for bloodstain pattern analysis but they are not used frequently. They still lack advancement and more work has to be done on software-based programs for interpreting the bloodstain pattern evidence (Blake and Sensabaugh 1976).

4.4 Characteristics of Bloodstain

Some several factors and characteristics influence the size and shape of the bloodstains. Blood is a complex composition of blood cells and plasma. Four types of blood cells are found in humans such as red blood cells, white blood cells, platelets, and plasma (Boward and Wilson 2013). These cells of the blood are suspended in a liquid. Red blood cells are the main component of the solid portion of blood. RBC does not have the genetic material—deoxyribonucleic acid (DNA). The other solid portion is the white blood cells, which contain DNA.

Since it contains genetic material, it is very crucial for forensic analysis. Blood also has clotting factors present in the plasma, which is the liquid portion. The liquid that remains after the blood has clotted is called serum. Like other liquids, blood has no shape of its own and hence adjusts according to its surrounding. The physical characteristics of the blood such as specific weight, viscosity, and surface tension are very crucial in the interpretation of its morphological characteristics (Boward and Wilson 2013). Blood also has some biological, serological, and immunological characteristics but has minor importance in bloodstain pattern analysis (Chapman et al. 1989). Hence, it will not be discussed in this chapter.

When blood passively strikes off a surface and falls onto a smooth hard horizontal surface, we have a very even round pattern. As the blood that hit, is put in motion, those blood droplets form oscillating sphere shapes. If blood droplets are fallen due to gravitational force alone, then droplets would be intact and not broken. However, any other external forces to the mass of blood would result in smaller droplets. Factors other than droplet volume, drop distance, and the blood source surface properties also affect the size, shape, and pattern of the bloodstains (Chen and Hortin 2000).

Higher effective energy would destroy surface tension and this would result in smaller droplets. The diameter of the bloodstains is generally dependent upon two things, one is the distance from which the droplet is falling to the surface and the other is the volume. Distance and volume are both directly proportional to the bloodstain. More the distance of the fall, more the diameter of the bloodstain. However, once it reaches the maximum height, i.e., more than 2.5 m (7 ft.), a very slight change would be seen in bloodstain diameter (Allery et al. 2001). The larger the volume of bloodstain, the more likely to be the diameter. The average volume of blood droplets is known to be 0.05 ml (Allen 1995). Further studies have shown that the average amount of passively dripping droplets depends on the surface properties of the object on which the blood drips. The falling of blood droplets from a surface is due to increasing its volume to a level where the earth's gravity dominates the viscosity, thus resulting in poor surface tension, causing blood droplets to fall from the surface.

The physical properties such as viscosity, specific density, and surface tension of blood make it difficult to break down into droplets. When an external force is applied to the static blood pool, some of the blood reacts and is thrown into the air. The distance a droplet travels in the air depends on the amount of force used to create the droplet, the size of the blood droplet, and the air resistance. If the exerted force is low then this would create a lesser number of blood droplets and also their size of them would be larger. Most droplets are over 3 mm in diameter, but very few are less than 3 mm in diameter. As the level of force decreases, the distance travelled by these droplets would be short. If the exerted force is high then this would create a greater number of blood droplets. As the level of force increases, the distance travelled by these droplets would be increased. The more the force, the less the volume and diameter of the individual droplets, and many blood droplets are 1 mm in diameter. Large droplets are still present, but these droplets tend to move farther from the blood source than smaller droplets having a size of 1–2 mm in diameter.

Small blood droplets lack the physical weight to resist airflow and friction, so they travel a shorter distance from the blood source and dissipate energy quickly. Most small droplets travel within 1 m (39 in.) of the blood source. In case of extremely high force such as gunshot spatter, thousands of blood droplets less than 1 mm in diameter are generated and propelled into the air. The larger the distance from the blood source, the larger the droplet size, but these droplets are smaller (2–3 mm) compared to the larger droplets produced by the low-force blood spatters. Many of the larger droplets produced by these high-force events are the result of smaller droplets that collide during flight and coalesce into larger volumes of blood droplets. The shape of the bloodstain could be known by the angle between the trajectory of the droplet and its surface.

When a drop of blood hits the horizontal plane from a vertical angle, the resulting blood stains are rounded. Lesser the angle between the blood droplet's flight path and the target surface, the more the length and less the width of bloodstains. We could also say, as the size of the angle decreases, the bloodstains become longer and narrower. Bloodstain coming from different angles looks different. Suppose a blood droplet has hit the target surface from a 15-degree angle, then the shape of this bloodstain would be long and the width would be narrower. The presence of a tail could be observed on these long narrow stains. When a blood droplet hits the target surface, the body of the droplet attaches to the surface. A small portion of a drop of blood breaks at the top and continues in the forward direction and gives the shape of a tail. In other words, the primary spatter sticks to the surface and the secondary spatter would appear like a tail (Fig. 4.1).

The tail is an important tool in reconstructing the bloodstain pattern. The tail points in the direction the droplet was moving when it hit the surface of the target. After an initial crime scene when photography and documentation have been completed, the crime scene investigator places a reference scale near the blood spatter and selects the individual stain for analysis. The investigator draws a line extending along a path or length axis of each select stain. The area where this line intersects is called the area of convergence which indicates the approximate height of the incident that produce bloodshed.

The examiner uses a loop to measure the length and width of each stain and then calculated the angle at which the droplet impacted the surface. The length and width of the specific bloodstain could be used to calculate the angle of impact (Fig. 4.2). The most frequently used formula for this calculation: Angle of impact = arc sin w/l. Once the investigator knows the area of convergence and impact angle for a group of

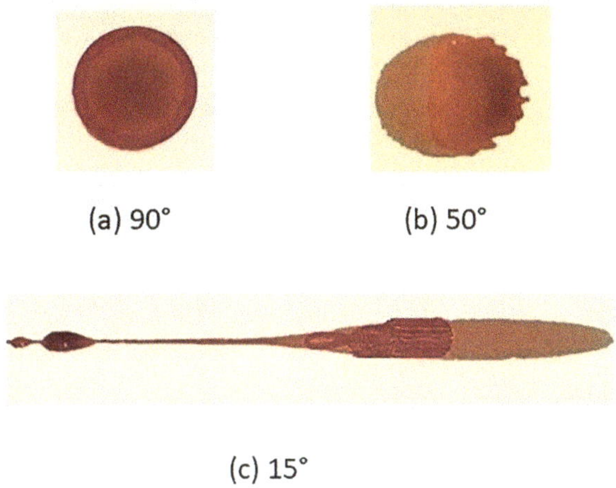

Fig. 4.1 Shape of bloodstain projected from different angles

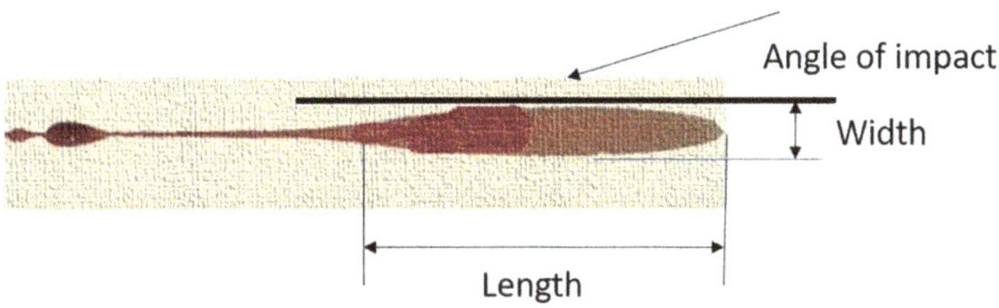

Fig. 4.2 Schematic representation of angle of impact

stains, the string process begins. The investigator attaches one end of the string to each stain's location at the point where the blood drop contacts the surface. Then, using your protractor hold the strings along the angle of impact and secure them to something stable. Once several strings are emplaced, the area of origin becomes visible. This complex time-consuming processing creates the 3-dimensional model that indicates where the victim was located at the time of the event.

The shape of the bloodstain and its width-to-length ratio are also dependent upon the type of surface on which blood strikes. The reconstruction and interpretation of the crime scene would be easier if the blood strikes a smooth, hard, nonporous surface. If the surface has properties that distort the shape of the bloodstain, then reconstruction and interpretation get difficult. The appearance, size, and shape of the bloodstains are also dependent upon the type of surfaces such as absorbent or non-absorbent. In the case of an absorbent surface, the blood gets soaked inside and thus resulting in altered width-to-length ratios.

These alterations may result in an inaccurate angle of impact determination; this information could prevent an accurate pattern reconstruction for locating the pattern blood source. In case of a rough surface, the blood droplet would break apart. This would lead to an inaccurate angle of impact.

Bloodstains found at the scene of crime have various shapes and pattern that helps in identifying the nature of the wound from which the blood originated. Every bloodstain has its unique geometrical properties that are very crucial in differentiating these bloodstain patterns. To maintain an objective overview, bloodstain patterns need to be described in a specific and unique way. This allows experts to understand the mechanism behind blood stain patterns while reading bloodstain patterns analysis reports. A standardized scientific approach to the interpretation of bloodstain patterns was defined by the "Working Group on Bloodstain Pattern Analysis" (SWGSTAIN) (Davies and Wilson 1974), which mainly complies with the terminology used by the German "Arbeitsgruppe für Blutspuren–Verteilungsanalyse." The three broad categories of bloodstain patterns are explained in Fig. 4.3.

4.4.1 Passive Drop

Passive bloodstains patterns, such as drops, flows, and pools, are created by the force of gravity alone. Such patterns can be found as a direct result of passive flow patterns on uneven surfaces (carpets, tar, etc.) with fairly high resistance (Fig. 4.4). This tread provides information about the process of movement. Drip stains are drops that fall undisturbed in the air. It physically maintains its spherical shape without breaking into small droplets.

Such a blood drop may descend from an exposed wound or any object containing a sufficient amount of blood to permit the formation of a drop. Flow pattern describes a change in the shape and direction of a bloodstain due to gravity and the angle of the surface on which it is formed. Flow patterns and pools could provide a piece of important information regarding the victim's movements during the attack as well as post-mortem movement or alteration of the body at the scene of death. These patterns can be found, for example, on the victim's body and clothing, and on the surface on which the victim lies. The type of surface and bloodstain angle are very crucial in interpreting the static findings. For example, the interruption of a flow pattern can aid in determining the sequence and duration of time between the flow and the interruption. If the direction of gravity does not match the flow of the blood on an object or body, it is conceivable that the object or body was relocated after the blood had dried.

A blood pool pattern is generally defined as an accumulation of blood found at the lowest point of flow patterns. Blood that pools on absorbent surfaces may be absorbed and diffused over the surface, resulting in a pattern that is larger than the initial pool. This is common in pools and on mattresses and sofas.

When any bloody object comes into contact with an unstained surface they leave a pattern known as transfer/contact stains. These are the result of compression or lateral movement. Examples include tool prints, bleeding fingerprints, footprints, palm prints, and other transfer patterns. Blood transfer patterns can be used to determine class or individual characteristics. These patterns aid in identifying the

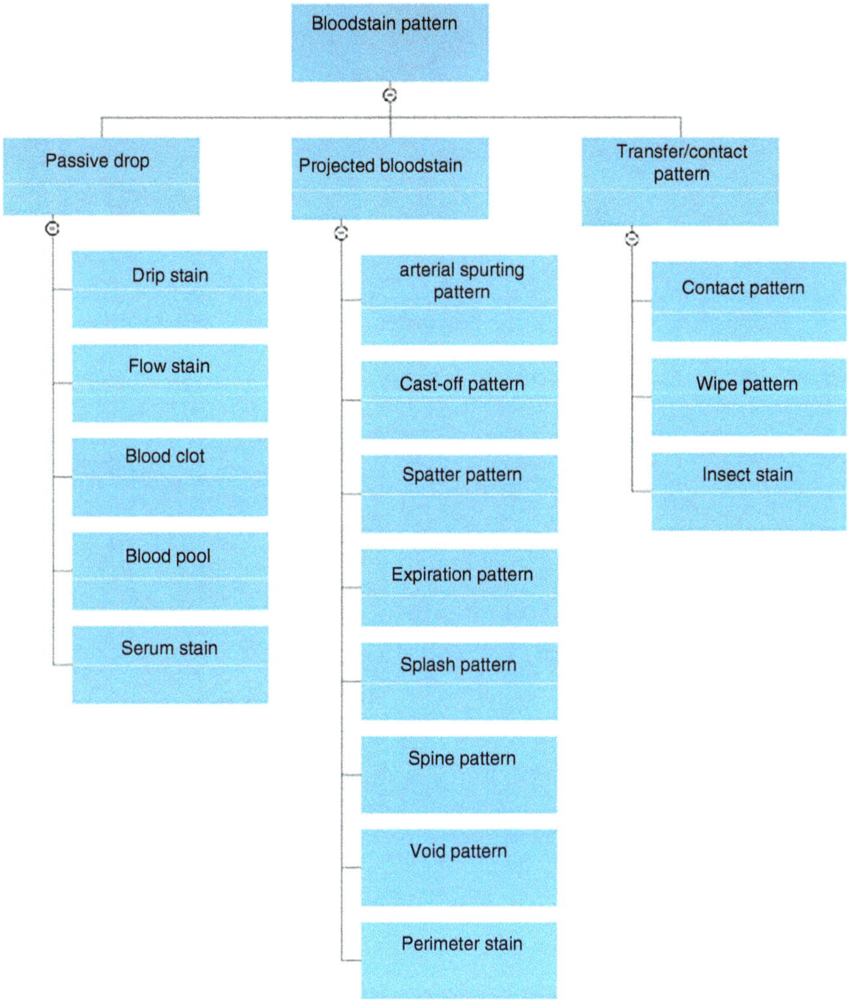

Fig. 4.3 Different types of bloodstain patter

victim's perpetrators or any other individuals present at the crime scene's directionality.

Blood has a strong adhesive force and due to this stain gets transferred from one surface to another. Due to its transferrable nature, the identification of the origin of the pattern might be unobtrusive. Contact patterns can give a good idea of the shape and size of the object in question (such as hard objects, fingerprints, hair, body parts, or soles). Depending on the condition of the sole (of the shoe), the ground, and the strength of the steps (slow, careful walking and high speed, intensive running), these contact patterns can also be seen at considerable distances (30–60 m) from the crime scene (Fig. 4.5).

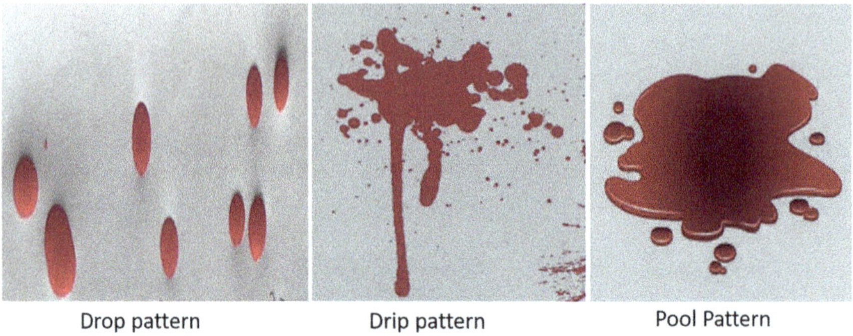

Fig. 4.4 Different types of passive bloodstain pattern

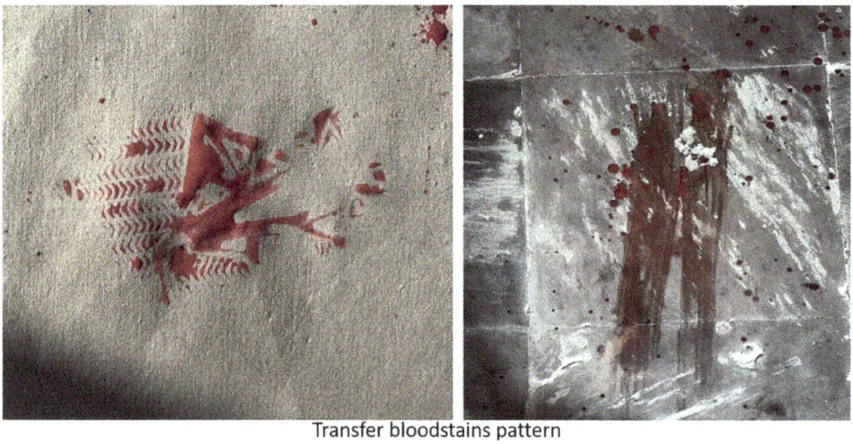

Fig. 4.5 Different types of transfer bloodstain pattern

In addition, you can distinguish between primary and secondary wipe patterns. The primary wipe pattern occurs when a bloody object moves tangentially over the surface and leaves a trail (called a wipe pattern). A secondary wipe pattern occurs when a clean, uncontaminated surface or object is tangentially dragged over an existing bloodstain (called a wipe pattern). Wipe patterns help in determining the chronological order of various incidents.

4.4.2 Projected Patterns

Projected patterns are blood spatter, which is caused by a force other than impact (Fig. 4.6). In arterial spray spatter, blood is projected from an arterial blood vessel in variable volumes when the artery is breached. Blood spurts out of the damaged area due to the pressure of continuous blood pumping. Arterial blood stain's size ranges

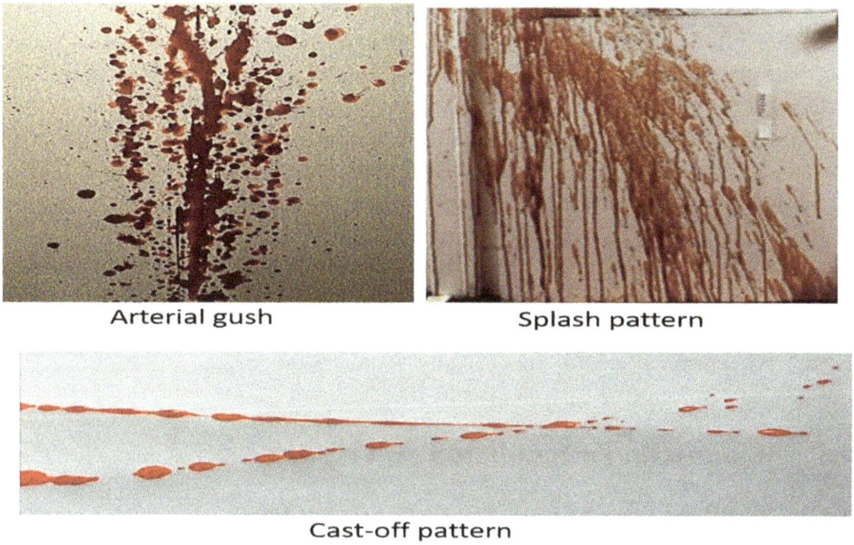

Fig. 4.6 Different types of projected bloodstain pattern

from massive gushing or spurting patterns each time the heart pumps to extremely little spray patterns. These patterns are easily distinguishable as the oxygenated blood spurting from an artery is bright red color than blood ejected from impact wounds.

A bloodstain pattern formed by a huge amount of liquid blood falling onto a surface. These patterns usually have a large central area with elongated peripheral bloodstains. They usually mean a cohesive impression, slightly separated into smaller lace patterns.

A splatter stain is the result of blood being pressed against a liquid blood source. You can distinguish between forward spatter caused by force directional movement and backward spatter caused by opposite directional movement. The effect in this context is any type of blunt trauma that is strong enough to break a homogeneous pattern by overcoming surface tension.

Expirated bloodstain patterns are generated by the blood that is evacuated through the mouth or nose as a result of an internal injury. As a result of being diluted by saliva, expirated blood may look lighter in color than impact spatters. The stains may contain visible air bubbles created by blood combining with air from the respiratory tracts or lungs if the blood was recently evacuated. There are many reasons for expirated blood such as injuries in skull, neck, or lung bleeding injury.

Cast-off spatter happens when blood is hurled from a blood-bearing object such as a hammer or knife. The size of the spatter is proportional to the size of the object's point; for example, the spatter formed by a bat will be greater than the spatter created by a sharp knife. The number of blows can also be calculated using the number of cast-offs spattered. The size of cast-off patterns is commonly smaller (6–7 mm) than passive drop patterns.

Gunshot spatter can cause mist-like dispersions (typical of high-velocity spatter), which are little blood spatters less than 0.1 mm in diameter. It can also create forward and backwards spatter from an entrance incision and exit wound as well. The distance between the firearm and the victim determines the amount of back spatter on a firearm or a shooter. The size of the spatter is governed by a number of parameters, including the caliber of the weapon, the amount of available blood, the position, and the number of bullets and impediments such as hair, clothing, and so on.

A void pattern is the absence of bloodstains in otherwise continuous staining patterns. The empty space on the surface or object could reveal information about the size and shape of the missing thing or person. This aids in establishing a crime scene's sequence and identifying changes.

4.5 Bloodstain Pattern Based on Velocity

4.5.1 Low-Velocity Impact Blood Spatter (LVIS)

A force or energy corresponding to normal gravitational pull up to a force or energy of 5 ft./s is called low-velocity (5 ft./s). The stain that results is frequently rather big, measuring 4 mm or more in diameter (Fig. 4.7). Low-velocity impact spatter has the following characteristics:

- the majority of spatters are larger than 3 mm in diameter;
- the formation of long, spiny projections is possible;
- the tail of the splatter frequently points toward the source rather than the direction of travel.

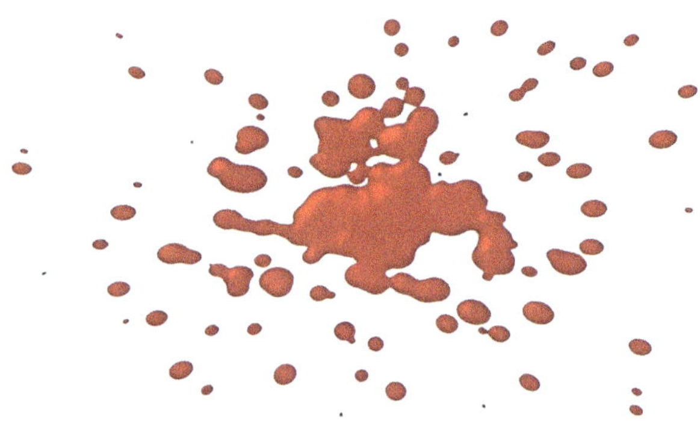

Fig. 4.7 Low-velocity impact blood spatter

4.5.2 Medium-Velocity Impact Blood Spatter (MVIS)

Medium velocity is defined as a force of 5–25 ft./s that is applied to a blood supply. The stains formed will be in the size range of 1–4 mm (Fig. 4.8). Medium velocity impact spatter has the following characteristics:

- the majority of spatters are 3 mm in diameter or smaller;
- spatter can travel long distances from its source.

4.5.3 High-Velocity Impact Blood Spatter (HVIS)

These types of bloodstains form when the blood source is exposed to a force greater than 100 ft./s. The resulting stain is typically less than 1 mm in diameter, though both smaller and larger stains may be visible (Fig. 4.9). These stains are commonly caused by gunshot wounds, explosions, and mechanical catastrophes. High-velocity impact spatter has the following characteristics:

- larger drops can reach greater distances than 3–4 ft.;
- atomized spatters only move 3–4 horizontally from their origin;
- gravity pushed them down fast due to their small size;
- entrance wounds were distributed in a cone-shaped pattern from the impact site in the direction of the original bullet;
- because blood is carried out with the projectile, exit wounds produce more spatter than entrance wounds.

Fig. 4.8 Medium-velocity impact blood spatter

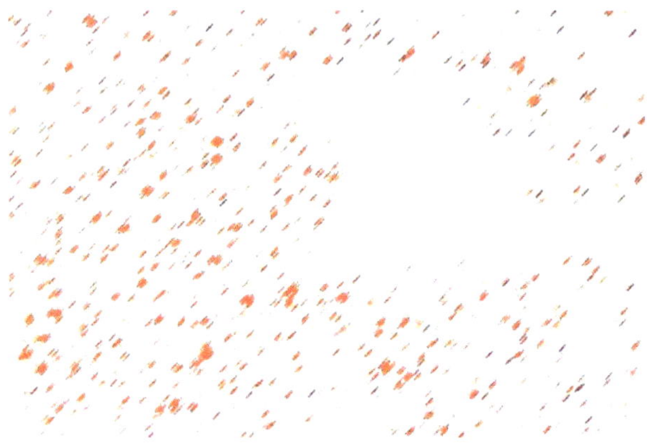

Fig. 4.9 High-velocity impact blood spatter

4.6 Blood Drying Times

The drying time of the bloodstain pattern is affected by many environmental and physical factors. Blood volume is the most important physical factor that affects the time it takes for a stain to dry. When exposed to the same environmental conditions, large amounts of blood dry more slowly than small amounts of blood. Environmental factors such as temperature, humidity, surface properties, and airflow affect the rate at which blood samples dry. Warm temperatures promote drying and speed up the drying of blood. Cold or very cold usually impedes the drying time of blood. Humidity also affects the drying time of blood.

When the same amount of liquid blood is deposited in environments with different humidity levels, blood in a high-humidity environment dries more slowly than blood in a low-humidity environment. Drying occurs when the moisture in the dirt evaporates into the surrounding air. In a high-humidity environment, the air is saturated with water, which suppresses evaporation and prolongs the drying time of blood. The properties of the surface on which the blood is deposited also affect the drying time of the blood. If the surface protects the blood by limiting the number of stains that are exposed to the environment, the blood will be less dry and the drying time will be longer.

Blood deposits on surfaces that maximize the surface area of stains exposed to the environment reduce drying time. An environment that provides good airflow to the exposed surface of the blood reduces the time required for the stain to dry. There are several factors and characteristics that influence the size and shape of the bloodstains. Blood is a complex composition of blood cells and plasma. Four types of blood cells are found in humans such as red blood cells, white blood cells, platelets, and plasma (Boward and Wilson 2013). These cells of the blood are suspended in a liquid. Red blood cells are the main component of the solid portion of blood. RBC does not have the genetic material—deoxyribonucleic acid (DNA). The other solid portion is the white blood cells, which contain DNA. Since it contains genetic material, it is very crucial for forensic analysis. Blood also has clotting factors present in the plasma which is the liquid portion.

4.7 Bloodstain Pattern Documentation

Documentation plays a very important role in reconstructing the bloodstain pattern evidence analysis. If the crime scene has not been documented properly, it always led to some subjective conclusions. The advancement in technology has helped a lot in forensic analysis. Some of those techniques are highly useful documentation and reconstruction of the bloodstain pattern evidence. The common technique used in the documentation of bloodstain pattern analysis is physical evaluation. This technique requires multiple analysts. Some of the forensic experts would engage in examining the bloodstained evidence while other experts would take scene notes, photographs, and sketches of the crime scene. The other technique used by forensic experts is the computer or digital method. This would evaluate data from the bloodstain patterns and crime scene and determine the blood source locations. Thorough and complete notes should accompany documentation and reconstruction of bloodstain pattern evidence.

The crime scene should be documented properly; it should contain all the required information and observations. Since the documentation of the crime scene would help further in reconstruction. Some other reports should be carefully preserved such as crime scene reports, police reports, medical examiner reports, and other forensic evaluations that may have been reviewed before the development of the analyst's conclusions. Bloodstain evidence should be photographed and videographed properly. This is one of the most important documents that help in the reconstruction of the crime scene. Photographs taken from the scene of the crime should be of high quality. Photos should always include a standardized scale and a fixed object about the bloodstains documented.

The proper lighting should be maintained, and overview shots, as well as detailed photographs, should be taken, detailed photographs should easily be recognized within overview shots, an upright position of the camera about the surface being photographed, and the use of a 50 mm objective lens, if possible. If required, additional lighting should be provided for photography. Once the bloodstain area has been identified, the photograph should be taken with proper measurements and also keeping some reference points. Every step of the reconstruction should be photographed. Once this process is complete, blood samples should be taken and sent to the Institute of Criminology for analysis.

Analysis of blood samples can help in the final interpretation of the pattern by helping the analyst connect the pattern to the individual involved in the blood spatter event. Drying time of the bloodstain pattern is affected by many environmental and physical factors. Blood volume is the most important physical factor that affects the time it takes for a stain to dry. When exposed to the same environmental conditions, large amounts of blood dry more slowly than small amounts of blood. Environmental factors such as temperature, humidity, surface properties, and airflow affect the rate at which blood samples dry. Warm temperatures promote drying and speed up the drying of blood. Cold or very cold usually impedes the drying time of blood. Humidity also affects the drying time of blood.

4.8 Research and Development of BPA/Advancing BPA Techniques

Blood is, as we all know, one of the most typical types of evidence found at a crime scene. Under laboratory circumstances, an experiment was undertaken to better understand bloodstain production by using Awlata dye. Awlata (Alta), an Indian dye used for women's grooming, was utilized to make fake bloodstains to better understand how bloodstains form at different heights and how they relate to spines and satellite stains. It was discovered that as the height of the fake blood drop increased, so did the distance of satellite stains originating from the false bloodstains. Satellite stains were directly proportional to the height of the bloodstain, while spines were inversely proportional, according to the experimental findings.

Although various countries have developed effective blood substitute products, they still have issues such as high cost, insufficient validation of physical properties through experimental comparison with human blood, and difficulties in manufacture and sale due to the separate capabilities of pattern reproduction and luminol reaction. Sang Yoon Lee et al. (Di Martino et al. 2004) produced a more practical blood substitute since the components used in its creation are freely available in the market and do not contain chemicals that are toxic to the human body.

These blood substitutes also have luminol reaction functionality as well as pattern transfer bloodstain functionality (bloodstain fingerprint, bloodstain fingerprint) dyeing capabilities. The viscosity and viscoelasticity were adjusted with "hyaluronate acid," "hemoglobin from bovine blood," and "potassium ferricyanide," and the pattern transfer bloodstain dyeing functionality was adjusted with "amino acid solution" and "bovine serum albumin." To make the NFBS, the selected components were blended in the required composition ratios. Another study conducted by Ravi Varma attempted to use artificial intelligence algorithms to help analysts analyze bloodstain patterns. BPA currently uses a manual analysis process, thus having forensic analysts who can reliably produce dependable results is critical. Human mistake, on the other hand, is unavoidable, and analyst error can lead to erroneous findings, jeopardizing casework. The angle of impact from simulated crime scene samples was estimated using artificial intelligence. As a result, it was discovered that the artificial intelligence-assisted approach was correct for 78.64% of the data examined (Engelbertz et al. 2010). Jadhav et al. (Herman et al. 2018) conducted one study to show that computational intelligence methodologies might be effectively combined with image processing techniques to help forensic investigators improve their performance in studying bloodstains, both in terms of time and accuracy of the analysis. Preliminary research using fuzzy clustering was conducted to corroborate these findings and encourage the computational intelligence community to take on this new issue of developing a formal definition of forensic intelligence.

Multiple Choice Questions

1. A bloodstain pattern is created by the force of gravity acting alone?
   (a) Blood clot pattern
   (b) Medium velocity spatter
   (c) Drip pattern
   (d) Passive drop
   Answer: (d)
2. What bloodstain pattern occurs when an object moves through an existing stain, or object tangentially dragged over an existing bloodstain?
   (a) Wipe pattern
   (b) Spatter pattern
   (c) Drip pattern
   (d) Splash pattern
   Answer: (a)
3. What type of blood spatter is typically created from a gunshot?
   (a) Cast-off
   (b) Low velocity
   (c) High velocity
   (d) Drip
   Answer: (c)
4. What would be the shape of a blood droplet at an angle of impact of about 90°?
   (a) Circular
   (b) Spherical
   (c) Oval
   (d) Tail
   Answer: (a)
5. Which weapon would create cast-off patterns consisting of small droplets in a linear pattern?
   (a) Knife
   (b) Baseball bat
   (c) Wooden plank
   (d) None of the above
   Answer: (b)

References

Ablett P (1983) The identification of the precise conditions for seminal acid phosphatase (SAP) and vaginal acid phosphatase (VAP) separation by isoelectric focusing patterns. J Forensic Sci Soc 23:254–256

Allen SM (1995) An enzyme linked immunosorbent assay (ELISA) for detection of seminal fluid using a monoclonal antibody to prostatic acid phosphatase. J Immunoassay 16(3):297–308

Allery JP, Telmon N, Mieusset R, Blanc A, Rougé D (2001) Cytological detection of spermatozoa: comparison of three staining methods. J Forensic Sci 46(2):349–351

Anoruo B, van Oorschot R, Mitchell J, Howells D (2007) Isolating cells from non-sperm cellular mixtures using the PALM microlaser microdissection system. Forensic Sci Int 173(2–3):93–96

Baxter SJ (1973) Immunological identification of human semen. Med Sci Law 13(3):155–165

Blake ET, Sensabaugh GF (1976) Genetic markers in human semen: a review. J Forensic Sci 21(4): 785–796

Boward ES, Wilson SL (2013) A comparison of ABAcard(®) p30 and RSID™-Semen test kits for forensic semen identification. J Forensic Leg Med 20(8):1126–1130

Chapman RL, Brown NM, Keating SM (1989) The isolation of spermatozoa from sexual assault swabs using proteinase K. Journal Forensic Sci Soc 29(3):207–212

Chen JT, Hortin GL (2000) Interferences with semen detection by an immunoassay for a seminal vesicle-specific antigen. J Forensic Sci 45(1):234–235

Davies A, Wilson E (1974) The persistence of seminal constituents in the human vagina. Forensic Sci 3(1):45–55

Di Martino D, Giuffrè G, Staiti N, Simone A, Le Donne M, Saravo L (2004) Single sperm cell isolation by laser microdissection. Forensic Sci Int 146(suppl):S151–S153

Engelbertz F, Korda JB, Engelmann U, Rothschild M, Banaschak S (2010) Longevity of spermatozoa in the post-ejaculatory urine of fertile men. Forensic Sci Int 194(1–3):15–19

Herman Y, Feine I, Gafny R (2018) Acid phosphatase test on Phadebas® sheets - an optimized method for presumptive saliva and semen detection. Forensic Sci Int 288:218–222

Identification of Blood

5

Shivam Chourasiya, Varsha Rani Patel, Himani Sharma, Moumita Sinha, and Tilak Ram Chandrakar

Abstract

One of the most important types of biological evidence that can be obtained for forensic investigation is blood. Because of the probability of a mixture of body fluids or other types of fluids can be present which resemble blood on items therefore conclusive human blood analysis is required in many cases, thus this chapter covers the composition, presumptive, and confirmatory assays for blood identification found at the crime scene. In order for justice to prevail, forensic investigators and scientists must be able to obtain accurate and complete results of the evidence at hand, as well as ensure that the integrity of the evidence has not been compromised.

Keywords

Blood · Presumptive assays · Confirmatory assays

5.1 Composition of Blood

Blood is a tissue that consists of a cellular component and a non-cellular liquid component (Fig. 5.1). Blood serves several functions, including transporting different materials throughout the body. In addition to carrying oxygen and nutrients to all

S. Chourasiya (✉) · V. R. Patel · H. Sharma · T. R. Chandrakar
Faculty of Science, Department of Forensic Science, Medi-Caps University, Indore, Madhya Pradesh, India
e-mail: shivam.chourasiya@medicaps.ac.in; varsharani.patel@medicaps.ac.in; himani.sharma@medicaps.ac.in; tilak.ram@medicaps.ac.in

M. Sinha
Department of Forensic Science, Kalinga University, Raipur, Chhattisgarh, India

A. Puri et al. (eds.), *Fundamentals of Forensic Biology*,
https://doi.org/10.1007/978-981-99-3161-3_5

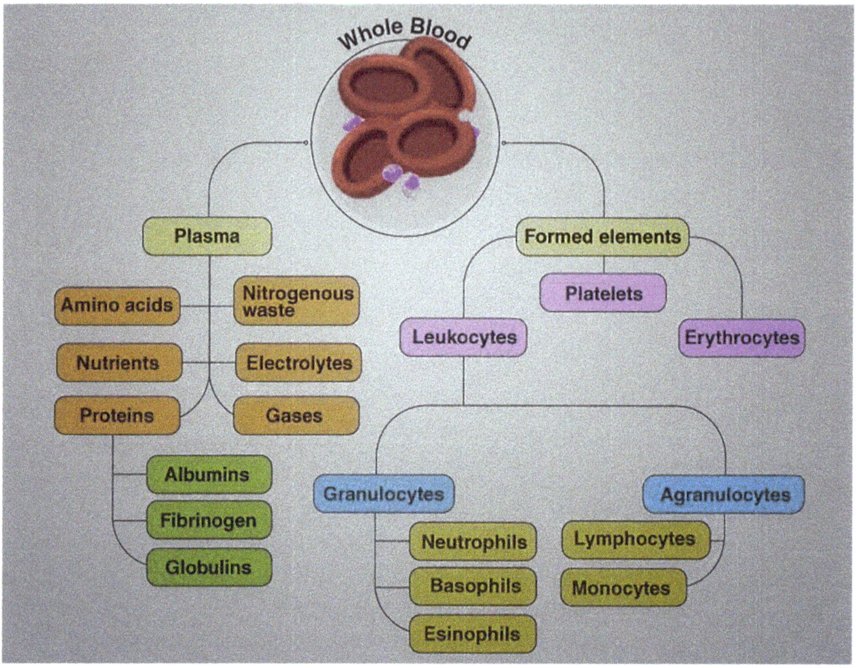

Fig. 5.1 Composition of blood

parts of the body, blood also distributes heat produced by the body's actively respirating tissues (such as the liver and the muscles) and removes waste products from the system (carbon dioxide and urea) (Psionica 2005).

5.1.1 Red Blood Cells

Erythrocytes are another name for these cells. In humans, they have a life span of about 3–4 months (Li 2015).

Previously, red blood cells (RBCs) were thought to be only oxygen and nutrient transporters to tissues. In addition to transporting oxygen and nutrients around the body, recent experiments have shown that RBCs play an important role in regulating systemic nitric oxide metabolism, redox regulation, blood rheology, and viscosity. Erythrocytes are made up of hemoglobin, which are proteins that transport oxygen. A heme molecule is made up of a protoporphyrin IX organic component and a ferrous (Fe^{2+}) iron ion (Fig. 5.2). A heme molecule is also known as ferroprotoporphyrin. Heme groups are also found in the blood of various animals as well as other proteins such as myoglobin in muscles and neuroglobin in the brain, which serves as the foundation for identifying blood found at a crime scene (Kuhn et al. 2017).

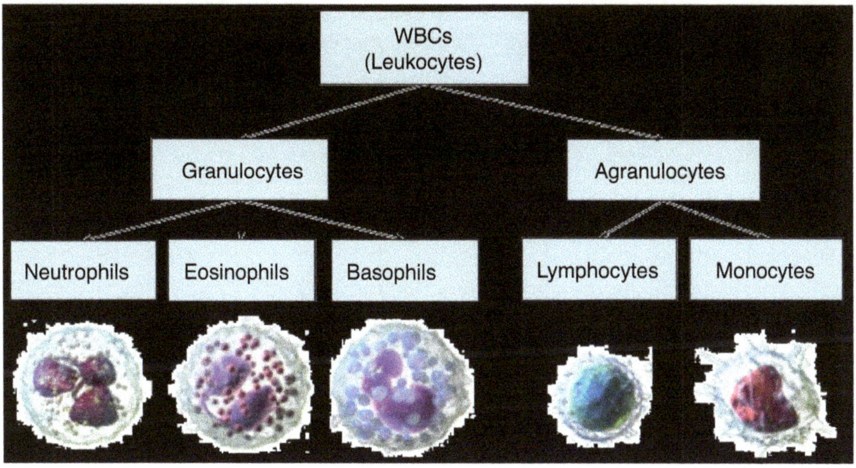

Fig. 5.2 Composition of WBC

5.1.2 White Blood Cells

White blood cells, also known as leucocytes, can be further categorized into two major groups: granulocytes, and agranulocytes. Granulocytes are classified as neutrophils, eosinophils, and basophils. Agranulocytes are further classified as lymphocytes and monocytes (Fig. 5.2). In the fight against disease and illness, white blood cells play a key role and because of their nuclei, they are the most common carriers of nuclear DNA in the circulatory system.

5.1.3 Platelets

Blood platelets, also known as thrombocytes, are tiny, colorless cell fragments that clump together to form clots and slow or stop bleeding. Platelets are produced in the bone marrow, which is a sponge-like tissue found within our bones. Red blood cells, white blood cells, and platelets all originate from stem cells in the bone marrow (Kuhn et al. 2017) (Fig. 5.3).

5.1.3.1 What Is Serum

The serum is plasma devoid of clotting factors, most notably fibrinogen. As a result, the serum does not clot when left alone. Typically, to obtain serum, all clotting agents in plasma are removed via progressive centrifugation or a blood sample is obtained and allowed to clot before the supernatant is extracted. The serum contains all of the other electrolytes, non-clotting proteins, drugs, and toxins. Human serum is typically used for diagnostic purposes. Other animal sera are utilized as anti-venom, antitoxins, and vaccines (Fig. 5.4).

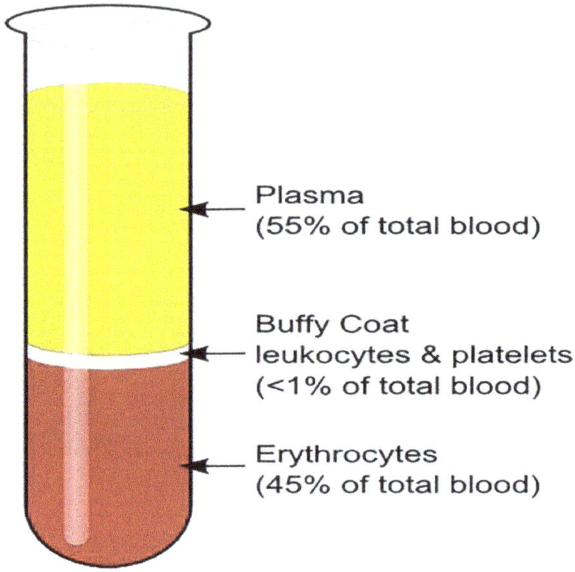

Fig. 5.3 In the presence of an anticoagulant, blood can be separated into two phases. Plasma, the liquid portion of blood, accounts for approximately 55% of blood volume. Erythrocytes, leucocytes, and platelets account for 45% of total blood volume (© Li 2015)

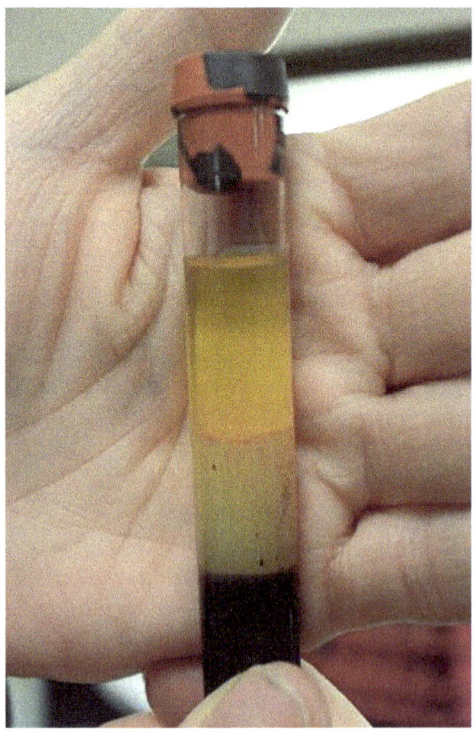

Fig. 5.4 Photograph of centrifuged blood (© Li 2015)

5.2 Presumptive Assays for Identification

5.2.1 Principle of Presumptive Assays

The purpose of presumptive blood tests is to detect traces of blood. These tests are predicated on the fundamental principle of the oxidation–reduction reaction catalyzed by the heme portion of hemoglobin. Consequently, colorless substrates catalyzed by heme undergo an oxidation reaction, resulting in chemiluminescence, fluorescence, or a color change. These tests are very sensitive and can find blood in samples that have been diluted up to 105–106 times. The presence of blood may be indicated by a positive reaction. Additionally, the majority of these tests do not impede forensic DNA analysis.

5.2.2 Oxidation: Reduction Reaction

In an oxidation–reduction reaction, the state of oxidation of a molecule changes. In particular, the oxidation of a molecule indicates that it has lost electrons, whereas the reduction of a molecule indicates that it has gained electrons.

Oxidants: Oxidants are substances that are chemically capable of being reduced, and as a result can acquire electrons from other molecules.

Reductants: Chemicals that are capable of being oxidized and can therefore donate their electrons to other molecules are referred to as reductants.

Biochemical reactions often lose hydrogen during oxidation. Figure 5.6 depicts an oxidation–reduction reaction for blood identification. In presumptive assays, heme serves as the catalyst and hydrogen peroxide serves as the oxidant. A colorless substrate is oxidized in the presence of heme, producing a product with color, chemiluminescence, or fluorescence.

5.2.3 Structure of Heme

Four pyrroles, small pentagon-shaped molecules with four carbons and one nitrogen, make a heme. Tetrapyrrole is formed by combining four pyrroles. A porphyrin is a tetrapyrrole with side chain substitutions that allow it to hold a metal ion. Heme is thus a porphyrin-containing iron. Heme has many forms [(a) Protoporphyrin IX]. (b) Heme (ferroprotoporphyrin). Hemochromagen (c), R = pyridine (pyridine ferroprotoporphyrin). (d) Hematin hydroxide, R = OH (ferriprotoporphyrin hydroxide); hematin chloride, R = Cl (ferriprotoporphyrin chloride)], which correspond to the numerous functions that it must perform in an organism (Li 2015) (Fig. 5.5).

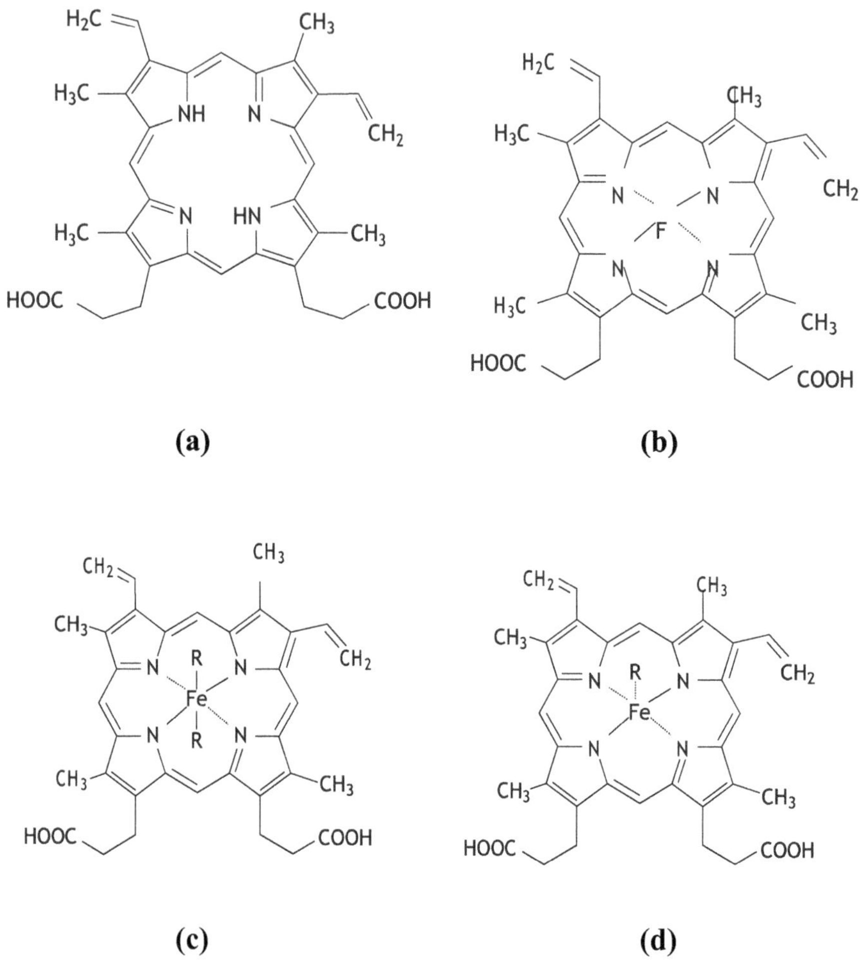

Fig. 5.5 Structure of the heme molecule (© Li 2015)

5.2.4 Types of Presumptive Assays for Blood

Presumptive assays are of three types: colorimetric assays, chemiluminescence, and fluorescence assays.

5.2.5 Colorimetric Assays

There are numerous methods for detecting heme in blood via color reactions. Phenolphthalein, leucomalachite green, and benzidine derivatives are the most commonly used agents. The color reactions produced by these assays can be seen with the naked eye right away.

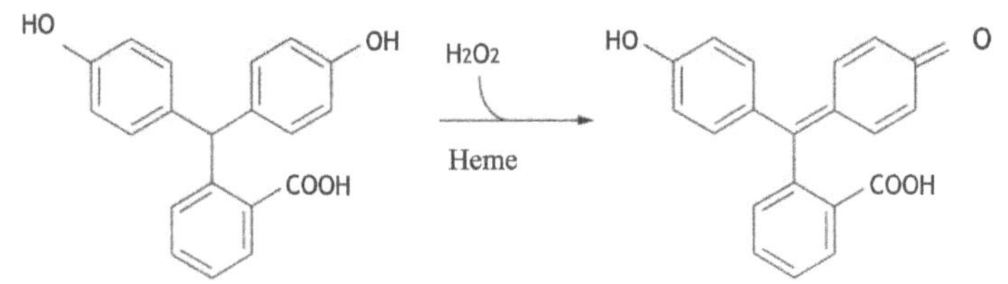

Reduced Phenolphthalein (Colourless)

Pink Colour Oxidized Phenolphthalein

Fig. 5.6 Chemical reaction of the phenolphthalein assay (© Li 2015)

5.2.5.1 Phenolphthalein Assay

Phenolphthalein, an indicator, and dye, is used in titrations of mineral and organic acids, as well as most alkalis. The phenolphthalein assay for blood identification is also known as the Kastle–Meyer test. Kastle published a study in 1901 that presented the results of a reaction in which phenolphthalein, a colorless compound, is catalyzed by heme with hydrogen peroxide as the oxidant (Fig. 5.6). Under alkaline conditions, the oxidized derivative is phenolphthalein, which appears pink.

5.2.5.2 Reagent Preparation

Stock Solution

Phenolphthalein: 2.0 g
Potassium hydroxide: 20.0 g
Distilled water: 100 ml
Zinc dust: 20.0 g

Mix, add a few boiling chips, and boil under reflux for 2–3 h or until the solution has lost its pink color. Cool and decant into a bottle containing some zinc to keep in the reduced form.

Working Solution

Solution # 1:	Ethanol	10 ml
Solution # 2:	Phenolphthalein stock	2 ml
	Distilled water	10 ml
	Ethanol	2 ml
Solution # 3:	3% Hydrogen peroxide	10 ml

Procedure

1. A small cutting, swabbing, or extract of the suspected bloodstain is placed on a filter paper or spot test paper.
2. Two or three drops of ethanol are placed on the stain.
3. Two drops of working phenolphthalein solution are added to the stain.

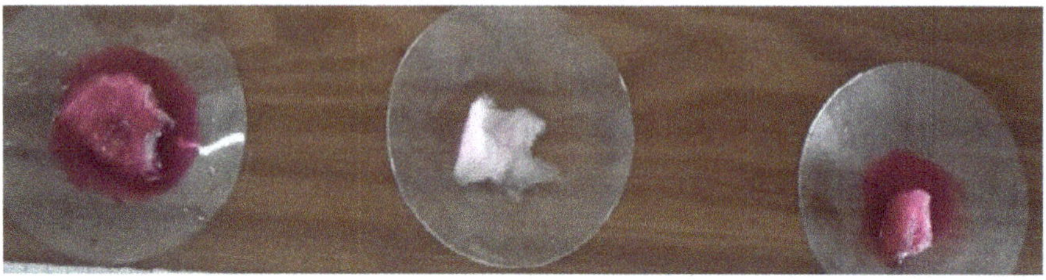

Fig. 5.7 Photograph of phenolphthalein assay results. Positive (left), negative (center), and control (right)

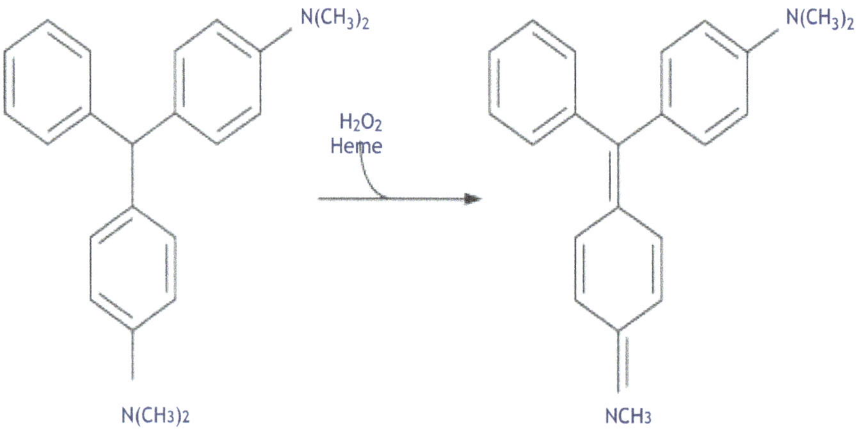

Fig. 5.8 Chemical reaction of the leucomalachite green assay (© Li 2015)

4. After waiting to insure that no color develops at this stage, two or three drops of 3% hydrogen peroxide are added.
5. An intense pink color is a positive test for peroxidase activity, indicative of hemoglobin. This is not a confirmatory test for blood (Fig. 5.7).

5.2.5.3 Leucomalachite Green Assay

The green color of malachite is a triphenylmethane dye. The colorless leuco base form of malachite green can be oxidized by heme catalysis to produce a green color. Hydrogen peroxide is used as the oxidant and the reaction is carried out under an acidic environment (Cox 1991) (Fig. 5.8).

5.2.5.4 Reagent Preparation

Stock Solution

Leucomalachite green: 0.25 g
Glacial acetic acid: 100 ml

Distilled water: 150 ml
Zinc dust: 5 g

Mix, add a few boiling chips, and boil under reflux for 2–3 h or until the solution has lost all its color. Cool and decant into a bottle containing some zinc to keep it in reduced form.

Hydrogen peroxide 3%.

1. Measure out 10 ml of 30% hydrogen peroxide.
2. Add 90 ml of deionized water.
3. May be stored at room temperature or refrigerated—expiration date 1 year.

5.2.5.5 Procedure

1. Swab the suspected blood stain with a clean filter paper or a swab, which may be moistened, if necessary, with deionized water, ethanol, or saline.
2. Apply 1–2 drops of the LMG reagent.
3. Note any blue-green color change. A blue-green color change at this step indicates a chemical oxidant and the test should be considered inconclusive. If there is no color change, proceed to the next step.
4. Add 1–2 drops of 3% hydrogen peroxide.
5. Note any immediate blue-green color change.
6. An immediate blue-green color change indicates a positive result. No color change indicates a negative result. A negative result indicates that either no blood is present or is below the limit of detection of the test (Fig. 5.9).

5.2.5.6 Benzidine and Its Derivatives

In the past, benzidine was used as a step in the process of making dyes (Fig. 5.10). It was later used as a blood-presumptive assay after it was discovered that heme can catalyze the oxidation of benzidine, resulting in a blue to dark blue color (carried out

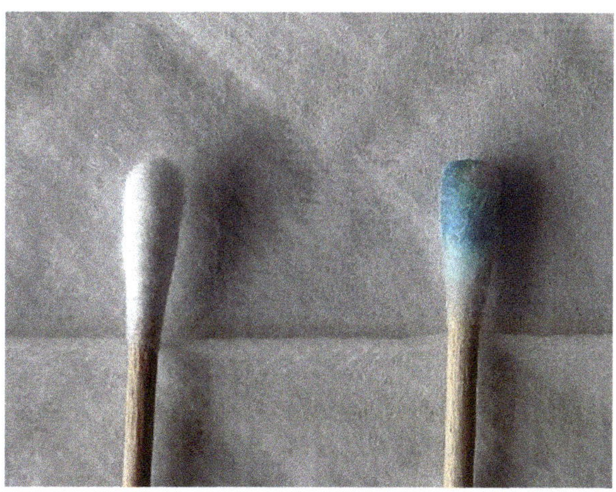

Fig. 5.9 Photograph of the leucomalachite green assay results. Negative (left) and positive (right) reactions (© Li 2015)

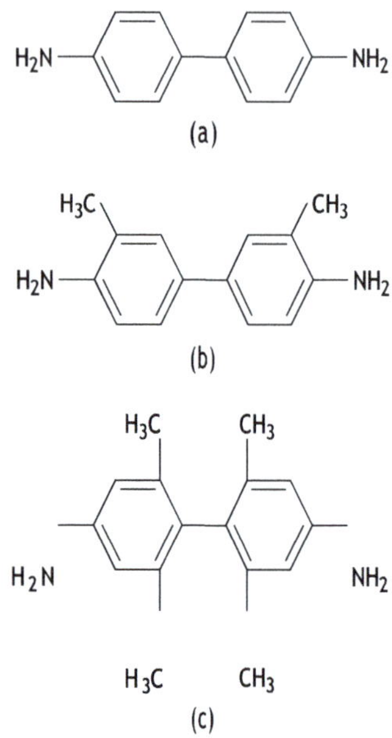

Fig. 5.10 Chemical structures of benzidine and derivatives: (**a**) benzidine, (**b**) orthotolidine, and (**c**) tetramethylbenzidine (© Li 2015)

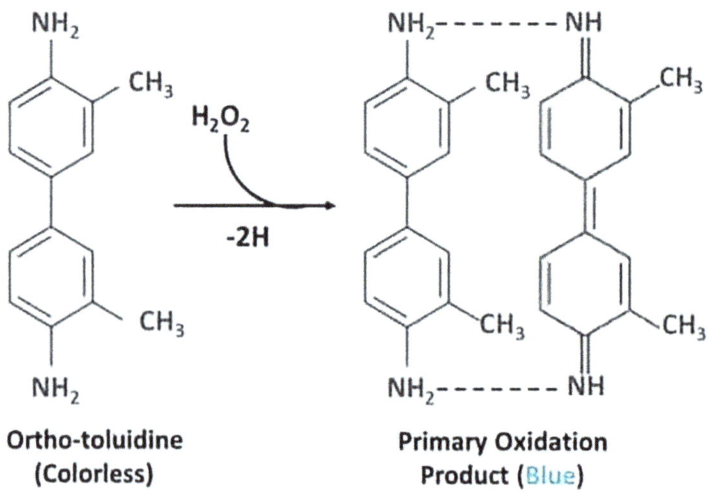

Fig. 5.11 Chemical reaction of the ortho toluidine assay (© Li 2015)

in an acid solution). The reaction must be read instantly because the blue color could change to brown over time. Since it was found to cause cancer, forensic testing no longer uses benzidine. Orthotolidine is made from dimethylated benzidine. Under acidic conditions, its oxidation reaction can be catalyzed by heme, resulting in a blue color reaction (Fig. 5.11). Based on studies with animals, ortho toluidine is also thought to be a possible carcinogen. Because of this, it has been replaced with tetramethylbenzidine (TMB). TMB is a benzidine tetramethyl derivative. Under acidic conditions, the oxidation of TMB can be catalyzed by heme to produce a green to blue-green color. TMB is still being used. Miles Laboratories' Hemastix® assay kit is a TMB-based assay that employs a TMB-containing strip device. A moistened sample is applied to a Hemastix® strip for testing. The presence of blood is indicated by the appearance of a green or blue-green color (Garner et al. 1976).

5.2.5.7 Reagent Preparation

Acetate Buffer
Sodium acetate: 5.0 g
Glacial acetic acid: 43.0 ml
Deionized water: 50.0 ml

Working Solution
TMB: 0.4 g
Acetate buffer: 20.0 ml

Mix, filter, and store in a brown-colored bottle in the refrigerator.

Procedure
1. Place a cutting or swabbing of the stain on a filter paper or spot test paper.
2. A drop of TMB solution is placed on the stain, followed by a drop of 3% hydrogen peroxide.
3. An immediate blue-green color is a positive test for peroxidase activity, indicative of hemoglobin. This is not a confirmatory test for blood (Fig. 5.12).

5.2.5.8 Chemiluminescence and Fluorescence Assays

Chemiluminescence assay: The emission of light occurs as a byproduct of a chemical reaction in the chemiluminescence assay. When there is blood present, luminol produces chemiluminescence.

Fluorescence assays: For a fluorescence assay, an oxidized product like fluorescein needs to be exposed to a certain wavelength of an excitation light source. The

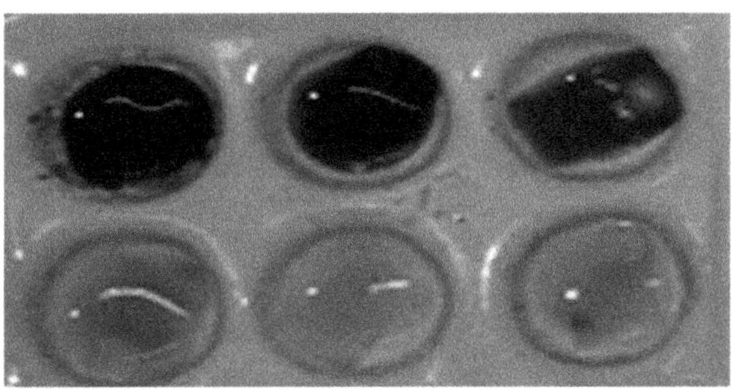

Fig. 5.12 Photograph of the benzidine assay results. Changes in color throughout various dilutions of blood can be seen

fluorescence is then emitted at longer wavelengths than the light source that caused it.

Advantages of Chemiluminescence and Fluorescence Assays

Chemiluminescent and fluorescent reagents have the advantage of being sprayed over large areas where latent bloodstains may be found. A positive reaction shows that there was blood and shows how bloody marks, like footprints and fingerprints, are made. These techniques are so sensitive that even trace amounts of blood can be located. They can be used to find signs of blood at crime scenes even after they have been cleaned.

Disadvantages of Chemiluminescence and Fluorescence Assays

It is necessary to take precautions with chemiluminescent and fluorescent reagents if stains are very small or have been washed. It may be challenging to isolate sufficient DNA for forensic DNA analysis if reagents are sprayed onto a sample because it is further diluting the sample. The use of a reagent is restricted to situations with a low or non-existent light source (Li 2015).

Luminol Test

Luminol is most commonly used as a chemiluminescent reagent. In the presence of an oxidant, the oxidation reaction of luminol catalyzed by heme produces light (Figs. 5.13–5.14). The light emitted by a positive reaction can only be seen in the dark, limiting luminol's applications. A luminol-enhanced pattern should be photographed as soon as possible before it fades away (Creamer et al. 2003).

Reagent Preparation

1. Solution 1: Add 2 ml of 3% H_2O_2 to approximately 50 ml of distilled water.
2. Solution 2: Mix 0.05 g luminol into 10 ml of 5% NaOH.
3. Mix Solution 1 and Solution 2 and bring the final volume up to 100 ml with distilled water.
4. Pour the reagent into the spray bottle.

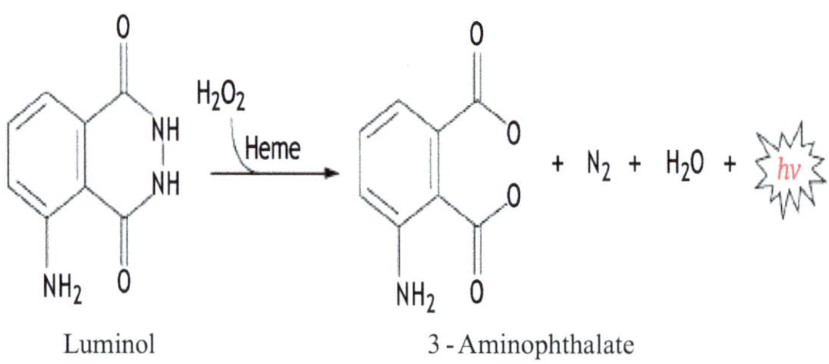

Fig. 5.13 Chemical reaction of luminol (© Li 2015)

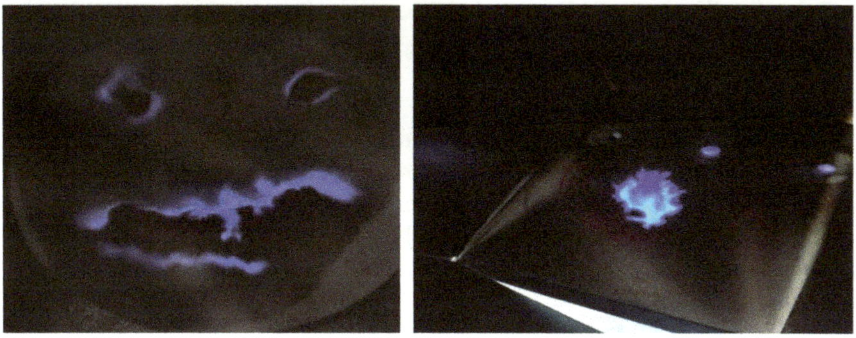

Fig. 5.14 Photograph of the luminol test: a blue chemiluminescence indicates the presence of blood

Fig. 5.15 Photograph of the fluorescence assay (© Li 2015)

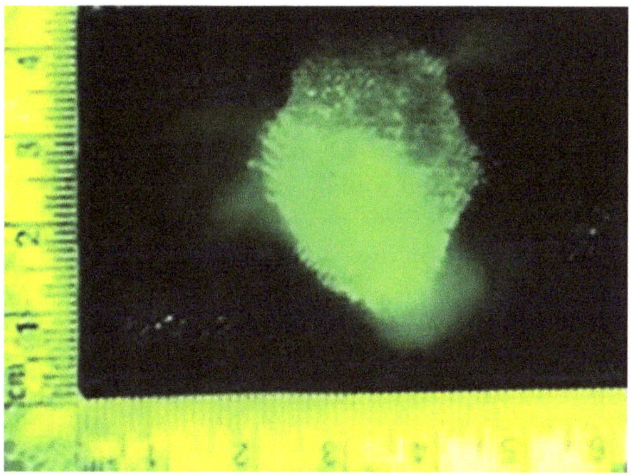

Procedure

1. Spray the luminol reagent on the suspected area where blood stains can be present.
2. Appearance of blue light indicates (Fig. 5.14) the presence of blood in the suspected area.

Fluorescein Test

Another reagent used to detect the presence of bloodstains at a crime scene is fluorescein (Fig. 5.15). Fluorescein exhibits fluorescent properties when oxidized and catalyzed by heme. Fluorescein-sprayed stains are typically exposed to light in the 425–485 nm range using an alternate light source device. When oxidized fluorescein is present, it emits an intense yellowish-green fluorescent light, indicating the presence of a bloodstain. Fluorescein-sprayed stains emit more light than luminol-sprayed stains (Cheeseman and Tomboc 2001).

5.3 Confirmatory Assays for Identification

5.3.1 Microcrystal Assays

In microcrystal assays, chemicals are used to treat bloodstains, resulting in the formation of crystals of heme molecules. Crystals formed as a result of this process have distinct morphologies that are characteristic of heme. These morphologies can be examined under a microscope and can be compared to a known standard. The presence of blood is strongly indicated by a positive microcrystal assay. However, confirmatory assays are typically less sensitive than presumptive assays. In addition, there is no way to tell the difference between human and animal blood using these methods.

5.3.1.1 Hemochromagen Crystal Assay

Hemochromagens are a type of heme derivative in which the heme's ferrous iron forms two bonds, one with nitrogenous bases and one with other heme molecules (Fig. 5.16). In the year 1864, the process of producing hemochromagen crystals was first documented. Since then, reports have surfaced of a number of different modifications. Since its publication in 1912, the Takayama crystal assay has been the method of choice for many forensic laboratories. In alkaline conditions, a bloodstain is treated with pyridine and glucose, which is a reducing sugar that can reduce ferric ions. This procedure results in the formation of crystals of pyridine ferroprotoporphyrin (Fig. 5.16).

5.3.1.2 Hematin Crystal Assay

The Hematin Crystal Analysis: The Teichmann crystal assay is another name for this particular analysis. Teichmann published a method for the formation of crystals from blood specimens in the year 1853. Hematin chloride, also known as

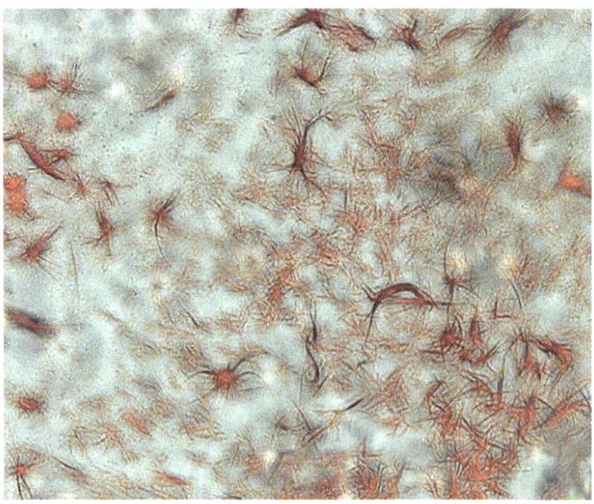

Fig. 5.16 Microcrystal assays using the Takayama method (© Li 2015)

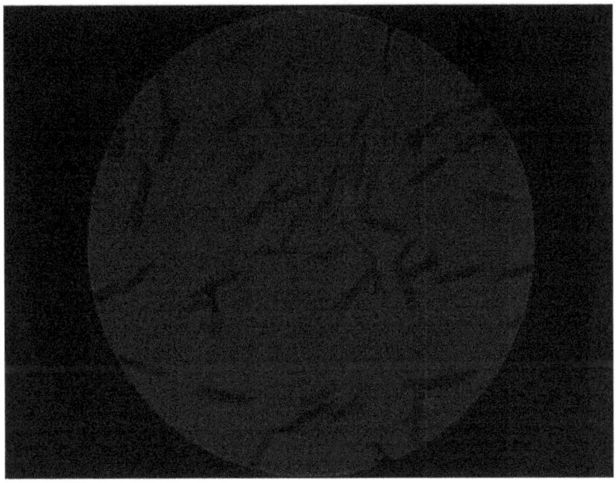

Fig. 5.17 Microcrystal assays using the Teichmann method (© Li 2015)

ferriprotoporphyrin chloride, is a brown crystal with a prismatic appearance that is produced from blood samples after they have been subjected to treatment with glacial acetic acid and salts (Fig. 5.17). Hematin is a heme derivative and its iron is in the ferric (Fe^{+3}) state (Fig. 5.17). The sensitivity and specificity of this hematin assay are not significantly different from those of hemochromagen assays. For older blood samples, the hematin assay is recommended because it has a higher reliability rate than the hemochromagen assay.

5.4 Conclusion

In traditional crimes, blood is often found at the crime scene. The testimony of forensic scientists is crucial in bringing about fairness in the courtroom, and this is accomplished through the examination of blood samples found at the crime scene. In a wide variety of criminal cases, blood is often detected using presumptive and confirmatory assays. Confirmatory assays involve the use of chemicals to treat bloodstains, which ultimately results in the formation of crystals composed of heme molecules. The heme crystals that are formed as a result of this process have distinct morphologies that are typical of the substance. Presumptive assays are based on the principle of an oxidation–reduction reaction, in which colorless substrates that are catalyzed by heme undergo an oxidation reaction, which results in chemiluminescence, fluorescence, or a change in color.

References

Cheeseman R, Tomboc R (2001) Fluorescein technique performance study on bloody foot trails. J Forensic Identif 51(1):12

Cox M (1991) A study of the sensitivity and specificity of four presumptive tests for blood. J Forensic Sci 36(5):1503–1511

Creamer JI, Quickende TI (2003) A comprehensive experimental study of industrial, domestic and environmental interferences with the forensic luminol test for blood. Wiley Analytical Sci 18(4). https://doi.org/10.1002/bio.723

Garner DD et al (1976) An evaluation of tetramethylbenzidine as a presumptive test for blood. J Forensic Sci 21(4):816–821

Kuhn V, Diederich L et al (2017) Red blood cell function and dysfunction: redox regulation, nitric oxide metabolism, anemia. In: Antioxidants & redox signaling, vol 26, issue 13. Mary Ann Liebert, Inc. https://doi.org/10.1089/ars.2016.6954

Li R (2015) Forensic biology. CRC Press Taylor & Francis Group

Serology Concept and Techniques

6

Arjun Rao Isukapatla, Mehar Chadha, and Moumita Sinha

Abstract

Forensic serology is the study of bodily fluids in the context of forensic investigations. Antigens and antibodies are important components of forensic serology because they can be used to identify the presence of bodily fluids, determine the source of the fluids, and link a suspect to a crime scene. An antigen is a molecule that can elicit an immune response in an organism. In the context of forensic serology, antigens found in bodily fluids can be used to identify the type of fluid present, such as blood, semen, saliva, or urine. The use of antigen–antibody interactions in forensic science is a valuable tool for identifying and analyzing bodily fluids and other substances found at crime scenes, which can help to establish the identity of suspects and provide important evidence in criminal investigations. In forensic serology, techniques such as enzyme-linked immunosorbent assays (ELISAs) and immunofluorescence assays (IFAs) are commonly used to detect the presence of antigens and antibodies in bodily fluids. These techniques can be used to identify the type of bodily fluid present, determine the blood group of an individual, and link a suspect to a crime scene based on the presence of bodily fluids.

A. R. Isukapatla
Faculty of Forensic Science, Department of Life Sciences, Christ University, Bangalore, Karnataka, India
e-mail: arjun.rao@christuniversity.in

M. Chadha
Department of Life Sciences, Christ University, Bangalore, Karnataka, India
e-mail: mehar.chadha@mfs.christuniversity.in

M. Sinha (✉)
Department of Forensic Science, Kalinga University, Raipur, Chhattisgarh, India
e-mail: moumita.sinha@kalingauniversity.ac.in

A. Puri et al. (eds.), *Fundamentals of Forensic Biology*,
https://doi.org/10.1007/978-981-99-3161-3_6

Keywords

Forensic serology · Antigens · ELISA · Immunoassay · Agglutination

6.1 Forensic Serology

Forensic serology is a branch of forensic science that deals with the identification and analysis of body fluids, such as blood, semen, saliva, and sweat, found at a crime scene. Serology involves the use of immunological techniques to detect and analyze antigens and antibodies present in bodily fluids. The analysis of bodily fluids can provide valuable information for criminal investigations. For example, the presence of blood can help determine the cause and manner of death, while the presence of semen or saliva can help identify potential suspects. Forensic serologists use a variety of techniques to analyze bodily fluids, including blood typing, DNA analysis, and enzyme immunoassays. They may also use microscopic examination to identify characteristics of the fluid or use chemical tests to identify the presence of certain compounds. Overall, forensic serology plays an important role in helping investigators piece together the events surrounding a crime and can provide critical evidence in criminal cases. Antigens and antibodies are important components of forensic serology, as they help to identify and analyze bodily fluids found at a crime scene. Antigens are substances that trigger an immune response in the body, while antibodies are proteins produced by the immune system in response to an antigen. In forensic serology, antigens and antibodies can be used to identify and analyze bodily fluids such as blood, semen, saliva, and sweat. For example, the ABO blood group system is based on the presence of specific antigens on the surface of red blood cells, which can be identified using specific antibodies. This information can be used to match blood samples found at a crime scene with the blood type of a suspect or victim. Similarly, antibodies can be used to detect the presence of specific antigens in bodily fluids.

The presence of the prostate-specific antigen (PSA) in semen can be detected using antibodies specific to PSA. This information can be used to identify the presence of semen at a crime scene and can be used to match the semen sample to a potential suspect. Blood antigens are proteins found on the surface of red blood cells and are used to determine an individual's blood type. The most commonly used blood group systems in forensic serology are the ABO system and the Rh system. Saliva antigens can also be used in forensic science to identify and analyze bodily fluids found at a crime scene (Old et al. 2009). Saliva contains a variety of proteins and enzymes, including alpha-amylase, which is an enzyme that breaks down starches. The detection of alpha-amylase is commonly used to identify the presence of saliva in a sample. This can be done using a variety of techniques, including colorimetric assays, immunological tests, and DNA analysis. Sweat antigens can also be used in forensic serology to help identify and analyze bodily fluids found at a crime scene (Schittek et al. 2001). Sweat contains a variety of proteins and electrolytes, including eccrine sweat gland-specific proteins such as dermcidin and sweat gland-specific antigen (SGS). The detection of sweat antigens can help identify the presence of sweat on a sample. This can be done using immunological tests, such as enzyme immunoassays, which can detect the presence of specific antigens in a sample.

6.2 Immunogen and Antigens

Immunogens and antigens are related concepts in immunology, but they are not exactly the same. An antigen is any substance that is recognized by the immune system as foreign or non-self and that can elicit an immune response. Antigens can be naturally occurring, such as the surface proteins of a virus or bacteria, or they can be synthetic or recombinant, such as those used in vaccines. On the other hand, an immunogen is a type of antigen that is capable of inducing an immune response in an organism by triggering the production of antibodies or immune cells specific to that antigen. In other words, not all antigens are immunogens, but all immunogens are antigens. Immunogens are usually larger, more complex molecules than other antigens, and they often require additional processing and presentation by antigen-presenting cells (APCs) to stimulate an immune response. Antigens are any substances that can be recognized by the immune system, while immunogens are a specific subset of antigens that can induce an immune response by themselves.

6.2.1 Use of Immunogen and Antigen in Forensic Serology

Immunogens and antigens are used in forensic serology to identify and analyze biological fluids found at crime scenes or in other types of forensic investigations. When a fluid sample is collected, it is typically tested to determine whether it contains antigens or antibodies that are specific to a particular body fluid, such as blood, semen, or saliva. One common method used in forensic serology is the enzyme-linked immunosorbent assay (ELISA), which relies on the specific binding of antibodies to antigens (Konstantinou 2017). For example, an ELISA test may be used to detect the presence of blood by looking for the antigens present in red blood cells. Another technique commonly used in forensic serology is the ABO blood typing system, which is based on the presence of specific antigens on the surface of red blood cells. By analyzing the antigens present in a blood sample, forensic scientists can determine the individual's blood type, which can provide valuable information for investigations. In addition to blood typing and ELISA, immunogens and antigens can also be used in DNA analysis to identify specific individuals from biological samples. This technique relies on the detection of unique DNA sequences, which can be amplified using polymerase chain reaction (PCR) and compared to DNA samples from known individuals (Table 6.1).

Table 6.1 Classification of some common types of immunogens

Sl. No.	Types of Immunogens	Characteristics
1	Proteins and peptides	Most common types of immunogens include many viral and bacterial surface proteins, as well as proteins found in allergens
2	Polysaccharides	Complex carbohydrates are found in bacterial and fungal cell walls, and they are often used in vaccines
3	Lipids	Some lipids can act as immunogens, such as glycolipids found on the surface of some bacteria
4	Nucleic acids	DNA and RNA can act as immunogens, especially when they are exposed to the immune system due to infection or cell death
5	Haptens	Small molecules that can elicit an immune response only when they are bound to a larger carrier protein, such as penicillin
6	Synthetic and recombinant immunogens	Immunogens that are artificially created in the laboratory, such as recombinant proteins used in some vaccines
7	Autoantigens	Self-antigens that are normally tolerated by the immune system, but can become immunogens in autoimmune diseases

6.3 Antibodies

Antibodies, also known as immunoglobulins, are proteins produced by the immune system in response to the presence of foreign substances in the body, such as bacteria, viruses, or other pathogens. Antibodies recognize and bind to specific targets, called antigens, on the surface of these foreign substances, thereby marking them for destruction by other cells of the immune system. Antibodies are composed of two heavy chains and two light chains, which are held together by disulfide bonds (Hamers-Casterman et al. 1993). Each antibody has a unique amino acid sequence in its variable regions, which determines its specificity for a particular antigen. The production of antibodies is a key component of the adaptive immune response, which is the body's defense mechanism against infection and disease. Vaccines work by stimulating the immune system to produce specific antibodies against a particular pathogen, without causing illness. There are five classes of antibodies, each with different biological properties and functions, including IgA, IgD, IgE, IgG, and IgM.

6.3.1 Types of Antibodies

There are five main types of antibodies, also known as immunoglobulins (Ig):

IgG: This is the most common antibody found in the blood and tissues. It provides long-term protection against bacteria, viruses, and toxins. IgG can cross the placenta from mother to fetus and provide passive immunity to the newborn.

IgM: This is the first antibody produced by the body in response to an infection. It is present in the blood and lymphatic system and is effective in clearing pathogens from the body.

IgA: This antibody is found in bodily secretions such as saliva, tears, breast milk, and mucus membranes. It plays a critical role in defending against infections in the respiratory and digestive systems.

IgE: This antibody is associated with allergies and plays a role in defending against parasitic infections. It triggers the release of histamine, which causes the symptoms of an allergic reaction.

IgD: This antibody is found in low levels in the blood and on the surface of B cells, where it helps to trigger the production of other antibodies. Its exact function is not well understood.

6.3.2 Production and Mechanism of Antibody

Antibodies or immunoglobulins (Ig) are Y-shaped proteins produced by B cells of the immune system in response to the presence of foreign substances called antigens. The production of antibodies is a complex process that involves several steps which are further explained in Fig. 6.1.

The mechanism of antibody production involves a process called somatic hypermutation (Neuberger and Milstein 1995), which introduces random mutations into the genes encoding the antibody receptors on B cells. These mutations can result in the production of antibodies with higher affinity for the antigen, which increases the effectiveness of the immune response. A small subset of B cells undergoes a process called class switching (Stavnezer et al. 2008), which changes the type of antibody produced. This allows the immune system to produce different types of antibodies, each with unique properties, to target different types of antigens.

6.3.3 Preparation of Antibodies

The preparation of antibodies typically involves immunization of an animal with the antigen of interest, followed by the collection of serum containing the antibodies produced by the animal's immune system. The general steps involved the preparation of antibodies specific to nature and types of uses.

Choosing appropriate animal: Typically, animals such as rabbits, mice, and goats are used for antibody production. The choice of animal depends on factors such as the desired antibody class and the size of the animal.

Process of immunization: The animal is immunized with the antigen of interest, either alone or conjugated to a carrier protein to enhance the immune response.

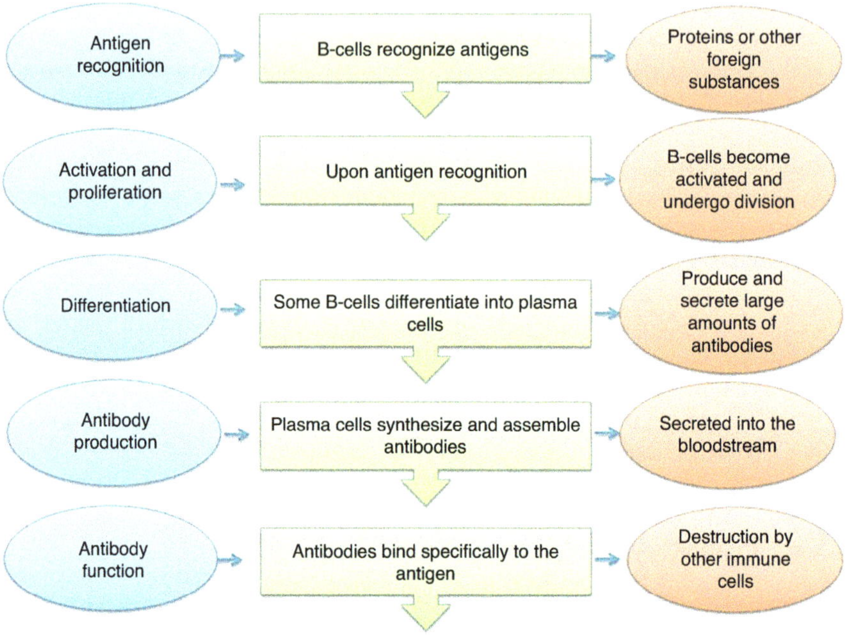

Fig. 6.1 Production process of antibodies

The antigen is usually injected multiple times over several weeks to boost the immune response.

Serum collection: Blood is then collected from the desired animal and allowed to clot, and the serum is separated from the clot by centrifugation. The serum contains the antibodies produced by the animal's immune system in response to the antigen.

Antibody purification: The antibodies can be purified from the serum using various methods, such as affinity chromatography, protein A or G chromatography, or precipitation with ammonium sulfate.

Antibody characterization: The purified antibodies can be characterized by methods such as enzyme-linked immunosorbent assay (ELISA) or Western blotting to confirm their specificity and affinity for the antigen.

Antibody storage: The purified antibodies can be stored in a suitable buffer at −20 ° C or −80 °C for long-term storage.

Antibodies are highly sensitive and specific detection tools that can recognize and bind to specific antigens with high affinity. The sensitivity of an antibody refers to its ability to detect low levels of the target antigen (Silverstein 1995), often expressed as the lowest concentration of antigen that can be reliably detected by the assay. The sensitivity of an antibody depends on several factors, including the specificity and affinity of the antibody for the target antigen, the quality of the antigen preparation, and the detection method used in the assay. Generally, the higher the affinity and specificity of the antibody, the more sensitive the assay will be. Many modern antibody-based assays, such as ELISA, fluorescence-based assays, and immunohistochemistry, can detect antigens in the low picogram to femtogram range (Ito et al. 2021). Some newer technologies, such as single-molecule imaging, can detect even lower concentrations of antigens. The sensitivity of an antibody-based assay is not the only factor that determines its accuracy and reliability. Other factors, such as assay specificity, reproducibility, and robustness, should also be considered when evaluating the performance of an antibody-based assay.

6.4 Antigen–Antibody Binding

Antigen–antibody reaction, also known as antigen–antibody interaction, is a specific chemical reaction that takes place during an immune response between antigens and antibodies produced by B cells of white blood cells. Antigens and antibodies are combined during agglutination (Reverberi and Reverberi 2007). The body employs this fundamental biological mechanism to protect itself from numerous outside invaders, such as viruses and their poisonous compounds. When antibodies attach to antigens in a precise and powerful manner, an antigen–antibody complex is created in the blood. Following that, the immune complex is sent to cellular systems where it can be neutralized or inactivated (Tang et al. 2021).

6.4.1 Antigen

Antigen is a material capable of triggering the immune system's activation of lymphocytes, the body's white blood cells that fight infections. Foreign antigens (also known as heteroantigens) and autoantigens are the two primary classifications of antigens (or self-antigens). The source of foreign antigens is external to the body. Examples include components of or compounds created by viruses, microorganisms (such as bacteria and protozoa), substances in snake venom, specific proteins in meals, and serum and red blood cell components from other people (Mohamed Abd El-Aziz et al. 2019). On the other hand, autoantigens come from within the body. Natural biological processes allow the body to distinguish between self and nonself, but in people with autoimmune illnesses, these processes trigger an immune reaction that results in the production of autoantibodies. An immunogen is an antigen that stimulates the production of antibodies by lymphocytes or the direct attack of the antigen by the immune system.

Antigenic determinants are parts of antigens that fit and bind to receptor molecules on the surface of lymphocytes that have a complementary structure. Initiating an immune response against the antigen, such as antibody formation, cytotoxic cell activation, or both, is stimulated by the lymphocytes' receptors binding to the antigens' surface molecules (Chaplin 2010). The type and quantity of antigen present, the method of entry into the body, and the unique properties of the host all affect how much antibody is produced in response to stimulation.

6.4.2 Antibody

An antibody is a protein-based element of the immune system that travels through the blood, detects, and destroys foreign entities like bacteria and viruses. Antibodies, also known as immunoglobulins, are Y-shaped proteins produced by the immune system in response to antigens. They have two main regions: the variable region and the constant region. The variable region is responsible for binding to the specific antigen that triggered its production, while the constant region is responsible for activating other components of the immune system to destroy the antigen. Antibodies can be classified into different types based on their structure and function. Antibodies remain in circulation in the blood after exposure to an antigen, acting as defense against additional exposures to that antigen.

6.4.3 Chemical Bonds Responsible for the Antigen–Antibody Reaction

Similar to how proteins connect to their cellular receptors or enzymes bind to their substrates, the interaction between the Ab-binding site and the epitope only includes noncovalent bonding. High ionic strength or extremely low pH can inhibit or dissociate the binding, which is reversible. Ag-Ab binding involves the following intermolecular forces:

Electrostatic bonds: They are formed when two protein side chains with opposing charges, such as an ionized amino group (NH^{4+}) on a lysine in the Ab and an ionized carboxyl group (COO^{-}) on an aspartate residue in the Ag, attract one another.

Hydrogen bonding: Hydrophilic groups, such as OH and C=O, NH and C=O, and NH and OH groups, can form relatively weak hydrogen bonds when the Ag and Ab are in close proximity.

Hydrophobic interactions: Hydrophobic groups, such as the side chains of valine, leucine, and phenylalanine, have a tendency to connect owing to Van der Waals bonding and aggregate in an aqueous environment, keeping water molecules out of their surroundings. As a result, their separation grows smaller, intensifying the attraction energies at play. Up to 50% of the Ag-Ab bond's overall strength is thought to come from interactions of this kind.

Bonds of Van der Waals: The interactions between the "electron clouds" that surround the molecules of Ag and Ab are what drive these forces. The contact has been compared to that which would occur between two molecules with alternating dipoles, alternating such that at any one time, oppositely orientated dipoles will be present in close-proximal regions of the Ag and Ab molecules.

6.4.4 Strength of Ag-Ab Interaction

6.4.4.1 Affinity

The degree of contact between an epitope and the antigen-binding site of an antibody is measured by affinity. The same fundamental thermodynamic rules that apply to all reversible biomolecular interactions characterize it:

$$K_A = [\mathrm{Ab} - \mathrm{Ag}] \div [\mathrm{Ab}][\mathrm{Ag}]$$

Affinity constant $= K_A$
[Ab] is the amount of open binding sites on the antibody in molar form.
[Ag] is the amount of vacant binding sites on the antigen in molar form.
Molar concentration of the antibody–antigen combination is (Ab-Ag)

To put it another way, K_A defines the quantity of antibody–antigen complex at the time equilibrium is established. Each antibody's rate of diffusion affects how long it takes for this to happen. High-affinity antibodies, on the other hand, bind more antigens in less time than low-affinity antibodies. As a result, K_A for antibodies can range significantly between below 105 mol^{-1} and over 1012 mol^{-1}, depending on a variety of variables such as pH, temperature, and buffer composition. Ab's affinity for an epitope is determined by the sum of all non-covalent contacts between that epitope and its specific Ag-binding site.

Low-affinity Ab: Weakly binds Ag and easily dissociates.
Strong affinity Ab: Tightly binds Ag and maintains the bond for a longer time.

6.4.4.2 Avidity

Antibodies and antigens are multivalent, which means they have several binding sites. Avidity is a measurement of an antibody's overall ability to bind to each binding site. The functional affinity is another name for avidity. Three elements determine avidity.

- The affinity for binding: the degree to which a particular binding site has a strong connection.
- The valency: The total number of binding sites involved.
- The structure of the involved antigen and antibody is the structural arrangement. All antibodies are multivalent. IgMs are decavalent, while IgGs are bivalent.
- A more stable antibody–antigen complex might result from an antibody and antigen structure that works well together.
- Avidity is the strength of the many interactions between multivalent Ab and Ag. The ability of an antibody to bind is better measured by avidity than by affinity. Low affinity can be made up for by high avidity.

6.4.4.3 Cross Reactivity

If two Ags have the same epitope or have chemically similar features, they may cross-react when an antibody is induced by one Ag.

6.4.5 Stages of Antigen–Antibody Interaction

The interactions between antigen and antibody undergo three phases.

1. The creation of the Ag-Ab complex is a need for the first phase of the reaction.
2. Agglutination, precipitation, and other observable phenomena are the outcome of the second stage.
3. The third stage includes neutralizing or destroying the Ag.

6.4.6 Types of Ag-Ab Interaction

6.4.6.1 Agglutination

When a certain Ag and its Ab combine in the presence of electrolytes at the right temperature and pH, the particles cluster or agglutinate. Agglutinins are the aggregates of cellular Ag produced by the serum's Ab. The aggregated particle antigens are known as agglutinogens.

1. Agglutination of slides: This is a quick and practical method of determining whether agglutinating antibodies are present.
2. Tube agglutination: This method is frequently used to calculate the Ab concentration. The Ab-containing serum is serially diluted with saline in tiny test tubes, and then a consistent volume of the Ag solution is added. A control tube that does not have any antiserum is kept. As soon as there is visible agglutination, the tubes are incubated. The titer is the test tube with the highest level of agglutination.
3. The passive agglutination test is analogous to the hemagglutination test, but the reaction's physical features are distinct. An Ag-coated carrier particle helps to convert a precipitation process into an agglutination reaction, increasing the sensitivity of the reaction. As carrier particles, RBC, latex particles, or bentonite might be employed. Tanned RBC (RBC coated with polystyrene) may occasionally be utilized.

6.4.6.2 Precipitation

When a soluble Ag and its Ab interact in the presence of an electrolyte (NaCl) at a particular pH and temperature, an insoluble precipitate of the Ag-Ab complex is created. The molecule that causes precipitation is known as precipitin, and the reaction is known as a precipitation reaction. Precipitation reactions can happen in both the gel and liquid media.

Liquid precipitation: To carry out an antigen–antibody interaction, increasing amounts of antigen are added to tubes containing a fixed amount of antibody. The combination response of the antigen and antibody causes precipitation.

Gel precipitation: These techniques use Petri plates or plates with agar gel or a similar gel. Both Ag and Ab rapidly diffuse throughout the gel system. Depending on the diffusion rate and concentration of the reactants, a zone of equivalence will develop at a particular place and be observable as precipitation. With intricate preparations of Ag or Ab, several bands form. They can be classified as either single diffusion or double diffusion techniques.

6.4.6.3 Complement Fixation

One technique to show the presence of antibodies in patient serum is complement fixation. The traditional technique for proving the presence of an antibody in patient serum is complement fixation. Two elements make up the complement fixation test. A combination of sheep red blood cells, a complement-fixing antibody such as immunoglobulin G made against the sheep red blood cells, and an external supply of complement—typically guinea pig serum—makes up the initial component of an indicator system. The anti-sheep antibody attaches to the surface of red blood cells when these components are combined under ideal circumstances. After this antigen–antibody combination is produced, a complement attaches to it, causing the red blood cells to lyse.

In addition to complementing, the second component consists of a recognized antigen and patient serum added to a suspension of sheep red blood cells. The complement fixation method's two components are put to the test one after the other. Prior to adding a complement to the mixture, the patient serum is added to the known antigen. The ensuing antigen–antibody complexes will bind every component if the serum has antibodies to the antigen. Afterwards, the antibody is introduced together with sheep red blood cells. Complement is accessible to bind to the sheep cell indicator system and anti-sheep antibody if it has not already been bound by a known antigen. A negative complement fixation test and the absence of antibodies in the patient's serum are both indicated by the lysis of the indicator sheep red blood cells. The absence of red blood cell lysis will be an indicator of a favorable outcome if the patient's serum does contain a complement-fixing antibody.

6.4.6.4 Hemagglutination

A serological assay called hemagglutination is used to pinpoint a potential virus and find related antibodies. Red blood cells are used as a source of antigens in the experiment. Red blood cells clump together as a result of a process called hemagglutination when some enveloped viruses, such as the influenza virus, are present. Hemagglutinin, a glycoprotein on the viral surface, interacts with red blood cells to cause them to group together and create a lattice. The hemagglutination inhibition assay (HIA) is used to assess the amount of antiviral antibodies that have been produced. In HIA, the absence of hemagglutination is used to determine whether antibodies are present in a particular sample.

6.4.6.5 Neutralization

By blocking or neutralizing any biological effects an antigen or infectious agent might have, a neutralizing antibody protects a cell from harm. Many viral infections can be avoided with the help of an antibody response, which may also aid in the healing of an infection. When a virus infects a vertebrate, antibodies are made against several epitopes of various virus proteins (Neurath 2008) via a process known as neutralization, a fraction of these antibodies can prevent viral infection. Virus–antibody complex development typically occurs in this situation.

6.5 Primary Binding Assays

Primary binding assays are experimental techniques used to measure the interaction between two or more molecules, such as a ligand and a receptor, and determine the strength and specificity of the binding. These assays are used to determine the affinity, stoichiometry, and kinetics of the interaction, as well as to identify potential drug candidates or evaluate the efficacy of a drug (Findlay 2008). Examples of primary binding assays include radioligand binding assays, fluorescence-based assays, and surface plasmon resonance assays. These assays are widely used in drug discovery and development, as well as in basic research to understand the mechanisms of biological processes. Some of the routinely used primary binding tests/assays in the laboratories are:

6.5.1 Radio Immunoassay

Radioimmunoassay (RIA) is a laboratory technique used to measure the concentration of a specific antigen (a substance that stimulates the production of an antibody) or antibody (a protein produced by the immune system in response to an antigen) in a given sample. It involves the use of a radioactive isotope, typically iodine-125, as a tracer to label the antigen or antibody being measured. In RIA, a known amount of the antigen or antibody to be measured is mixed with a fixed amount of a corresponding antibody or antigen that has been labeled with a radioactive isotope (Lowenstein et al. 2006). The mixture is then allowed to incubate, during which time the labeled and unlabeled molecules compete for binding to the corresponding antibody or antigen. After the incubation period, the mixture is separated, typically using a solid-phase adsorbent, such as a bead or a plate, which binds the antibody or antigen. The amount of radioactive isotope bound to the solid-phase adsorbent is then measured using a scintillation counter or other radiation detector. The amount of radioactivity detected is proportional to the amount of labeled antigen or antibody in the sample. By comparing the amount of radioactivity in the sample to that of a known standard, the concentration of the antigen or antibody in the sample can be determined. RIA is a sensitive and precise technique that has been widely used in clinical and research laboratories to measure a variety of antigens and antibodies, including hormones, drugs, and infectious agents.

There are several methods of radioimmunoassay (RIA) used in laboratory practices, but the basic principle involves the use of a labeled antigen or antibody and a corresponding unlabeled antigen or antibody. The labeled antigen or antibody is usually a radioactive isotope, such as iodine-125 or tritium, while the unlabeled antigen or antibody is the one being measured in the sample. Some common methods of RIA include:

Competitive RIA: In this method, a fixed amount of labeled antigen or antibody is mixed with a known amount of unlabeled antigen or antibody and a fixed amount of corresponding antibody or antigen. The mixture is allowed to incubate, and the proportion of labeled antigen or antibody that remains bound to the antibody is inversely proportional to the concentration of unlabeled antigen or antibody in the sample.

Sandwich RIA: This method involves the use of two antibodies that bind to different parts of the same antigen. The antigen is sandwiched between the two antibodies, one of which is labeled with a radioactive isotope. The amount of radioactivity detected is proportional to the amount of antigen in the sample.

Direct RIA: In this method, the antigen or antibody is directly labeled with a radioactive isotope, and the amount of radioactivity detected is proportional to the amount of antigen or antibody in the sample.

Indirect RIA: This method involves the use of a secondary antibody that is labeled with a radioactive isotope and binds to the primary antibody–antigen complex. The amount of radioactivity detected is proportional to the amount of the primary antibody–antigen complex in the sample. The choice of the RIA method depends on the specific antigen or antibody being measured and the sensitivity and specificity required for the particular type of assay.

6.5.2 Immunofluorescence Assay

Immunofluorescence assay (IFA) is a highly sensitive and specific technique for detecting and visualizing antigens or antibodies in a variety of samples. It is commonly used in medical diagnosis, research, and quality control testing. The principle of immunofluorescence assay (IFA) is based on the use of fluorescent-labeled antibodies to detect and visualize specific antigens or antibodies in a sample (Im et al. 2019). The assay utilizes the specific binding of antibodies to their target antigens. In an IFA, the sample is first fixed onto a glass slide or a membrane and then blocked to minimize non-specific binding. A primary antibody specific to the antigen of interest is added to the sample and allowed to bind to the target antigen. A fluorescent-labeled secondary antibody is then added, which binds to the primary antibody, allowing for visualization of the antigen or antibody of interest. The sample is then washed to remove any unbound antibodies, and examined under a fluorescence microscope. When the fluorescent-labeled secondary antibody binds to the primary antibody that is bound to the antigen, it produces light of a specific color, which can be detected and visualized using a fluorescence microscope. The presence or absence of fluorescence at specific locations in the sample can be interpreted to determine the presence or absence of the antigen or antibody of interest.

Immunofluorescence assay (IFA) is a highly sensitive and specific technique for detecting and visualizing antigens or antibodies in a variety of samples. This technique is a highly sensitive technique that can detect even small amounts of antigens or antibodies in a sample (Abraham et al. 1994). This makes it useful for detecting low-level infections or for monitoring immune responses. IFA is a highly specific technique that can differentiate between closely related antigens or antibodies. This is important in distinguishing between different strains of viruses or bacteria. The IFA technique can also be used to quantify the amount of antigen or antibody present in a sample. This is useful for monitoring changes in the immune response over time and can be used to detect a wide range of antigens or antibodies in various types of samples, including cells, tissues, and body fluids. IFA can be performed in two easy ways.

Direct immunofluorescence assay (DFA): This is a type of immunofluorescence assay (IFA) used to detect the presence of a specific antigen in a sample. It involves the use of a fluorescent-labeled primary antibody that binds directly to the antigen of interest, allowing visualization of the location of the antigen under a fluorescence microscope.

Indirect immunofluorescence assay (IFA): This is a type of immunofluorescence assay used to detect the presence of a specific antibody in a sample. It involves the use of a fluorescent-labeled secondary antibody that binds to the primary antibody that was already bound to the antigen of interest, allowing visualization of the location of the antibody under a fluorescence microscope.

6.5.3 Immunoprecipitation (IP)

This technique is used to isolate and purify a specific protein or complex of proteins from a complex mixture of biological samples, such as cell lysates or body fluids. It involves the use of an antibody that specifically recognizes and binds to the target protein, followed by the precipitation of the antibody–protein complex using a suitable matrix, such as protein A/G beads or magnetic beads. The basic steps involved in immunoprecipitation are as follows:

1. Sample preparation: The sample is first lysed to release the proteins of interest. The lysate is then clarified by centrifugation to remove cellular debris and other insoluble materials.
2. Antibody incubation: The clarified lysate is incubated with a specific antibody that recognizes and binds to the target protein.
3. Immunoprecipitation: The antibody–protein complex is then precipitated by adding a suitable matrix, such as protein A/G beads or magnetic beads, which specifically bind to the antibody. The matrix is then washed to remove any non-specifically bound proteins.

4. Elution: The protein of interest is then eluted from the matrix using a suitable buffer or reagent, and can be further analyzed by methods such as Western blotting or mass spectrometry.

Immunoprecipitation is a powerful technique that allows specific isolation and purification of target proteins from complex biological samples. It is widely used in a range of applications, such as protein–protein interaction studies, identification of post-translational modifications, and analysis of protein complexes.

6.5.4 Enzyme-Linked Immunosorbent Assay (ELISA)

ELISA is a widely used laboratory technique used to detect and quantify the presence of specific proteins or antibodies in a biological sample, such as blood or urine. ELISA involves the use of a capture antibody that specifically recognizes and binds to the target protein or antibody of interest, followed by detection with a secondary enzyme-linked antibody that produces a colorimetric or fluorescent signal. ELISA is a powerful and widely used technique with several advantages (Engval 2010), including its high sensitivity and specificity, ease of use, and ability to detect a wide range of proteins and antibodies in a variety of biological samples. It is used in a variety of applications, such as clinical diagnosis, vaccine development, and drug discovery. The basic steps involved in an ELISA are as follows:

1. Coating: The wells of a microtiter plate are coated with a capture antibody that specifically recognizes and binds to the target protein or antibody of interest.
2. Blocking: The wells are then blocked to prevent non-specific binding of other proteins or antibodies.
3. Sample incubation: The biological sample, such as blood or urine, is added to the wells and allowed to incubate with the capture antibody.
4. Detection antibody incubation: A secondary antibody, which is linked to an enzyme such as horseradish peroxidase, is added to the wells and allowed to bind to the target protein or antibody of interest.
5. Substrate incubation: A substrate that is specific for the enzyme is added to the wells, causing a color change that indicates the presence and amount of the target protein or antibody.

In ELISA, the enzymes used are typically linked to the secondary antibody used for detection. The most commonly used enzyme for this purpose is horseradish peroxidase (HRP), which can be conjugated to the secondary antibody via a chemical coupling reaction. Other enzymes that can be used in ELISA include alkaline phosphatase, beta-galactosidase, and glucose oxidase. The enzyme-linked secondary antibody is added to the ELISA plate after the primary antibody has bound to the antigen or protein of interest. The substrate specific to the enzyme is then added, causing a color change or fluorescence that is proportional to the amount of antigen or protein present in the sample. The detection system is often optimized to ensure high sensitivity and low background signal.

In ELISA, the colorimetric substrates used to visualize the enzyme-linked detection antibody are typically chromogenic substrates that produce a visible color change upon reaction with the enzyme. The most commonly used colorimetric substrates for ELISA include:

1. TMB (3, 3′, 5, 5′-tetramethylbenzidine): TMB is a widely used substrate that produces a blue color when oxidized by horseradish peroxidase (HRP). The reaction can be stopped with the addition of an acid, which changes the color to yellow. TMB is often used in sandwich ELISAs for the detection of proteins.
2. ABTS (2, 2′-azino-bis (3-ethylbenzothiazoline-6-sulfonic acid)): ABTS is a soluble chromogenic substrate that produces a green color when oxidized by HRP. The reaction can be stopped with the addition of an acid, which changes the color to yellow. ABTS is often used in competitive ELISAs for the detection of small molecules.
3. PNPP (*p*-nitrophenyl phosphate): PNPP is a colorless substrate that produces a yellow color when cleaved by alkaline phosphatase (AP). PNPP is often used in phosphatase-based ELISAs for the detection of proteins.
4. BCIP/NBT (5-bromo-4-chloro-3-indolyl phosphate/nitroblue tetrazolium): BCIP/NBT is a substrate that produces a blue/purple color when cleaved by alkaline phosphatase (AP). BCIP/NBT is often used in ELISAs for the detection of nucleic acids or protein–protein interactions.

The choice of the substrate depends on the enzyme used and the desired colorimetric signal. Colorimetric ELISA substrates are widely used due to their simplicity, affordability, and compatibility with standard laboratory equipment.

There are several types of ELISA, including direct, indirect, sandwich, and competitive ELISAs, each with its own unique applications.

6.5.4.1 Direct ELISA

Direct ELISA is a type of ELISA that is used to detect the presence of a specific antigen in a sample. In a direct ELISA, the antigen is immobilized on a solid support, such as a microtiter plate, and a labeled primary antibody specific to the antigen is used for detection (Shah and Maghsoudlou 2016). The principle of direct ELISA involves the following steps:

1. Coat the solid support: The first step in a direct ELISA is to coat the solid support with the antigen of interest. This can be done by adding a known amount of the antigen to the wells of a microtiter plate and allowing it to bind to the surface of the plate.
2. Block non-specific binding: After the antigen is coated on the solid support, the plate is treated with a blocking solution to prevent non-specific binding of the primary antibody.

3. Add the primary antibody: A labeled primary antibody specific to the antigen is added to the coated plate and allowed to bind to the immobilized antigen.
4. Wash the plate: The plate is washed to remove any unbound primary antibody.
5. Detect the signal: The signal is detected by adding the appropriate substrate for the enzyme linked to the primary antibody, and measuring the resulting colorimetric or fluorescent signal.

The intensity of the signal is proportional to the amount of antigen present in the sample. Direct ELISAs are commonly used for the detection of viral or bacterial antigens, as well as other proteins or small molecules. The main advantage of a direct ELISA is that it is a simple and straightforward assay, with fewer steps and potential sources of error compared to other ELISA formats. However, it may have lower sensitivity than other types of ELISAs.

6.5.4.2 Indirect ELISA

Indirect ELISA is a type of ELISA that is used to detect the presence of antibodies against a specific antigen in a sample. In an indirect ELISA, the antigen of interest is immobilized on a solid support, such as a microtiter plate, and a primary antibody specific to the antigen is used to capture the antigen (Kohl and Ascoli 2017). A labeled secondary antibody, specific to the primary antibody, is then used for detection. The intensity of the signal is proportional to the amount of antibodies against the antigen present in the sample. Indirect ELISAs are commonly used for the detection of antibodies in serum or other biological fluids, as well as for the quantification of antigens that do not have suitable antibodies available for direct detection. The main advantage of an indirect ELISA is that it is more sensitive than a direct ELISA, due to the amplification of the signal by the secondary antibody. However, it requires an additional step and can be more time consuming than a direct ELISA.

6.5.4.3 Sandwich ELISA

Sandwich ELISA (enzyme-linked immunosorbent assay) is a type of ELISA that is used to detect the presence of a specific antigen in a sample. In a sandwich ELISA, two specific antibodies are used to sandwich the antigen of interest, which is captured in between them (Kohl and Ascoli 2017). The intensity of the signal is proportional to the amount of antigen present in the sample. Sandwich ELISAs are commonly used for the detection and quantification of antigens in biological fluids, such as serum or plasma. The main advantage of a sandwich ELISA is its specificity and sensitivity, as it uses two antibodies to capture and detect the antigen, resulting in a highly specific and sensitive assay.

6.6 Secondary Binding Assay

Secondary binding assays, also known as precipitation assays, are a type of immunoassay used to detect and quantify the presence of specific antigens in a sample. These assays rely on the principle of antigen–antibody binding, which is the basis of the immune response. In a precipitin-based secondary binding assay, a sample containing the antigen of interest is mixed with an antibody specific to that antigen. If the antigen is present in the sample, it will bind to the antibody to form an antigen–antibody complex. The complex then reacts with a secondary antibody, which recognizes and binds to the antigen–antibody complex.

The secondary antibody is usually labeled with a detectable marker, such as an enzyme or a fluorescent dye, which enables the detection and quantification of the complex. The antigen–antibody complex formed in the first step can also cause the precipitation of the complex out of solution, hence the name precipitation assay. The amount of precipitation can be used as an indirect measure of the amount of antigen in the sample. Precipitin-based secondary binding assays are commonly used in medical and veterinary diagnostics, food safety testing, and forensic investigations. Examples of these assays include the radial immunodiffusion assay, the Ouchterlony double diffusion assay, and the immunoelectrophoresis assay. These assays are highly sensitive and specific, making them a valuable tool in a wide range of applications.

6.6.1 Precipitin-Based Assays

6.6.1.1 Immunodiffusion

Immunodiffusion is a type of immunological technique used to detect and quantify the presence of specific antigens or antibodies in a sample. It involves the diffusion of an antigen and an antibody toward each other in a semi-solid medium, such as agar, and the formation of a visible precipitate at the point where the antigen and antibody react. There are two main types of immune diffusion techniques used in forensic samples analysis.

6.6.1.2 Radial Immunodiffusion (RID)

This is a quantitative technique used to determine the concentration of a specific antigen in a sample. In this technique, a known amount of antibody is incorporated into an agar gel and a well is cut into the gel. The sample containing the antigen is then added to the well, and the antigen diffuses out from the well and reacts with the antibody in the gel. As the antigen diffuses outwards, a circular ring of precipitation forms, and the diameter of the ring is proportional to the concentration of the antigen in the sample.

6.6.1.3 Double Immunodiffusion

This technique, also known as the Ouchterlony method, is a qualitative technique used to detect the presence of a specific antigen or antibody in a sample. In this technique, two wells are cut into an agar gel, and a sample containing the antigen is added to one well, while a sample containing the antibody is added to the other well. The antigen and antibody diffuse outwards from their respective wells and react when they meet in the gel, resulting in the formation of a visible precipitin line between the two wells. The position and shape of the precipitin line can be used to determine the specificity of the reaction and the presence of the antigen or antibody in the sample. Immunodiffusion is a versatile and widely used technique in medical and biological research, clinical diagnostics, and quality control testing in industries such as food and pharmaceuticals. It is relatively simple to perform, requires minimal equipment, and can be adapted for high-throughput screening of large numbers of samples.

6.6.2 Immunoelectrophoresis (IEP)

IEP is a type of immunological technique used to separate and identify proteins based on their size, charge, and antigenic properties. It combines electrophoresis, a technique that separates molecules based on their charge, with immunological detection using antibodies. In a typical IEP assay, a mixture of proteins is separated by electrophoresis on a gel and then transferred to a membrane. Antibodies specific to a particular protein of interest are then added to the membrane, and any proteins that bind to the antibody are detected using a detection system, such as an enzyme-linked immunosorbent assay (ELISA). Immunoelectrophoresis can be used to identify and quantify specific proteins in a sample, and can be used in clinical diagnostics, research, and quality control testing. It is a relatively simple and cost-effective technique, but requires some specialized equipment and expertise. Immunoelectrophoresis is a powerful technique for separating and detecting specific proteins based on their antigenic properties (Shinomiya et al. 1978). When combined with secondary binding assays such as ELISA, it can be used to detect and quantify very small amounts of proteins in a sample. IEP can be performed using various techniques, including single and double radial immunodiffusion, rocket immunoelectrophoresis, crossed immunoelectrophoresis, and electroimmunodiffusion. These techniques differ in the way the proteins are separated and the way the antibodies are applied.

6.6.2.1 Rocket Immunoelectrophoresis

This is a variation of immunoelectrophoresis, a technique used to separate and identify proteins based on their size, charge, and antigenic properties. In rocket immunoelectrophoresis, the antigen and antibody are separated by electrophoresis in a gel, and the resulting precipitin line takes on the shape of a rocket. To perform rocket immunoelectrophoresis, a sample containing the antigen of interest is mixed with a specific antibody, and the mixture is loaded into a well cut into an agarose gel (Walker 1996). An electric field is then applied to the gel, causing the antigen and antibody to migrate toward the anode. As they migrate, the antigen and antibody form a complex, which precipitates out of the solution to form a visible line. The precipitin line takes on the shape of a rocket due to the way the antigen and antibody are distributed in the gel. The antigen is located in the center of the rocket, while the antibody is distributed around the periphery of the rocket. The height of the rocket is proportional to the concentration of the antigen in the sample. The rocket can be visualized by staining the gel with a suitable dye, and the height of the rocket can be measured to determine the concentration of the antigen in the sample. Rocket immunoelectrophoresis is a relatively simple and inexpensive technique, and can be used to detect and quantify specific antigens in a wide range of biological samples. It has been used in clinical diagnostics, research, and quality control testing, and is particularly useful for detecting small changes in antigen concentration over time.

6.6.2.2 Crossed Immunoelectrophoresis

Crossed immunoelectrophoresis is another variation of immunoelectrophoresis, a technique used to separate and identify proteins based on their size, charge, and antigenic properties. In crossed immunoelectrophoresis, two-dimensional separation of the antigen and antibody is achieved by performing electrophoresis in two different directions. To perform crossed immunoelectrophoresis, a sample containing the antigen of interest is loaded onto a gel and allowed to electrophorese in one direction (Tuller and Saunders 2012). Once the antigen has been separated from other proteins in the sample, an antibody specific to the antigen is added to the top of the gel. Electrophoresis is then performed perpendicular to the initial direction of separation, causing the antigen–antibody complex to migrate toward the cathode. As the antigen–antibody complex migrates across the gel, it forms a visible line of precipitation, which can be visualized by staining the gel with a suitable dye. The position and intensity of the line are used to identify and quantify the antigen in the sample. Crossed immunoelectrophoresis is a highly sensitive technique that can detect small amounts of antigens in a sample. It is also relatively simple and inexpensive, making it a popular technique in clinical diagnostics, research, and quality control testing. However, it requires specialized equipment and expertise to perform and can be time consuming.

6.6.2.3 Electro Immunodiffusion

Electroimmunodiffusion is a technique used to identify and quantify specific proteins in a sample based on their antigenic properties. It is a variation of immunodiffusion, a technique that involves the diffusion of antigens and antibodies through a semisolid medium such as agarose gel. In electroimmunodiffusion, a sample containing the protein of interest is loaded onto a well cut into an agarose gel, and an antibody specific to the protein is added to the gel. An electric field is then applied to the gel, causing the antigen and antibody to diffuse through the gel and form a precipitin line (Sweet and Elvins 1976). The precipitin line forms at the point where the antigen and antibody are in equivalence, and its shape and size are used to identify and quantify the antigen in the sample. The line can be visualized by staining the gel with a suitable dye, and its position and intensity can be measured to determine the concentration of the antigen in the sample. Electroimmunodiffusion is a relatively simple and cost-effective technique that can be used to detect and quantify specific proteins in a wide range of biological samples. It is particularly useful for detecting and characterizing antigens with low molecular weights, and has been used in clinical diagnostics, research, and quality control testing. However, it requires some specialized equipment and expertise to perform and can be time consuming.

6.6.3 Agglutination-Based Assays

Agglutination-based assays are a type of immunoassay that detects the presence or absence of a specific antigen in a sample by causing agglutination or clumping of particles. These assays are based on the principle of antigen–antibody interactions, where specific antibodies bind to their corresponding antigens and cause agglutination. Agglutination assays can be performed in a variety of formats, such as slide agglutination, tube agglutination, and microplate agglutination. In slide and tube agglutination assays, a sample containing the antigen is mixed with a specific antibody and the mixture is allowed to settle. If the antigen is present in the sample, it will bind to the antibody and cause clumping or agglutination of the particles. In microplate agglutination assays, a sample is added to a microplate containing wells coated with a specific antibody. The antigen in the sample binds to the antibody and causes agglutination of the particles, which can be visualized by turbidity or color change.

Agglutination assays are commonly used in clinical and diagnostic settings to detect and quantify a wide range of analytes, including infectious agents, hormones, and blood group antigens. They are also used in food safety and environmental testing to detect the presence of contaminants or toxins. The advantages of agglutination assays include their simplicity, speed, and low cost. They can be performed using minimal equipment and do not require complex instrumentation or specialized expertise. However, their sensitivity and specificity can be affected by factors such as the concentration and purity of the sample, the quality of the antibody, and the presence of interfering substances in the sample. There are several types of agglutination-based assays, each with different formats and applications. Some of the most common agglutination-based assay types are as follows:

6.6.3.1 Direct Agglutination Assay

A direct agglutination assay is a type of agglutination-based assay that involves the direct binding of an antibody to the target antigen. The antibody used in this assay is specific to the target antigen, and when the antigen is present in the sample, it binds to the antibody, leading to visible agglutination or clumping of the particles. In a direct agglutination assay, the sample is mixed with a specific antibody, and the mixture is allowed to settle. If the antigen of interest is present in the sample, it will bind to the antibody, causing visible agglutination or clumping of the particles. The degree of agglutination is typically proportional to the concentration of the antigen in the sample.

6.6.3.2 Indirect Agglutination Assay

An indirect agglutination assay is a type of agglutination-based assay that involves the use of a secondary particle coated with the antigen or antibody of interest. The primary antibody or antigen is then added to the mixture, causing agglutination or clumping of the secondary particles. In an indirect agglutination assay, the secondary particle, such as latex beads or erythrocytes, is coated with the antigen or antibody of interest. The sample containing the complementary antibody or antigen is added to the mixture, causing agglutination of the secondary particles. The degree of agglutination is proportional to the concentration of the target antigen or antibody in the sample.

6.6.3.3 Hemagglutination Assay

Hemagglutination assay is a type of agglutination-based assay that uses red blood cells (RBCs) as the secondary particle. Hemagglutination is the clumping of RBCs caused by the binding of antibodies to specific antigens on the surface of the cells. In a hemagglutination assay, RBCs are mixed with a known antigen or antibody. If the complementary antibody or antigen is present in the sample, it will bind to the RBCs, causing visible agglutination or clumping of the cells. The degree of agglutination is typically proportional to the concentration of the antigen or antibody in the sample.

6.6.3.4 Particle Agglutination Assay

Particle agglutination assay is a type of agglutination-based assay that uses particles such as latex beads, gold nanoparticles, or magnetic beads as the secondary particle. The particles are coated with a specific antigen or antibody and mixed with the sample containing the complementary antibody or antigen. If the target analyte is present in the sample, it will bind to the coated particles, causing visible agglutination or clumping. The particle agglutination assay is simple, rapid, and highly sensitive, making it a useful tool in clinical and diagnostic settings. It does not require specialized equipment or expertise, and can be performed using a small amount of sample.

6.7 Conclusion

Forensic serology plays a crucial role in criminal investigations by providing valuable information about the bodily fluids found at the crime scene. The conclusions drawn from forensic serology can help identify suspects, confirm or refute alibis, and provide crucial evidence for the prosecution in court. Immunoassays are analytical methods that use antibodies to detect the presence of specific molecules in bodily fluids. In forensic science, immunoassays have been widely used for the detection of drugs, poisons, and other substances in biological samples. ELISA is a type of immunoassay that is widely used in forensic science for the detection and quantification of drugs, hormones, and other biomolecules in biological samples.

References

Abraham P, Babu PG, John TJ (1994) Comparison of the indirect immunofluorescence assay with western blot for the detection of HIV antibody. Indian J Med Res 99:143–148

Chaplin DD (2010) Overview of the immune response. J Allergy Clin Immunol 125(2):S3–S23

Engvall E (2010) The ELISA, enzyme-linked immunosorbent assay. Clin Chem 56(2):319–320

Findlay JWA (2008) Specificity and accuracy data for ligand-binding assays for macromolecules should be interpreted with caution. AAPS J 3:433–434. https://doi.org/10.1208/s12248-008-9047-2

Hamers-Casterman CTSG, Atarhouch T, Muyldermans SA, Robinson G, Hammers C, Songa EB, Hammers R (1993) Naturally occurring antibodies devoid of light chains. Nature 363(6428): 446–448

Im K, Mareninov S, Diaz MFP, Yong WH (2019) An introduction to performing immunofluorescence staining. Biobanking 1897:299–311

Ito E, Iha K, Yoshimura T, Nakaishi K, Watabe S (2021) Early diagnosis with ultrasensitive ELISA. Adv Clin Chem 101:121–133

Kohl TO, Ascoli CA. Immunometric double-antibody sandwich enzyme-linked immunosorbent assay. Cold Spring Harb Protoc. 2017 2017(6):pdb.prot093724

Konstantinou GN (2017) Food allergens. Methods molecular biology. In: Lin J, Alcocer M (eds) Enzyme-linked immunosorbent assay (ELISA), vol 1592. Humana Press, New York, NY

Lowenstein JM, Reuther JD, Hood DG, Scheuenstuhl G, Gerlach SC, Ubelaker DH (2006) Identification of animal species by protein radioimmunoassay of bone fragments and blood-stained stone tools. Forensic Sci Int 159(2–3):182–188

Mohamed Abd El-Aziz T, Soares AG, Stockand JD (2019) Snake venoms in drug discovery: valuable therapeutic tools for life saving. Toxins 11(10):564

Neuberger MS, Milstein C (1995) Somatic hypermutation. Curr Opin Immunol 7(2):248–254

Neurath AR (2008) Immune response to viruses: antibody-mediated immunity. Encyclopedia Virol 56–70. https://doi.org/10.1016/B978-012374410-4.00591-4. Epub 2008 Jul 30. PMCID: PMC7149652

Old JB, Schweers BA, Boonlayangoor PW, Reich KA (2009) Developmental validation of RSID™-saliva: a lateral flow immunochromatographic strip test for the forensic detection of saliva. J Forensic Sci 54(4):866–873

Reverberi R, Reverberi L (2007) Factors affecting the antigen-antibody reaction. Blood Transfus 5(4):227

Schittek B, Hipfel R, Sauer B, Bauer J, Kalbacher H, Stevanovic S, Garbe C (2001) Dermcidin: a novel human antibiotic peptide secreted by sweat glands. Nat Immunol 2(12):1133–1137

Shah K, Maghsoudlou P (2016 Jul) Enzyme-linked immunosorbent assay (ELISA): the basics. Br J Hosp Med (Lond) 77(7):C98–C101

Shinomiya T, Muller M, Muller PH, Lesage R (1978) Immunoelectrophoresis used for identification of blood stains in forensic medicine. Forensic Sci Int 12(2):157–163

Silverstein AM (1995) The historical origins of modern immunology. In: Immunology. Academic Press, London, pp 5–20

Stavnezer J, Guikema JE, Schrader CE (2008) Mechanism and regulation of class switch recombination. Annu Rev Immunol 26:261–292

Sweet GH, Elvins JW (1976) Studies by crossed electroimmunodiffusion on the individuality and sexual orgin of bloodstains. J Forensic Sci 21(3):498–509

Tang AF, Enyindah-Asonye G, Hioe CE (2021) Immune complex vaccine strategies to combat HIV-1 and other infectious diseases. Vaccine 2021(9):112

Tuller H, Saunders R (2012) The use of crossover immunoelectrophoresis to detect human blood protein in soil from an ambush scene in Kosovo. J Forensic Sci 57(4):873–879

Walker JM (1996) The protein protocols handbook. Springer protocols handbooks. In: Walker JM (ed) Rocket Immunoelectrophoresis. Humana Press. https://doi.org/10.1007/978-1-60327-259-9_137

Identification of Semen

7

Anuj Bharadwaj and Tanya Chauhan

Abstract

Semen is very important evidence in sexual assault cases. It is produced in post-pubescent males and is ejaculated in response to sexual stimulation. Clothes, beddings, and the victim's body are common places where semen stains can be found in sexual assault cases. Its detection on evidence can help in confirming the involvement of a male perpetrator in the crime. It can also help in establishing penile penetration in sexual assault cases. Semen is a rich source of DNA as spermatozoa are the source of genetic information from one generation to another generation. Therefore, if DNA is successfully isolated, it can help in the individual identification of the perpetrator.

Keywords

Spermatozoa · Alkaline phosphatase · Presumptive · Confirmative test

Semen is very important evidence in sexual assault cases. It is produced in post-pubescent males and is ejaculated in response to sexual stimulation. Clothes, beddings, and the victim's body are common places where semen stains can be found in sexual assault cases. Its detection on evidence can help in confirming the involvement of a male perpetrator in the crime. It can also help in establishing penile penetration in sexual assault cases. Semen is a rich source of DNA as spermatozoa are the source of genetic information from one generation to another generation.

A. Bharadwaj (✉)
CFSL Chandigarh, Chandigarh, India

T. Chauhan
Department of Forensic Science, National Forensic Science University Delhi Campus (LNJN NICFS), New Delhi, Delhi, India

A. Puri et al. (eds.), *Fundamentals of Forensic Biology*,
https://doi.org/10.1007/978-981-99-3161-3_7

Therefore, if DNA is successfully isolated, it can help in the individual identification of the perpetrator.

Therefore, it is very important to identify the semen stain. Various presumptive and confirmatory tests are performed to identify the semen stain based on different components of semen. First, the stain is located and identified visually by naked eyes or using an alternate light source. If not visible, the stain is subjected to a presumptive test like the prostatic acid phosphatase test. If the result for the presumptive test is positive, then the stain is subjected to confirmatory tests. The presence of spermatozoa in the questioned stain will confirm the stain to be semen stain. Other methods like immunochromatographic methods and ELISA are helpful if the perpetrator has low or no sperm present in his semen. RNA-based methods have also been developed to detect semen.

7.1 Composition

Semen is a thick viscous liquid. A healthy human male can release 2–6 mL of semen per ejaculation. It is heavier than water, and its density ranges between 1.027 and 1.037. It is weakly alkaline. Semen is composed of sperm cells, amino acids, sugars, lipids, enzymes, proteins, and flavonoids. Semen can be divided into sperm cells and seminal fluid. Seminal fluid is a mixture of glandular secretion. It helps in providing nourishment to sperm cells. The contributors to seminal fluids are as follows:

- **Seminal vesicle fluid**—It is 60% of seminal fluid. **Semenogelin** is present in it, which plays role in the coagulation of semen. It also consists of flavin which causes fluorescence under UV in semen.
- **Prostatic fluid**—It is 30% of seminal fluid. It consists of prostatic acid phosphatase and prostate-specific antigen. These two are very useful for semen identification in a forensic lab.
- **Epididymis** and **Bulbourethral** or **Cowper's gland** each contribute 5% to the seminal fluid (Fig. 7.1).

The normal sperm cell count in a healthy human male is 10^7–10^8 spermatozoa per mL of semen. But there could be a reduction in the number of sperm cells produced. Multiple factors affect the sperm cell count. Some of them are as follows:

- Certain medical conditions like ***azoospermia*** (no spermatozoa) and ***oligospermia*** (abnormally low sperm count). It must be noted that seminal fluid is still secreted in these individuals.
- Genetic disorder.
- Alcohol and drug abuse.

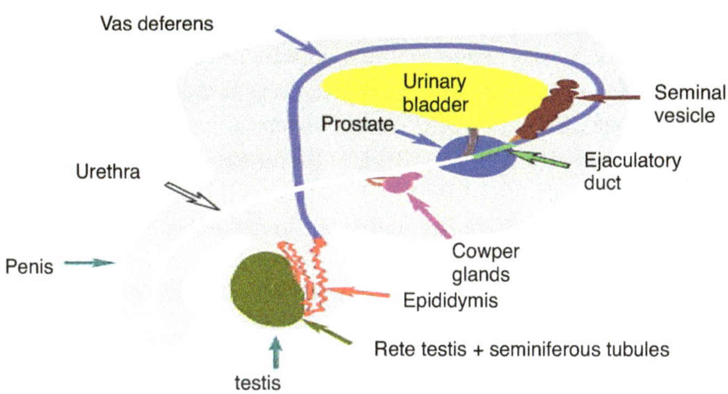

Fig. 7.1 Human male reproductive tract

- Prolonged exposure to certain chemicals.
- Surgical procedure—***vasectomy.***[1]

7.1.1 Acid Phosphatase (AP)

These are a group of phosphatases that are active at an acidic pH. Acid phosphatases are ubiquitous enzymes; they are also present in mammalian liver and cauliflower stem juice. Semen has a higher concentration of acid phosphatase than other body parts and plant tissues. Acid phosphatase released by the prostate gland has a major contribution to acid phosphatase activity in semen. At the time of puberty, a large-scale synthesis of acid phosphatase is done by epithelial cells of the prostate gland. After the age of 40, the acid phosphatase level starts decreasing. The level of acid phosphatase in semen is unaffected by the vasectomy procedure. Also, there is no correlation between acid phosphatase level and the number of spermatozoa in the semen. There is no variation between the level of prostatic acid phosphatase in a male with normal sperm count and a clinically infertile male. The half-life of acid phosphatase is 6 months at 37 °C, but in wet conditions, the half-life will decrease. If a semen stain is dried and stored at −20 °C, AP can be detected for 1 year. It is also used in the screening of prostatic carcinoma as a high level of prostatic acid phosphatase is detected in the serum of a person suffering from prostatic carcinoma.

7.1.2 Prostate-Specific Antigen (PSA)

Prostate-specific antigen is the major protein present in seminal fluid, approximately 0.5–2.0 mg/mL. It is produced by androgen stimulated epithelial cells of the prostate

[1]In vasectomy, bilateral segment of vas deference is removed which prevent spermatozoa from reaching distal parts of male reproductive tract.

gland and secreted into semen. It is also found in paraurethral, perianal, apocrine sweat, and mammary glands. Therefore, a small quantity of PSA is also detected in urine, faecal material, sweat, and milk. It is also present in the bloodstream in a lower concentration. An elevated level of PSA in plasma indicates prostate cancer.

The molecular weight of PSA is **30 kDa**; therefore, it is also known as **P30**. At room temperature, the half-life of PSA is 3 years which is greatly reduced if the stain is in wet condition. The PSA is responsible for hydrolysing ***Semenogelin.***[2]

7.1.3 Seminal Vesicle-Specific Antigen (SVSA)

There are two types of SVSAs: Semenogelin I (SG I) and Semenogelin II (SG II). It forms a coagulum which is hydrolysed in minutes by PSA. Both SG I and SG II are present in the following tissue of the male reproductive system:

- Seminal vesicles
- Ductus deference
- Prostate gland
- Epididymis

These are also present in the following tissues of the following organs:

- Skeletal muscles
- Colon
- Trachea
- Kidney

Their presence in sera of lung indicates lung cancer in the individual. In seminal fluid, concentration of SVSA is higher than PSA. Also, unlike PSA, it is not present in urine, milk, and sweat. Therefore, it is a better indicator of the presence of semen than PSA.

7.2 Spermatozoa

Spermatozoa (spermatozoon *sin.*) are male reproductive cells. They are flagellated structures approximately 55 μm in length. They are produced in seminiferous tubules of testes of post-pubescent males. The process is known as spermatogenesis. The approximate length of spermatozoa is 50 μm. In spermatogenesis, spermatogonia differentiate into spermatids and these spermatids differentiate into spermatozoa. These spermatozoa get transferred to the epididymis and undergo maturation.

[2] Semenogelin is responsible for gel formation in semen.

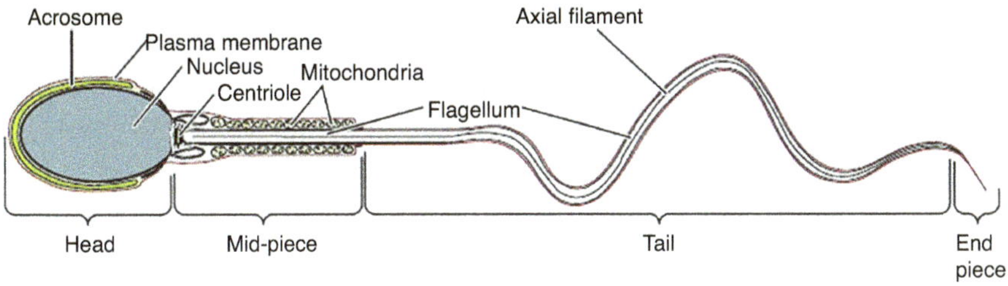

Fig. 7.2 Spermatozoon

During ejaculation, they travel through ductus deference or vas deference, ejaculatory duct, and ultimately prostatic urethra.

A spermatozoon has three morphologically distinct structures:

- Head—It is an ovoid which is 4.5 μm in length, 2.5 μm in width, and 1.5 μm in thickness. It contains a haploid nucleus at its centre and is capped with acrosome. Acrosome contains enzymes required for fertilization.
- Middle piece—It contains mitochondria, these mitochondria generate the energy required for moving the tail of a sperm cell.
- Tail—It is 90% of the total length of a spermatozoon. It provides motility to the spermatozoon. The tail loses its motility within 3–6 h of ejaculation and easily fall off (Fig. 7.2).

Certain cell organelles like endoplasmic reticulum, Golgi apparatus, lysosome, and peroxisome are absent in a mature spermatozoon. In an ejaculate, approximately 60% of spermatozoa are morphologically normal but morphological abnormalities are also very common. It must be noted that the human sperm cell is morphologically distinct from the animal sperm cell. Therefore, it is possible to identify species of semen through microscopical examination (Fig. 7.3).

7.3 Presumptive Assay for Identification

7.3.1 Visual Identification

It is easy to identify semen stains on a piece of fabric if there is a high volume of semen, and the fabric is unwashed because the stain exhibits a stiff and crusty appearance. Depending upon the colour and thickness of substrate, the semen stain appears translucent or opaque spots with a yellowish tint to the naked eyes. The dried semen stain can also be visualised under an alternating light source (ALS) at 450–495 nm with orange filtered goggles. Semen stains fluoresce more intensely than other body fluids. The fluorescence is due to flavins present in human semen. The intensity of fluorescence can be affected by the substrate. Sometimes, the semen

Fig. 7.3 Spermatozoa of different organisms

stain does not fluoresce because of certain dyes present in the fabric. If suspected stain fluoresces under UV, then it is subjected to an AP test (Figs. 7.4 and 7.5).

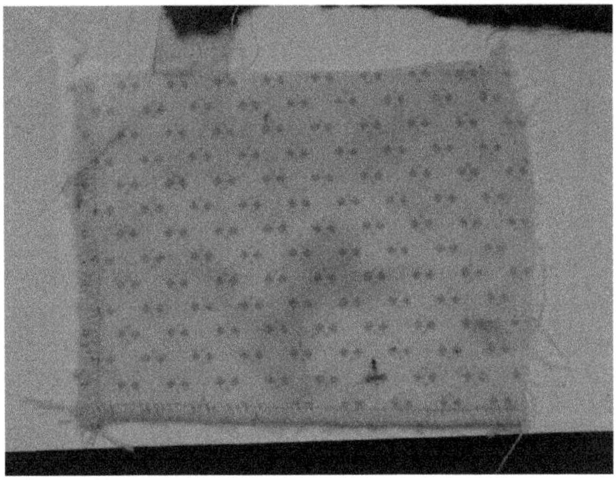

Fig. 7.4 Semen stain on light colour fabric

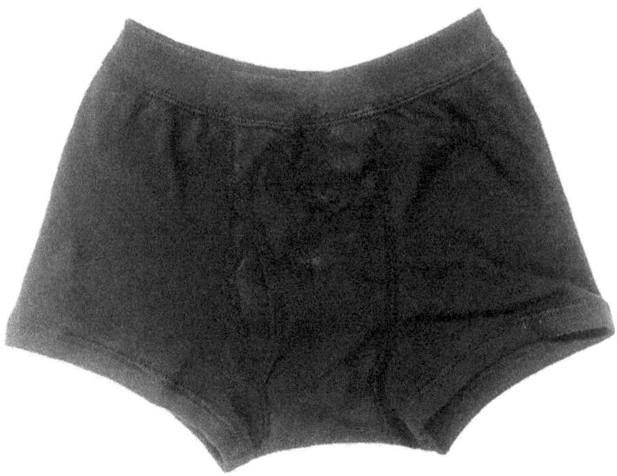

Fig. 7.5 Semen stain on dark colour fabric

7.3.2 AP Detection Through Colorimetric Assay

Acid phosphatase can hydrolyse a variety of esters. It catalyses the removal of a phosphate group from the substrate. The enzyme shows optimal activity at pH 7. In presence of stabilized diazonium salt, an insoluble coloured precipitate is formed at the site of the AP activity. To reduce the interference from non-prostatic acid phosphatase isoenzymes, prostatic acid phosphatase specific substrates like alpha naphthyl phosphate or thymolphthalein monophosphate are used. These substrates are readily hydrolysed by prostatic acid phosphatase and slowly hydrolysed by other acid phosphatase isoenzymes.

$$\text{Phosphate monoester} + \text{Acid Phosphatase} \rightarrow \text{Phosphate group} + \text{Alcohol}$$

If alpha naphthyl phosphate is used as the substrate, then the following reaction takes place:

$$\alpha \text{ naphthyl phosphate} + \text{Prostatic Acid Phosphatase} \rightarrow \text{Phosphate group} + \alpha \text{ naphtol}$$

This alpha naphthol reacts with brentamine fast blue B (stabilized diazonium salt) and forms a purple colour azo dye:

$$\alpha \text{ naphtol} + \text{Brentamine Fast Blue B} \rightarrow \text{Azo dye(Purple colour)}$$

In a forensic lab, the test is performed in the following way:

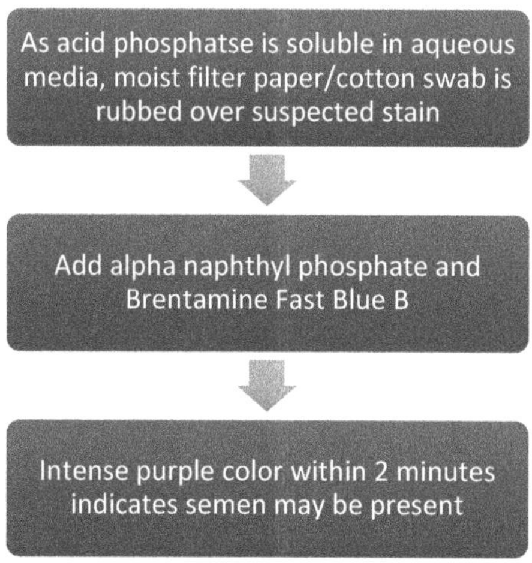

Result in discussion: If there is no purple colouration within 2 min, it indicates a negative result. While performing the test, one must note the intensity and time taken to develop the colour. Other body fluids like vaginal secretion, perspiration, faeces, or urine may give a positive reaction after a minute or so. The combination of these fluids can be found in the crotch area of undergarments.

7.3.3 AP Detection Through Fluorometric Assay

This method is more sensitive than the colourimetric method. It is also used for mapping the dimensions and orientation of semen stain. The substrate used in this method is 4-methylumbelliferonne phosphate (MUP).

4-methylumbelliferone phosphate + Acid Phosphatase $\rightarrow$ 4-methylumbelliferone + Phosphate group

The 4-methylumbelliferone fluoresce under long-wave UV radiation. This test is performed in the following way:

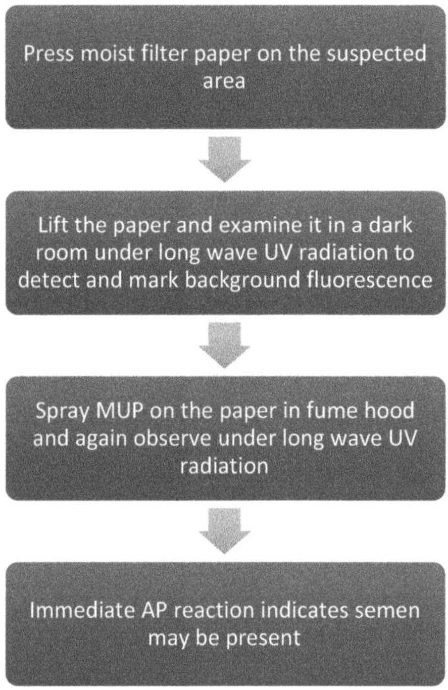

7.4 Confirmatory Assay for Identification

7.4.1 Microscopic Examination

Finding spermatozoa in the suspected stain under the microscope is a confirmation that the questioned stain is of semen. For microscopic examination, first, the stain is extracted out of the fabric on the glass slide and suitable staining methods is used to differentiate spermatozoa from the background.

There are two ways to prepare the semen extract. One method is to damp the stained area with sterile water and rub or roll it over a microscopic slide. Another method is to cut the stained area, dip it in sterile water, gently vortex it and transfer the liquid on a microscopic slide, evaporate the liquid at room temperature, or fix the slide at low heat.

Fig. 7.6 Christmas tree staining of spermatozoa

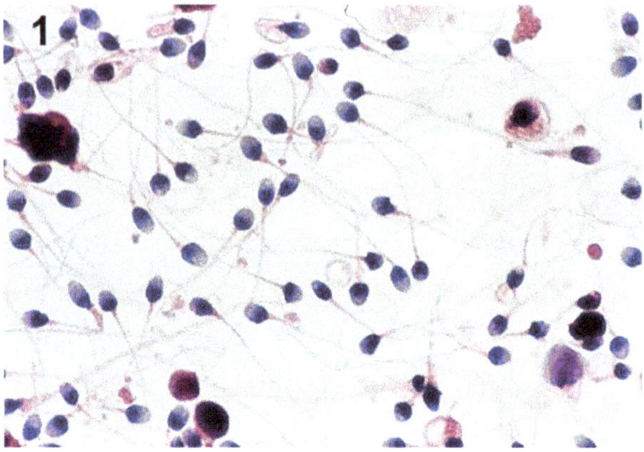

Christmas tree staining is the most popular staining method for staining spermatozoa in forensic labs. In this method, nuclear fast red (NFR) and picroindigocarmine (PIC) are used. After staining, the slide is observed under the microscope, if spermatozoa are present, acrosomal cap and nucleus will appear pink-red, the tail and midpiece will appear blue-green, and epithelial cells will appear blue-green with red nuclei (Fig. 7.6).

SPERM HY-LITER Fluorescent Staining Kit is another method used to fluorescently detect spermatozoa.

Laser Capture Microdissection (LCM) is an advanced technique used to isolate spermatozoa from epithelial cells of the victim on a glass slide. In this technique, a thin layer of thermosensitive polymer is used. This layer is present on the tip of LCM. After spermatozoa are identified under the microscope, the LCM tip is placed over the area and the polymer layer is melted by an infrared laser. The sealed spermatozoa are lifted off the slide and placed in separate tubes for DNA analysis.

7.4.2 Identification Through PSA

7.4.2.1 Immunochromatographic Assay

Commercial kits like PSA-Check-1 (VED-LAB), Seratec PSA Semiquant (Seratec Diagnostica), and One-step ABAcard PSA (Abacus Diagnostic) are used for immunochromatographic identification of PSA. These kits use anti-human PSA antibodies. In the ABAcard method, there are three zones named sample well, test zone, and control zone present on a nitrocellulose membrane. In the sample well, labelled monoclonal anti-human PSA antibodies are present. In the test zone, immobilized polyclonal anti-human PSA antibodies are present. And in the control zone, antiglobulin is present.

A drop of the semen stain extract is loaded in the sample well, the antigen of semen positive sample binds to antibody present in the sample well and make an antigen-antibody complex. This complex diffuses through the membrane and

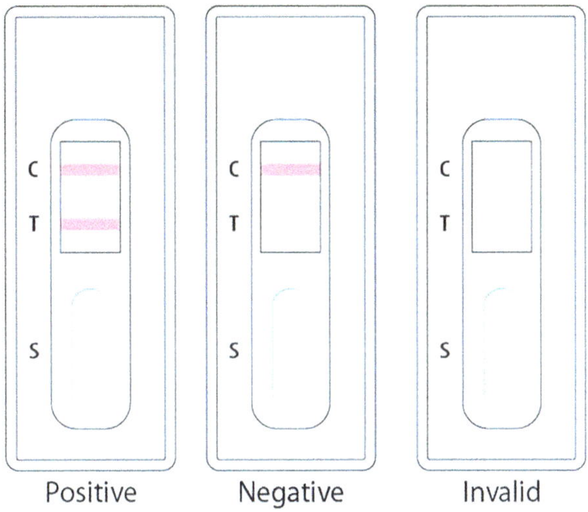

Fig. 7.7 ABA card for P30 test

reaches the test zone. In the test zone, this complex binds to monoclonal anti-human PSA antibody and make another complex of antibody-antigen-antibody. The ABAcard uses pink dye for a positive result. Therefore, a pink line is formed at the test zone for the positive sample. When unbound labelled monoclonal antihuman PSA antibodies reach the control region, they bind to antiglobulin. Here also, a pink line is formed (Fig. 7.7).

If a sample contains PSA, then a pink line will be formed at the test zone as well as at the control zone within a minute. A pink line only at the control zone indicates a negative result. Artefacts may appear if a high quantity of seminal fluid is tested.

7.4.2.2 ELISA

Enzyme-Linked Immunosorbent Assay (ELISA) technique is used to detect PSA with the help of anti-PSA antibodies. Here, sandwich ELISA is the most commonly used. Multiple samples can be tested at the same time using a 96-well titre plate. Wells of this plate are coated with polystyrene, and anti-PSA antibodies are immobilised in these wells. When a sample containing PSA is added to the well, it forms a complex with immobilised anti-PSA antibody. In the next step, secondary anti-PSA-antibody is added to the mix which results in an antibody-antigen-antibody sandwich. In the final step, labelled antiglobulin which is specific to secondary anti-PSA antibody is added. Detection can be made through either fluorometric or colorimetric techniques based on the label used with antiglobulin (Fig. 7.8).

The amount of PSA can also be quantified in this method by comparing a standard with a known concentration. This method is very specific and sensitive but quite time-consuming.

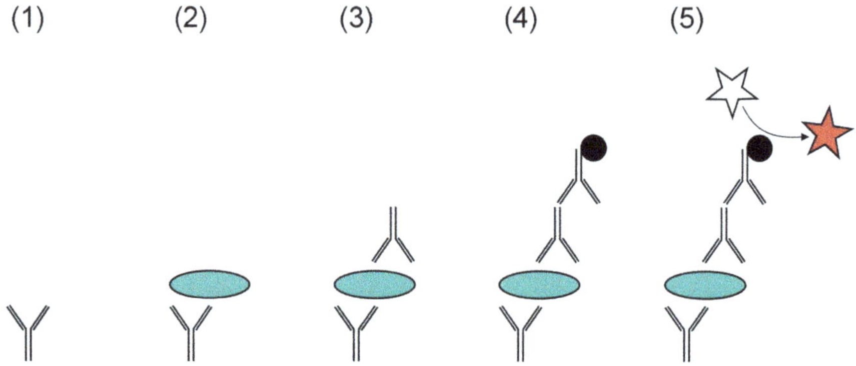

Fig. 7.8 Antibody-Antigen-Antibody Sandwich ELISA technique for P30

7.4.3 Identification Through SVSA

7.4.3.1 Immunochromatographic Assay

Commercial kits like RSID Semen Test (Independent Forensics) and Nanotrap Sg are available for SVSA detection. The RSID semen test is commonly used. The method is similar to ABAcard. But in the RSID method, the sample well contains monoclonal anti-SG antibodies, the test zone also contains immobilized monoclonal anti-Sg antibodies but for a different epitope of SVSA, and the control zone contains antiglobulin. A positive sample of semen will result in a pink line at the test zone and the control zone in 10 min.

The advantages of using this method are that it is highly sensitive, no species cross-reactivity has been observed, not responsive to other body fluids, and the result is not affected by condom lubricants or spermicides. But similar to the ABAcard method for PSA, a high dose hook effect occurs at a concentration above 3 μL.

7.4.3.2 ELISA

Similar to ELISA for PSA, the sandwich ELISA technique can also be used for the detection of SVSA. Anti-Sg antibodies are used. In ELISA for SVSA, anti-Sg antibodies are immobilised on polystyrene coated tubes. A semen positive sample when added to the tube, the SVSA in semen forms a complex with immobilised anti-Sg antibody. In the next step, secondary anti-Sg-antibody which is specific for an epitope different from the primary anti-Sg antibody is added to the mix which results in an antibody-antigen-antibody sandwich. Finally, a labelled antiglobulin which is specific to the secondary anti-Sg antibody is added. Based on the label used on antiglobulin colorimetric or fluorometric signals are generated which are detected using a spectrophotometer. The intensity of the signal detected by the spectrophotometer is proportional to the quantity of the antigen. The antigen can also be quantified by comparing the standard with known concentration (Fig. 7.9).

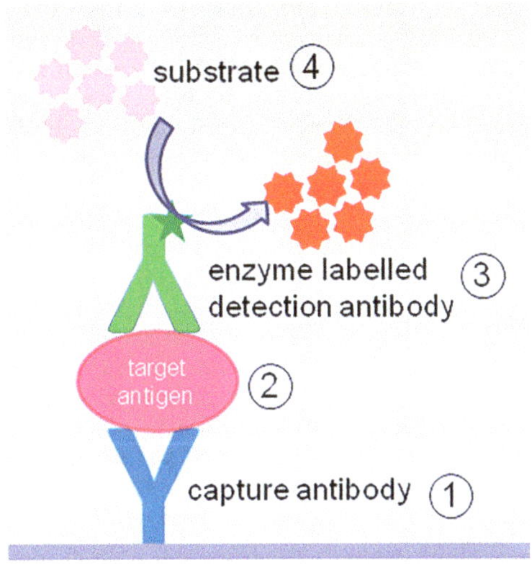

Fig. 7.9 Antibody-Antigen-Antibody Sandwich ELISA technique for SVSA

Table 7.1 Genes used for RNA-based semen detection

Gene	Description
Protamine 1 (PRM1)	DNA binding protein involved in condensation of spermatozoa chromatin
Protamine 2 (PRM2)	DNA binding protein involved in condensation of spermatozoa chromatin
Semenogelin 1 (SEMG1)	Seminal vesicle specific antigen

7.4.4 RNA-Based Detection

It is a new method of semen detection. In RNA-based method, expression of specific mRNA is detected using the RT-PCR technique. These mRNAs are exclusive to spermatozoa and cells of the male reproductive tract. The RNA-based method is highly specific and can be automated. But the only limitation is that RNA is easily degraded by endogenous ribonucleases. Following is the list of tissue specific genes which are utilised for RNA-based detection of semen (Table 7.1):

Further Reading

Ablett P (1983) The identification of the precise conditions for seminal acid phosphatase (SAP) and vaginal acid phosphatase (VAP) separation by isoelectric focusing patterns. J Forensic Sci Soc 23:254–256

Allen SM (1995) An enzyme linked immunosorbent assay (ELISA) for detection of seminal fluid using a monoclonal antibody to prostatic acid phosphatase. J Immunoassay 16(3):297–308

Allery JP, Telmon N, Mieusset R, Blanc A, Rougé D (2001) Cytological detection of spermatozoa: comparison of three staining methods. J Forensic Sci 46(2):349–351
Anoruo B, van Oorschot R, Mitchell J, Howells D (2007) Isolating cells from non-sperm cellular mixtures using the PALM microlaser micro dissection system. Forensic Sci Int 173(2–3):93–96
Baxter SJ (1973) Immunological identification of human semen. Med Sci Law 13(3):155–165
Blake ET, Sensabaugh GF (1976) Genetic markers in human semen: a review. J Forensic Sci 21(4): 785–796
Boward ES, Wilson SL (2013) A comparison of ABAcard(®) p30 and RSID™-Semen test kits for forensic semen identification. J Forensic Leg Med 20(8):1126–1130
Chapman RL, Brown NM, Keating SM (1989) The isolation of spermatozoa from sexual assault swabs using proteinase K. J Forensic Sci Soc 29(3):207–212
Chen JT, Hortin GL (2000) Interferences with semen detection by an immunoassay for a seminal vesicle-specific antigen. J Forensic Sci 45(1):234–235
Davies A, Wilson E (1974) The persistence of seminal constituents in the human vagina. Forensic Sci 3(1):45–55
Di Martino D, Giuffrè G, Staiti N, Simone A, Le Donne M, Saravo L (2004) Single sperm cell isolation by laser microdissection. Forensic Sci Int 146(suppl):S151–S153
Engelbertz F, Korda JB, Engelmann U, Rothschild M, Banaschak S (2010) Longevity of spermatozoa in the post-ejaculatory urine of fertile men. Forensic Sci Int 194(1–3):15–19
Herman Y, Feine I, Gafny R (2018) Acid phosphatase test on Phadebas® sheets - an optimized method for presumptive saliva and semen detection. Forensic Sci Int 288:218–222
Hueske EE (1977) Techniques for extraction of spermatozoa from stained clothing: a critical review. J Forensic Sci 22(3):596–598
James SH, Nordby JJ, Bell S (2014) Forensic science an introduction to scientific and investigative techniques. CRC Press, Boca Raton
Lewis J, Baird A et al (2013) Improved detection of semen by use of direct acid phosphatase testing. Sci Justice 53:385–394
Li R (n.d.) Forensic biology. CRC Press, Boca Raton
Nolan A, Speers SJ, Murakami J, Chapman B (2018) A pilot study: the effects of repeat washing and fabric type on the detection of seminal fluid and spermatozoa. Forensic Sci Int 289:51–56
O'Leary KR, Glynn CL (2018) Investigating the isolation and amplification of microRNAs for forensic body fluid identification. MicroRNA (Shariqah, United Arab Emirates) 7(3):187–194
Parikh D (2014) Parikh's textbook of medical jurisprudence forensic medicine and toxicology. CBS, New Delhi
Park JL, Park SM, Kim JH, Lee HC, Lee SH, Woo KM, Kim SY (2013) Forensic body fluid identification by analysis of multiple RNA markers using nanostring technology. Genomics Inform 11(4):277–281
Rao NG (2010) Textbook of forensic medicine & toxicology. JAYPEE, New Delhi
Saferstein R (2018) Criminalistics an introduction to forensic science. Pearson, London
Sakurada K, Ikegaya H, Fukushima H, Akutsu T, Watanabe K, Yoshino M (2009) Evaluation of mRNA-based approach for identification of saliva and semen. Leg Med (Tokyo) 11(3):125–128
Sakurada K, Watanabe K, Akutsu T (2020) Current methods for body fluid identification related to sexual crime: focusing on saliva, semen, and vaginal fluid. Diagnostics (Basel, Switzerland) 10(9):693
Sehgal V (2018) Forensic science in criminal investigation & court evidence. Selective & Scientific Books, New Delhi
Sharma B (2016) Forensic science in criminal investigation. Universal Law Publishing, New Delhi
Singh I (2011) Textbook of human histology. JAYPEE, New Delhi
Suttipasit P (2019) Forensic spermatozoa detection. Am J Forensic Med Pathol 40(4):304–311
Tian H, Lv M, Li Z, Peng D, Tan Y, Wang H, Li Q, Li F, Liang W (2018) Semen-specific miRNAs: Suitable for the distinction of infertile semen in the body fluid identification? Forensic Sci Int Genet 33:161–167

Tian H, Huang S, Bai P, Xiao X, Peng D, Zhao H, Liu Y, Feng Q, Liao M, Li F, Liang W (2021) The effect of infertile semen on the mRNA-based body fluid identification. Electrophoresis 42(16):1614–1622

Virkler K, Lednev IK (2009) Analysis of body fluids for forensic purposes: from laboratory testing to non-destructive rapid confirmatory identification at a crime scene. Forensic Sci Int 188(1–3): 1–17

Wang Z, Zhang SH, Di Z, Zhao SM, Li CT (2013) Messenger RNA profiling for forensic body fluid identification: research and applications. Fa Yi Xue Za Zhi 29(5):368–374

Westring CG, Wiuf M, Nielsen SJ, Fogleman JC, Old JB, Lenz C, Reich KA, Morling N (2014) SPERM HY-LITER™ for the identification of spermatozoa from sexual assault evidence. Forensic Sci Int Genet 12:161–167

Introduction to Saliva and Its Forensic Analysis

8

Aditya Kumar Kar, Radha Patel, and Ankit Debnath

Abstract

Saliva is a complex biological, extracellular fluid secreted by acinar cells of the major and minor salivary glands. Saliva has been an important piece of forensic evidence in cases of molestation, and drug/alcohol intake, for DNA profiling without the suspect knowing because of its ease of availability and non-invasive collection method. The analysis of saliva is also quite cost-effective, adding to the reason for saliva being a very preferred medium of choice in forensic/criminal investigation among all bodily fluids.

Tests for identification include preliminary and confirmatory tests. A preliminary test can provide possibilities of the presence of saliva at the crime scene while a confirmatory test can identify a specific genetic material.

Analysis of saliva can shed light on a person's physiological and psychological conditions, providing information about eating habits, behaviors, abnormalities, and even sex and age.

Keywords

Extracellular · Salivary glands · Amylases · Phadebad reagent

Aditya Kumar Kar, Radha Patel and Ankit Debnath contributed equally with all other contributors.

A. K. Kar (✉) · R. Patel · A. Debnath
Department of Forensic Science, Adamas University, Barasat, Kolkata, India

A. Puri et al. (eds.), *Fundamentals of Forensic Biology*,
https://doi.org/10.1007/978-981-99-3161-3_8

8.1 Introduction

Forensic investigators have confirmed the fact numerous times that saliva has helped solve high-profile cases (Kassin et al. 2013). The case of Geovanni Borjas reported on 31 May 2017 was one such example, where police arrested Geovanni, for the murder of 22-year-old Bree' Anza Guzman and 17-year-old Michelle Lozano. The detectives used DNA extracted from the saliva which matched with a DNA sample in the statewide database that was identified to belong to a close relative of the suspect they were looking for, who turned out to be Geovanni's father. Later, personal identification helped strengthen the alibi.

A crime scene can contain salivary evidence in the form of spit thrown on the sides, saliva on discarded chewing gum, and bite marks on the victim. It can also be obtained from used cutlery, glass bottles, masks, or any objects that show marks of being bitten (Fig. 8.1).

Saliva has been an important piece of evidence in cases of molestation, and drug/alcohol intake, for DNA profiling without the suspect knowing because of its ease of availability and non-invasive collection method. The analysis of saliva is also quite cost-effective, adding to the reason for saliva being a very preferred medium of choice in forensic/criminal investigation amongst all bodily fluids.

Advancements in biotechnology have improved the isolation of human DNA from dried saliva stains obtained from lip prints and bite marks at a crime scene. But in case there is a lack of DNA samples for comparative identification, it can still be used for biological profiling. It can include the age, gender, and personal characteristics of an individual for reconstructive identification. This utilizes the study of the human microbiome, microbial DNA of microorganisms present in salivary secretion, combined with salivary biomarkers to shed light on the lifestyle, medical condition, and personal habits aiding in forensic investigation (Kapoor and Chowdhry 2018) (Fig. 8.2).

8.1.1 What Is saliva?

Saliva is a complex biological, extracellular fluid secreted by acinar cells of the major and minor salivary glands. Seventy percent of saliva is produced by the submandibular salivary gland, 25% from parotids, and 5% from sublingual salivary glands. It is an indicator of various plasma constituents (Proctor and Carpenter 2007) (Fig. 8.3).

8.1.2 Function

1. Lubrication
2. Breakdowns starch into simpler sugars
3. Creates food bolus

Fig. 8.1 Graphical representation of use of saliva in various disciplines

4. Important in sense of taste
5. Maintains pH of the mouth

8.2 Activity of Amylases

Amylases are enzymes that cleave polysaccharides such as starches, which are composed of D-glucose units connected by α1–4 linkages.

Two types of amylases characterized are as follows:

1. B-amylases found in plants and bacterial sources
2. Human α-amylases

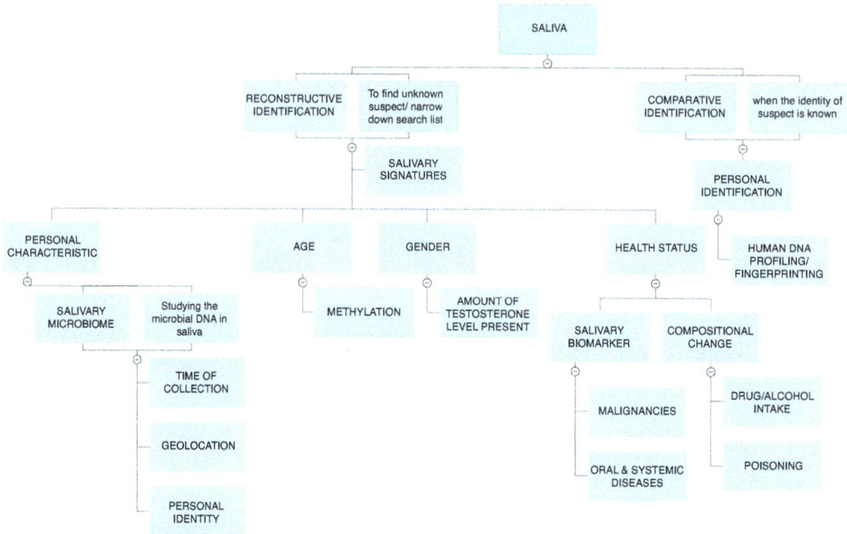

Fig. 8.2 Flowchart represent forensic significance of saliva. (Adopted and modified from Kapoor and Chowdhry 2018)

Human α-amylases have two major isoenzymes, i.e., multiple forms that differ in their amino acid sequences.

1. Human salivary α-amylase (HSA), encoded by the *Amy1* locus, is synthesized at the salivary glands and secreted into the oral cavity.
2. Human pancreatic α-amylase (HPA), encoded by the *Amy2* locus, is synthesized by the pancreas and secreted into the duodenum (Li et al. 2015).

The identification of saliva is largely based on detecting the presence of amylases in a sample. As we know, starch reacts with iodine to give black-blue color but in the presence of amylase, starch molecules break down into maltose which does not show a reaction with iodine (Figs. 8.4 and 8.5).

8.3 Identification of Saliva

The identification of saliva is largely based on detecting the presence of amylase in the sample, for that two types of assay techniques can be utilized. The first can detect the enzymatic activity of any type of amylase, unable to distinguish between HSA; the second being highly specific to HSA. This technique along with RNA-based assay is more confirmatory.

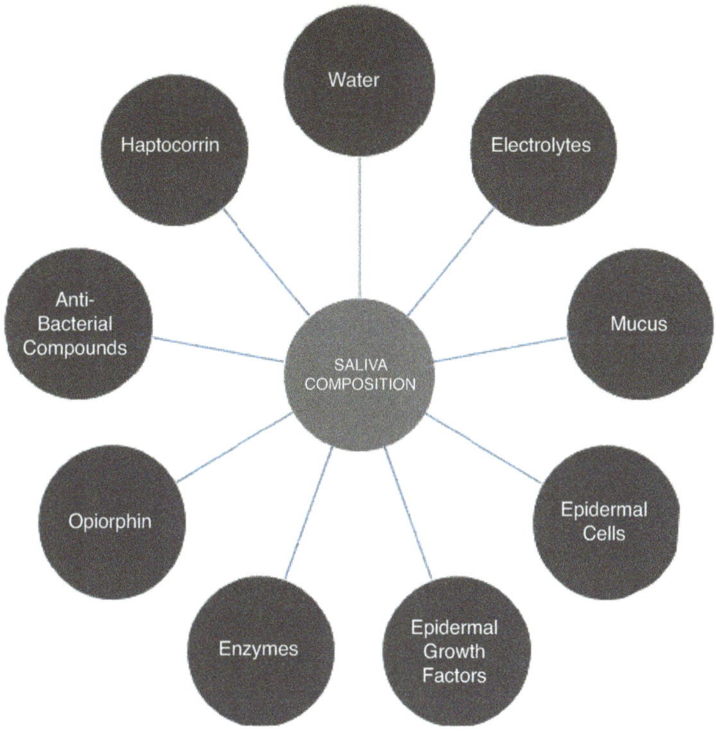

Fig. 8.3 Representative diagram for different components of saliva

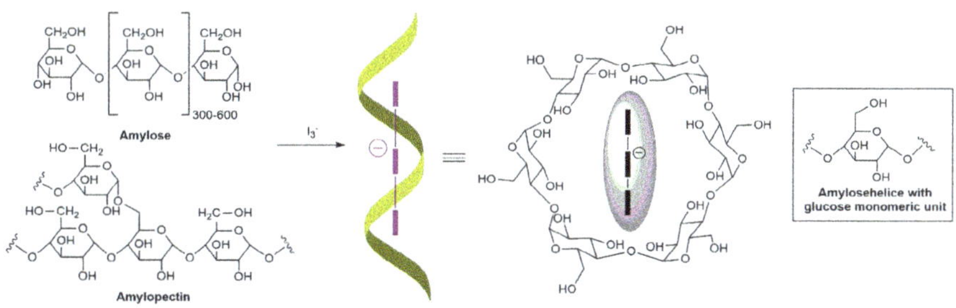

Fig. 8.4 Reaction shows formation of starch-iodine complex

8.3.1 Preliminary Examination

8.3.1.1 Visual Examination

The use of alternate light sources facilitates the search and visualization of saliva stains. However, 470 nm excitation wavelength can be used with orange goggles to allow fluorescence, which is usually less intense than seminal stains.

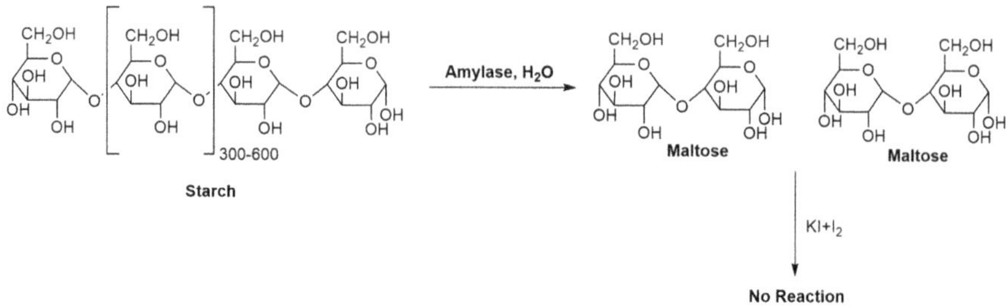

Fig. 8.5 Reaction shows action of amylase on starch

8.3.1.2 Starch-Iodine Assay

This assay is based on the activity of amylase enzyme present in saliva on starch.

8.3.1.3 Radial Diffusion Assay

Principle

Starch reacts with iodine to form a dark blue complex. In the presence of amylase, starch breaks down into monosaccharides or disaccharides, which does not give any coloration when iodine is added.

Procedure

1. An agar gel containing starch is prepared. A sample well is created by punching a hole in it.
2. Extract of the questioned sample is placed into the well. If amylase is present, it diffuses into the gel and hydrolyzes the starch present in it.
3. The gel is then stained using an iodine solution.
4. A clear area in the gel indicates amylase activity, and the size of the clear area is proportional to the amount of amylase present in the sample.

Note: This assay is not specific to HSA and can produce false-positive results (Li et al. 2015).

8.3.1.4 Colorimetric Assay

The principle of this assay is based on the dye-labeled amylase substrates such as dye-conjugated amylose or amylopectin. These substrates are not soluble in water. In presence of amylase, the dye-containing moieties are cleaved and are soluble in water to produce a color. Most such assays are not HSA specific, although their specificity can be tested by using inhibitors that preferentially inhibit HAS.

Phadebas Test

If it is suspected that the stain to be tested is a weak saliva stain, or if testing the supernatant from an extracted stain or swab, the Phadebas reagent used as a tube test is more sensitive than Phadebas paper. When testing swab(s) for the presence of saliva, they may need to be extracted according to a separate protocol. The supernatant from the extraction procedure can then be tested using the Phadebas tube test protocol below:

1. Prepare suitable positive (saliva) and negative (distilled water) control samples, in tubes (for example Eppendorf tubes).
2. Extraction of samples:
   (a) *Stain*: Cut out a small portion of the stain (approx. 3 × 3 mm) and transfer it to a sterile tube. Add 0.5–1 ml of sterile distilled water and leave to soak for 1 min. Agitate vigorously using a mechanical shaker for 30 s.
   (b) *Swab*: Follow internal protocol for extracting swabs.
3. Pipette 0.5–1 ml of supernatant from the stain or swab extraction to another tube.
4. Add 1 (one) Phadebas tablet to each test and control tube.
5. Top up the sample and control tubes to 1 ml with sterile distilled water and agitate using a mechanical shaker.
6. Incubate the tubes at 37 °C for 30 min. Remove and agitate each tube.
7. Centrifuge the tubes at 10,000 × *g* for 1 min.
8. A positive amylase reaction will produce a blue-colored supernatant solution, the depth of color depending upon the amylase concentration. A negative reaction will result in a clear supernatant (Whitehead and Kipps 1975).

8.3.2 Confirmatory Test

8.3.2.1 Immunochromatographic Assay

Immunochromatographic assay is a rapid and simple test which is used to detect presence and absence of analyte in a sample.

RSID-Saliva Test (Rapid Stain Identification of Saliva)

Principle

Commercially produced immune-chromatographic kits include the RSID-saliva kit.

1. Sample containing amylase is loaded in a sample well.
2. Antigen binds to a labeled anti-HSA Ab to form a labeled Ab-HSA complex.
3. At the test zone, the labeled complex binds to an immobilized antihuman HSA Ab to form a labeled Ab-HSA-Ab sandwich.
4. At the controlled zone, the labeled anti-HAS Ab binds to an immobilized antiglobulin and is captured at the controlled zone.

Procedure

Forensic samples obtained on cotton swabs should be extracted in 300–400 μL of RSID™-Universal Buffer: shake for 10 s, longer incubation times are optional. Alternatively, a portion of a swab may be used, and sufficient RSID™-Universal Buffer should be added to easily cover the sample. Stains on fabric or paper should be sampled by taking a punch or cutting (≈20 mm^2) of the item. The punch or

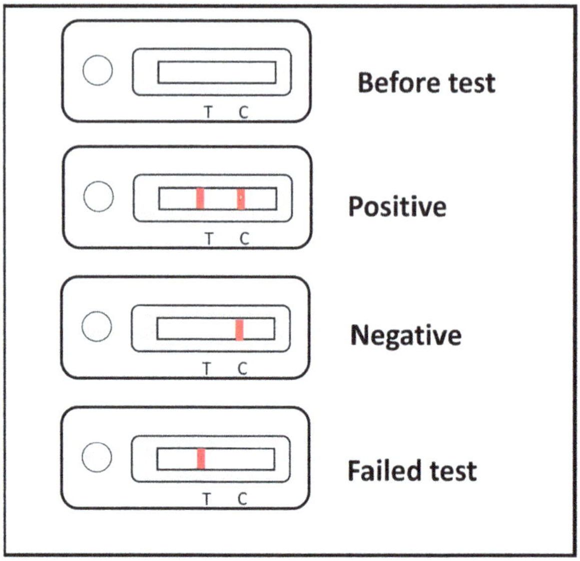

Fig. 8.6 Representative diagram of various results obtained by immunochromatographic test through RSID Kit

cutting should be extracted in 100 μL of RSID™-Universal Buffer for 10 s or longer (Fig. 8.6).

Assays should be performed at room temperature. It is recommended that a positive and negative control be included with every assay.

1. Remove the cassette from the foil pouch. Discard silica gel desiccant.
2. Combine extract aliquot (max of 20 μL) with RSID™-Universal Buffer to bring the test sample to a total volume of 100 μL.
3. Add sample in RSID™-Universal Buffer to sample window. Start timing.
4. At 10 min, score and record results as shown in the scoring results diagram shown in Fig. 8.6 (Casey and Price 2010).

Enzyme-Linked Immunosorbent Assay (ELISA)

This method can be used for the detection and quantification of a sample using anti-HAS antibodies, utilizes reporting enzymes to produce colorimetric or fluorometric signals. The intensity of the signals can be detected spectrophotometrically and is proportional to the amount of bound antigen. The number of HAS can be quantified by comparing it to a standard known concentration. This method is specific and highly sensitive in detecting HAS; it is very time-consuming.

8.4 Examination of Saliva for Various Biological and Chemical Key Points

8.4.1 RNA-Based Assay

This technique is recently developed for the identification of saliva. It is based on the expression of certain genes in certain cells/tissue. Thus, the technique uses the detection of mRNA present in certain cells of the oral cavity for the identification of saliva. These assays utilize the reverse transcription PCR method to detect gene expression. RNA-based assays present higher specificity and are amenable automation. However, one limitation is that RNA is unstable because of degradation by endogenous ribonucleases.

8.4.2 Detection of Psychotropic Drugs

Detection of psychotropic drugs and ethanol levels in saliva has been a major boost in traffic safety for its easy roadside procedure. Overall sensitivity and specificity are 91–98%. Drugs that can be detected include amphetamine, cocaine, opiates, and cannabinoids. THC is only detected in saliva as a result of long-duration smoking. SCN (thiocyanate) concentration is analyzed by HPLC to distinguish non-smokers from smokers (Youso et al. 2012). Saliva is not ideal evidence for post-mortem identification of ethanol due to contamination and decomposition by microorganisms.

8.4.3 ABO Grouping Using Saliva

Bite marks and spits being a potential source of salivary transfer can contain around 80% of ABO blood group antigens. Ninety percent of ABO DNA profiles can be obtained from epithelial cells, in saliva by DNA profiling. However, the degradation of genetic material over time is a huge hindrance.

8.4.4 Gender Determination Through Saliva

The presence of Barr-bodies in epithelial cells in saliva samples can be a significant parameter in the identification of sexual dimorphism. Barr-bodies are extra inactivated X-chromosomes found exclusively in females. It can be identified as a small, dark-stained body near a nuclear envelope.

8.4.5 Microbiome Study of Saliva for Personal Identification

Oral microflora, commonly mutans Streptococci, Actinomyces, and Veillonella, are also transferred through bite marks. Streptococcus mutans is the golden standard in forensic identification as S-mutans strain are unique to every individual acquired during birth and persists through life. Hence, the subtyping of mutans streptococci species has definite implications in forensic microbiology (Leake 2014).

8.5 Conclusion

Saliva has become a key piece of evidence in forensic science, becoming more popular for its cost-effectiveness, less time-consuming, and noninvasive methods. Salivary analysis has provided aid in both postmortem and antemortem analysis.

Saliva can shed light on both the physiological and psychological conditions of a person, providing information about eating habits, behaviors, abnormalities, and even sex and age.

Tests for identification include preliminary and confirmatory tests. A preliminary test can provide possibilities of the presence of saliva at the crime scene while a confirmatory test can identify a specific genetic material. The presence of human salivary amylase or HSA in the saliva is utilized for many of the popular confirmatory tests. Screening tests are used to determine the possible presence of hormones, genetic material, and controlled samples.

It is a fairly new field of research and needs to be explored further for it to be optimal for both adjunct and confirmatory sources of personal identification.

References

Casey DG, Price J (2010) The sensitivity and specificity of the RSID™-saliva kit for the detection of human salivary amylase in the Forensic Science Laboratory, Dublin, Ireland. Forensic Sci Int 194(1–3):67–71

Kapoor P, Chowdhry A (2018) Salivary signature in forensic profiling: a scoping review. J Forensic Dent Sci 10(3):123

Kassin SM, Dror IE, Kukucka J (2013) The forensic confirmation bias: Problems, perspectives, and proposed solutions. J Appl Res Mem Cogn 2(1):42–52

Leake S (2014) Human identification through analysis of the salivary microbiome: proof of principle. Université de Lausanne, Faculté de droit et des sciences criminelles

Li R. (2021) '15.2.1.2.1 starch iodine assay', in Forensic biology. 2nd edn. Boca Raton: CRC Press, Taylor & Francis Group

Proctor GB, Carpenter GH (2007) Regulation of salivary gland function by autonomic nerves. Auton Neurosci 133(1):3–18

Whitehead P, Kipps AE (1975) A test paper for detecting saliva stains. J Forensic Sci Soc 15:39–42

Youso SL, Rockwood GA, Logue BA (2012) The analysis of protein-bound thiocyanate in plasma of smokers and non-smokers as a marker of cyanide exposure. J Anal Toxicol 36(4):265–269

Identification of Vaginal Secretions and Menstrual Blood

9

Arjun Rao Isukapatla, Mehar Chadha, Nisha Kaushik, and Sunanda Dhenge

Abstract

Vaginal stains and menstrual stains are two types of bodily fluid stains that can be important in forensic science investigations. Vaginal stains refer to bodily fluids left on clothing, bedding, or other surfaces as a result of sexual activity, while menstrual stains refer to bloodstains left on surfaces during menstruation. Identifying these stains is significant in forensic science because they can provide critical evidence in sexual assault investigations. The presence of these stains can also help identify a suspect by linking their DNA to the stain. Methods for identifying vaginal and menstrual stains may include visual inspection, alternate light sources, chemical or DNA analysis. The identification of vaginal and menstrual stains is a critical part of forensic investigation in sexual assault cases, and can provide important evidence to support the prosecution of perpetrators and help ensure justice for victims.

Keywords

Vaginal stains · Menstrual stains · mRNA profiling · RT-PCR · D-Dimer · LDH Assay

A. R. Isukapatla (✉) · M. Chadha
Department of Life Sciences, Christ University, Bengaluru, Karnataka, India
e-mail: arjun.rao@christuniversity.in

N. Kaushik · S. Dhenge
Forensic Science Education Society, Raipur, Chhattisgarh, India

A. Puri et al. (eds.), *Fundamentals of Forensic Biology*,
https://doi.org/10.1007/978-981-99-3161-3_9

9.1 Introduction

The identification of vaginal stains is an important aspect of a sexual assault investigation. These stains can provide valuable evidence in determining whether sexual contact occurred, as well as in identifying the perpetrator. The process of identifying vaginal stains typically involves collecting samples from the victim's clothing, bedding, or body using a rape kit. The samples are then sent to a forensic laboratory, where they are analyzed for the presence of semen, sperm, or other biological materials. In cases where the stains are visible to the naked eye, they may be examined under a microscope to determine their composition and origin. The vaginal microbiota is a complex ecosystem consisting of a diverse array of microorganisms, including bacteria, viruses, fungi, and protozoa (Smith and Ravel 2017; Gajer et al. 2012). In vaginal stains, various types of bacteria may be present, depending on the individual's vaginal microbiome and any potential infections or imbalances such as bacterial vaginosis (BV), sexually transmitted infections (STIs), and vulvovaginal candidiasis (VVC) (Chee et al. 2020). Menstrual blood stains can be used to establish a link between a suspect and a victim, as the presence of menstrual blood on the suspect's clothing or other belongings may indicate contact with the victim during or after menstruation. In criminal investigations, menstrual blood stains can be used as evidence in cases involving sexual assault or other crimes where bodily fluids may be involved (Holtkötter et al. 2018). Forensic scientists can differentiate between menstrual blood and other types of blood based on the presence of certain proteins or enzymes.

9.2 Identification of Vaginal Acid Phosphatase

Vaginal acid phosphatase (VAP) examination is a forensic test used to detect the presence of acid phosphatase enzyme in vaginal secretions. Acid phosphatase is an enzyme found in high levels in semen, and therefore, its presence in vaginal secretions can indicate recent sexual activity. The VAP examination is a simple and inexpensive test that can provide valuable information in cases of sexual assault or rape (Schumann et al. 1976). The test involves swabbing the vaginal area to collect a sample of secretions, which is then analyzed for the presence of acid phosphatase. If the enzyme is present in high levels, it can indicate recent sexual activity, although it cannot determine whether the sexual activity was consensual or non-consensual. The reliability of the VAP test can be affected by several factors, such as the time elapsed since sexual activity, the use of certain medications, and the presence of certain medical conditions (Schiff 1978). Consequently, the results of the VAP test should always be interpreted in conjunction with other forensic evidence and clinical findings. VAP examination can be a useful tool in forensic investigations involving sexual assault or rape, but it should be used in conjunction with other tests and evidence to ensure accurate and reliable results.

9.2.1 Brentamine Fast Blue Test

Acid phosphatase is an enzyme found in high levels in seminal fluid, and therefore, its presence in vaginal stains can indicate recent sexual activity. Acid phosphatase activity in vaginal stains can be detected using a variety of forensic tests, including the Brentamine fast blue test and the acid phosphatase stain test (Byard and Payne-James 2015). The acid phosphatase stain test involves the use of a solution containing a substrate that reacts with acid phosphatase to produce a color change. A sample of the vaginal stain is collected and mixed with the solution, and the mixture is then observed under a microscope. If acid phosphatase is present in the stain, it will react with the substrate in the solution, producing a brown color. Acid phosphatase activity in vaginal stains can be affected by several factors, such as the time elapsed since sexual activity, the use of certain medications, and the presence of certain medical conditions. Therefore, the results of acid phosphatase tests should always be interpreted in conjunction with other forensic evidence and clinical findings. The detection of acid phosphatase activity in vaginal stains can be a useful tool in forensic investigations involving sexual assault or rape, but it should be used in combination with other tests and evidence to ensure accurate and reliable results.

9.2.2 p-Nitrophenyl Phosphate Technique

The p-nitrophenyl phosphate technique is a biochemical staining method that can be used to detect alkaline phosphatase activity in vaginal secretions (Bowen and Ryder 1976). Alkaline phosphatase is an enzyme that is produced by a variety of cells, including those in the vaginal epithelium, and can be used as a marker for the presence of vaginal fluid. To perform the staining, a sample of vaginal secretions is collected using a swab or other collection device. The swab is then placed into a solution containing Sigma p-nitrophenyl phosphate, which is a substrate that is converted to a yellow product by alkaline phosphatase. The sample is incubated for a short period of time, typically a few minutes, and then, the color of the solution is observed. A yellow color indicates the presence of alkaline phosphatase activity, and therefore the presence of vaginal secretions. The Sigma p-nitrophenyl phosphate technique is a relatively simple and inexpensive method for detecting vaginal secretions, and it can be useful in forensic investigations or medical examinations. It is also important to note that the technique only indicates the presence of vaginal fluid, and cannot be used to determine the timing or circumstances of its deposition.

9.2.3 Sodium Thymolphthalein

The sodium thymolphthalein technique is another biochemical staining method that can be used to detect the presence of vaginal secretions in forensic investigations or medical examinations (Seiden and Duncan 1983; Elkins 2012). This method also relies on the detection of alkaline phosphatase activity in the vaginal fluid. To perform the staining, a sample of vaginal secretions is collected using a swab or other collection device. The swab is then placed into a solution containing sodium thymolphthalein, which is a pH indicator that changes color in response to changes in pH. The sample is incubated for a short period of time, typically a few minutes, and then the color of the solution is observed. A blue color indicates the presence of alkaline phosphatase activity and therefore the presence of vaginal secretions. The sodium thymolphthalein technique is also a relatively simple and inexpensive method for detecting vaginal secretions, and it can be used in conjunction with other staining methods for the confirmation of results. Similar to the Sigma p-nitrophenyl phosphate technique, the sodium thymolphthalein technique indicates the presence of vaginal fluid and cannot be used to determine the timing or circumstances of its deposition.

9.3 Identification of Vaginal Bacteria

In forensic science, identifying the presence of bacteria in vaginal fluid can provide important information about a sexual assault or other crime. Bacteria in vaginal fluid can be detected using various techniques, including culture-based methods and molecular biology techniques (Xia et al. 2016). Culture-based methods involve growing bacteria in a laboratory setting on special media that promote the growth of specific bacterial species. This method is useful for identifying the type of bacteria present in a vaginal fluid sample and determining if an infection is present. However, this method can take several days or even weeks to obtain results, which may not be suitable for forensic investigations. Molecular biology techniques, such as polymerase chain reaction (PCR) and next-generation sequencing (NGS), can be used to rapidly detect and identify the presence of specific bacterial species in vaginal fluid. These techniques can be used to detect bacteria associated with sexually transmitted infections (STIs), such as Chlamydia trachomatis, Neisseria gonorrhoeae, and Trichomonas vaginalis. They can also be used to detect bacterial species associated with bacterial vaginosis, such as Gardnerella vaginalis, Atopobium vaginae, and Prevotella species (Castro et al. 2021). The identification of specific bacterial species in the vaginal fluid can provide important information about a sexual assault, as it may indicate the presence of an STI or other health concern. It may also help to identify potential sources of the bacteria, such as a suspect or other individuals involved in the investigation.

9.3.1 Types of Bacteria Present in Vaginal Stains

The most common bacteria found in vaginal stains are typically lactobacilli, which are beneficial bacteria that help to maintain a healthy vaginal environment by producing lactic acid and other substances that help to regulate the pH and prevent the growth of harmful microorganisms. Other bacteria may also be present in vaginal stain, which includes *Gardnerella vaginalis*, *Escherichia coli* (*E. coli*), *Streptococcus agalactiae*, and *Staphylococcus aureus* (Lindahl et al. 2005; Money 2005).

9.3.1.1 *Gardnerella vaginalis*

It is a type of bacteria that is commonly found in the vaginal microbiota of women. While it is not typically associated with criminal investigations or forensic science, it may be of interest in cases involving sexual assault or other types of sexual violence (Srinivasan and Fredricks 2008). If Gardnerella vaginalis is identified in a forensic sample, such as a swab taken from the vaginal area of a sexual assault victim, it may provide valuable information about the perpetrator. For example, if the DNA of the bacteria is collected and analyzed, it may be possible to determine whether the bacterium matches the DNA of a suspect or suspects. However, it is important to note that the presence of Gardnerella vaginalis in a forensic sample does not necessarily indicate sexual assault or other criminal activity. This bacteria is a normal part of the vaginal microbiota and can be present in the absence of any criminal activity. Therefore, while Gardnerella vaginalis identification may be a useful tool in forensic investigations, it should always be interpreted in the context of the specific case and other available evidence.

9.3.1.2 *Escherichia coli* (*E. coli*)

Escherichia coli (*E. coli*) is a type of bacteria that is commonly found in the gastrointestinal tract and is not normally associated with the vaginal microbiota. Therefore, the identification of *E. coli* in vaginal stains could potentially be a cause for concern, as it may indicate contamination from fecal matter or improper sample handling (Castro et al. 2019). If *E. coli* is detected in vaginal stains, it is important to consider the possibility of contamination during sample collection or handling. It may also be useful to confirm the presence of *E. coli* with additional testing, such as genetic analysis, to rule out any false positives. However, it is important to note that the presence of *E. coli* in vaginal stains does not necessarily indicate sexual assault or other criminal activity. In fact, the majority of *E. coli* strains are not pathogenic and do not cause illness. Therefore, while the identification of *E. coli* in vaginal stains may be a cause for further investigation, it should always be interpreted in the context of the specific case and other available evidence. Additional testing and analysis may be needed to determine the significance of *E. coli* in the sample and whether it is relevant to the investigation.

9.3.1.3 *Streptococcus agalactiae*

Streptococcus agalactiae (group B streptococcus or GBS) is a type of bacteria that is commonly found in the vaginal and rectal microbiota of healthy women. GBS identification in vaginal stains may be of interest in forensic investigations, particularly in cases involving infant deaths or cases of sexual assault. In cases involving infant deaths, GBS can be a potential cause of infection and may be identified in samples collected from the infant or mother. In cases of sexual assault, GBS may be identified in samples collected from the victim, as it can be transmitted during sexual contact. The presence of GBS in vaginal stains may be used as evidence of sexual contact, although it is important to note that the presence of GBS alone does not necessarily indicate non-consensual sexual activity (Smith et al. 2018). To identify GBS in vaginal stains, standard microbiological techniques such as culturing and biochemical testing can be used. More advanced techniques such as genetic testing may also be used for confirmation or to obtain more detailed information about the strain of GBS. Overall, the identification of GBS in vaginal stains can be a useful tool in forensic investigations, particularly in cases involving infant deaths or sexual assault. However, as with any forensic evidence, it should always be interpreted in the context of the specific case and other available evidence.

9.3.1.4 *Staphylococcus aureus*

Staphylococcus aureus (*S. aureus*) is a type of bacteria that is commonly found on the skin and in the nose of healthy individuals, and can sometimes cause infections. While it is not typically associated with the vaginal microbiota, it may be present in vaginal stains in certain cases. *S. aureus* identification in vaginal stains may be of interest in forensic investigations, particularly in cases involving sexual assault or other types of sexual violence. In cases involving sexual assault, *S. aureus* may be identified in samples collected from the victim, as it can be transmitted during sexual contact. The presence of *S. aureus* in vaginal stains may be used as evidence of sexual contact, although it is important to note that the presence of *S. aureus* alone does not necessarily indicate non-consensual sexual activity (Ellem et al. 2017). To identify *S. aureus* in vaginal stains, standard microbiological techniques such as culturing and biochemical testing can be used. More advanced techniques such as genetic testing may also be used for confirmation or to obtain more detailed information about the strain of *S. aureus* (Deng et al. 2019). Overall, the identification of *S. aureus* in vaginal stains can be a useful tool in forensic investigations, particularly in cases involving sexual assault or other types of sexual violence. However, as with any forensic evidence, it should always be interpreted in the context of the specific case and other available evidence. The presence of this bacteria in vaginal stains can provide important information for forensic investigations, as it may help to identify potential sources of infection or other health concerns.

9.3.2 Composition of Different Level Proteins Present in Vaginal Stains

Proteins are an important component of vaginal stains, as they can provide valuable information about the source and timing of the stain. In sexual assault investigations, proteins found in vaginal swabs or stains can be used to identify the presence of semen or other bodily fluids, and to help determine the timing of the assault. Proteins can also provide information about the health status of the individual. Elevated levels of certain proteins may indicate the presence of an infection or inflammation in the genital area (Dobay et al. 2019). In addition, the specific types of proteins found in vaginal stains can provide clues about the individual who deposited the stain.

Proteins found in semen can be used to identify the individual's genetic profile, which can be useful in identifying a suspect in cases of sexual assault. The analysis of proteins in vaginal stains is an important tool in sexual assault investigations and can provide valuable information for forensic analysis. The composition of proteins in vaginal stains can vary depending on a number of factors, such as the individual's menstrual cycle, sexual activity, and vaginal infections or conditions (Raffi et al. 1977). However, some of the proteins that have been found in vaginal stains include lactoferrin, serum albumin, cervical mucus glycoproteins, immunoglobulins, and prostate-specific antigen (PSA) (Amabebe and Anumba 2018).

9.3.2.1 Lactoferrin

Lactoferrin is a protein that is found in various bodily fluids, including breast milk, tears, and vaginal fluid. In the context of sexual assault investigations, the detection of lactoferrin in vaginal stains can be used as a biomarker for the presence of vaginal fluid (Jarosik and Land 2000). Lactoferrin is produced in the female genital tract, and its presence in vaginal fluid can be used to determine whether a stain is of vaginal origin. This information can be helpful in distinguishing between vaginal fluid stains and other bodily fluids, such as blood or saliva. However, it is important to note that the presence of lactoferrin in vaginal stains does not necessarily indicate sexual activity or sexual assault. Other factors, such as menstrual bleeding or vaginal infections, can also result in the presence of lactoferrin in vaginal fluid. Additionally, the absence of lactoferrin does not necessarily rule out the possibility of sexual activity or sexual assault, as not all vaginal fluid samples will contain detectable levels of lactoferrin (Rönnqvist et al. 2006).

9.3.2.2 Serum Albumin

Serum albumin is a protein found in blood plasma, and it can be present in vaginal stains due to contamination from menstrual blood or other sources. Its presence in a vaginal stain can be used to suggest that the stain may contain blood, but it does not necessarily indicate the presence of sexual activity or sexual assault (Joshi et al. 1990). In cases where the presence of semen is being investigated, the detection of serum albumin in a vaginal stain may suggest the presence of male blood, and therefore, semen. The presence of serum albumin alone is not sufficient evidence to conclude that sexual activity has occurred or that a sexual assault has taken place (Sakurada et al. 2020). It must be evaluated in conjunction with other evidence such as DNA analysis, physical examination, and witness statements. It is also worth noting that serum albumin can be present in vaginal fluid from non-sexual sources, such as vaginal infections or medical procedures. Therefore, the presence of serum albumin alone does not prove the presence of semen or sexual activity, and its detection should always be considered in conjunction with other factors in the investigation.

9.3.2.3 Cervical Mucus Glycoproteins

Cervical mucus glycoproteins are a major component of cervical mucus, which is produced by the cervix and lines the vaginal canal. The properties of cervical mucus change throughout the menstrual cycle in response to hormonal changes, and these changes can be used to track fertility (Moncla et al. 2016). Vaginal stains may contain cervical mucus glycoproteins, which can be detected through laboratory testing. This testing is commonly used in sexual assault investigations to identify the presence of semen or other bodily fluids, and to help determine the timing of the assault. Cervical mucus can be present in the vaginal canal for several days after intercourse, and may also be present due to natural vaginal secretions (Adnane et al. 2018; Zegels et al. 2010). Therefore, additional testing and evidence must be considered in order to determine whether sexual activity occurred and, if so, when it occurred.

9.3.2.4 Immunoglobulins

Immunoglobulins may be present in vaginal stains, particularly if there has been recent sexual activity or if there is an infection or inflammation in the genital area (Paterson et al. 2006). In sexual assault investigations, testing for the presence of immunoglobulins in vaginal swabs or stains can help to confirm recent sexual activity or the presence of an infection. It is also to be noted that the presence of immunoglobulins in vaginal stains does not necessarily indicate sexual assault or any other criminal activity (Cauci et al. 1998; Breedveld et al. 2022). Immunoglobulins may also be present due to natural vaginal secretions or as a result of non-sexual contact with bodily fluids, such as during medical procedures or contact sports (Hein et al. 2005). Additional testing and evidence must be considered in order to determine the source and timing of immunoglobulins in vaginal stains, and to determine whether any criminal activity has occurred.

9.3.2.5 Prostate-Specific Antigen (PSA)

Prostate-specific antigen (PSA) is a protein that is produced by the prostate gland in males. It is commonly used as a marker for prostate cancer, but it can also be found in semen and in smaller amounts in other bodily fluids. PSA should not normally be present in vaginal stains from females, as females do not have a prostate gland (Kulczycki et al. 2017). In cases of sexual assault, the presence of PSA in vaginal swabs or stains from females may indicate that semen from a male with a prostate gland was deposited in the vagina during the assault. The presence of PSA in vaginal stains is not definitive evidence of sexual assault, as there may be other explanations for its presence (Hochmeister et al. 1999; Kamenev et al. 1989). Some medical procedures can result in the accidental transfer of PSA-containing fluids to the vaginal area. Additional testing and evidence must be considered in order to determine the source and timing of PSA in vaginal stains, and to determine whether any criminal activity has occurred.

9.3.3 Genetic Markers Identification

9.3.3.1 mRNA Profiling

mRNA profiling has a wide range of applications in forensic science. It can be used for various purposes, including identifying the tissue origin of biological fluids found at crime scenes (Juusola and Ballantyne 2005), assessing the postmortem interval (PMI), and detecting the presence of body fluids that are difficult to visualize. One of the primary uses of mRNA profiling in forensic science is the identification of the tissue origin of biological fluids. mRNA profiling can be used to identify the presence of specific genes or gene expression patterns in a given sample, including body fluids (Jakubowska et al. 2013). Body fluids such as blood, urine, saliva, and semen can be identified using mRNA profiling by analyzing the expression of specific genes that are unique to each type of fluid. Specific genes such as hemoglobin or immunoglobulin genes are expressed in blood, while prostate-specific antigen (PSA) gene is expressed in semen. By analyzing the expression of these genes using mRNA profiling techniques such as reverse transcription-polymerase chain reaction (RT-PCR), it is possible to identify the type of body fluid present in a sample (Haas et al. 2009). mRNA profiling for body fluids identification can be useful in various forensic applications, such as identifying the source of a bodily fluid found at a crime scene or determining the presence of a disease or infection based on the expression of certain genes in a body fluid sample.

mRNA markers can be used for the identification of vaginal stains, as different cells in the body have unique gene expression patterns. It is of importance that mRNA markers alone may not be sufficient for accurate identification of vaginal stains as there may be variations in gene expression patterns between individuals. Additionally, environmental factors and the age of the sample may also affect the accuracy of mRNA marker identification. Some potential mRNA markers that have been suggested for the identification of vaginal stains include mucin 1 (MUC1), human beta-defensin 1 (HBD1), and lactoferrin (LTF). These markers are typically expressed in vaginal epithelial cells and secretions and can be detected through techniques such as reverse transcription polymerase chain reaction (RT-PCR) or microarray analysis. The use of mRNA markers should be complemented by other traditional methods of stain identification such as microscopy and DNA analysis to improve the accuracy of results. It is possible to distinguish biological material using certain mRNA markers since every tissue and bodily fluid has a unique and individual mRNA profile. Usage of the PCR methodology improves the test's sensitivity when compared to traditional confirmatory procedures. The steps involved in the RNA analysis are RNA isolation, quantification, reverse transcription, amplification of specific transcripts, and amplicon identification.

9.3.3.2 mRNA Markers Used in Vaginal Stain Identification

There are different types of mRNA markers that can be used to identify vaginal stains. These markers can be used to detect the presence of vaginal epithelial cells, which are shed from the lining of the vagina and can be present in vaginal secretions. One commonly used mRNA marker is the human beta-defensin 1 (hBD1) gene

(Valore et al. 1998). This gene is expressed in the vaginal epithelium and is not present in other tissues, making it a specific marker for vaginal cells. The hBD1 gene is a human beta-defensin gene that produces a protein with antimicrobial properties. It is expressed in a variety of human tissues, including the skin, oral cavity, and reproductive tract, including the vagina. The identification of hBD1 gene in vaginal stains can be used as a marker for vaginal epithelial cells. This marker alone may not be sufficient for positive identification of vaginal stains as other factors, such as the presence of spermatozoa or other cellular material, must also be taken into consideration (Hanson and Ballantyne 2013). To identify hBD1 gene in vaginal stains, DNA extraction and analysis techniques such as PCR (polymerase chain reaction) and DNA sequencing can be used. PCR amplifies the DNA sequence of interest, while DNA sequencing confirms the presence of the hBD1 gene in the sample. Another mRNA marker that can be used is the mucin 4 (MUC4) gene. This gene is also expressed in the vaginal epithelium and can be used to identify vaginal cells in stains (Hadžić et al. 2011). The MUC4 gene is a human gene that encodes for the mucin 4 protein, which is a glycoprotein expressed in various tissues, including the vagina (Cossu et al. 2009). The MUC4 gene is a potential marker for vaginal epithelial cells, and its identification in vaginal stains can be used in forensic investigations for identifying the presence of vaginal secretions.

Other potential mRNA markers for vaginal stain identification include the human glyceraldehyde-3-phosphate dehydrogenase (GAPDH) gene, the human lactate dehydrogenase C (LDHC) gene, and the human lactoferrin (LTF) gene. However, the use of these markers may not be as specific as hBD1 and MUC4. The GAPDH gene is a human gene that encodes for the glyceraldehyde-3-phosphate dehydrogenase protein, which is involved in the glycolysis pathway and is expressed in many tissues, including the vagina. The GAPDH gene is a potential marker for vaginal epithelial cells, and its identification in vaginal stains can be used in forensic investigations for identifying the presence of vaginal secretions. The LDHC gene is a human gene that encodes for the lactate dehydrogenase C protein, which is involved in the conversion of pyruvate to lactate in various tissues, including the reproductive tract, including the vagina. The LDHC gene has been proposed as a potential marker for the identification of vaginal epithelial cells in forensic investigations. The LTF gene is a human gene that encodes for lactotransferrin, which is an iron-binding glycoprotein found in various bodily fluids, including vaginal secretions (Salzmann et al. 2021). The LTF gene has been proposed as a potential marker for the identification of vaginal epithelial cells in forensic investigations.

9.3.3.3 RT-PCR

RT-PCR (reverse transcription-polymerase chain reaction) is a molecular technique used to amplify RNA sequences into complementary DNA (cDNA) and subsequently amplify the cDNA sequences using PCR. It is a powerful tool for detecting and quantifying specific RNA molecules, including those expressed by genes in vaginal epithelial cells. RT-PCR can be used in forensic investigations for the identification of vaginal stains. To use RT-PCR for the identification of vaginal stains, the RNA must first be extracted from the stain and converted into cDNA using reverse transcription (Ballantyne and Juusola 2007). The cDNA can then be amplified using PCR with primers specific to the gene of interest, such as hBD1, MUC4, GAPDH, LDHC, or LTF. The PCR products can be visualized and analyzed using gel electrophoresis or other methods, such as sequencing or melting curve analysis. Multiplex PCR is a molecular technique that allows for the simultaneous amplification of multiple DNA or RNA targets in a single reaction. It is a useful tool for forensic investigations, as it allows for the identification of multiple genes of interest in a single sample. Multiplex PCR can be used for the identification of vaginal stains by targeting multiple genes expressed in vaginal epithelial cells, such as hBD1, MUC4, GAPDH, LDHC, and LTF (Gipson et al. 1999). For identification of vaginal stains using the multiplex PCR, DNA or RNA must first be extracted from the stain. The extracted DNA or RNA is then used as a template for PCR with multiple sets of primers, each targeting a specific gene of interest. The PCR products are then visualized and analyzed using gel electrophoresis or other methods.

9.3.3.4 Protein Identification

Protein estimation in vaginal fluid can be helpful in identifying various conditions or diseases. An increased level of protein in vaginal fluid can indicate inflammation or infection. There are several methods for protein estimation in vaginal fluid, including biochemical assays, immunological assays, and electrophoresis. Biochemical assays involve measuring the total protein content using colorimetric or spectrophotometric techniques. Immunological assays involve measuring specific proteins using antibodies that bind to the protein of interest. Electrophoresis separates proteins based on their size and charge and can provide a more detailed analysis of the protein composition in vaginal fluid (Igoh et al. 2015). The specific method used for protein estimation in vaginal fluid will depend on the purpose of the analysis and the available resources. Protein markers can be used to identify the presence of specific proteins in a vaginal stain, which may provide additional information about the source of the stain. Certain proteins are specific to semen or vaginal secretions, which can help to determine whether the stain came from a male or female individual. There are several protein markers that can be used for this purpose. Prostate-specific antigen (PSA) is a protein marker that is specific to semen (Wen et al. 2012). If PSA is detected in a vaginal stain, it is a strong indication that the stain came from a male individual. Other protein markers that can be used to identify vaginal secretions include human cervicovaginal fluid (CVF) proteins such as lactoferrin, lysozyme, and mucin. These proteins are specific to vaginal secretions and can be used to identify the presence of such secretions in a stain.

9.4 Menstruation: D-Dimer Assay and LDH Assay

9.4.1 Menstrual Blood

Menstruation is the regular flow of blood and the removal of the endometrium, which has deteriorated, from the uterus of non-pregnant women. The uterus is crucial in preparing the uterine endometrium for the potential implantation of an embryo in the future. The myometrium and the endometrium make up the uterus' linings. The uterus's muscular fibers make up the myometrium. The stroma and simple columnar epithelium make up the endometrium. Sexual assault is a serious offense and the identification of body fluids originating from sexual activity is a crucial aspect of forensic investigations (Bagwe 2018). The discrimination between peripheral and menstrual blood can be highly relevant for police investigations because it provides potential evidence regarding the issue of consent. In situations where blood may be present on a surface, such as clothing, an accurate distinction between menstrual blood and peripheral blood could be a crucial finding in figuring out whether the bloodstain had come from trauma or menses, potentially revealing information about the consent issue. Menstrual blood can be identified using a variety of methods, including microscopy, the identification of lactate dehydrogenase isozymes, messenger RNA (mRNA) and micro-RNA (miRNA) profiling, and the detection of fibrinolysis by-products.

The PMB test is a duplex test combining human hemoglobin and D-dimer detection, developed for the identification of blood and menstrual fluid, both at the crime scene and in the laboratory. Duplex D-dimer/hemoglobin assay reliably detects the presence of human hemoglobin and identifies samples containing menstrual fluid by detecting the presence of D-dimers. Post-mortem blood, which also shows a high D-dimer concentration, can be distinguished from menstrual blood by the analysis of myoglobin levels. Post-mortem blood is unlikely to be present in a typical sexual offense case.

9.4.2 D-dimer Assay

D-dimer assay is a blood test used to detect the presence of D-dimer, a small protein fragment that is produced when a blood clot is degraded in the body. While D-dimer assay is not typically used to identify vaginal stains directly, it can be used to support the presence of a vaginal stain and to indicate the possibility of sexual assault (Baker et al. 2011). In cases of sexual assault, it is common to perform a sexual assault related tests, which includes the collection of vaginal swabs to test for the presence of semen, sperm, or other biological materials. If a vaginal stain is identified, a D-dimer assay may be performed on the vaginal swab to detect the presence of D-dimer. The presence of D-dimer in a vaginal stain sample can indicate the presence of blood, which can be consistent with sexual assault or trauma. Although the presence of D-dimer in a vaginal stain does not necessarily indicate sexual assault, as it can also be produced by other medical conditions or injuries.

Fibrin D-dimer is the breakdown byproduct of cross-linked fibrin. It stands for the marker of intravascular thrombogenesis. During menstruation, the fibrinolytic and coagulation pathways are active. D-dimer fragment assays can be used forensically to identify menstrual blood (Tsai et al. 2022). In an enzyme-linked immunosorbent assay, antibodies bind to D-dimer antigens on the solid phase. The D-dimer-antibody complex can then be examined using an antibody-based detection method. This method takes a long time, but it is very delicate. In latex agglutination tests, the interaction of antibodies and D-dimers that is present on carrier's results in the formation of aggregates during the agglutination process. D-dimer testing can positively identify menstrual blood samples. These assays do not respond well to peripheral blood, despite the low levels of D-dimer present there. Menstrual blood can be distinguished from peripheral blood using D-dimer assays. Although post-mortem blood is not usually detected in sexual assault cases, its discovery would make it more difficult to interpret the results.

9.4.3 Lactate Dehydrogenase (LDH) Assay

Lactate dehydrogenase (LDH) assay can be used in forensic cases to identify body fluids, such as vaginal fluid, semen, saliva, and blood, in crime scene samples. LDH is a stable enzyme and can remain detectable for several days after deposition, making it a useful marker for the identification of body fluids. Forensic scientists can use LDH assay in combination with other tests, such as acid phosphatase (AP) and prostate-specific antigen (PSA), to differentiate between various body fluids. The presence of LDH, AP, and PSA can indicate the presence of semen, while the presence of only LDH and AP can indicate the presence of vaginal fluid (Ben et al. 2007). LDH assay can also be used to determine the age of a stain. As time passes, the amount of LDH in a stain decreases, so measuring the LDH levels can provide an estimate of the age of the stain. The use of LDH assay in forensic cases can provide valuable information in identifying body fluids and estimating the age of stains. Lactate dehydrogenase (LDH) is an enzyme found in many body tissues, including the vagina. LDH levels can be measured in vaginal fluid to help diagnose conditions such as bacterial vaginosis, which is a common vaginal infection. To perform an LDH assay to identify vaginal stains, a sample of vaginal fluid is collected and analyzed for LDH levels using a commercial LDH assay kit. The kit contains reagents that react with LDH to produce a measurable signal, which is typically measured using a spectrophotometer. If the LDH level in the vaginal fluid is elevated, it may indicate an infection or other abnormal condition. It is important to note that LDH levels can also be elevated in response to non-infectious factors, such as trauma or inflammation.

Lactate dehydrogenase (LDH) is a family of enzymes that catalyze the conversion of lactate to pyruvate. There are five different isoenzymes of LDH, which are named LDH-1, LDH-2, LDH-3, LDH-4, and LDH-5 (Virkler and Lednev 2009). These isoenzymes differ in their composition and distribution in different tissues. In the identification of vaginal stains using LDH assay, the LDH isoenzymes present in vaginal fluid are typically LDH-1 and LDH-2. LDH-1 is mainly found in heart muscle and red blood cells, while LDH-2 is found in the liver and other tissues. In vaginal fluid, the LDH-1 and LDH-2 isoenzymes are typically found in approximately equal amounts. The ratio of LDH-1 to LDH-2 can be used to differentiate between vaginal fluid and semen. Semen contains a higher proportion of LDH-5, while vaginal fluid contains higher levels of LDH-1 and LDH-2. A low ratio of LDH-1/LDH-2 in a stain is indicative of vaginal fluid, while a high ratio is indicative of semen. The LDH isoenzymes used in vaginal stain identification are mainly LDH-1 and LDH-2, and the ratio of LDH-1 to LDH-2 can be used to differentiate between vaginal fluid and semen.

9.5 Conclusion

The identification of vaginal stains in criminal cases is significant because it can provide critical voice in sexual assault investigations. By identifying and analyzing these stains, forensic examiners can determine whether sexual activity occurred, and if so, whether it was consensual or non-consensual. The presence of vaginal stains can also help identify a suspect by linking their DNA to the stain. The significance of different identification methods may vary depending on the specific case and circumstances. Some highly sophisticated technological methods for identifying vaginal stains include enzyme assays, Protein identification, mRNA profiling, and DNA analysis. These methods can help confirm the presence of bodily fluids and identify the source of the stain, which can be crucial in linking a suspect to the crime.

References

Adnane M, Meade KG, O'Farrelly C (2018) Cervico-vaginal mucus (CVM)–an accessible source of immunologically informative biomolecules. Vet Res Commun 42(4):255–263

Amabebe E, Anumba DO (2018) The vaginal microenvironment: the physiologic role of lactobacilli. Front Med 5:181

Bagwe K (2018) Evaluation of recently developed methods for the forensic detection of menstrual blood (Doctoral dissertation, Boston University)

Baker DJ, Grimes EA, Hopwood AJ (2011) D-dimer assays for the identification of menstrual blood. Forensic Sci Int 212(1-3):210–214

Ballantyne J, Juusola J (2007) Messenger RNA profiling: body fluid identification using multiplex reverse transcription-polymerase chain reaction (RT-PCR). Research Foundation of The University of Central Florida Inc., Orlando

Ben SQ, Ni SS, Shen HH, Shi YX, Huang SB, Xu JH, Huang JF (2007) The dynamic changes of LDH isoenzyme 3 and D-dimer following pulmonary thromboembolism in canine. Thromb Res 120(4):575–583

Bowen ID, Ryder TA (1976) Use of the p-nitrophenyl phosphate method for the demonstration of acid phosphatase during starvation and cell autolysis in the planarian Polycelis tenuis Iijima. Histochem J 8:319–329

Breedveld AC, Schuster HJ, van Houdt R, Painter RC, Mebius RE, van der Veer C, van Egmond M (2022) Enhanced IgA coating of bacteria in women with Lactobacillus crispatus-dominated vaginal microbiota. Microbiome 10(1):1–11

Byard R, Payne-James J (2015) Encyclopedia of forensic and legal medicine. Academic Press, Cambridge

Castro J, Machado D, Cerca N (2019) Unveiling the role of Gardnerella vaginalis in polymicrobial bacterial vaginosis biofilms: the impact of other vaginal pathogens living as neighbors. ISME J 13:1306–1317

Castro J, Rosca AS, Muzny CA, Cerca N (2021) Atopobium vaginae and Prevotella bivia are able to incorporate and influence gene expression in a pre-formed Gardnerella vaginalis biofilm. Pathogens 10(2):247

Cauci S, Monte R, Driussi S, Lanzafame P, Quadrifoglio F (1998) Impairment of the mucosal immune system: IgA and IgM cleavage detected in vaginal washings of a subgroup of patients with bacterial vaginosis. J Infect Dis 178(6):1698–1706

Chee WJY, Chew SY, Than LTL (2020) Vaginal microbiota and the potential of Lactobacillus derivatives in maintaining vaginal health. Microb Cell Factories 19(1):203

Cossu C, Germann U, Kratzer A, Bär W, Haas C (2009) How specific are the vaginal secretion mRNA-markers HBD1 and MUC4? For Sci Int Genet Suppl Ser 2(1):536–537

Deng L, Schilcher K, Burcham LR, Kwiecinski JM, Johnson PM, Head SR, Doran KS (2019) Identification of key determinants of Staphylococcus aureus vaginal colonization. MBio 10(6): e02321

Dobay A, Haas C, Fucile G, Downey N, Morrison HG, Kratzer A, Arora N (2019) Microbiome-based body fluid identification of samples exposed to indoor conditions. Forensic Sci Int Genet 40:105–113

Elkins KM (2012) Forensic DNA biology: a laboratory manual. Academic Press, Cambridge

Ellem JA, Kovacevic D, Olma T, Chen SA (2017) Rapid detection of Group B streptococcus directly from vaginal–rectal specimens using liquid swabs and the BD Max GBS assay. Clin Microbiol Infect 23(12):948–951

Gajer P, Brotman RM, Bai G, Sakamoto J, Schütte UME, Zhong X et al (2012) Temporal dynamics of the human vaginal microbiota. Sci Transl Med 4:132–152

Gipson IK, Spurr-Michaud S, Moccia R, Zhan Q, Toribara N, Ho SB, Gargiulo AR, Hill JA (1999) MUC4 and MUC5B transcripts are the prevalent mucin messenger ribonucleic acids of the human endocervix. Biol Reprod 60:58–64

Haas C, Klesser B, Maake C, Bär W, Kratzer A (2009) mRNA profiling for body fluid identification by reverse transcription endpoint PCR and realtime PCR. Forensic Sci Int Genet 3(2):80–88

Hadžić G, Lukan A, Drobnič K (2011) Practical value of the marker MUC4 for identification of vaginal secretion in penile swabs. For Sci Int Genet Suppl Ser 3(1):e222–e223

Hanson EK, Ballantyne J (2013) Highly specific mRNA biomarkers for the identification of vaginal secretions in sexual assault investigations. Sci Justice 53(1):14–22

Hein M, Petersen AC, Helmig RB, Uldbjerg N, Reinholdt J (2005) Immunoglobulin levels and phagocytes in the cervical mucus plug at term of pregnancy. Acta Obstet Gynecol Scand 84(8): 734–742

Hochmeister MN, Budowle B, Rudin O, Gehrig C, Borer U, Thali M, Dirnhofer R (1999) Evaluation of prostate-specific antigen (PSA) membrane test assays for the forensic identification of seminal fluid. J For Sci 44(5):1057–1060

Holtkötter H, Dias Filho CR, Schwender K, Stadler C, Vennemann M, Pacheco AC, Roca G (2018) Forensic differentiation between peripheral and menstrual blood in cases of alleged sexual assault—validating an immunochromatographic multiplex assay for simultaneous detection of human hemoglobin and D-dimer. Int J Legal Med 132:683–690

Igoh A, Doi Y, Sakurada K (2015) Identification and evaluation of potential forensic marker proteins in vaginal fluid by liquid chromatography/mass spectrometry. Anal Bioanal Chem 407:7135–7144

Jakubowska J, Maciejewska A, Pawłowski R, Bielawski KP (2013) mRNA profiling for vaginal fluid and menstrual blood identification. Forensic Sci Int Genet 7(2):272–278

Jarosik GP, Land CB (2000) Identification of a human lactoferrin-binding protein in Gardnerella vaginalis. Infect Immun 68(6):3443–3447

Joshi S, Sharma VK, Sarin AR (1990) Characterisation of vaginal secretion stains in forensic cases—a critical review. In: 13th congress of the International Society for Forensic Haemogenetics (Internationale Gesellschaft für forensische Hämogenetik eV) New Orleans, October 19–21, 1989. Springer, Berlin, pp 355–357

Juusola J, Ballantyne J (2005) Multiplex mRNA profiling for the identification of body fluids. Forensic Sci Int 152:1–12

Kamenev L, Leclercq M, Francois-Gerard C (1989) An enzyme immunoassay for prostate-specific p30 antigen detection in the postcoital vaginal tract. J For Sci Soc 29(4):233–241

Kulczycki A, Brill I, Snead MC, Macaluso M (2017) Prostate-specific antigen concentration in vaginal fluid after exposure to semen. Contraception 96(5):336–343

Lindahl G, Sta M, Areschoug T (2005) Surface proteins of Streptococcus agalactiae and related proteins in other bacterial pathogens. Clin Microbiol Rev 18:102–127. https://doi.org/10.1128/CMR.18.1.102-127.2005

Moncla BJ, Chappell CA, Debo BM, Meyn LA (2016) The effects of hormones and vaginal microflora on the glycome of the female genital tract: cervical-vaginal fluid. PLoS One 11(7): e0158687

Money D (2005) The laboratory diagnosis of bacterial vaginosis. Can J Inf Dis Med Microbiol 16(2):77–79

Paterson SK, Jensen CG, Vintiner SK, McGlashan SR (2006) Immunohistochemical staining as a potential method for the identification of vaginal epithelial cells in forensic casework. J Forensic Sci 51(5):1138–1143

Raffi RO, Moghissi KS, Sacco AG (1977) Proteins of human vaginal fluid. Fertil Steril 28(12): 1345–1348

Rönnqvist PD, Forsgren-Brusk UB, Grahn- Håkansson EE (2006) Lactobacilli in the female genital tract in relation to other genital microbes and vaginal pH. Acta Obstet Gynecol Scand 85(6): 726–735

Sakurada K, Watanabe K, Akutsu T (2020) Current methods for body fluid identification related to sexual crime: focusing on saliva, semen, and vaginal fluid. Diagnostics 10(9):693

Salzmann AP, Bamberg M, Courts C, Dørum G, Gosch A, Hadrys T, Haas C (2021) mRNA profiling of mock casework samples: results of a FoRNAP collaborative exercise. Forensic Sci Int Genet 50:102409

Schiff AF (1978) Reliability of the acid phosphatase test for the identification of seminal fluid. J Forensic Sci 23(4):833–844

Schumann GB, Badawy S, Peglow A, Henry JB (1976) Prostatic acid phosphatase: current assessment in vaginal fluid of alleged rape victims. Am J Clin Pathol 66(6):944–952

Seiden HOWARD, Duncan GT (1983) Presumptive screening test for seminal acid phosphatase using sodium thymolphthalein monophosphate. J Assoc Off Anal Chem 66(1):207–209

Smith SB, Ravel J (2017) The vaginal microbiota, host defence and reproductive physiology. J Physiol 595:451–463

Smith A, Kaczmar A, Bamford RA, Smith C, Frustaci S, Kovacs-Simon A, O'Neill P, Moore K, Paszkiewicz K, Titball RW et al (2018) The culture environment influences both gene regulation and phenotypic heterogeneity in Escherichia coli. Front Microbiol 9:1–13

Srinivasan S, Fredricks DN (2008) The human vaginal bacterial biota and bacterial vaginosis. Interdiscip Perspect Inf Dis 2008(2008):750479

Tsai LC, Liu KL, Lin WY, Lin YC, Huang NE, Lee JCI, Hsieh HM (2022) Evaluation of three commercial kits effective identification of menstrual blood based on the D-dimer. Forensic Sci Int 338:111389

Valore EV, Park CH, Quayle AJ, Wiles KR, McCray PB Jr, Ganz T (1998) Human beta-defensin-1: an antimicrobial peptide of urogenital tissues. J Clin Invest 101:1633–1642

Virkler K, Lednev IK (2009) Analysis of body fluids for forensic purposes: from laboratory testing to non-destructive rapid confirmatory identification at a crime scene. Forensic Sci Int 188(1-3): 1–17

Wen Y, Whitin J, Yu T, Cohen H, Polan ML, Chen B (2012) Identification of protein marker in vaginal wall tissues of women with stress urinary incontinence by protein chip array. J Obstet Gynaecol Res 38(1):89–96

Xia Q, Cheng L, Zhang H, Sun S, Liu F, Li H, Diao Y (2016) Identification of vaginal bacteria diversity and it's association with clinically diagnosed bacterial vaginosis by denaturing gradient gel electrophoresis and correspondence analysis. Infect Genet Evol 44:479–486

Zegels G, Van Raemdonck GA, Tjalma WA, Van Ostade XW (2010) Use of cervicovaginal fluid for the identification of biomarkers for pathologies of the female genital tract. Proteome Sci 8:1–23

Urine

10

Rudresh Channe, H. Manderiya, Tanya Chauhan, Alka Mishra, Preeti Bhatnagar, Avinash Puri, and Nithyanandam Mahalakshmi

Abstract

Urine is a liquid waste product excreted by the kidneys through a process known as urination, and it exits the body via the urethra. Typically, urine appears as a clear, pale yellow solution, but its color can vary widely, ranging from colorless to light yellow. The color of urine is primarily determined by the presence of specific pigments, compounds, waste product concentrations, and hydration levels.

Urine is a valuable substance for drug detection and monitoring drug exposure because it generally lacks a significant amount of protein. It contains high concentrations of metabolites, which allows for a longer detection window for most drugs. As a result, urine is considered one of the most reliable sources for

R. Channe · H. Manderiya
Department of Forensic Science, Institute for Excellence in Higher Education, Bhopal, Madhya Pradesh, India

T. Chauhan
Department of Forensic Science, Institute of Excellence, Bhopal, Madhya Pradesh, India

Department of DNA, Forensic Sciences, Chennai, Tamil Nadu, India

A. Mishra
Department of Forensic Science, Medi-Caps University, Indore, India

P. Bhatnagar
Department of Biology, Regional Forensic Science Laboratory, Gwalior, Madhya Pradesh, India

A. Puri (✉)
Department of DNA, Regional Forensic Science Laboratory, Bhopal, Madhya Pradesh, India

N. Mahalakshmi
Department of DNA and Forensic Biology, Regional Forensic Science Laboratory, Indore, Madhya Pradesh, India

A. Puri et al. (eds.), *Fundamentals of Forensic Biology*,
https://doi.org/10.1007/978-981-99-3161-3_10

drug detection or monitoring drug usage. However, it's worth noting that some drugs may not be detectable at low levels in urine.

Keywords

Urine · Drug · Crystal · Nephrons

Urine is a liquid waste secreted by the kidneys by a process called as urination and is excreted through the urethra. Usually urine is a transparent pale yellow solution, but it can widely range from colourless to pale yellow. The colour of urine is primarily influenced by the presence of certain pigments and compounds, as well as the concentration of waste products and hydration level.

Urine can be extracted directly as urine is relatively free of protein. Urine contains high concentrations of metabolites and provides longer detection time for most drugs. Hence urine is considered as one of the best evidence for detection of drugs or monitoring drug exposure. There are certain drugs that cannot be detected in low levels in urine.

The specific gravity of urine is 1.005–1.030. The normal pH of urine is between 4.6 and 8.0. The colour of normal urine is that of clear straw, and consists mild odour and the presence of some crystals and small cells from the tissues which line the urinary bladder. Such components as sugars, proteins, bacteria, yeast cells, and parasitic organisms are not present in normal urine. But such components are detected in urine due to certain medical issues. Antibiotics such as penicillin and foods such as red beets and asparagus may affect the odour and colour of urine.

10.1 Identification of Urine

The identification of urine can be crucial in forensic investigation particularly in cases involving sexual assault. The stains of urine identification can be useful in the investigation of homicidal cases involving ligature or manual strangulation.

Urine is performed less frequently by forensic laboratories because of the insensitivity of the tests and low success rate with DNA profiling. Identifying or detecting urine has been achieved through UV visual examination. Stains may fluoresce or luminesce under alternate light sources, but diluted stains are harder to detect. There are several methods for the identification of urine and have relied on identifying inorganic ions and organic compounds that concentrate in urine. Urine detection relies on identifying two organic compounds such as Urea and Creatinine. Both these components are found in other body fluids such as blood, saliva, and semen. Urea and creatinine are found at relatively high levels in liquid urine. As liquid urine is absorbed into fabric surfaces, it spreads across the surface, effectively diluting the test components.

10.2 Constituents of Urine

Normal constituents: Sodium chloride, uric acid, hippuric acid, potassium, oxalate, amino acid, ammonia, traces of glucose, ketose, vitamins, hormones, calcium, sulphate, and creatinine.

Abnormal constituents: Sugar, protein, ketone bodies, bile salts, etc.

10.3 Composition of Urine

The principal function of the urinary system is to maintain volume and composition of body fluids within normal limits. The urinary system has a major role in excretion. The urinary system, also known as urinary tract or renal system, consists of kidneys, ureters, bladder, and urethra.

The primary route in which body eliminates substances is through the kidneys. The kidney excretes body wastes and harmful chemicals into the urine. The formation of urine takes place in the kidneys, in the nephrons. The basic structural and functional unit of the kidney is nephron. Each kidney consists of about one million nephrons. The nephron is composed of a glomerulus (Bowman's capsule) and a renal tubule.

The first step in the urine formation is filtration. The blood flows through glomeruli, much of its fluid, except cells and large molecules and filters through capillaries into Bowman's capsule. The preliminary form of urine is glomerular filtrate, which consists of water, salts (sodium and potassium ions), glucose, and urea (waste product). The filtrate then enters the renal tubule where the process of reabsorption takes place. During the reabsorption process, water, glucose, nutrients, and ions such as sodium are reabsorbed back into the blood. The secretion process is the final step in urine formation. The secreted ions combine with the remaining filtrate and become urine. The urine flows out of the nephron tubule into a collecting duct. It passes out of the kidney through the renal pelvis, into the ureter, and down to the bladder.

The nephrons of the kidney process blood and create urine through a process of filtration, reabsorption, and secretion. Urine is about 95% water and 5% waste products. Nitrogenous wastes excreted in urine include urea, creatinine, ammonia, and uric acid. Ions such as sodium, potassium, hydrogen, and calcium are also excreted (Figs. 10.1 and 10.2).

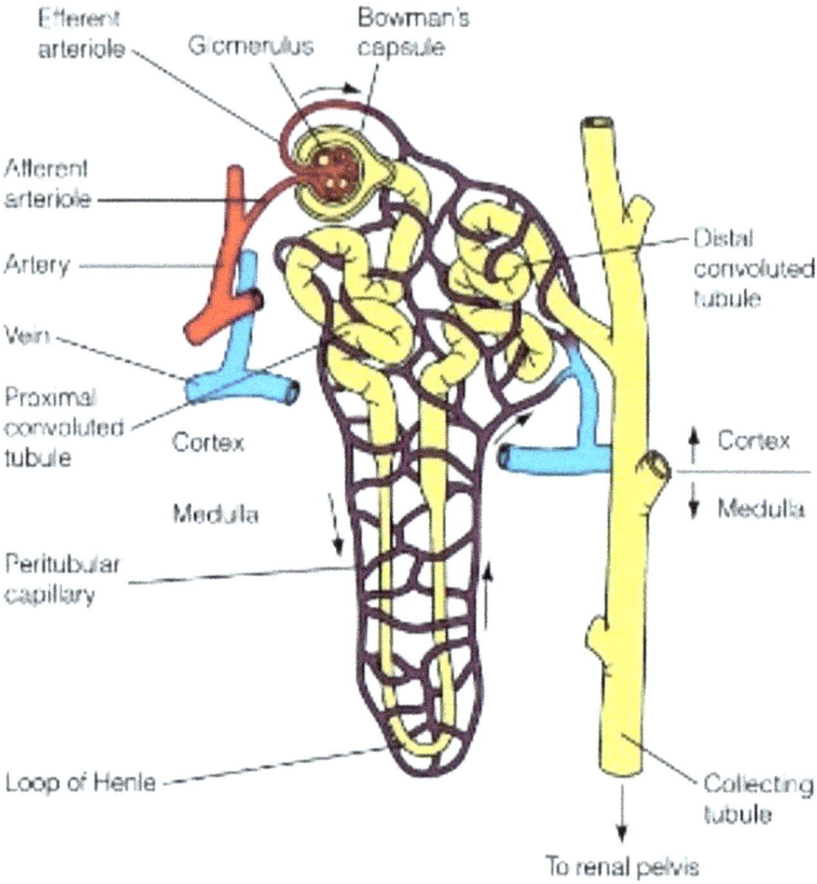

Fig. 10.1 The nephron

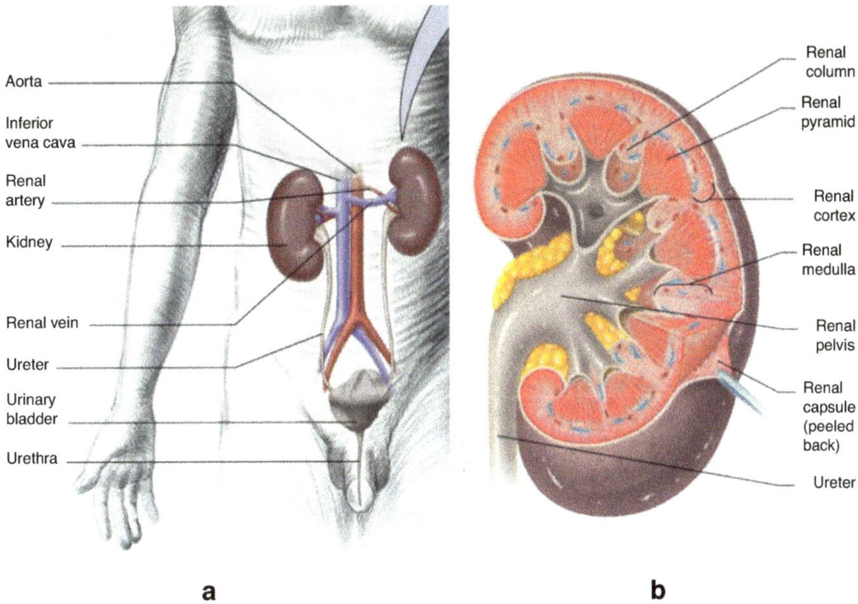

Fig. 10.2 Human urinary system. (**a**) Anterior view of urinary system showing relationship of kidneys, ureters, urinary bladder, and urethra. (**b**) A cross section of the human kidney showing the cortex, medulla, and renal pelvis

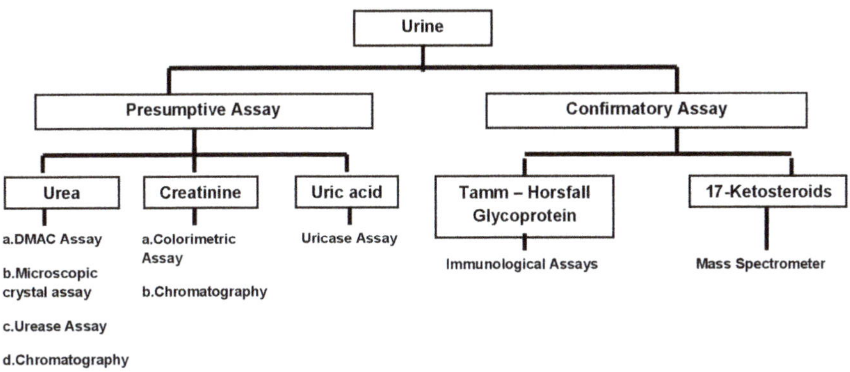

10.4 Presumptive Assay

10.4.1 Identification of Urea

Urea is a waste product of many living organisms and is one of the major organic components of human urine. Urea is a major waste product formed as a result of breakdown of amino acids, particularly in the metabolism of proteins. These amino

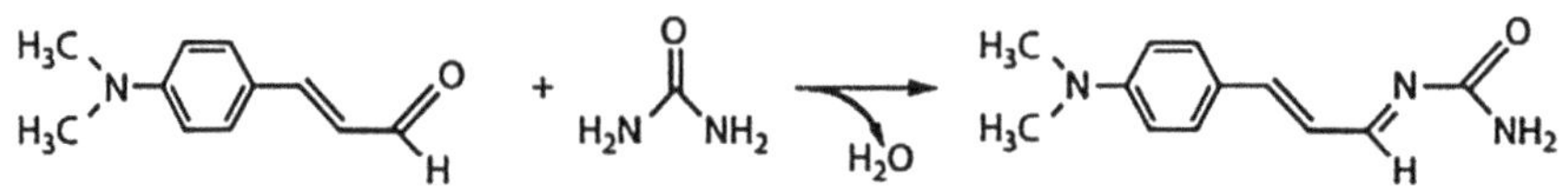

Fig. 10.3 Chemical reaction of DMAC assay

acids are metabolized and converted to ammonia, carbon dioxide, water, and energy in the liver. Since ammonia is toxic, the liver converts it into a non-toxic compound called urea, which is then safely transported in the blood to the kidneys. The normal range of urea is 2.6–6.5 mm in human blood. Abnormal urea levels cause diseases such as liver disease, real disease, hereditary urea cycle abnormalities, and dietary problems.

10.5 DMAC Assay

The most commonly used assay for the forensic identification of urine stains is the para-dimethylaminocinnamaldehyde (DMAC) assay. This is a simple and rapid assay. This assay can be performed using colorimetric and fluorometric methods (Fig. 10.3).

Colorimetric method is a colour producing method. In this method, a small (−1 cm^2) portion of stain is cut and is extracted with distilled water (1 ml). This extraction is then transferred into a piece of filter paper and is allowed to dry. After drying add 0.1% DMAC solution. DMAC reacts with urea and it doesn't react with creatinine, ammonia, or uric acid. If urea is present then there is production of pink-coloured (or magenta coloured) product. The appearance of pink colour after or within few minutes applying the DMAC reagent is considered as positive reaction. No colour change is considered as negative reaction (Fig. 10.4).

Fluorometric method is another method for the detection of urea and for locating urine stains on bedding or clothing. In this method, a large piece of evidence to be examined such as clothing is covered by moistened filter paper sheet containing DMAC solution. The clothing (evidence) and filter are then wrapped in aluminium foil sheet and are left overnight in a press. It should be ensured that the clothing (evidence) is in close contact with filter paper. Alternatively, the layers can be heated using an iron so that there is transferring of a small amount of stain into the paper. The paper is then lifted, and the dyes and pigment can inhibit the fluorescence (Fig. 10.5).

10.6 Urease Assay

The urease assay is another assay for the identification of urea. The urease assay identifies those organisms that are capable of hydrolysing urea to produce ammonia and carbon dioxide. In this method, ammonia is detected using an acid-base

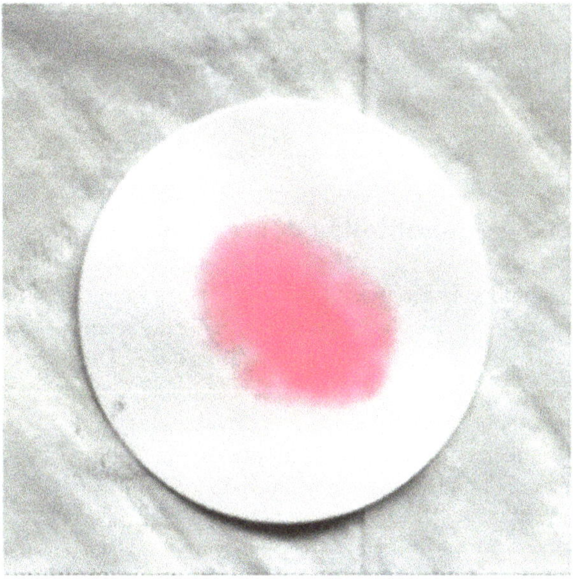

Fig. 10.4 DMAC colorimetric assay

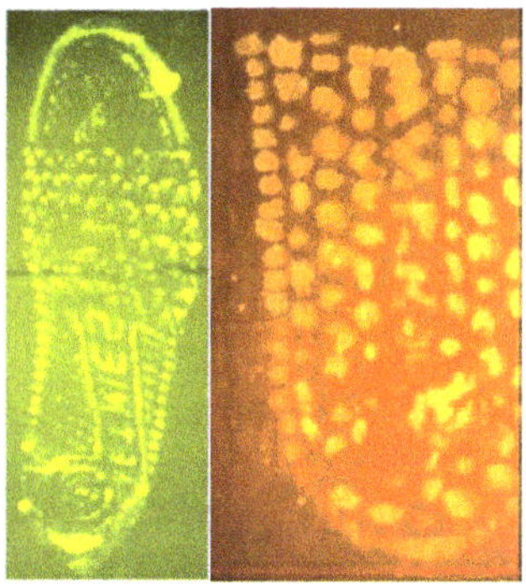

Fig. 10.5 Fluorometric assay

Fig. 10.6 Biochemical reaction of urease

indicator. Bromothymol blue initially appears blue but undergoes a color change to black when it reacts with manganese and silver nitrates (Fig. 10.6).

10.7 Microscopic Crystal Assay

Microscopic examination of urea crystals and urea nitrate crystals converted from urea and microscopic examination of crystals using xanthydrol have been used for identification.

10.7.1 Identification of Creatinine

Creatinine is a waste product that comes from the normal wear and tear on muscles of the body. During this process, phosphocreatine undergoes a breakdown in muscle cells, resulting in the formation of creatine. This breakdown releases stored energy. Creatine is then metabolized to creatinine. Creatinine is released from muscle cells into blood. A small amount of creatinine is secreted by renal distal tubules whereas serum creatinine is largely filtrated by renal glomeruli. Creatinine is present in urine, blood, and also in semen (Fig. 10.7).

10.8 Colorimetric Assay (Jaffe's Method)

The creatinine present in urine can be detected by Jaffe colour test. In this method, creatinine produces quantitatively a red colour with picnic acid in alkaline medium. Under alkaline conditions picnic acid is used to convert creatinine to creatinine picrate. The product formed is bright red (Fig. 10.8).

10.8.1 Identification of Uric Acid

Uric acid has distinctive, strong absorbance at 293 nm and the quantification of the absorbance at 293 nm is the simplest way for uric acid assay. Measuring the disappearance of uric acid absorption at 293 nm after treating it with uricase is a method to quantitatively determine uric acid levels. Uricase converts uric acid to 5-hydroxyisourate, causing a decrease in absorption at 293 nm. This change in absorbance is directly related to uric acid concentration and is used for clinical diagnoses and biochemical studies.

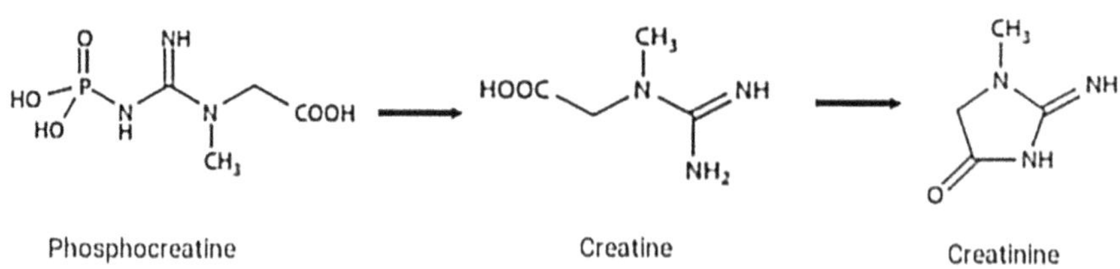

Fig. 10.7 Schematic view of the formation of creatinine

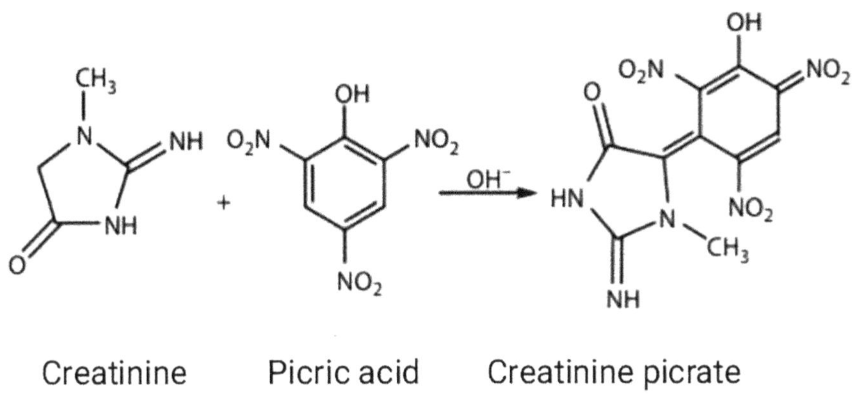

Fig. 10.8 Chemical reactions of Jaffe's assay

10.9 Confirmatory Assay

10.9.1 Tamm–Horsfall Protein

Tamm–Horsfall protein is the most abundant urinary protein. It is also known as "Uromodulin". THP is exclusively produced by renal tubular cells of the distal loop of Henle. THP is secreted into lumen from the apical plasma membrane of epithelial cells. THP excretion decreases in stone formers, and therefore it was proposed as a potential biomarker for kidney stone disease. A number of assays have been developed to detect THP including electroimmunodiffusion, gel electrophoresis, radio immunoassay, and enzyme- linked immunosorbent assay (ELISA).

RSID urine utilizes a polyclonal rabbit antibody that is specific to THP.

These methods are specific to THP that is a protein specific to urine and not typically found in other body fluids like plasma, saliva, semen, sweat, or vaginal secretion or fluids.

10.9.2 Identification of 17-Ketosteroids

The group of compounds known as 17-Ketosteroids are metabolites of hormones produced by the adrenal cortex, the tests, and to a limited extent, the ovaries. These are groups of metabolites derived from the breakdown of hormones, including androgens, produced by adrenal cortex, testes, and ovaries. The derivatives of 17-Ketosteroids in human urine include androsterone, dehydroepiandrosterone (DHEA), and etiocholanolone.

Androsterone is derived from sex hormones and it is a steroid metabolite. It is produced as a result of the metabolism of testosterone, primarily in the liver. Androsterone is a steroid hormone with weak potency of testosterone.

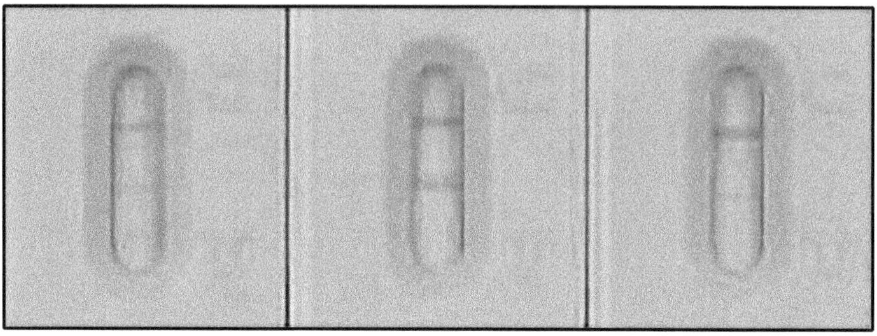

Fig. 10.9 Results obtained using immunochromatographic assays for identification of THP in urine. From left to right negative control, undiluted urine samples, diluted (1:10) urine samples, and diluted (1:100) urine samples. A visible line in Test and Control zone indicates a positive result. The visible line in Control zone only indicates a negative result

Dehydroepiandrosterone (DHEA) is an endogenous steroid hormone, produced by adrenal gland, gonads, and brain. DHEA is also known as androstenolone. It is synthesized from pregnenolone and is also related to androstenedione, testosterone, and estrogens. This compound serves as a metabolic intermediate in the biosynthesis of gonadal steroids.

Etiocholanolone (5-Androsterone) is an androgenically inactive metabolite of testosterone and androstenedione and 5 epimer of androsterone.

17-Ketosteroids, such as androsterone, dehydroepiandrosterone (DHEA), and etiocholanolone, can undergo further modification in the liver through glucuronidation and sulfation. These compounds are present in urine as conjugates. In glucuronidation, glucuronic acids are added to 17-Ketosteroids and in sulfation, sulphates are transferred to 17-Ketosteroids. These conjugates are more soluble in water than the non-conjugated 17-Ketosteroids as they contain charged moieties. A new method for identifying human urine stains utilizes High-Performance Liquid Chromatography (HPLC) analysis of five major 17-Ketosteroids conjugates.

The five major components of 17-Ketosteroids conjugates present in human urine are androsterone glucuronide, androsterone sulfate, DHEA sulfate (Dehydroepiandrosterone sulfate), etiocholanolone glucuronide, and etiocholanolone sulfate. The analysis of 17-Ketosteroid conjugates is useful for the identification of human urine stains. Thus these five conjugates of 17-Ketosteroids can be identified and quantified by Liquid Chromatography–Mass Spectrometry (LC-MS) (Fig. 10.9).

Multiple-Choice Questions

1. In urease assay, urease hydrolyses urea to produce
   (a) Ammonia and carbon monoxide
   (b) Ammonia and carbon dioxide
   (c) Calcium and carbon dioxide
   (d) Phosphate and carbon dioxide

2. Jaffe test for urine sample is reaction between
   (a) Urea and ammonium hydroxide
   (b) Creatinine and ammonium hydroxide
   (c) Urea and picric acid
   (d) Creatinine and picric acid
3. Which of the following is considered as simple and rapid assay
   (a) Immunological assay
   (b) DMAC assay
   (c) Urease assay
   (d) Uricase assay
4. Which of the following is abnormal constituent
   (a) Sugar
   (b) Oxalate
   (c) Amino acid
   (d) Ketone
5. Which of the following is not involved in the urinary tract
   (a) Kidney
   (b) Urethra
   (c) Lungs
   (d) Ureters

Answers

1. (a)
2. (d)
3. (b)
4. (a)
5. (c)

Introduction to Sweat and Its Forensic Analysis

11

Aditya Kumar Kar, Radha Patel, and Tanima Nandi

Abstract

Sweat analysis in forensic science is an important aspect in the field where the results obtained can be used to determine the presence of toxic substances in the body such as drugs, metals, toxins, etc. Advanced analysis of this bio-fluid can also help us determine the number of persons present in the scene of crime due to the determination of the composition of constituents present in the sweat. The most important advantage of sweat evidence is that because of having unique feature, it can help in distinguishing individuals easily. This is possible due to the fact that sweat contains a large number of metabolites and amino acids that can be detected on the surface, which tend to fluctuate over time due to the change in lifestyle. Because of this reason, sweat stains have become important biological evidence in cases like kidnapping, rape cases, and molestation which can be found in the clothes, on the carpet, used handkerchiefs, or on the victim's/suspect's body, etc. at various crime scenes.

Keywords

Thermoregulation · Secretion · Sweat · Fluids

Aditya Kumar Kar, Radha Patel and Tanima Nandi contributed equally with all other contributors.

A. K. Kar (✉) · R. Patel · T. Nandi
Department of Forensic Science, Adamas University, Barasat, Kolkata, India

A. Puri et al. (eds.), *Fundamentals of Forensic Biology*,
https://doi.org/10.1007/978-981-99-3161-3_11

11.1 Introduction

Identification of body fluids is an important aspect of forensic science and is based on the principle of individuality and principle of comparison (Hanssen et al. 2017). Sweat present in a crime scene is a vital source of evidence. An average square inch of skin contains 650 sweat glands, meaning our bodies leave tiny amounts of sweat on everything that we touch, be it our phones or the dining table or the knife that we use (Jain and Bhattacharya 2020). Thus, sweat is an important evidence for the forensic investigators, analyzing sweat left behind at a crime scene can be useful in determining the number of people who were there thus narrowing down the list of potential suspects (Jain and Bhattacharya 2020).

11.1.1 What is Sweat?

Sweat is a bodily secretion whose main function is to prevent overheating of the body. It is a clear, salty liquid, and perspiration/sweating is the process by which our body cools itself. The sweat comes out of the body through tiny holes known as pores. It is a normal occurrence, when it is hot, or when we exercise, or excited or nervous. It is produced by the glands present in the dermis, namely, eccrine, apocrine, and apoeccrine. On average, a person can have 2–4 million sweat glands present in the body, with most of them on the palms of hands and soles of the feet (Taylor and Machado-Moreira 2013). Excessive sweating is known as hyperhidrosis while a lack of it is known as hypohidrosis.

Sweat is normally a transparent biofluid with low tonicity and a slightly acidic nature with a pH between 4 and 6. It has been found out that basic drugs accumulate in sweat then blood based on the pH partition theory, all the same making it important evidence for analysis and study. Although sweat evidence is very hard to find and collect and has a limited scope, examination and analysis of sweat evidence or samples is very vital from the forensic point of view (Jadoon et al. 2015). Sweat forms the basis of analysis of latent fingerprints in crime scenes. The examination of sweat can help us determine a specific odour or an individual's blood group, for an instance through the Gee's urea nitrate test, which we will study later in this chapter.

11.1.2 What Causes Sweat?

Sweating is caused by the sweat glands present in our body. When our body temperature rises, the glands, namely, eccrine, and apocrine release fluids cool our body as they evaporate (Bovell 2015). It is regulated by the sympathetic division of the autonomic nervous system and we have little control over it. Some of the simple causes of sweating are listed below:

- Emotion and stress, for instance, anger, fear, embarrassment, anxiety, etc.
- Elevated temperature
- Any illness such as cancer, fever, hypoglycemia, or specific medications like painkiller
- Specific type of foods such as spicy foods, caffeinated drinks, alcoholic beverages, etc.
- Menopause

11.1.3 Functions of Sweat

While it has already been discussed that the most essential function of sweat is thermoregulation, some of the other functions of sweat which are equally as vital are given below (Bovell 2015).

- Prevents overheating of the body
- Excrete waste products
- Emotional response
- Keeping the stratum corneum moist to ensure better tactile skills and elasticity of the palms and soles
- Electrolyte balance

11.2 Biology of Perspiration

11.2.1 Sweat Glands

The sweat producing glands also known as sudoriferous glands are widely distributed in our skin. A person on average can have more than 2.5 million sweat glands. The three main types of sweat glands are as follows:

11.2.1.1 Eccrine Glands

They are the most numerous and are widely distributed all over the body. They are found especially in higher concentration on the palms, soles of feet, and the forehead. Eccrine glands may be seen in various stages of development at a time. Eccrine sweat glands first appear on the palms and soles during the fourth month of gestation. By the eighth fetal month, they resemble adult sweat glands, but are not fully functional until about 2 years of age. It should be noted that no new sweat glands are formed after birth (Bovell 2015).

Structure of Eccrine Sweat Gland

Eccrine glands are simple, coiled tubular glands. The secretory part lies in the dermis and the glands open to the skin surface through ducts or pores. To understand about this a bit more, take a look at the picture given below.

The secretory part of the gland displays a single-layered structure, consisting of three types of cells:

1. Clear cells
2. Dark cells
3. Myoepithelial cells

The cells rest on the basal lamina and are arranged in the form of pseudostratified epithelium.

The ductal part of the gland is divided into two parts:

1. Intradermal
2. Intraepidermal

This part extends from half of the basal coil in dermis to opening at the surface. It consists of basal cell, ductal lumen, and luminal cell. It lacks myoepithelium, clear cells and dark cells. The cells are spread out in the form of stratified cuboidal epithelium (Bovell 2015).

11.2.1.2 Apocrine Glands

Apocrine sweat glands are associated with hair follicles and are found mostly in the armpits, the groin, and the area around the nipples of the breast. Apocrine glands in the skin are scent glands, and their secretion usually has an odour. Apocrine glands have a low secretory output and hence no significant effect on temperature regulation of the body (Fortney and Vroman 1985). Apocrine glands are larger than eccrine glands, with a diameter 10 times greater. They are made up of two primary parts: a coiled secretory structure and an accompanying straight duct. They first appear during the fourth to fifth month of gestation. They are dormant postnatally until they develop their secretory portion and are functional at around puberty under the influence of ongoing hormonal activity (Fortney and Vroman 1985). They can occasionally be found on the eyelids, abdomen, and scalp too.

Structure of Apocrine Sweat Glands

The apocrine sweat glands are divided into two main parts which are subdivided into the following:

1. Ductal part: intradermal & intraepidermal
2. Secretory part: coiled (lower dermis)

The secretory part where the gland is located lies in the lower dermis or the subcutaneous fat and is made up of only secretory cells. The secretory portion of the apocrine gland contains several large granules. The gland is made up of a single layer of columnar or cuboidal cells, that rest on a layer of myoepithelial cells (Groscurth 2002). The ductal part is made up of a layer of basophilic cells with a periluminal eosinophilic cuticle. When tip of the epithelial cord has reached the level of the sebaceous gland, the intradermal ductal lumen begins to form. The secretion and cellular detritus move mostly through the apocrine duct into the pilosebaceous follicle present in the infundibulum, but they can be found opening right into the epidermal surface (Saga 2002).

11.2.1.3 Apoeccrine Gland

Apoeccrine glands are mixed glands that are found in the adult human axillae and share some of the functional and morphological features of both the eccrine and apocrine gland. Apoeccrine sweat glands secrete fluid continuously somehow comparable to eccrine sweat glands. They represent less than 10% of all glands seen in human axilla. Apoeccrine glands can be found in the anogenital region. They are larger than eccrine glands but smaller than apocrine glands and are found in all levels of the dermis. Apoeccrine glands develop from eccrine like precursor glands that underwent 'apocrinization' during puberty due to the nearby growth factors. They consist of myoepithelial cells, apocrine secretory, and eccrine secretory cells (Bovell 2015).

Structure of Apoeccrine Gland

A typical morphological characteristic of this gland is the irregular structure of the tubule and as well as the cells. The structure of apoeccrine sweat glands are somehow similar to that of the eccrine gland. The secretory segment is irregularly dilated, and the duct is long and thin and comparable to the eccrine gland. The duct opens into the skin surface.

11.2.2 How Sweating Occurs?

Sweating is mainly caused by three different physiological stimuli such as:

- Temperature
- Gustatory
- Emotional

11.2.2.1 Thermoregulatory Sweating

Sweating caused due to temperature or heat induced sweating starts at the forehead. When the temperature of the body rises due to fever or exercise or the environment, the body's sweat gland releases a salty liquid known as sweat which when evaporates leaves behind the sensation of cooling. This type of sweating helps dissipate excess heat and prevents overheating of the body (Sawka and Wenger 1988).

11.2.2.2 Gustatory Sweating

Eating very spicy food can cause physiologic gustatory sweating in many people due to a reflex known as trigeminovascular reflex. Gustatory sweating is characterized by profuse sweating immediately after or during the ingestion of food. The trigeminovascular reflex naturally occurs symmetrically on the face or scalp and predominantly over the nose, lips, neck, and forehead. Gustatory sweating is typically done by eccrine sweat glands (Wilke et al. 2007).

11.2.2.3 Emotionally Induced Sweating

Emotionally induced sweating is induced by emotional stimuli such as nervousness, fright, anxiety, or anger, etc. Although this type of sweating can occur all over the body, this type of sweating is most evident on the face, axilla, palms, soles of feet, and armpits. This is basically a result of high eccrine sweat gland densities at this body sites.

11.3 Composition of Sweat

The overall pH of sweat ranges between 4 and 6. While 99% of the content of sweat is water, it has other important components too. The various constituents of sweat are listed below (Fig. 11.1)

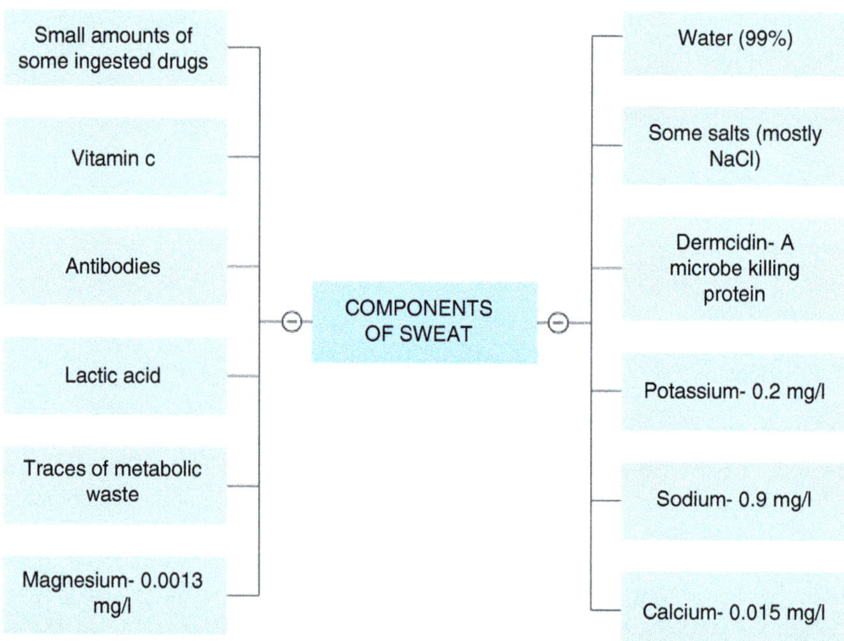

Fig. 11.1 Flowchart represent composition of sweat

11.4 Identification of Sweat

Primarily, the examination that should be done upon finding a suspected sweat sample in the crime scene is to see is that the particular sample in question is sweat or not? Sweat is rich in chemicals and can indicate the physiological state of the body, and sometimes can be the indicator of the emotions if a certain individual is/was going through. Sweat contains metabolites (glucose, lactose, etc.), trace elements, electrolytes, and small macromolecular components. Sweat analysis can substitute blood analysis. Though the chemical composition of sweat samples is complex, the collected amount is usually small (Yu and Sun 2020). It is very important to differentiate the components of sweat and to also improve the susceptibility of sweat detecting tests or sensors for sweat analysis.

11.4.1 Preliminary Examination of Sweat

While sweat at times can be identified through odour and its consistency, it is important to perform sweat identification examinations to get surety of the sample in question. Although there are not many preliminary tests prescribed for sweat analysis, the most common preliminary tests done to see if a given sample is sweat not are Odour test and Gee's Urea test.

11.4.1.1 Odour Test

A tiny amount of the suspected sweat sample is collected and heated. The particular scent that it gives off is noted and can be used for the purpose of investigation if it provides any certain clue pertinent to the case.

11.4.1.2 Gee's Urea Test

The sweat sample is at first tested or examined for the presence of urea using the Gee's urea nitrate test.

Procedure

- The questioned sample is extracted using acetone and the acetone extract with the help of evaporation is concentrated.
- The concentrated is then filtered and evaporated until it is dried.
- The residue is then treated with acetone solution and then the mixture is mixed with the help of a glass rod.
- A drop of the solution is taken on a slide and then it is allowed to dry.
- Nitric acid solution with the help of glass rod is added and then the slide is covered with coverslip and observed under the microscope for urea nitrate crystals for the indication of the presence of urea in the sample.

11.4.2 Confirmatory Examination of Sweat

There are many tests that are performed for the analysis of sweat stain sample with the Dermcidin test being one of the most reliable and efficient ones. Let us get an overview on the various confirmatory examinations done on sweat sample for the aid of investigation.

11.4.2.1 Dermcidin Test Using Elisa

Dermcidin present in sweat plays a vital role by protecting the epithelial barrier from infections. In the field of forensic, it can be used as a biomarker for confirmatory analysis of sweat identification. The detection of dermcidin in sweat sample can be performed by utilizing the ELISA technique. This method is highly sensitive and is able to detect dermcidin in sweat samples that are diluted 10,000 times. Dermcidin is encoded by the DCD gene (Schittek et al. 2001). The detection of DCD by ELISA was examined in 10 mL of sweat. It produced very high values compared to the other body fluids, even in the 10,000-fold dilution.

11.4.2.2 Instrumental Analysis of Sweat

SEM-EDX (Scanning Electron Microscope-Energy Dispersive X-ray Analysis)

Scanning electron microscope (SEM) has been acquired to investigate the morphology, microstructure, and composition of the constituents in the sweat sample. The same instrument, coupled with an energy-dispersive spectrometer (EDX), is used to carry out the energy-dispersive X-ray spectroscopy to determine the presence and distribution of the chemical elements in samples. SEM paired with EDX helps to establish the relative concentrations of metal, phosphorus, sulfur, chlorine, potassium, calcium, and different metal traces. The sweat analysis showed that chlorine and metal were the sole systematically clear peaks among the various samples, and metallic element was generally visible. The massive chlorine peak is employed as the basis of comparison and identification (Quinton 1978).

Fluorescence Spectroscopy

The fluorescence sensing technique can be used to detect sweat by comparing the levels of electrolytes and metabolites found in the sample collected from a scene of crime. This technique depends on the link between analyte concentration and light intensity. Once the fluorescent material is exposed to specific excitation light source, it generates light. Once the fluorescent material undergoes a particular reaction with the target analyte, the fluorescent signal changes. The benefits of fluorescence sensing embody high sensitivity, strong acuteness, and easy use. It can be used to detect metabolites (such as lactate and urea) and electrolytes (such as Cl^-, Na^+, Cu^{2+}) in sweat. This technique can also help us to find and detect the levels of specific metals (possibly harmful and present in risky levels) present in the body which are excreted by sweat, for comparison and analysis (Yu and Sun 2020).

GC-MS (Gas Chromatography–Mass Spectrometry) for Drug Detection in Sweat

Sweat can be used as a biological matrix for the testing of drugs subject to abuse. There are many tests that are prescribed to detect the presence of sweat of which the most utilized ones are GC-MS electron ionization. The basic principle of this instrument involves the patch test technology for qualitative detection of previously used drugs by individuals. There are various types of drugs that can be detected with the help of instrumental analysis of sweat sample such as such as codeine, heroin, fentanyl, morphine, etc. (Jadoon et al. 2015).

11.5 Conclusion

However, sweat as studied above proves to be an essential biofluid for investigation purposes. Forensic experts claim that the investigator often tend to overlook the presence of sweat in crime scene, which is a miscalculation on their part. Sweat analysis plays an important role in testing the presence of drugs in an individual's body and also for the presence of poisons and various harmful metals. It has been found that upon intensive analysis of sweat sample, DNA can be extracted (provided skin cells are included in the oils and fluids of sweat) which as we all know serves as the pinnacle for the purpose of investigation and for the means of individual identification (Vandewoestyne et al. 2013). Sweat stains have become important biological evidence in cases like kidnapping, rape cases, molestations, etc. It can be present on any kind of clothing or, for instance, on the victim's/suspect's body which can associate us to suspect. It can also be present on surfaces/objects touched by victim or suspect (Lee et al. 2001). Sweat, which is a skin secretion, contains various metabolites and the levels of those are extremely distinctive to each of us. Three of those metabolites are lactate, urea, and glutamate. Lactate appears in higher concentration in our sweat and varies greatly based on a person's lifestyle. Glutamate and urea are found in different parts of sweat and are also highly concentrated. Studies show that the chances of any two people having identical levels of all these above-mentioned metabolites is almost nil, thus proving to be one of the various reasons to study sweat and its effectiveness in aiding an investigation. Although as already been discussed sweat can be used to find out how many individuals were present at the scene of crime, it has been advised that it should not be held as the only basis of comparison. Contrary to DNA, which always remains constant, a person's sweat construct is always changing and based on their lifestyle. Therefore, while these tests related to sweat can determine the number of personnel present at the crime scene, it is not reliable for determining who was at the crime scene. Sweat as evidence regarding to these types of cases should be utilized in concurrence with traditional DNA testing and investigation procedures rather than as a replacement.

References

Bovell D (2015) The human eccrine sweat gland: structure, function and disorders. J Local Glob Health Sci 2015(1):5

Fortney SM, Vroman NB (1985) Exercise, performance and temperature control: temperature regulation during exercise and implications for sports performance and training. Sports Med 2:8–20

Groscurth E (2002) Anatomy of sweat glands. Curr Probl Dermatol 30:1–9

Hanssen EN, Avershina E, Rudi K, Gill P, Snipen L (2017) Body fluid prediction from microbial patterns for forensic application. Forensic Sci Int Genet 30:10–17

Jadoon S, Karim S, Akram MR, Kalsoom Khan A, Zia MA, Siddiqi AR et al (2015) Recent developments in sweat analysis and its applications. Int J Anal Chem 2015:164974

Jain M, Bhattacharya B (2020) Examining the significance of saliva and sweat in forensic science. J Legal Stud Crim Justice 1(1):6–15

Lee HC, Palmbach T, Miller MT (2001) Henry Lee's crime scene handbook. Academic Press, Cambridge

Quinton PM (1978) Techniques for microdrop analysis of fluids (sweat, saliva, urine) with an energy-dispersive X-ray spectrometer on a scanning electron microscope. Am J Physiol Renal Physiol 234(3):255–259

Saga K (2002) Structure and function of human sweat glands studied with histochemistry and cytochemistry. Prog Histochem Cytochem 37(4):323–386

Sawka MN, Wenger C (1988) Physiological responses to acute exercise-heat stress. Army Research Inst of Environmental Medicine, Natick

Schittek B, Hipfel R, Sauer B, Bauer J, Kalbacher H, Stevanovic S et al (2001) Dermcidin: a novel human antibiotic peptide secreted by sweat glands. Nat Immunol 2(12):1133–1137

Taylor NA, Machado-Moreira CA (2013) Regional variations in transepidermal water loss, eccrine sweat gland density, sweat secretion rates and electrolyte composition in resting and exercising humans. Extreme Physiol Med 2(1):1–30

Vandewoestyne M, Van Hoofstat D, Franssen A, Van Nieuwerburgh F, Deforce D (2013) Presence and potential of cell free DNA in different types of forensic samples. Forensic Sci Int Genet 7(2): 316–320

Wilke K, Martin A, Terstegen L, Biel S (2007) A short history of sweat gland biology. Int J Cosmet Sci 29(3):169–179

Yu H, Sun J (2020) Sweat detection theory and fluid driven methods: a review. Nanotechnol Precis Eng 3(3):126–140

Forensic Significance and Examination of Faecal Matter

12

Varsha Rani Patel, Himani Sharma, Shivam Chourasiya, and Tilak Ram Chandrakar

Abstract

Faecal matter is an often overlooked but possibly significant piece of evidence. It is found in a variety of casework scenarios, ranging from trace amounts to whole stool deposits. Small amounts of faeces can be transmitted to the sodomite's penis or other objects introduced anally in circumstances of sexual assault. In contrast, a criminal may purposely or unintentionally defecate at the site of a crime, leaving behind an entire bowel movement. In many situations, the identification and individualization of faeces can establish a relationship between the victim and the perpetrator.

Animal metabolism produces faeces as an excretory by-product. It is made up of bilirubin, muscle fibres, and undigested cellular and vegetable material. Its identification may occasionally be quite important, particularly in sodomy situations. For more than a century, faeces have been examined as part of criminal investigations. In particular, faecal analysis was used in 1948 to connect a suspect's shoes to a crime scene in a burglary case. Finding a connection between the reference sample and the faecal evidence is one of the goals of faecal analysis, which could help connect a suspect to a crime scene. The identification of faecal matter is valuable in providing important information for a criminal investigation. For example, the presence of faecal matter may corroborate a sexual assault involving sodomy, assault with faecal matter, vandalism, bestiality, and burglary during which the perpetrator defecated at the scene, faeces can be found as evidence on the penile swab and in the clothing. If faeces are already present at the crime scene, faeces may be transferred to the suspect's shoes or clothing. In

V. R. Patel (✉) · H. Sharma · S. Chourasiya · T. R. Chandrakar
Faculty of Forensic Science, Medi-Caps University, Indore, Madhya Pradesh, India
e-mail: varsharani.patel@medicaps.ac.in; himani.sharma@medicaps.ac.in; shivam.chourasiya@medicaps.ac.in; tilakram.chandrakar@medicaps.ac.in

A. Puri et al. (eds.), *Fundamentals of Forensic Biology*,
https://doi.org/10.1007/978-981-99-3161-3_12

sodomy and bestiality, the suspect's faeces play a key function in determining his or her drug habit. It has also been noticed that the ingestion of certain medicines causes faeces to turn green, black, and red.

Alkaline phosphatase, IgA immunoglobulins, intestinal parasites, pancreatic-amylase, urobilinogen and urobilin, and vegetable remains have been analysed as part of forensic methods for the characterization of human faeces. It can also be utilised for blood grouping, hence aiding in the individualization of individuals. Today, the forensic DNA analysis of sloughed intestinal epithelial cells seen in faecal debris may efficiently determine the unique properties of a faecal sample as it enables nuclear DNA-based discrimination and CODIS comparison.

Keywords

Faecal matter · Urobilinogen · Urobilin · Stercobilin

12.1 Formation of Faeces

Food that has been broken down by the digestive system results in a specific sort of waste material called faeces. Undigested food items, shed intestinal epithelial cells, intestinal bacteria, bile pigments, electrolytes, and water can all be found in human faeces. The final stage of digestion causes the intestines to produce faeces. Initially, liquid faeces travel to the colon. On the small intestine's surface, the majority of the nutrients are absorbed. The surface of the lumen in the large intestine is where water, sodium, and chloride are absorbed. Faeces are formed out of the leftover luminal contents. The stomach retains food for between 2 and 6 h. The small intestine and large intestine each add an additional 3–5 h and 12–24 h, respectively, to the total time required for transit (Fenwick et al. 1983).

12.2 What is in Human Faeces?

It has been determined that, in general, human faeces is made up of 75% water and 25% solid material, which is made up of 30% dead bacteria, 10–20% fat, 10–20% inorganic material, 2–3% proteins, and 30% of residual food that hasn't been digested. In addition to bile pigments (bilirubin) and dead leukocytes (white blood cells), cell debris shed from the intestinal mucosa is also excreted in the waste material (Shleenkova et al. 2015). The number of excretions per person per day ranges from 135 to 270 g, and 88 to 97% of the compounds are organic after drying. Skatole and thiols contribute to the characteristic odour of faeces. These substances are composed of sulphur, amines, and carboxylic acids. Skatole is generated from the amino acid tryptophan. The brown colour of faeces results from the activity of bacteria on bilirubin, the end product of haemoglobin breakdown (red blood cells) (Nakonechna et al. 2017).

12.3 Urobilinogen: Its Formation

The source of urobilinogen is bilirubin. Bilirubin is created by your body as it normally breaks down old red blood cells. Bile, a substance that aids in food digestion in your intestines, is created by your liver using bilirubin. Bilirubin needs to be processed by the body before it can be eliminated. In this context, the term "metabolism" refers to the production and disintegration of bilirubin for elimination. Bilirubin metabolism occurs in three stages:

1. Initial, pre-hepatic
2. Hepatic
3. Post-hepatic

12.3.1 The Pre-hepatic Phase

The pre-hepatic phase involves

1. The breakdown of old red blood cells to release haemoglobin.
2. The pigment heme and globin are produced from haemoglobin. Globin is disassembled into its component amino acids and then recycled by the body.
3. Iron and the green pigment biliverdin are produced when heme is broken down.
4. The enzyme biliverdin reductase transforms the substance biliverdin into unconjugated bilirubin.
5. Blood albumin binds to unconjugated bilirubin and carries it to the liver.

12.3.2 Hepatic (Liver) Phase

Unconjugated bilirubin is changed by UGT into its conjugated form during the **hepatic (liver) phase**. Now that bilirubin is soluble, it can be expelled.

12.3.3 Post-hepatic (Post-liver) Phase

Conjugated bilirubin is discharged into the small intestine as a component of bile during the **post-hepatic (post-liver) phase**. In the small intestine, bile breaks down fat before continuing on through the digestive system. When it gets to the colon, also known as the large intestine, it is deconjugated and changed into urobilinogen. Urobilinogen is converted to stercobilin, which gives faeces their brown colour, in the stomach to an extent of about 80%. Urobilinogen's remaining 20% is either recycled back to the liver via re-absorption into the blood or it is transported to the kidneys where it is transformed to urobilin and then eliminated as urine. A portion of the reason why urine is yellow is due to urobilin.

12.4 Identification Assays

The forensic study of faeces frequently requires the identification of faecal stains on swabs and clothing, from which only minute amounts of sample may be obtained. For the analysis of faeces stains, macroscopic, microscopic, and chemical testing can be helpful. The identification of urobilinogen is by far the most prevalent chemical test conducted on faeces. Urobilinogen, including urobilin and stercobilin, is derived from the breakdown of heme and is expelled in faeces. Apart from these, DNA identification and analysis of faecal microbiota plays a crucial role in individualisation (Barbosa 2012).

12.4.1 Macroscopic Analysis

Macroscopic examination of the general appearance and properties of faeces majorly include colour and odour, conducted primarily through sight and smell. The general qualities of scent and hue are as follows:

- Colour
  Faecal matter is often dark due to urobilinogen; however, in newborns, it is yellow due to unaltered bilirubin and a milk-based diet.
    Abnormal Colour: yellow, green, blood spot, brilliant red, and black
- Odour
  Aromatic scent attributable to indole and skatole.
    Increased smell is a result of excessive protein intake.
    Common in infants and adults, sour milk digestion is normal in infants.
    Gross—Extreme diarrhoea (Maclagan 1946).

12.4.2 Microscopic Analysis

During the microscopic examination, the probable stained regions are softened with normal saline or distilled water for about 30 min. Scrapping of soften stained region is taken on a sterile slide followed by adding 1–2 drops of normal saline and 1 drop of Lugol's Iodine (aqueous iodine) are added to a little scraping of the stain placed on a microscope slide. The microscopic slide is covered with a cover slip and viewed under a microscope to identify undigested food particles such as vegetable residues and muscle fibres. In addition to huge numbers of bacteria, pus cells, epithelial cells, Entamoeba histolytica cysts, and giardia intestinal ova of intestinal helminths are occasionally detected (Muwonge et al. 2010).

12.4.2.1 Why Lugol's Iodine?

Lugol's Iodine is a fast, non-specific contrast dye that is applied to direct wet mounts of the faecal material to help differentiate parasitic cysts from host white blood cells, which aids in the investigation of microorganisms that vary from person to person.

Lugol's Iodine stain is designed for use with wet mount preparations and concentration procedures to detect intestinal protozoa and helminth ova and larvae. Numerous protozoa and cysts absorb the dye and look brown, while the remainder of the sample remains transparent. Iodine discolours protozoan nuclei and intracytoplasmic organelles brown, hence facilitating their identification. Lugol's Iodine is a very concentrated substance that must be diluted before usage. Strong iodine solutions tend to agglomerate faecal particles and disrupt the refractility of other microorganisms, necessitating dilution. A 100 ml of Lugol's Iodine requires 5.0 g of crystalline iodine and 10.0 g of potassium iodide World Health Organization (2019).

12.4.2.2 Suggested Procedure

(a) Before usage, dilute Lugol's Iodine with sterile deionized water at a ratio of 1:5 (it is recommended that working solution should be prepared fresh approximately every 3 weeks).
(b) Mix a little amount (2 mg) of faeces with a drop of sterile physiological (0.85%) saline on a clean glass slide to create a direct smear of the material.
(c) Cover the sample with a coverslip and inspect the wet mount preparation for the presence of organisms (motile protozoa and others). The organisms are exceedingly pale and translucent, making them easier to view in dim light.
(d) After a thorough examination of the wet mount, a drop of Lugol's Iodine (working solution) may be inserted at the edge of the coverslip, or a new mount may be produced using iodine alone. If desired, the prepared slide can be sealed.
(e) Examine the microscope slide for undigested materials or any brown organism entities World Health Organization (2019).

12.4.3 Chemical Test

Faecal stains can be analysed with the aid of chemical assays. Urobilinogen detection is by far the most frequent chemical analysis done on faeces. Urobilinogen, such as urobilin and stercobilin, is produced by the breakdown of heme and expelled in faeces. Using the Schlesinger and Edelman tests, urobilinogen can be identified.

12.4.3.1 Schlesinger Test

In the Schlesinger test, a sample is combined with saturated zinc acetate in ethanol solution to generate aurobilinoid–zinc chelation complex, which emits a distinctive green fluorescence under UV light.

The standard implementation of the Schlesinger's test, however, is relatively insensitive to minute urobilin. A new process that has been updated and utilised for the past 2 years has proven to be significantly more sensitive. The modification consists primarily of extracting the zinc-urobilin complex with chloroform and observing the chloroform solution under filtered ultraviolet light for a golden-yellow fluorescence.

Method: Add a few crystals of ammonium persulphate to 5 ml of urine in a small separator and shake vigorously to dissolve. Mix in 5 ml of zinc acetate saturated with alcohol. Add 10 ml of chloroform, mix gently, allow to separate, and transfer the chloroform layer to a test tube. Add a few drops of pure alcohol to clear, then blend, and analyse under ultraviolet light filtration. The presence of urobilin is indicated by the observation of a golden-yellow fluorescence (Attwood et al. 2006).

12.4.3.2 Edelman Test

Edelman's test is an altered version of Schlesinger's test. A sample is treated with a solution of mercuric salt to produce a pink compound. Fluorescence is produced by further zinc salt treatment. However, in the Edelman test, less fluorescence is recorded than in the Schlesinger test.

With these tests, it is common to produce inconclusive and inconsistent findings, as faeces do not always glow visibly. Furthermore, the intensity of the detected fluorescence differs between samples. However, a spectrometric measurement of the fluorescence detection of faecal urobilinogen based on the principle of the Schlesinger test can boost the reliability and selectivity of the tests (Lloyd and Weston 1982).

12.4.4 Spectrometric Study of Fluorescence Produced by Urobilinogen

Fluorescence spectra of faecal material extracts in the presence of zinc ion exhibit the well-known green fluorescence utilised in the Schlesinger test for urobilinogen, as well as various additional fluorescence. All of these can be confirmed in a single spectrum by the synchronous fluorescence approach, which, with new extraction conditions, permits the detection of urobilinogen fluorescence in as little as 50 ng of human faeces. This is a 1000-fold reduction in the detection limit of the original visual approach, and a significant gain in selectivity. On the fluorescence, numerous reagents' effects have been investigated. In Edelman's version of the test, the presence of mercuric ion significantly suppresses fluorescence.

Method: 1 ml of zinc acetate solution (1% zinc acetate methoxy ethanol solution and 0.2% Tris) is applied to a dry sample. The resulting suspension is then sonicated for 5 min, heated to 100°C for 10 min, cooled, and centrifuged. Excitation and emission peaks at 507 and 514 nm, respectively, can be used to determine the presence of urobilinogen (Abdalla et al. 2017).

12.4.5 DNA in Faeces

From the stomach's cardiac opening to the anal canal, the human digestive tract is lined with columnar epithelium. Human DNA extracted from faeces is derived from gastrointestinal epithelial cells.

In forensic investigations, the phenol-chloroform extraction method is a successful, general-purpose technique. The organic procedure comprised the following:

1. A cell digestion step (~12 h incubation at 56°C in 400 μl of stain extraction buffer [10 mM Tris, 10 mM EDTA, 100 mM NaCl, 39 mM dithiothreitol, 2% SDS, 20 μl of 10 mg/ml proteinase K, and pH 8.0]).
2. A phenol/chloroform/isoamyl alcohol extraction step.
3. A Centricon R© YM-100 (Amicon) concentration and recovery step. By vacuum-enhanced evaporation (Savant DNA SpeedVac R© DNA110) and/or dilution with sterile water, the faecal DNA retentates are adjusted to a final volume of roughly 45 μl.

The presence of degradative and inhibitory compounds, including plant polysaccharides and bile salts, in faeces that co-extract with the target DNA poses a significant challenge. In addition, these compounds may interfere with the study of other co-occurring physiological fluids using traditional techniques. However, efficacious methods for removing faecal inhibitory compounds have been devised. In one approach, DNA samples are diluted to lower the inhibitor concentration. In other techniques, hexadecyl trimethyl ammonium bromide or chromatographic cellulose fibre powder are utilised with an organic extraction (Ghatak et al. 2013).

12.4.5.1 Advancements in DNA Extraction from Faecal Matter

QIAGEN Inc. (Valencia, CA) has developed a commercially available kit for the extraction of nuclear DNA from human faeces using a reasonably rapid silica-based extraction method. Included in the package is a reagent that purportedly eliminates the faeces-derived PCR inhibitors. In addition, the method permits the preferred lysis of human cells over bacterial ones. Vandenberg and van Oorschot examined the effectiveness of the kit. Using the QuantiBlot Human DNA Quantitation Kit (Perkin Elmer, Foster City, CA), the D17Z1 primate-specific probe, and the SF Microfiltration Apparatus, extracted DNA samples can be quantified by slot blotting (Bio-Dot) followed by amplifying the faecal DNA using the AmpFSTR R© Profiler Plus TM PCR Amplification Kit (PE Biosystems, Foster City, CA). Further, STR typing can be performed with capillary electrophoresis and laser-induced fluorescence with the ABI PRISM 310 Genetic Analyzer.

12.4.6 Identification of Microorganisms in Faeces

Using molecular biological approaches, the identification of faeces bacteria can be one of the sensitive and straightforward methods for identifying faeces. There are more than 4000 bacterial species in the human intestinal microbiota. Traditional methods for detecting faeces use the cultivation of faecal indicator bacteria like Escherichia coli or Enterococci spp. However, these faecal indicator bacteria only make up a small part of the faecal microbiota, so they may not be good enough for forensic identification. Bacteroides makes up around 30% of faecal microbiota, which are the most prevalent bacteria in human faeces. Bacteroides could be utilised to identify faeces for forensic purposes. Bacteroides is a genus of anaerobic, Gram-negative rod-shaped bacteria. The diet of the host can change the populations of microorganisms in faeces. Bacteroides species predominate in the faeces of individuals whose diets are rich in saturate fats and proteins, which are prevalent in Western diets. Prevotella species, another genus of Gram-negative bacteria, are prevalent in the faeces of persons who follow a low-fat and carbohydrate-rich diet. By recognising certain DNA sequences of the rpoB gene, which codes for the subunit of bacterial RNA polymerase, Bacteroides can be identified. *B. uniformis*, *B. vulgatus,* and *B. thetaiosamicron* are the three chief species of Bacteroides found in faeces. *B. uniformis* and *Bacteroides thetaiosamicron* cannot be detected in blood, saliva, sperm, urine, vaginal secretions, or on the surface of the skin. *B. uniformis* and *Bacteroides thetaiosamicron* are therefore regarded as a particular indicator bacterium for faecal identification. Occasionally, *B. vulgatus* can also be found in samples of feminine fluid. Therefore, caution must be exercised while interpreting the results of a *B. vulgatus* screening. Real-time PCR can be used to detect bacterial genes by amplifying the RNA polymerase-subunit gene of *B. uniformis* and *B. vulgatus* and the α-1-6 mannanase gene of *B. thetaiotaomicron* using a minor groove binding probe. However, a number of studies have demonstrated that either *B. uniformis* or *B. vulgatus* was present in every sample. Therefore, *B. uniformis* and *B. vulgatus* are more ideal target species for the identification of faecal material than *B. thetaiotaomicron*. Upon detection of *B. vulgatus* and/or *B. uniformis*, it is likely that the sample contains faeces.

12.5 Conclusion

The excretory material performs the same vital role as other biological fluids in the legal procedure of suspect and victim identification. It is as dependable as blood and other body fluids since it contains undigested food, nucleated epithelial cells, traces of blood and its other cells, bile pigments, drug metabolites, and many stomach-dwelling bacteria. Analysis of the aforementioned evidence can assist link a suspect to a crime. In addition to distinguishing between adult and juvenile faeces, macroscopic analysis can also be used to determine the species to some extent. Microscopic analysis leads to the screening of individuals based on the presence of undigested materials, which can reveal a person's eating habits. Blood and its corpuscles and other specific cells can also reveal specific diseases, which can be used to distinguish between the innocent and the main perpetrator. As it is unique to human faeces and missing from animal faeces, urobilin can be regarded a beneficial agent. Some modern technologies, such as spectrophotometry investigation of fluorescence-produced faeces in the Schlesinger and Edelman test and DNA profiling, have also been implemented to assess the molecular specificity of stool matter. Analysing the microbiota in faeces is a new field, as many microorganisms have yet to be discovered. These microorganisms are as closely associated with a person as their DNA.

References

Barbosa MR (2012) Chemical composition and formation of human feces–problems and solutions of large mergers demographics in developing countries. In Tenth International Symposium on Recent Advances in Environmental Health Research 39

Fenwick GR, Heaney RK, Mullin WJ, VanEtten CH (1983) Glucosinolates and their breakdown products in food and food plants. CRC Crit Rev Food Sci Nutr 18(2):123–201

Maclagan NF (1946) Faecal urobilinogen: clinical evaluation of a simplified method of estimation. Br J Exp Pathol 27(3):190

Nakonechna, O., Stetsenko, S., Popova, L., & Tkachenko, A. (2017). Metabolism of proteins and nucleic acids

Muwonge, R., da Ganda Manuel, M., Filipe, A. P., Dumas, J. B., Frank, M. R., & Sankaranarayanan, R. (2010). Visual screening for early detection of cervical neoplasia in Angola. Int J Gynecol Obstet, 111(1), 68–72

World Health Organization (2019) Bench aids for the diagnosis of intestinal parasites. World Health. Organization

Attwood T, Cammack R, Atwood T, Campbell P, Parish H, Smith T et al (2006) Oxford dictionary of biochemistry and molecular biology

Lloyd JBF, Weston NT (1982) A spectrometric study of the fluorescence detection of fecal urobilinoids. J Forensic Sci 27(2):352–365

Abdalla K, Zuhailawati H, Rahmat A, Azizan A (2017) Characteristics of zinc phosphate coating activated by different concentrations of nickel acetate solution. Metall Mater Trans A 48:771–779

Ghatak S, Muthukumaran RB, Nachimuthu SK (2013) A simple method of genomic DNA extraction from human samples for PCR-RFLP analysis. Journal of biomolecular techniques: JBT 24(4):224

Shleenkova, A., Luta, E., Bogun, L., & Yabluchansky, M. (2015). Hypothyroidism & Thyroiditis.

Introduction to Vomitus and Its Forensic Analysis

13

Aditya Kumar Kar, Chitrita Chakraborty, and Pooja Uppal

Abstract

Vomiting is a gastrointestinal response to poisoning or drug overdose and also it can occur due to trauma faced in violent assault like in case of strangulation, sexual assault, etc. Thus, the identification of gastric fluid in a crime scene can help in framing the crime scene and can corroborate to a criminal act. Vomitus contains gastric fluid which contains HCl, mucus, pepsinogen, and various gastric-specific protein biomarkers like MUC5AC, GAST, etc. along with epithelial cells and food residue. The detection of which in a suspected stain will help to identify vomitus.

Keywords

Nausea · Gastric secretions · Glands · Ulcers · Assay

13.1 Introduction

Vomiting and nausea are gastrointestinal problems that are generated by varying emetic stimuli through the nervous system. Vomiting is an involuntary act of forceful expulsion of gastrointestinal contents through the mouth (Li 2021). They are regarded as defense mechanism of the body when dangerous toxins,

---

Aditya Kumar Kar, Chitrita Chakraborty and Pooja Uppal contributed equally with all other contributors.

A. K. Kar (✉) · C. Chakraborty · P. Uppal
Department of Forensic Science, Adamas University, Barasat, Kolkata, India

A. Puri et al. (eds.), *Fundamentals of Forensic Biology*,
https://doi.org/10.1007/978-981-99-3161-3_13

microorganisms, or drugs gain entry into the body through different routes (gastrointestinal track or parenteral route).

The process of vomiting begins with the parasympathetic nervous getting stimulated that in turn increases salivation. Then, there is deep inhalation followed by closure of glottis to protect the lungs from aspiration of the vomit. This event is subsequently followed by contractions of diaphragm and then contraction of abdominal muscles that compresses the stomach. Due to relaxation of sphincter and esophagus, the gastric contents are then forced upward and expelled out though the mouth (Li 2021). There is also excessive sweating and increased heart rate during vomiting due to activation of sympathetic nervous system.

Vomitus is sometimes found in cases of strangulation, poisoning or drug overdose, sexual abuse, and child abuse. The positive identification of vomit stain may help to prove or disapprove a suspect and victim's involvement and further help in framing the scene of crime (Akutsu et al. 2016).

13.2 Composition of Vomiting

Vomit contains gastric secretions, which consists of secretions from oxyntic glands and pyloric glands of stomach mucosa. Gastric secretion consists of HCl, pepsinogen, mucus, gastrin hormone, and other intrinsic factors. Gastrin hormone is secreted by 'G' cells, the function of which is to stimulate stomach to release gastric acid. It contains Mucin 5AC (MUC5AC) which is secreted from epithelial goblet cells of stomach which is a gel-forming glycoprotein. Vomit also contains epithelial cells that come from the lining of the gastrointestinal track. It also contains digested, semi-digested, and un-digested food residue.

13.3 Characteristics Feature of Vomitus

1. The pH of vomitus is in general highly acidic. Its optimal pH value is 2, but it is influenced by the type of food residue it contains (Akutsu et al. 2016).
2. Malodorous.
3. Color-Color of vomit may indicate underlying condition like it is foamy or whitish in case of acid reflux, gastritis; it is clear in appearance in case of brain injury, gastric outlet obstruction, cyclic vomiting disorder, migraine, morning sickness, gastroenteritis; green or yellow colored in case of bile reflux, food poisoning, gastroenteritis, liver failure; pink or red colored in case of amyloidosis, damage to throat, mouth or gums, peptic ulcer, stomach cancer; it is brown in color in case of blocked intestine, severe constipation, stomach cancer. Brownish or blackish color vomit indicates internal bleeding particularly in the lower gastrointestinal track region that undergoes oxidation and appears brownish or blackish (Li 2021).

13.4 Examination of Vomitus

1. **Mucus**: The inner surface of the stomach is lined by two-layer mucus system, inner attached mucus layer and outer loose/unattached mucus layer which protects the inner wall of stomach from digestive enzymes and acidic environment (Johansson et al. 2013). Some amount of mucus comes out while vomiting. Mucus shows opalescence appearance while adding acetic acid (33%). By identifying the presence of mucus, we can determine whether the stain is vomit or not.
2. **Free HCL**: Some amount of HCL is present in our stomach, which is secreted by oxyntic glands of parietal cells of mucosa. HCL maintains the low pH in stomach and protects it from bacteria, microbes and helps in digestion (Beasley et al. 2015). No other body fluid contains HCL other than gastric fluid. For testing of free HCL in vomit samples, Gunzburg's reagent is used. Gunzburg's reagent is prepared by mixing phloroglucinol, vanillin, and alcohol. It shows brilliant red color in the presence of free HCL.
3. **Microscopic Examination of Food Residue**: Vomit may contain digested or semi-digested food material, which are examined under a microscope. This method is not reliable because many times vomit does not contain any visible substance. So, this method cannot verify whether the sample is vomit or not.
4. **Microscopic Examination of Endothelial Cells**: Vomit may contain few endothelial cells. For viewing the cells under microscope, the extract is first centrifuged for 10 min and then a thin film of the residue part is made on the slide and observed under the microscope.
5. **Pepsin Assay**: Pepsin is the most important endopeptidase present in gastric fluid. It is secreted by gastric chief cells as an inactive form pepsinogen, which is activated in the presence of HCL. The main function of this enzyme is break down of peptide bonds present in the middle of proteins. The preferred amino acid targets of pepsin for cleavage reaction are tryptophan, phenylalanine, and tyrosine. This proteolytic activity of pepsin is used as an assay to identify the vomit samples. In pepsin assay, an insoluble protein dye complex called fibrin blue is used, which is colorless. If the sample contains vomit, the pepsin present in vomit cleaves the fibrin blue and releases a chromophore (soluble in water) which shows blue color. This assay is carried out in fibrin blue containing agarose gel. The sample is loaded in the gel and incubated for 4–6 h at 37°C. If the sample contains vomit, a concentric, blue and translucent ring is observed around the sample. This assay is most reliable, as no other proteolytic enzyme and no other body fluid shows the positive reaction. It also shows positive reaction for old vomit stains up to 3–6 months. The gel plate can be photographed, dried, and preserved for the purpose of record (Yamada et al. 1992).

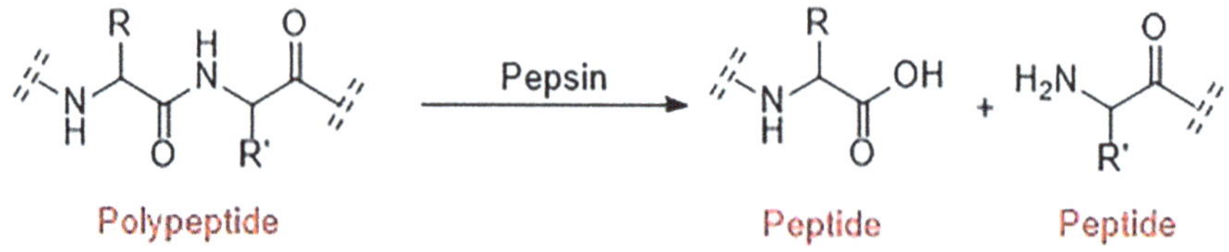

Cleavage of polypeptide by pepsin

6. **ELISA for the Detection of Gastric Mucosa-Expressing Proteins**: Tomoko et al. conducted a study to identify vomit samples by ELISA in which they focused on four gastric mucosa-expressing proteins (gastric-specific protein biomarkers), pepsinogen I (PGA), pepsinogen II (PGC) which are precursors for pepsin A and Pepsin C, secreted from chief cells of stomach. Gastrin (GAST) is a peptide hormone which is secreted from G cells and mucin 5AC (MUC5AC) is glycosylated protein which is secreted from epithelial goblet cells of stomach. Results showed that PGC and PGA are reliable but show some cross-reactivity with other body fluids. However, pepsinogen I show cross-reactivity with urine and seminal stains due to its expression in these fluids. MUC5AC is only reliable when used along with the other markers. GAST is not included for the identification of vomit because it is a gastrointestinal hormone and not secreted into the cavity of stomach. PGA and PGC also show positive results for old vomit stains. Although ELISA is simple to perform and is a robust technique, it has a drawback of non-specific antibody recognition, especially if a complex matrix is present and in those cases, it gives false positive or negative results. Also, information obtained from ELISA is limited to specific detection of only previously designated proteins. Apart from this, no other information like food consumed before vomiting can be obtained. So, ELISA is found to be a satisfactory technique, especially for the detection of pepsinogen I (PGA), thus it is an effective technique for identification of vomitus (Akutsu et al. 2016).
7. **Proteomic Approach Using Mass-Spectroscopy**: Proteomics is a technology to link protein expression patterns to its original tissue or to the biological fluid. This approach fulfills the requirement of forensic scientist through the identification of tissue-specific or organ-specific protein expression patterns. Thus, proteomics can assess the nature of body fluids. So, in case of detection of vomitus, mass spectroscopy-based proteomics not only help in identification of vomiting stain through the detection of tissue-specific proteins but also it can detect food-derived proteins which will help to identify the nature of meal, which might help to frame the crime scene. It must be considered that type of proteins and peptides found in vomit is greatly affected by force used for its expulsion and the nature of ingested food. If duodenal proteins are present in vomit stain, it indicates failure of pylorus that could be related to some pathological condition. If the force used for expulsion is more, then there will be increased intra-abdominal pressure because of retching and contraction, which sometimes induces small intestine reverse

peristalsis, so in those cases enzymes from small intestine (pancreatic enzymes) can be traced in vomit samples. Also, characterization of "digestomes" can further help to infer about the physiological decomposition of food proteins inside the human body.

For purification of proteins from vomiting stain for proteomic investigation, the proteins are solubilized by washing vomiting stain with 400 μl of reducing buffer consisting of 6 M guanidine HCl, 0.3 M Tris, 1 mM EDTA, 10 mM DTT, pH 8.0, it is then incubated at 56°C for 1 h. After this, it is alkylated with threefold iodoacetamide (IAA) for 40 min in a dark room at room temperature. Then, 100 μl aliquot containing reduced proteins are diluted with 25 mM Ammonium bicarbonate pH 7.8 containing Dithiothreitol. Then, a micro-lowery modified method kit is used for the quantization of polypeptides. It is further digested overnight at 37°C with trypsin. After trypsinolysis, peptides are again purified using prepacked C18 Sep-pak cartridge washed with 0.1% TFA (trifluoroacetic acid) and then eluted with 70% acetonitrile. It is finally concentrated, lyophilized, and then re-constituted in 0.1% formic acid (v/v) at a concentration of 0.5 μg/μl for LC-MS/MS analysis. For LC-MS/MS analysis, Eluent A should consists of 0.1% formic acid (v/v) in LC-MS grade water and Eluent B should consists of 0.1% formic acid (v/v) in acetonitrile. For the purpose of separation, 2–45% gradient of B for over 150 min at 30 nL/min constant flow rate is used (Pieri et al. 2019).

References

Akutsu T, Saito H, Iwase H, Watanabe K, Takamura A, Sakurada K, Miyasaka S (2016) The applicability of ELISA detection of gastric mucosa-expressing proteins for the identification of vomit. Int J Legal Med 131(2):359–364. https://doi.org/10.1007/s00414-016-1409-1

Beasley DE, Koltz AM, Lambert JE, Fierer N, Dunn RR (2015) The evolution of stomach acidity and its relevance to the human microbiome. PLoS One 10(7):e0134116

Johansson ME, Sjövall H, Hansson GC (2013) The gastrointestinal mucus system in health and disease. Nat Rev Gastroenterol Hepatol 10(6):352–361

Li R (2021) Identification of vomitus. In: Forensic biology essay. CRC Press, Boca Raton

Pieri M, Silvestre A, De Cicco M, Mamone G, Capasso E, Addeo F, Picariello G (2019) Mass spectrometry-based proteomics for the forensic identification of vomit traces. J Proteome 209: 103524. https://doi.org/10.1016/j.jprot.2019.103524

Yamada S, Hirata K, Tsugawa N, Bunai Y, Ohya I (1992) Vomit identification by a pepsin assay using a fibrin blue-agarose gel plate. Forensic Sci Int 52(2):215–221. https://doi.org/10.1016/0379-0738(92)90110-i

Nucleic Acids: DNA and RNA 14

Rutwik Shedge, Aditi Iyengar, Monisha Samuel, and Tanya Chauhan

Abstract

Nucleic acids are biomacromolecules that exist naturally in all the living organisms, and play fundamental roles in many biological processes such as catalysis of biochemical reactions and regulation of a plethora of cell activities (Cooper, The cell: a molecular approach. 2nd ed. Sinauer Associates, Sunderland, 2000). Various types of nucleic acids have tremendous potential in medicine, in form of treatments and sensors for detection of ions, small molecules, amino acids, proteins, and cells (Thatcher, Principles and applications of molecular diagnostics. Elsevier, Amsterdam, 2018). They are used for gene regulation, bioimaging, drug delivery, and disease therapy (Thatcher, Principles and applications of molecular diagnostics. Elsevier, Amsterdam, 2018). In forensic science, nucleic acids have applications in identification, estimation of post-mortem interval, paternity and maternity tests, genealogy, toxicology, and pathology to name a few (Li, Forensic biology: identification and DNA analysis of biological evidence. CRC Press, Boca Raton, 2021).

R. Shedge (✉)
School of Forenic Science, National Forensic Sciences University, Gandhinagar, Gujarat, India

A. Iyengar
Impact Science, Cactus Communications, Mumbai, India

M. Samuel
School of Forenic Science, National Forensic Sciences University, Gandhinagar, Gujarat, India

Department of Forensic Science, National Forensic Sciences University, Delhi Campus (LNJN NICFS), New Delhi, Delhi, India

T. Chauhan
Deparment of Forensic Science, National Forensic Sciences University, Tripura Campus, Agartala, Tripura, India

A. Puri et al. (eds.), *Fundamentals of Forensic Biology*,
https://doi.org/10.1007/978-981-99-3161-3_14

Keywords

Nucleic acids · DNA · RNA · Ribose · Nitrogen bases · Phospodiester bonds

14.1 Definitions

Nucleic acids: Biomacromolecules that exist naturally in all the living organisms, and play fundamental roles in many biological processes

DNA: Fundamental genetic material of humans

RNA: Biomacromolecule vital in multiple biological processes such as coding, decoding, and expression of genes

miRNA: micro-RNA or short non-coding RNA nucleotides that range between 15–30 bases

Pentose: A five carbon sugar that forms the nucleic acids

Purines: Nitrogenous bases derived from purine heterocyclic rings

Pyrimidines: Nitrogenous bases derived from pyrimidine heterocyclic rings

STR profiling: DNA typing using autosomal and sex chromosomal short tandem repeats

Chaotropic agents: Chemicals that deplete the hydration cell of DNA and cause it to reversibly bind to a substrate

14.2 Native Nucleic Acids

Native nucleic acids are DNA and RNA. DNA stands for deoxyribonucleic acid, and is the hereditary material in humans (Cooper 2000). It carries the genetic instructions necessary for growth, development, and reproduction of humans (Ha and Bhagavan 2023). RNA stands for ribonucleic acid, and is vital in multiple biological processes such as coding, decoding, and expression of genes (Grabmüller et al. 2015). Nucleic acid was first discovered by Friedrich Miescher in 1869, and was dubbed as nuclein, for its discovery within the nucleus (Dahm 2005).

14.2.1 Structure of Nucleic Acids

Nucleic acids are comprised of nucleotides, which are monomers made of three constituents such as a 5-carbon sugar (pentose), a phosphate group, and a nitrogen base (Fig. 14.1) (Thatcher 2018). Depending upon whether the pentose sugar is deoxyribose or ribose, the resultant nucleic acid is DNA or RNA. The presence of phosphate group (phosphoric acid) is why the macromolecules are called nucleic 'acids.' Nucleic acids consist of five different nitrogenous bases such as two purines and three pyrimidines. The nitrogenous bases are called purines or pyrimidines on the account of being derived from purine and pyrimidine heterocyclic rings (Fig. 14.2). Purine bases include adenine (6-aminopurine) and guanine (2-amino-

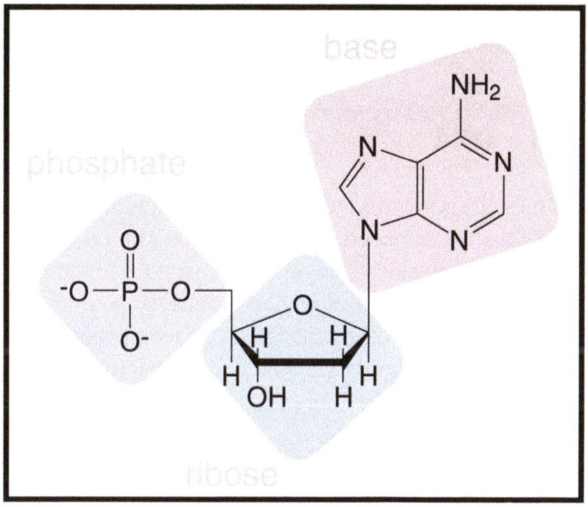

Fig. 14.1 Nucleic acid composition (reference image for the illustrator)

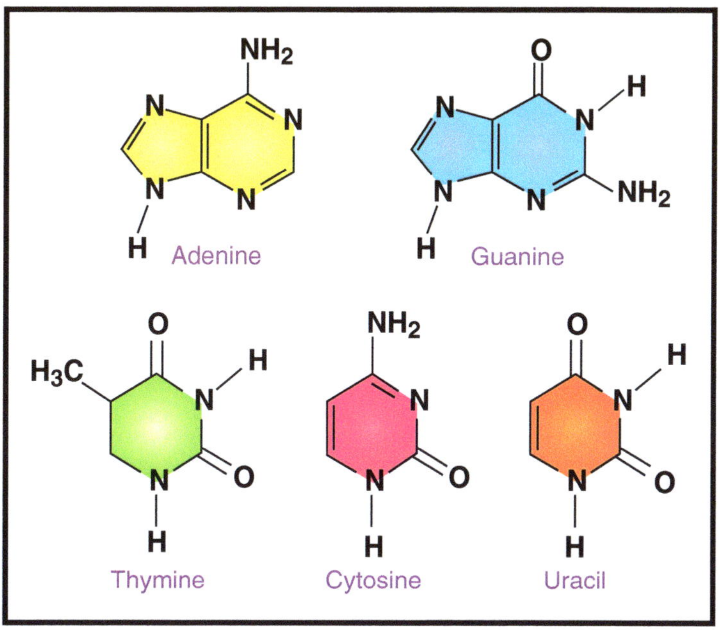

Fig. 14.2 Nitrogenous bases of nucleic acids (reference image for the illustrator)

6-oxypurine), while pyrimidine bases include thymine (5-methyl-2,4-dioxipyrimidine), cytosine (2-oxo-4-aminopyrimidine), and uracil (2,4-dioxoypyrimidine). Since all the nitrogenous bases are aromatic in nature, they can absorb radiation in the UV spectrum, and thus can be identified and quantified in an unknown sample (Butler 2015). In DNA, the nitrogenous bases found are adenine (A), guanine (G), thymine (T), and cytosine (C), while in RNA, thymine is replaced by uracil (U).

14.3 DNA

DNA or the deoxyribonucleic acid is a polymer that consists of two polynucleotide chains that are wrapped around each other to form a helical structure (Fig. 14.3). The two strands are said to be in an antiparallel orientation, and are connected to each other by means of hydrogen bonds between complimentary nitrogenous bases (Kyle and Shampo 1998). In the DNA, adenine binds to thymine by means of two hydrogen bonds, while guanine and cytosine bond to each other by means of three

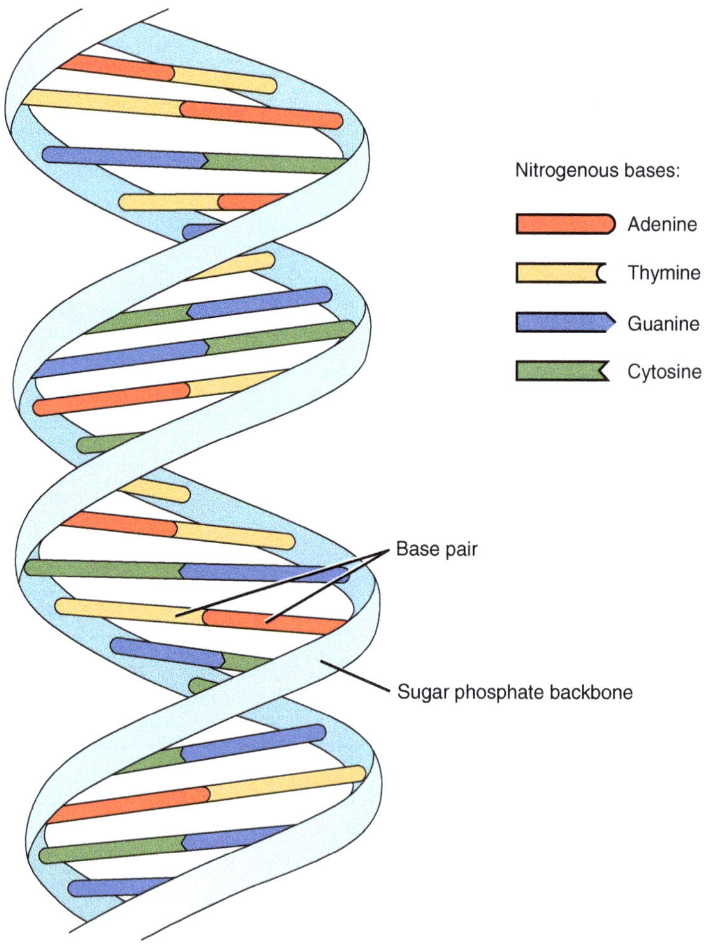

Fig. 14.3 Helical structure of DNA (reference image for the illustrator)

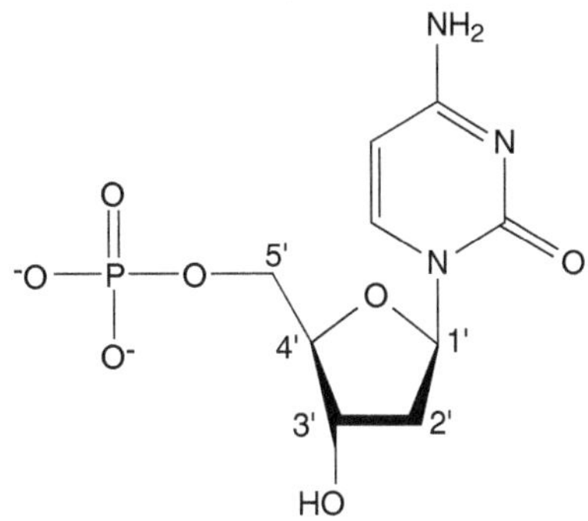

Fig. 14.4 Nucleotide with phosphate group connected to the 5′-end and hydroxyl group present at the 3′-end (reference image for the illustrator)

Table 14.1 Amount of DNA that can be extracted from commonly encountered forensic evidences

Forensic evidence	Amount of DNA
Liquid blood	20,000–40,000 ng/mL
Blood stain	250–500 ng/cm^2
Liquid semen	150,000–300,000 ng/mL
Post coital vaginal swab	10–3000 ng/swab
Liquid saliva	1000–10,000 ng/mL
Oral swab	100–1500 ng/swab
Plucked hair	1–750 ng/root
Shed hair	1–10 ng/root
Urine	1–20 ng/mL
Bone	3–10 ng/mg
Tissue	50–500 ng/mg

hydrogen bonds. Both the strands are coiled around the same axis, and have the same pitch of 3.4 nm. The pair of strands have a radius of 1 nm, and form two separate grooves such as a major groove which is 2.2 nm wide and a minor groove which is 1.2 nm wide.

The backbone of DNA is made up from alternating phosphate and deoxyribose molecules, joint together by means of phosphodiester bonds. The phosphodiester bonds are formed between the third and fifth carbon atoms of adjacent deoxyribose rings. These are known as the 3′-end and the 5′-end carbons. Any DNA strand has a 3′-end where the free hydroxyl group (–OH) is attached and a 5′-end where the phosphate group is attached (Fig. 14.4).

14.3.1 DNA Extraction: Principles

Extraction of DNA, its quantitation, and eventual profiling forms the crux of forensic molecular biology. Forensic scientists often have to deal with old, impure, degraded biological fluids as evidences, and are expected to isolate DNA from them for further downstream processing. This makes it imperative, that the forensic scientists use standardized methods of DNA isolation, that will ensure that sufficient amount of DNA is extracted for subsequent STR profiling. While the different methods of DNA extraction exist, the principle for them is somewhat similar. The amount of DNA that can be isolated from different forensic evidences are detailed in Table 14.1 (Butler 2015).

14.3.1.1 Cell and Tissue Disruption

DNA extraction begins with cell and tissue disruption. The cells present within the biological evidence found at a scene of crime or on the victim need to be disrupted so that the DNA present within their nuclei may be isolated. Cell or tissue disruption can be done by boiling, heating, treatment with alkalis, or by mechanical methods such as use of mortar and pestle, homogenizer, or a barocycler (Li 2021). Harder tissues such as the teeth and bones may be first subjected to liquid nitrogen, and grinded using any of the aforementioned mechanical methods. However, in case of teeth and bones, decalcification is a must, and is done by giving the bone or teeth sample repeated treatment with chelating agents such as EDTA (Ethylenediaminetetraacetic acid). The time required for successful decalcification may be as long as several days, and depends upon the size and source of the sample.

14.3.1.2 Lysis of Cellular and Organelle Membranes

After disruption of cells and tissues, the broken membranes of cells and their organelles need to by lysed to facilitate isolation of DNA. This lysis procedure is usually conducted in a Tris buffer (trisaminomethane) to maintain a pH where endogenous DNases remain inactive (Heathfield et al. 2021). In this Tris buffer, certain chaotropic agents and salts are added, and these carry out the lysis of cell membrane. Chaotropic agents such as guanidium chloride, 2-propanol, lithium perchlorate, etc. cause entropic disorder in the lipid bilayer of cell membranes, and cause loss of their hydrophobic properties. Detergents such as sodium dodecyl sulphate and sarkosyl are also added to the buffer, and they 'poke holes' in the cell membrane. Other enzymes such as proteinase-k are added which denature proteins and dissociate histone proteins from the DNA.

14.3.1.3 Removal of Cytoplasmic Constituents

After lysis, the isolated DNA still exists amongst the wreckage of lysed lipids, proteins, and other organic material. Removal of DNA from this mixture can be done using multiple rounds of phenol-chloroform extraction, or through repeated washing or through reversible binding of DNA to a solid material such as silica.

14.3.2 DNA Extraction Methods

14.3.2.1 Organic Extraction

Organic extraction, also known as phenol-chloroform extraction, is one of the most commonly used manual methods of DNA extraction in a forensic setting. Cell lysis is performed using a detergent such as sodium dodecyl sulphate (SDS) and proteinase-K. The removal of cytoplasmic contents is achieved by using a solution

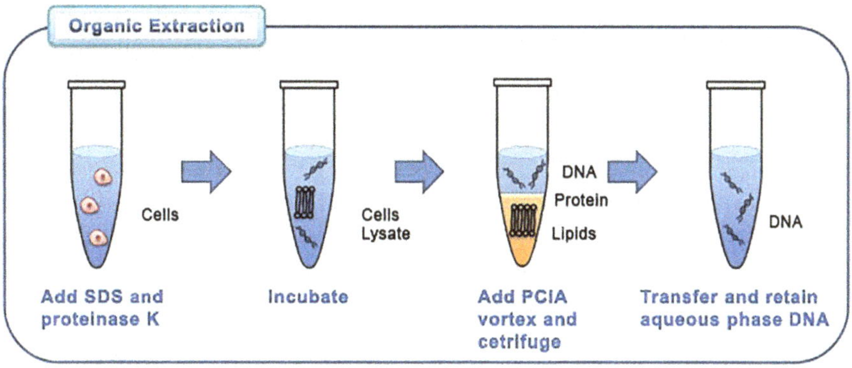

Fig. 14.5 Organic extraction of DNA (reference image for the illustrator)

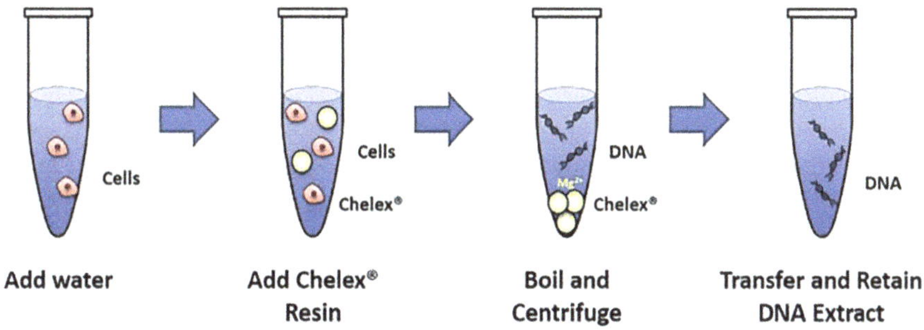

Fig. 14.6 Chelex®-based method of DNA extraction (reference image for the illustrator)

of phenol:chloroform:isoamyl alcohol in a proportion of 25:24:1 (Heathfield et al. 2021). Phenol is used to separate the proteins and lipids from the cytoplasmic contents. However, since the density of phenol is just slightly higher than that of water, separation of phenol from the aqueous phase is difficult. Hence, chloroform is added to the mixture as it has much higher density than phenol, and forms an organic layer at the bottom. Since addition of chloroform leads to foaming, isoamyl alcohol is added to reduce the foam. The aqueous layer rises to the top and contains the soluble DNA in it. The lipids are found in the organic layer, while the proteins are found in the interphase between organic and aqueous layers (Fig. 14.5).

The aqueous layer is separated into another tube, and precipitation of DNA is done. This is achieved by addition of ethanol, which depletes the hydration shell surrounding the dissolved DNA, causing it to precipitate. Ultrafiltration can also be used as an alternative for concentration of DNA.

14.3.2.2 Chelex®-Based Extraction

This technique is also known as the boiling and chelation method (Fig. 14.6). The first step in this method involves thorough washing of the sample to remove any inhibitors of the polymerase chain reaction. The sample is then incubated at 56°C for 20 min to soften the cell membranes and separate clumps of cells from each other. Lysis of the cell membrane is then achieved by boiling the sample. The DNA that gets released during this step is also accompanied by other cellular contents including DNases. If allowed to be active, the DNases may degrade the extracted DNA. To prevent their activity, a chelating resin composed of styrene divinylbenzene copolymers (Chelex® 100) is used (Ip et al. 2015). This ion exchange resin consists of paired iminodiacetate groups that bind to the divalent cations such as magnesium. Magnesium is a cofactor for the activity of DNases, and thus, its pairing with Chelex® 100 keeps the endogenous DNase inactive. The removal of Chelex® and other cytoplasmic constituents is done by centrifugation. It is important to completely separate the extracted DNA and Chelex® as during amplification, magnesium ions are required, and presence of any Chelex® molecules may inhibit the PCR reaction.

14.3.2.3 Silica-Based Extraction

Silica-based DNA extraction works on the principle of reversible adsorption of the DNA molecule onto silica surface in the presence of high concentrations of chaotropic salts (Fig. 14.7) (Vinueza-Espinosa et al. 2019). Normally, in an aqueous solution, the hydration shell surrounding the DNA molecule shields the negative charge of the DNA, and ma DNA hydrophilic. In presence of high concentration of chaotropic salts, the hydration shell of the DNA gets depleted and DNA becomes hydrophobic. This causes DNA to get reversibly adsorbed onto a silica surface.

The extraction procedure begins with cell disruption using proteinase-K. The DNA lysate is added to a stationary phase made up of silica (SiO_2) in the presence of chaotropic salts such as guanidinium thiocyanate and guanidinium hydrochloride. The presence of these salts causes the DNA to reversibly bind to the silica surface,

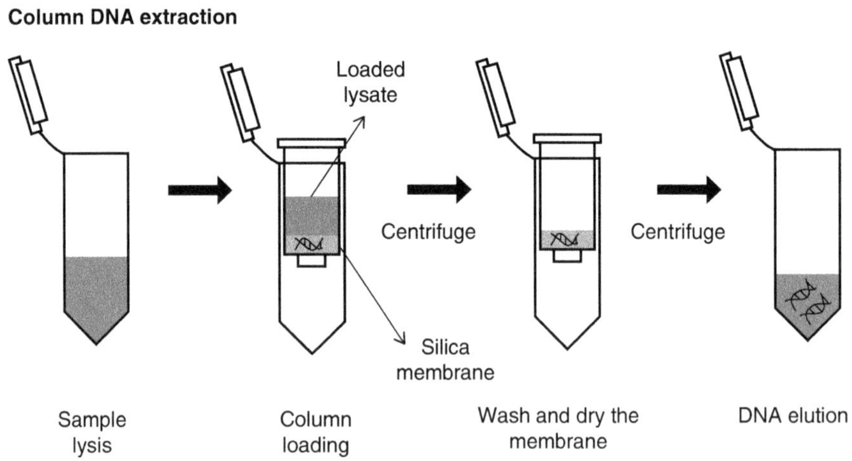

Fig. 14.7 Silica-based column method of DNA extraction (reference image for the illustrator)

while the other cytoplasmic constituents are not adsorbed. The cytoplasmic constituents, chaotropic agents, and other contaminants are washed away using an ethanol-based solution. The adsorbed DNA can be eluted using low salt water solutions. These solutions cause the DNA to rehydrate and thus elute.

Automated DNA Extraction

Silica membrane devices have been created for automated DNA extraction. Using 96-well silica membrane plates and a plethora of robotic platforms, DNA can be extracted without any manual input (Li 2021). Another modification of the silica-based method for automated DNA extraction is the use of silica-coated paramagnetic beads or particles (Witt et al. 2012). The sample lysate is simply added to the automated machine, where the DNA gets adsorped onto the magnetic beads. The cytoplasmic contents are washed away, and the magnetic beads are subjected to a method which rehydrates the bound DNA. These advances in DNA extraction technologies have allowed for development of parallel and high-throughput processing.

14.3.2.4 Differential Extraction

One of the more common cases dealt by forensic biologists are the sexual assault cases. In such cases, often, the evidences recovered are mixtures of semen of the perpetrator and the vaginal discharge of the victim. In such cases, different DNA extraction methods have to be utilized to separate DNA from spermatozoa of male contributor and from non-sperm cells of the victim such as the epithelial cells. If the DNA from these two different sources is not separated, complications may arise during interpretation of resultant DNA profiles.

The method used to differentially extract DNA from sperm cells and non-sperm cells has its principle based on the different cell membrane properties of these two types of cells. First, DNA is extracted from the epithelial cells using proteinase-K. This enzyme can denature the proteins of cell membrane of non-sperm cells with relative ease, while is ineffective in lysing cell membranes of the sperm cells. The plasma membrane of sperm cells contains proteins that are linked with disulphide bonds, which imparts extra mechanical stability and thus makes the plasma membrane resistant to the activity of proteinases. After the non-sperm is extracted, it gets released in the supernatant, which can be separated. The left-over sperm cells are separately lysed using a reducing agent called dithiothreitol (DTT). In presence of DTT and proteinase-K, the sperm cells get lysed and their DNA is extracted (Fig. 14.8) (Ng et al. 2017; Sinha et al. 2022).

14.4 RNA

RNA or ribonucleic acid is a polymer that is found in a single-stranded structure in the human body (Busa and Leung 2021). Similar to DNA, RNA molecule too has a sugar-phosphate backbone, except, in RNA ribose is the pentose sugar. Ribose sugar has a hydroxyl group attached in the 2′ position, which is absent in the DNA

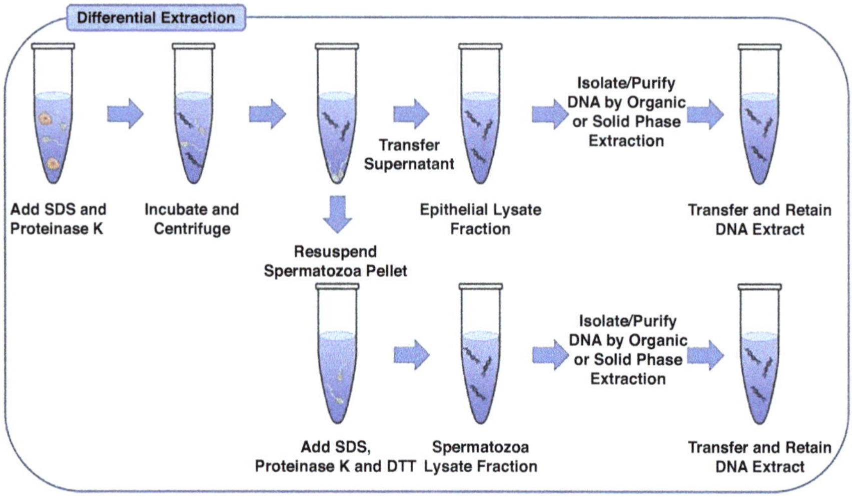

Fig. 14.8 Differential extraction of DNA (reference image for the illustrator)

molecule. The phosphate group is attached to the 3′-end of one ribose molecule, and 5′-end of the other ribose molecule (Fig. 14.9). As with DNA, nitrogenous bases form a part of the RNA molecule, but in the RNA, thymine (T) is replaced by uracil (U).

RNA synthesis happens by the means of DNA through a process known as transcription, while protein synthesis, or translation happens by the means of RNA (Cox and Arnstein 2003; Wise and Lou 2021). Three major types of RNA are involved in the process of translation: mRNA, rRNA, and tRNA (Fig. 14.10).

mRNA stands for messenger RNA, and is the RNA that is transcribed from DNA during the process of transcription. mRNA is a single-stranded RNA that carries information required for protein synthesis from nucleus to the cytoplasm where translation occurs (Liljas 2013). rRNA stands for ribosomal RNA, and is the primary component of ribosomes, which are cell organelles that perform protein synthesis (Joseph 2016). tRNA stands for transfer RNA, and is the molecule responsible for carrying amino acids to ribosomes for protein synthesis (Doherty and Guo 2016).

14.4.1 RNA Extraction

14.4.1.1 RNA-DNA Coextraction

RNA-DNA coextraction allows for simultaneous isolation of high-quality RNA and DNA extraction for forensic profiling and biological fluid identification (Fig. 14.11). Biological fluids are first lysed using a lysis buffer that contains high concentration of chaotropic salts such as guanidinium thiocyanate. These salts inactivate the endogenous DNases and RNases while also lysing the cell membranes. The lysate is passed through a silica membrane column, where due to the high concentration of

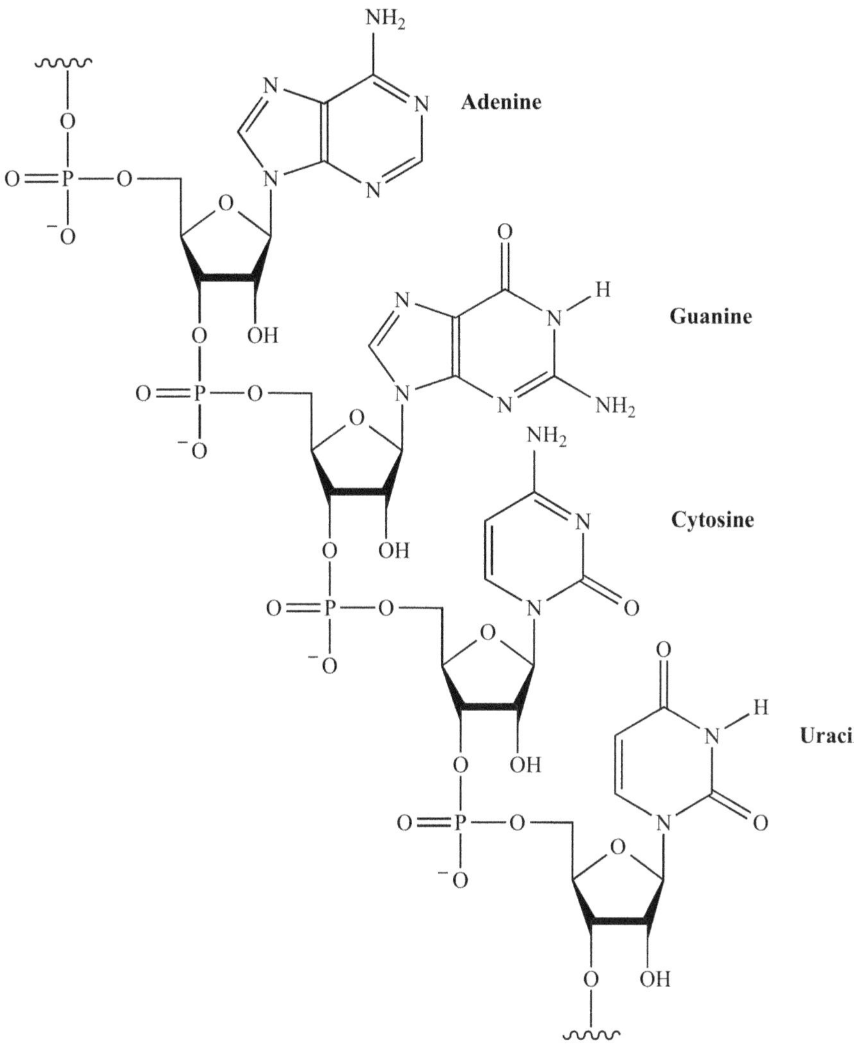

Fig. 14.9 Structure of RNA (reference image for the illustrator)

chaotropic salts, DNA gets selectively adsorped over RNA. The column is washed, and DNA gets eluted. The solution that comes out of this silica membrane column has lysate, chaotropic salts, and the RNA. Ethanol is then added to this liquid, and passed through another silica membrane column. Ethanol and chaotropic salts reduce the hydrophilic property of RNA to such an extent that it can now bind to the silica layer of this second column. RNA including mRNA gets eluted in this process (Majumdar et al. 2015; van den Berge and Sijen 2022).

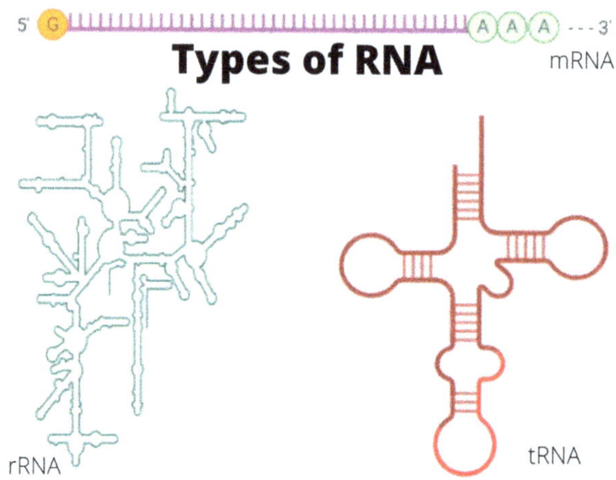

Fig. 14.10 mRNA, rRNA, and tRNA (reference image for the illustrator)

14.4.1.2 miRNA Extraction

miRNAs or microRNAs are small non-coding RNAs that range between 15 and 30 nucleotides in length. miRNAs have recently been utilized for the identification of biological fluids and body tissues. Since miRNAs are really small molecules, they cannot be successfully isolated using traditional RNA extraction methods. Instead, miRNA isolation happens in two phases: phase 1 involves organic solvent extraction, and phase 2 involves solid phase extraction (Fig. 14.12) (Corthell 2014; Tantray et al. 2023).

The cells of biological fluids or body tissues are first lysed. The lysate is then extracted using organic solvents such as phenol and chloroform, along with a high concentration of guanidinium thiocyanate. An acidic pH is maintained as at this pH, DNA partitions to the organic phase, while RNA gets dissolved in the aqueous phase.

The aqueous phase is then separated, and passed through a silica membrane filter at low ethanol concentration. Presence of low concentration ethanol and guanidinium thiocyanate depletes the hydration shell surrounding larger RNAs and causes them to bind to the silica membrane. The smaller RNAs (miRNAs) do not bind to the silica membrane, and pass off in the filtrate. This filtrate is passed through another silica membrane filter, but this time with higher ethanol concentration. This causes miRNAs to bind to the silica membrane, and can be eluted using low ionic strength solution.

MCQs

1. The backbone of nucleic acids is made by which bond between the phosphate and sugar molecules?
   (A) Hydrogen bonds
   (B) Disulphide bonds

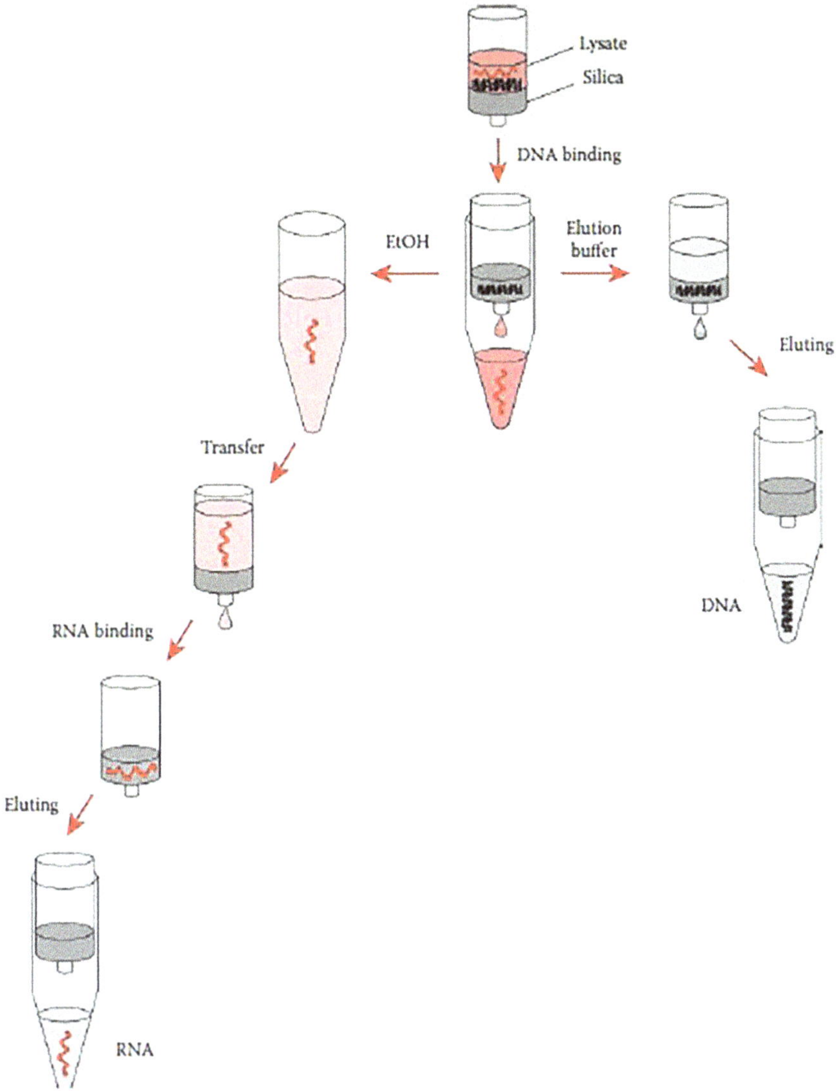

Fig. 14.11 Coextraction of RNA and DNA (reference image for the illustrator)

(C) phosphodiester bonds
(D) None of the above
Answer: C

2. In differential extraction of sperm and non-sperm DNA, lysis of sperm cells is done by addition of which of the following reducing agents?
(A) Dithiothreitol
(B) Sodium dodecyl sulphate

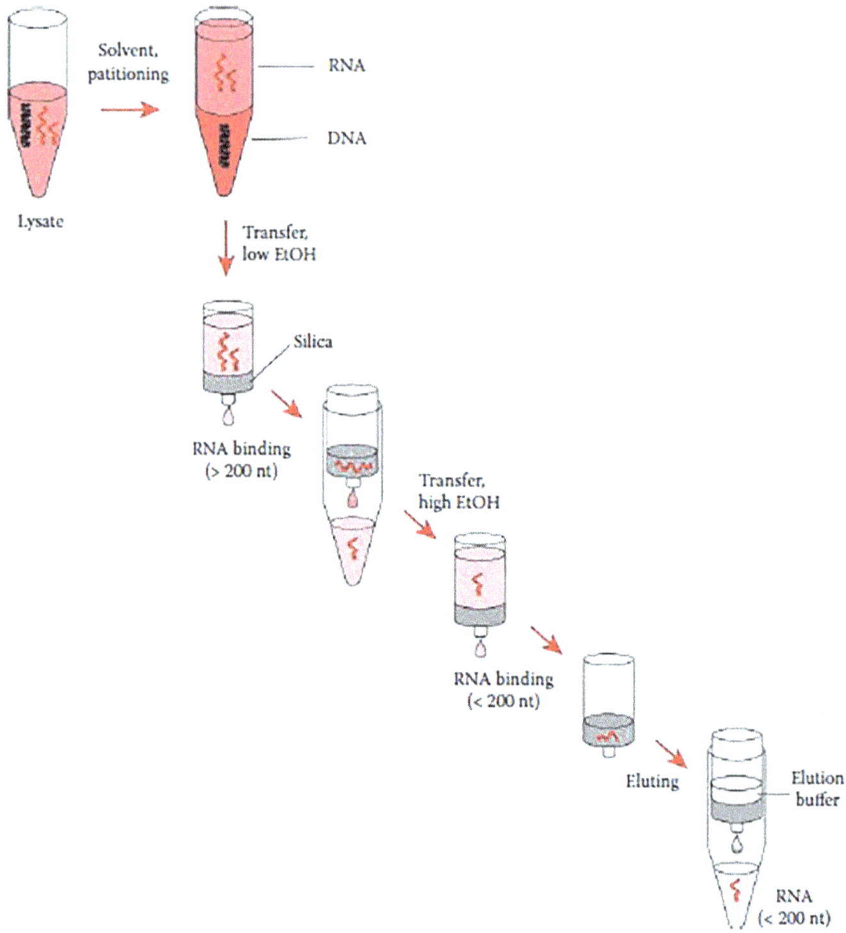

Fig. 14.12 miRNA extraction (reference image for the illustrator)

(C) Sarosyl
(D) Ethanol
Answer: A

3. Reversible adsorption of DNA happens on silica in presence of which of the following agents?
(A) Chelating agents
(B) Chaotropic agents
(C) Oxidising agents
(D) Stabilizers
Answer: B

4. The ideal proportion of phenol, chloroform, and isoamyl alcohol in organic extraction of DNA is _________

(A) 25:25:1
(B) 25:24:1
(C) 25:20:1
(D) 25:1:1
Answer: B

5. Which of the following nitrogenous bases is absent in RNA?
(A) Adenine
(B) Cytosine
(C) Guanine
(D) Thymine
Answer: D

References

Busa VF, Leung AKL (2021) Thrown for a (stem) loop: how RNA structure impacts circular RNA regulation and function. Methods 196:56–67. https://doi.org/10.1016/j.ymeth.2021.02.019

Butler JM (2015) The future of forensic DNA analysis. Philos Trans R Soc B Biol Sci 370: 20140252. https://doi.org/10.1098/rstb.2014.0252

Cooper GM (2000) Expression of genetic information. In: The cell: a molecular approach, 2nd edn. Sinauer Associates, Sunderland

Corthell JT (2014) Chapter 2 - DNA and RNA extraction protocols. In: Corthell JT (ed) Basic molecular protocols in neuroscience: tips, tricks, and pitfalls. Academic Press, San Diego, pp 11–19

Cox RA, Arnstein HRV (2003) Translation of RNA to protein. In: Meyers RA (ed) Encyclopedia of physical science and technology, 3rd edn. Academic Press, New York, pp 31–51

Dahm R (2005) Friedrich Miescher and the discovery of DNA. Dev Biol 278:274–288. https://doi.org/10.1016/j.ydbio.2004.11.028

Doherty J, Guo M (2016) Transfer RNA. In: Bradshaw RA, Stahl PD (eds) Encyclopedia of cell biology. Academic Press, Waltham, pp 309–340

Grabmüller M, Madea B, Courts C (2015) Comparative evaluation of different extraction and quantification methods for forensic RNA analysis. Forensic Sci Int Genet 16:195–202. https://doi.org/10.1016/j.fsigen.2015.01.006

Ha CE, Bhagavan NV (2023) Chapter 20 - Structure and properties of DNA. In: Ha CE, Bhagavan NV (eds) Essentials of medical biochemistry, 3rd edn. Academic Press, San Diego, pp 453–476

Heathfield LJ, Haikney TE, Mole CG et al (2021) Forensic human identification: investigation into tooth morphotype and DNA extraction methods from teeth. Sci Justice 61:339–344. https://doi.org/10.1016/j.scijus.2021.05.005

Ip SCY, Lin S, Lai K (2015) An evaluation of the performance of five extraction methods: Chelex® 100, QIAamp® DNA Blood Mini Kit, QIAamp® DNA Investigator Kit, QIAsymphony® DNA Investigator® Kit and DNA IQ™. Sci Justice 55:200–208. https://doi.org/10.1016/j.scijus.2015.01.005

Joseph S (2016) Ribosomal RNAs and protein synthesis. In: Bradshaw RA, Stahl PD (eds) Encyclopedia of cell biology. Academic Press, Waltham, pp 384–395

Kyle RA, Shampo MA (1998) Wilkins, Crick, and Watson: nobel prize for work on the structure of DNA. Mayo Clin Proc 73:362. https://doi.org/10.1016/S0025-6196(11)63703-8

Li R (2021) Forensic biology: identification and DNA analysis of biological evidence. CRC Press, Boca Raton

Liljas A (2013) Messenger RNA. In: Maloy S, Hughes K (eds) Brenner's encyclopedia of genetics, 2nd edn. Academic Press, San Diego, pp 369–370

Majumdar G, Vera S, Elam MB, Raghow R (2015) A streamlined protocol for extracting RNA and genomic DNA from archived human blood and muscle. Anal Biochem 474:25–27. https://doi.org/10.1016/j.ab.2014.12.021

Ng HH, Lim M-L, Hoe SY et al (2017) Modified differential DNA extraction to reduce processing time of sexual assault exhibits. For Sci Int Genet Suppl Ser 6:e252–e254. https://doi.org/10.1016/j.fsigss.2017.09.094

Sinha SK, Brown H, Holt H et al (2022) Development and validation of a novel method "SpermX™" for high throughput differential extraction processing of sexual assault kits (SAKs) for DNA analysis. Forensic Sci Int Genet 59:102690. https://doi.org/10.1016/j.fsigen.2022.102690

Tantray JA, Mansoor S, Wani RFC, Nissa NU (2023) Chapter 25 - RNA isolation from human tissue**Note: Handling Conditions Standard Precautions must be followed when handling all solid tissue samples. Samples can be stored in RNA later followed by storage in a −80°C freezer. In: Tantray JA, Mansoor S, Wani RFC, Nissa NU (eds) Basic life science methods. Academic Press, New York, pp 107–109

Thatcher SA (2018) 3 - Nucleic acid isolation. In: Rifai N, Horvath AR, Wittwer CT (eds) Principles and applications of molecular diagnostics. Elsevier, Amsterdam, pp 35–46

van den Berge M, Sijen T (2022) Development of a combined differential DNA/RNA co-extraction protocol and its application in forensic casework. For Sci Int Rep 5:100261. https://doi.org/10.1016/j.fsir.2022.100261

Vinueza-Espinosa DC, Santos C, Martínez-Labarga C, Malgosa A (2019) Assessing DNA recovery from highly degraded skeletal remains by using silica-based extraction methods. For Sci Int Genet Suppl Ser 7:810–812. https://doi.org/10.1016/j.fsigss.2019.10.204

Wise JA, Lou H (2021) Transcription|Messenger RNA processing in eukaryotes. In: Jez J (ed) Encyclopedia of biological chemistry III, 3rd edn. Elsevier, Oxford, pp 411–419

Witt S, Neumann J, Zierdt H et al (2012) Establishing a novel automated magnetic bead-based method for the extraction of DNA from a variety of forensic samples. Forensic Sci Int Genet 6:539–547. https://doi.org/10.1016/j.fsigen.2012.01.002

Quantitation and Quality Assessment of DNA

15

Monisha Samuel, Rutwik Shedge, and Tanya Chauhan

Abstract

The structure of DNA consists of two polynucleotide chains that run in opposite direction to one another and are held together by hydrogen bonds between the bases. The stacking of bases one upon another gives rise to grooves which run along the helix and thus makes up the double helical structure of DNA. These unique structural traits and complimentary nature of this biomolecule along with the pairing of bases has allowed researchers to create a wide range of sophisticated techniques for determining the quantity and quality of DNA present in a given sample (Nicklas and Buel, Anal Bioanal Chem 376:1160–1167, 2003a). Major objectives that have driven the development of DNA quantitative approaches throughout history have been raising the informational value of the result as well as increasing the analytical accuracy. Though it is very unique to the field of forensic science where due to the scarcity of sample along with its degraded nature, it becomes of paramount importance that the quantitation and quality assessment techniques used minimize the amount of sample requirement. Thus, DNA quantification approaches have transformed substantially during the last many years, as investigators are constantly refining the pre-existing methods and integrating new technologies.

M. Samuel (✉) · R. Shedge
Department of Forensic Science, National Forensic Sciences University, Tripura Campus, Agartala, Tripura, India

T. Chauhan
Department of Forensic Science, National Forensic Science University Delhi Campus (LNJN NICFS), New Delhi, Delhi, India

A. Puri et al. (eds.), *Fundamentals of Forensic Biology*,
https://doi.org/10.1007/978-981-99-3161-3_15

Keywords

Forensic science · DNA quantification · *Alu* sequences · Polymerase chain reaction · Real-time · Human · Florescence

15.1 Definitions

Deoxyribonucleic Acid: Deoxyribonucleic acid, DNA, is a biomolecule with a double helix structure that carries the sequences of genetic code which form the basis of all life.

Polymerase Chain Reaction: Polymerase chain reaction (PCR) is a laboratory technique for rapidly producing (amplifying) millions to billions of copies of a specific segment of DNA. PCR involves using short synthetic DNA fragments called primers to select a segment of the genome to be amplified, and then multiple rounds of DNA synthesis to amplify that segment.

Short Tandem Repeats: STRs are short tandem repeats located on the telomeric region of chromosome often known as microsatellite. The highly polymorphic regions of DNA repeated 5–50 times are called as the microsatellite. The STRs are short repeats of 1–6 bp and occurred up to 50 times, forming a nucleotide length of 100–120 bp.

Alu Sequences: Alu sequences are a heterogeneous group of primate-specific interspersed repetitive DNA elements with an estimated frequency of 500,000 to 1 million copies per genome.

15.2 Reason for DNA Quantitation

The quantity and quality of DNA in a forensic sample is very crucial, as many other techniques like effective DNA amplification and STR profiling are directly dependent on it. It is very commonly observed that the quantity, quality, and length of the extracted DNA can differ significantly depending on the source and extraction technique. This has thus direct impact on the downstream techniques since the amount of input DNA and the relative length of that DNA can influence the decision as to whether the traditional short tandem repeat (STR) typing, mini-STR typing, or mitochondrial DNA sequencing can be employed as the genotyping approach. Secondly, before using any destructive analytical technique, it is crucial to calculate the total amount of DNA accessible because as it is absolutely vital in forensic analysis to retain as much of the evidence as possible for retesting. Last but not least, the outcomes of preliminary quantitative and qualitative analyses enable a more knowledgeable interpretation of subsequent analytical outcomes. Moreover, this is especially crucial when working on cases where samples must be examined but analyses frequently include challenging specimen that contain minuscule amounts of DNA and have probably experienced environmental stress (DNA degradation). In addition to this, the substrate on which the DNA was deposited has a large degree of impact on both the quality and quantity of DNA (for example, blood on Denim or synthetic fabric). There is also higher possibility of trace substrate chemicals which may also be removed together with the DNA at the source. This could further affect the quantity as well as quality of the DNA. Because of this, the effectiveness and sensitivity of the extraction process are crucial factors in determining whether a given extraction method is appropriate for a given forensic material (Köchl et al. 2005).

15.3 Methods

The DNA quantitation technique can be broadly categorised into two approaches. One of the earliest approaches employed in forensic investigation has been quantitation of the total DNA obtained from the specimen irrespective of its origin. This method was quick but species selectivity was one of the primary drawbacks. Thus, whether the obtained DNA was human or non-human in nature could not be identified. On the advent of PCR methodology, the new approach was employing quantitative analytical techniques for selective identification of the amount of human DNA in a sample. For the longest time, there was just one method for selectively determining human DNA; today, there are multiple options available. The method section includes both non-specific DNA analysis as well as to those that provide human-specific (primate-specific) DNA quantification.

15.3.1 Non-Nucleic Acid-Based Quantitation Methods

Chemical, immunological, and/or microscopic investigations of the evidence is the primary technique used to screen for common biological fluids including blood, semen, and saliva. Even though the findings of these tests are primarily qualitative, they can also give a rough evaluation of the likely amount of DNA per unit area of a sample. Due to variations in cell counts per volume of biological fluid and the likelihood of uneven redistribution of cells during the deposition and drying of the biological fluid on its substrate, these procedures are limited in that the total number of cells that contain DNA within each sample may not always be equal. Also, this technique only superficially analyses the sample and further tests are required for confirmation. For example, microscopic examination of semen for the purpose of spermatozoa detection is a typical test used in forensic investigation. Since the quantity of DNA in each cell is known, this approach can help in estimating the predicted DNA yield from sexual assault evidence. This visual inspection may be used to determine the expected amount of male DNA and confirm the existence of spermatozoa in such a case (Saferstein 2002).

15.3.2 Nucleic Acid-Based Quantitation Methods

15.3.2.1 Spectrophotometric Quantitation

When the species of origin was unimportant, two instrumental techniques—ultraviolet and fluorescence spectroscopy—have been widely utilised to quantify DNA. A relatively straightforward approach known as ultraviolet spectroscopy has been around for a while and is still in use today, frequently serving as a benchmark for other methods. The principle behind it is that majority of biological molecules do not naturally absorb visible light, but they do absorb ultraviolet radiation. DNA's heterocyclic bases which are aromatic in nature absorb in the ultraviolet portion of the electromagnetic spectrum. It is notable that the λmax range of wavelengths for all Watson-Crick bases is from 250 to 280 nm. In double-stranded DNA (dsDNA), absorption of UV light at 260 nm is less than expected from the sum total of the bases present due to the stacking of the bases. They are stacked in such a manner that there is coupling between them and it thus reduces UV absorption. Thus, during denaturation, the UV absorption of a dsDNA sample rises by 20–30% as it becomes single stranded in nature. Samples analysed by UV spectroscopy are analysed at 260 nm and at 280 nm, and the optical density (OD) ratio at 260/280 is used to measure the purity. Samples with a 260/280 nm ratio of 1.8–2.0 are considered pure or relatively free or clean from various impurities. Samples with lesser values should be re-extracted to achieve accurate quantification since they may be contaminated by protein, phenol, or another UV-absorbing co-extracted material. At 260 nm, uncontaminated samples with an OD of 1 in a 1-cm path-length cell contain around 50 μg mL^{-1} double-stranded DNA, or 40 μg mL^{-1} single-stranded DNA or RNA. For applications where samples provide microgram quantities of DNA, UV spectroscopy is a quick and easy technique. The technique's main drawbacks are its inability to distinguish between ssDNA and dsDNA (unless denaturation studies are performed), absorption by RNA and other impurities, sensitivity of the technique, and the larger amount of the sample needed for accuracy (Nicklas and Buel 2003a).

15.3.2.2 Fluorescence-Based Quantitation

Limits of DNA detection using fluorescent-based quantitation techniques could be very low making them more sensitive than UV-based approaches. However, to analyse DNA using fluorescence, a suitable dye that will interact with the DNA must be added. The dye changes its emission spectrum from that in its unbound condition or fluoresces when it binds with the polymer. The fluorescence amplification that happens as a result of the dye's fixed orientation after binding to DNA is what allows intercalating dyes to detect DNA. Certain aspects are required to be taken into account while choosing a dye. Some colours may bind to DNA or RNA similarly or preferentially, whereas others may have a stronger affinity for ssDNA than for dsDNA. The instrument must be also equipped with filters that match or closely correspond to the dye's excitation and emission wavelengths once the dye has been chosen (Gallagher 1994). Many fluorescent dyes have been used to quantify DNA and as new dyes are becoming available greater sensitivity and selectivity in the quantification of DNA is achieved. Single-stranded and double-stranded DNA can be now detected using dyes like PicoGreen and OliGreen, respectively. PicoGreen is able to detect double-stranded DNA concentrations as low as 25 pg mL^{-1}, which is an improvement of almost 400-fold as compared to some of the dyes used previously (Singer et al. 1997). The fluorescence data must be compared to a standard curve that is normally created each time an assay is carried out since, unlike UV spectroscopy, they are relative. These techniques, while more sensitive than UV spectrophotometry, are not human-specific and could overstate the total amount of DNA present due to possible bacterial contamination.

15.3.2.3 Gel-Based Quantitation

The use of agarose gel electrophoresis (AGE) as a qualitative method to assess the size, purity, and integrity of DNA. The method is based on the fact that equal-sized fragments have similar migration paths. When spectrophotometric measurements are unavailable or sample yields are poor, AGE offers an alternative quantitative evaluation due to the correlation between the amount of DNA present and the intensity of dye. In this technique, the sample extract is put via gel electrophoresis process. Although various media can be used, 1–2% agarose is frequently used in gels (Volkin and Cohn 1954). Extracts that need to be quantified are placed into wells with samples having known levels of DNA. Following electrophoresis, the extracts are seen on a short-wavelength UV transilluminator, and comparisons with the standard samples helps quantifying the sample DNA (Barbisin and Shewale 2010). The DNA in gels can be seen using a variety of dyes. The popular is ethidium bromide (EtBr), which is frequently used in the gel and running buffer of the electrophoresis device. Others like DAPI attach to DNA firmly and remain attached to the molecule, and the DNA-dye complex is visualized after electrophoresis. The major drawback of this approach is that it has a lower limit of about 1ng of DNA and thus is not an option for samples with limited amounts of DNA (Sutherland and Shih 1983).

15.3.2.4 Slot-Blot Hybridization Technique

One of the simpler methods for immobilising large amounts of unfractionated DNA on a nylon or nitrocellulose membrane include the slot blotting technique. The relative abundance of the target sequences in the blotted DNA preparations can subsequently be determined using hybridization analysis. The blot produces a hybridization pattern that may be analysed using densitometric scanning. To elaborate it, the technique comprises hybridization of a biotinylated oligonucleotide probe to DNA samples immobilized on a nylon membrane and subsequent binding of streptavidin-horseradish peroxidase conjugate to the arrested biotin molecules. The peroxidase that is indirectly attached to the DNA samples reduces hydrogen peroxide when chemiluminescent detecting reagents are added. This reaction is associated with the oxidation of luminol, and the photons emitted are sensed using autoradiography film. The amount of DNA immobilised on the membrane in each spot is correlated with the size and density of the slots created on the film. Therefore, by comparing the dots made from a series of dilutions of a standard DNA sample, one may estimate the amount of DNA present in the sample extract slots (Saferstein

2002). Because the denaturation of samples is essential to the process, this technology can be applied to DNA that has been extracted using a Chelex. The assay is "human" DNA-specific as the oligonucleotide probe's sequence complements the D17Z1 satellite repeat region, which is only present in higher primates. This measurement technique is sensitive up to 0.1 ng of DNA (Andersen 1998). It is a simple technique and can be easily performed with limited resources. Even so, it was still not as accurate as downstream PCR-based typing techniques. Therefore, even if the quantification method showed that no DNA was present in the sample, the majority of laboratories would still amplify it. Additionally, slot blot data interpretation has been difficult, subjective, and time-consuming.

15.3.2.5 *Alu*-Element-Based Quantitation

To further develop the existing techniques, procedures that did not require sample hybridization on membranes were created which eventually helped in speeding up the quantification process. The AluQuant™ human quantification system is one example of this kind of system (Mandrekar et al. 2001). The high copy number Alu repeats in the human and primate lineages are the focus of the AluQuant assay. A family of repeated elements known as Alu sequences can be found in the human genome in quantities ranging from 500,000 to 1,000,000 copies, making up 6–13% of the haploid genome (Schmid 1996; Mighell et al. 1997). A-rich region connects two comparable monomers that make up the consensus Alu sequence, which is around 280 base pairs long. Alu sequences are thought to be retroposons descended from 7SL RNA, which is present in all species and has undergone significant amplification during the years of primate evolution (Schmid 1996). In this assay, the target DNA is hybridised with an Alu-based probe, starting a series of enzymatic processes that result in the creation of light that may be detected by a luminometer. The test is sensitive down to 50 pg of DNA and has a dynamic range of 0.1–50 ng (Mandrekar et al. 2001). The advantage of the AluQuant method over the slot blot method is that the DNA does not need to be blotted or bound to a fixed surface for this assay. The assay calls for two separate samples of each standard and crime scene sample, which can add to the operator's workload and material requirements while also requiring the use of larger quantity of forensic sample. Additionally, a luminometer is also required for this quantitation method. Alu element-based alternative approaches for the rapid identification and quantitation of human DNA are being now employed that use *Alu* PCR-based system to quantify human genomic DNA from 2.5 to 100 pg (Walker et al. 2003; Nicklas and Buel 2003b).

15.3.3 PCR-Based Quantitation

15.3.3.1 Real-Time PCR

One of the most potent instruments for quantitative nucleic acids analysis which has been made available to the scientific, medicinal, and diagnostic communities has been PCR in real-time. This new method is an improvement on the original PCR, which was created in the middle of the 1980s by Kary Mullis and colleagues. Real-time qPCR with TaqMan or 5′-nuclease fluorogenic assay is mostly commonly used where the amplification has to be monitored in real time. A fluorescent reporter is thus employed to observe the accumulation of amplified products during PCR. The fluorescence signals of the reporter molecule increase as amplified products gather with each cycle of PCR. Real-time qPCR is superior to other approaches in that it operates in a closed-tube system, severely lowering the risk of contamination or carryover, and it generates a linear response proportionate to the amount of input DNA up to 5 orders of magnitude. Other advantages include higher sensitivity with a large dynamic range (30 pg to 100 ng) as well as more precise measurements of smaller quantities of DNA in samples, lesser laboratory manipulations, flexibility towards automation and an ability to sense PCR inhibitors (Singh and Roy-Chowdhuri 2016).

There are commercial qPCR kits available for measuring the amount of Y chromosome and human DNA. In addition, the qPCR technique can be used to quantify mtDNA. qPCR amplifies particular DNA sequences while also measuring their amounts at the same time using commercially available fluorescence-detecting thermocyclers. A sequence-specific fluorescently labelled oligonucleotide probe or an intercalating double-stranded DNA-binding dye can serve as the fluorescent reporter. One of the most commonly used real-time PCR techniques is the TaqMan method.

A TaqMan assay uses standard PCR procedures. The probe, which has a fluorescent molecule (often FAM) on the 5′ end that is quenched by another dye (commonly TAMRA) on the 3′ end, attaches to the PCR product during the annealing phase of the reaction. Through the phenomenon of fluorescent resonance energy transfers (FRET), the quencher significantly lowers the fluorescence given off by the reporter while the probe is still intact. As the PCR primer is extended, this probe is broken down by the 5′ exonuclease activity of Taq polymerase, which allows the fluorophore to become free of the quencher and report fluorescence. More the amount of free fluorophore produced during each PCR cycle results in fluorescence that is directly proportional to the number of cycles and the amount of input DNA. For a given amount of amplified product, fewer PCR cycles are needed if there is the presence of higher initial concentration of target templates in the sample. The initial concentration of target templates can be shown using the cycle threshold (Ct). The cycle threshold (Ct) value is thus defined as the cycle number when the fluorescence of a PCR product can be detected above the background signal. Therefore, the Ct value increases with a decreasing amount of template. A plot of CT against the log10 of the initial concentration of a set of DNA standards will yield the standard curve (straight line) and target sequences in an unknown sample can be measured by comparing it to the standard curve. However, artifacts from the reaction mix or instrument can modify the fluorescence measurements linked with the Ct calculation and this will result in template-independent changes to the Ct value. Therefore, the Ct values from PCR reactions run under different conditions or with different reagents are not compared directly (Pryor and Wittwer 2006).

Further, PCR inhibition can also be quantified by examining the consequence of the inhibitor on the internal control sequence cycle threshold, curves and their characteristics. Variations that modify the slope may exhibit lowering of amplification efficiency and, therefore, a possible inhibition of the DNA polymerase. A shift of the curve to later cycles with no modifications in slope may direct towards competitive template binding inhibition (Lee et al. 2014).

15.4 Conclusion

The chapter describes several techniques for quantifying DNA for forensic studies. These methods were initially centred on the straightforward quantification of total DNA using UV or dyes, regardless of the species of origin. But with the advent of human-specific PCR techniques, exact quantification for STR studies is now possible with sensitivity reported down to picogram levels. Future multiplexing of assays focusses on simultaneous quantification of human nuclear, Y chromosome, and/or mtDNA as well as DNA quality evaluation in a single tube. This is only made possible by the development of new multiplex qPCR techniques which can target several loci of varied sizes. Along with this, the use of internal positive controls has also enabled the measurement of inhibition. All this together has opened up many different possibilities and will lead into generation of a plethora of useful information for the forensic scientist.

Multiple Choice Questions

1. Short fragments of DNA with radioactive tags are known as
   (A) Probes
   (B) Primers
   (C) STRs
   (D) SSRs
   Answer: A
2. Which of the following cells are not useful for DNA analysis
   (A) Spermatozoa
   (B) Vaginal cells
   (C) Red Blood Cells
   (D) Buccal Mucosal cells
   Answer: C
3. The first technique that was adopted for forensic DNA analysis was
   (A) PCR
   (B) STR
   (C) RFLP
   (D) HLA-DQAI
   Answer: C
4. TaqMan assay is a type of method used in
   (A) RT PCR
   (B) End-Point PCR
   (C) Real-time PCR

(D) Alu Quant
Answer: C

5. Ct value stands for
   (A) Cytochrome threshold
   (B) Cycle threshold
   (C) Cycle time
   (D) Cell threshold
   Answer: B
6. UV spectroscopy is a technique used for
   (A) DNA typing
   (B) DNA quantitation
   (C) Species specific DNA quantitation
   (D) DNA profiling
   Answer: B
7. Alu sequences are
   (A) Interspersed repetitive DNA elements
   (B) Plasmid
   (C) Cosmid
   (D) Vector
   Answer: A
8. EtBr (ethidium bromide) which is a carcinogen is frequently used in
   (A) PCR
   (B) Agarose gel electrophoresis
   (C) DNA profiling
   (D) Spectrophotometry
   Answer: B
9. Internal control sequences are used in PCR for examining
   (A) PCR inhibition
   (B) Annealing
   (C) Extension
   (D) Polymerase activity
   Answer: A
10. Streptavidin-horseradish peroxidase conjugate is used in which of the following quantitation technique.
   (A) Electrophoresis
   (B) UV spectroscopy
   (C) Reverse Transcriptase PCR
   (D) Slot Blot assay
   Answer: D

References

Andersen J (1998) Quantification of DNA by slot-blot analysis. Methods Mol Biol 98:33–38. https://doi.org/10.1385/0-89603-443-7:33

Barbisin M, Shewale J (2010) Assessment of DNA extracted from forensic samples prior to genotyping. For Sci Rev 22(2):199–214. https://doi.org/10.1201/B15361-10

Gallagher SR (1994) Quantitation of DNA and RNA with absorption and fluorescence spectroscopy. Curr Protoc Hum Genet 116:1–8. https://doi.org/10.1002/0471142905.HGA03DS00

Köchl S, Niederstätter H, Parson W (2005) DNA extraction and quantitation of forensic samples using the phenol-chloroform method and real-time PCR. Methods Mol Biol 297:13–30. https://doi.org/10.1385/1-59259-867-6:013/COVER

Lee SB, Mccord B, Buel E (2014) Advances in forensic DNA quantification: a review. Electrophoresis 35:3044–3052

Mandrekar MN, Erickson AM, Kopp K et al (2001) Development of a human DNA quantitation system. Croat Med J 42:336–339

Mighell AJ, Markham AF, Robinson PA (1997) Alu sequences. FEBS Lett 417:1–5. https://doi.org/10.1016/S0014-5793(97)01259-3

Nicklas JA, Buel E (2003a) Quantification of DNA in forensic samples. Anal Bioanal Chem 376: 1160–1167

Nicklas JA, Buel E (2003b) Development of an Alu-based, real-time PCR method for quantitation of human DNA in forensic samples. J Forensic Sci 48:2002414. https://doi.org/10.1520/jfs2002414

Pryor RJ, Wittwer CT (2006) Real-time polymerase chain reaction and melting curve analysis. Methods Mol Biol 336:19–32. https://doi.org/10.1385/1-59745-074-x:19

Saferstein R (2002) Forensic science handbook, vol 1. Prentice Hall, Hoboken

Schmid CW (1996) Alu: structure, origin, evolution, significance and function of one-tenth of human DNA. Prog Nucleic Acid Res Mol Biol 53:283–319. https://doi.org/10.1016/S0079-6603(08)60148-8

Singer VL, Jones LJ, Yue ST, Haugland RP (1997) Characterization of PicoGreen reagent and development of a fluorescence-based solution assay for double-stranded DNA quantitation. Anal Biochem 249:228–238. https://doi.org/10.1006/ABIO.1997.2177

Singh C, Roy-Chowdhuri S (2016) Quantitative real-time PCR: recent advances. Methods Mol Biol 1392:161–176. https://doi.org/10.1007/978-1-4939-3360-0_15

Sutherland BM, Shih AG (1983) Quantitation of Pyrimidine dimer contents of nonradioactive deoxyribonucleic acid by electrophoresis in alkaline agarose gels. Biochemistry 22:745–749. https://doi.org/10.1021/BI00273A006/ASSET/BI00273A006.FP.PNG_V03

Volkin E, Cohn WE (1954) Estimation of nucleic acids. Methods Biochem Anal 1:287–305. https://doi.org/10.1002/9780470110171.CH11

Walker JA, Kilroy GE, Xing J et al (2003) Human DNA quantitation using Alu element-based polymerase chain reaction. Anal Biochem 315:122–128. https://doi.org/10.1016/S0003-2697(03)00081-2

Polymerase Chain Reaction: An Indispensable Molecular Biology Tool of 21st Century

16

Parth Sharma, Aditi Mishra, Sarthak Misra, Ulhas Gondhali, and Tanya Chauhan

Abstract

Due to its universal applications, Polymerase Chain Reaction (PCR) became an essential technique in various scientific branches such as microbiology, biotechnology, oncology, clinical studies, forensic science, phylogenetic studies, migration studies, and genetic testing. It can selectively amplify a nucleotide sequence from a complex mixture of sequences with a reproducible and easily operatable protocols in a common laboratory. The modern PCR workflows and systems have advanced to a point where it can amplify a low abundant nucleotide sequence from a single cell and run high-throughput runs using nano well or chip-based systems. In this chapter we are taking a tour that explores the fundamentals of PCR and its components, mode of action of various PCR workflows and systems, foundation of primer designing, and optimizations of PCR protocols.

P. Sharma
School of Applied Sciences and Technology, Gujarat Technological University, Ahmedabad, Gujarat, India

A. Mishra
Kristu Jayanti College, Bangalore, India

S. Misra
Indira Gandhi Institute of Medical Sciences, Patna, India

U. Gondhali (✉)
Jindal Institute of Behavioural Sciences, O.P. Jindal Global University, Sonipat, India

T. Chauhan
Department of Forensic Science, National Forensic Science University Delhi Campus (LNJN NICFS), New Delhi, Delhi, India

A. Puri et al. (eds.), *Fundamentals of Forensic Biology*,
https://doi.org/10.1007/978-981-99-3161-3_16

Keywords

Polymerase chain reaction (PCR) · Primers · Nucleotides · Amplification · Gel electrophoresis

16.1 Definition of Terms

PCR	Polymerase chain reaction
Polymerases	Enzymes that extend nucleotide polymers such as DNA
Primers	Oligonucleotide that selectively binds to a target sequence and marks the start point of amplification for DNA polymerase
Template	Sequence of interest potentially containing a sequence region to be amplified
Amplicon	A target sequence amplified by PCR

16.2 Introduction to Polymerase Chain Reaction (PCR)

Central dogma of molecular biology (CDMB) is the core principle followed by every known life to manifest its features. Therefore, components of CDMB such as nucleic acid (DNA and RNA) can be used to infer sequence-specific features of the source organism via genetic analysis. But recovery of these nucleic acids is dependent on the type and condition of initial samples, as the samples may contain mixed nucleic acid from multiple organisms, and the nucleic acids are highly degradable in open environment. Even if nucleic acids are extracted in high quality and quantity, analysis at genome level becomes expensive and impractical for most of the cases. All above-mentioned hurdles can be overcome and reproducibility of the analysis can be improved by use of the Polymerase Chain Reaction (PCR) technique. Kary Mullis innovated the PCR method in the year 1983 from an idea he had during a weekend drive to his cabin in the woods with his girlfriend for which he was awarded with the Nobel Prize in Chemistry in the year 1993. (Rahman et al. 2013) Polymerase chain reaction in its essence is a protocol used for amplification of the nucleic acid strands at specific loci to exponentially multiply the existing copies for enhanced detection. Due to its wide applications, this technique is used around the globe routinely by scientists in various fields.

Since its birth, the technique has evolved into multiple generations of PCRs including traditional PCR wherein the region of interest in the DNA strand was amplified and the amplicon band was qualitatively visualized on the gel slab of Agarose or Polyacrylamide Gel Electrophoresis, a quantitative PCR (q-PCR) capable of quantifying the amplification in real time, and modern digital PCR (d-PCR) with higher sensitivity and amplification power. There are a lot of advantages of the PCR technique that has made it indispensable because of its ability to reproduce results, rapidly generate large concentrations of DNA for analysis, operate in multiplex samples in a single run easily and economically, and allow dynamic range of run cycles based on the choice and requirement of the researcher. For example, traditional PCRs become common laboratory utility used by researchers, whereas q-PCR gained spotlight during the Covid-19 pandemic where it was considered a Gold-standard for determining if a person was infected with the virus, and if yes, how much was the viral load present in the individual and up to what extent was the individual capable of spreading the virus in the surrounding environment.

With the advantages, there are few limitations associated with the PCR protocol like necessary prior knowledge of the sequences to be amplified, inhibitor contaminations can prevent the amplification process, increased number of cycles or sequence similarity in primer sequences cause generation of non-specific PCR products, and a smaller number of cycles or non-binding primers generate products in non-detectable limits. Non-specific binding also occurs with high GC content in the primer resulting in formation of unnecessary products along with the required amplified regions giving false positive results.

16.3 Components and Factors Affecting PCR

The PCR is executed with a 0.2–0.5 ml reaction tube placed in a thermocycler (machine that can modulate temperature change over time). The reaction mixture includes a target nucleic acid template, short oligonucleotide sequences called primers that selectively bind and assist in amplifying a sequence region between the binding sites from the template, thermostable polymerase with co-factor that can execute the amplification process between bound forward and reverse primer (forward primer attaches to anti-sense strand and reverse primer attaches to the sense strand), free nucleotides that are used by polymerase as a building block to synthesize the amplicon in 5′–3′ direction, and a buffer to maintain optimal pH range of 8.0–9.5. The quality of the amplification can depend upon quantity of template, effective binding of primers to targeted site, type of polymerase used, concentration of co-factors, and presence of any inhibitors (Roux 2009) (Fig. 16.1).

Next, the PCR workflow, a program instructing thermocycler to temporally change the temperature in controlled manner, is divided into three steps: Denaturation, Annealing, and Extension. The denaturation occurs in a temperature range of 94–95 °C, where the unwinding of the double-stranded nucleic acid strands such as DNA occur so that the primers and polymerase can bind to it in the next step of annealing where the temperature should be below the denaturing temperature and compatible with the physical properties of the primers used which can usually range between 50 and 60 °C for the commonly used universal bacterial 16S ribosomal RNA gene primers. The third and final step of the protocol is extension wherein the added dNTPs get attached to the strand to form a complementary strand to the template strand and the temperature range of this step depends on the type of polymerase used for the amplification which is 72 °C for commonly used Taq DNA Polymerase (Joshi and Deshpande 2010).

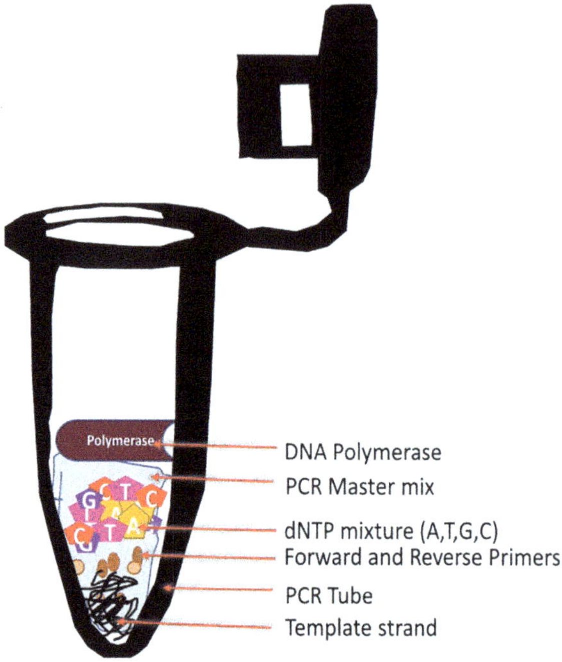

Fig. 16.1 Components and factors affecting reaction rate in PCR

16.4 Types of PCR

With the development of modern molecular tools, PCR technique also advanced to improve its detection sensitivity, detection resolution, amplification length, detection of nucleotide modifications, and its error rates. These advancements are usually attained by incorporation of polymerases with unique properties, use of special primer setups, chemical or physical modification of the template before or after the PCR, and/or use of template or probe binding dyes to measure amplification in real time.

16.4.1 Use of Various Enzymes in PCR

16.4.1.1 Reverse Transcriptase PCR (RT-PCR)

The protocol is based on amplification of RNA sequences by converting the RNA sequences to cDNA form by use of enzyme called reverse transcriptase. Primers for this first PCR step are selected based on the target RNA population to be amplified. For example, if sample source is eukaryotic and the objective is to study mRNA population, then poly-dT primers are used to selectively amplify RNAs with poly-A tail, whereas random hexamer primers are used to arbitrarily amplify RNA molecules from the samples. The cDNA is then subjected to second PCR allowing synthesis of the complementary DNA strand with primers designed based on the cDNA nucleotide sequence and the protocol is carried out using the steps of the standard PCR protocol. One of the applications of the DNA amplified using RNA as a precursor is that the amplified product is devoid of introns. Other applications of this technique include study of indels, genetic disease detection, oncological detection, and detection of expressed genes.

16.4.2 Amplified Fragment Length Polymorphism (AFLP)

In this modification of the PCR protocol, the template strand of interest is cut into fragments via restriction endonucleases that are further ligated with adapters to ensure primer binding prior to the amplification. The fragments created can vary in length based on the distribution of endonuclease-specific restriction sites on the template and hence the amplified products are of varied lengths that are visualized as separate bands on the agarose gel. The protocol is applicable in cases of assessing diversity and determination of and within species level organisms assessing migration patterns, phylogenetic studies, and genetic mapping of organisms (Fig. 16.2).

16.4.3 High-Fidelity PCR

This version of PCR uses proofreading DNA polymerase instead of standard Taq polymerase to improve the confidence of amplification. These high-fidelity polymerases commonly have 3′–5′ exonuclease activity to remove erroneous nucleotide and incorporate the correct nucleotide to lower the PCR error rate at a cost of

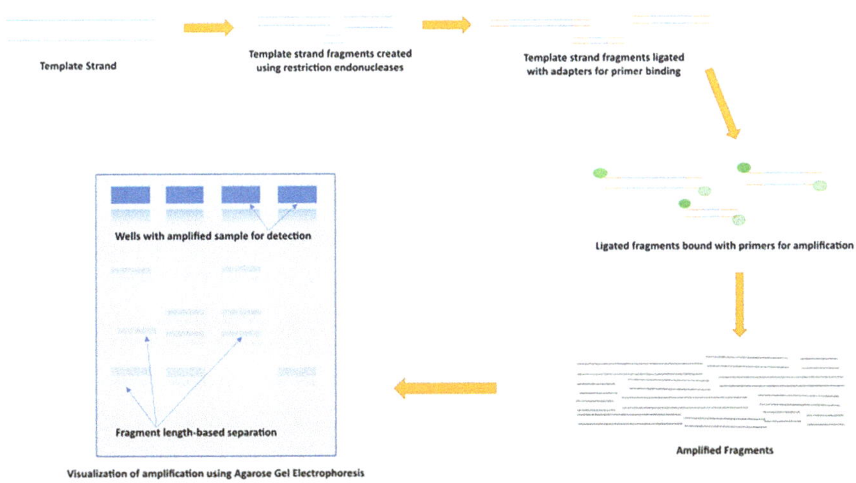

Fig. 16.2 Amplified fragment length polymorphism (AFLP)

longer run time. The high-fidelity PCR is essential for experiments where accuracy of amplification is required such as next-generation sequencing (NGS), SNP analysis, and cloning protocols. Eckert & Kunkel (1991)

16.4.4 Hot Start PCR

In case of challenging PCR protocol where amplification is carried out at room temperature, the reaction may reduce the amplification amplitude by developing primer-dimers. This issue can be solved by using antibody-blocked polymerase that is activated only after denaturation step, hence preventing formation of primer-dimers. In simple terms, this method reduces the non-specific amplification and potentially increases the amplification yield.

16.4.5 Ligation-Mediated PCR

In studies concerning unknown nucleotide sequences, known adapter or linker sequence is ligated to the unknown sequence of interest by DNA ligase before amplification using linker-specific primers. Usually, these adapters are short oligonucleotides and come handy during DNA foot printing, genome walking, and DNA sequencing. In NGS, ligation-mediated PCR is used to index a nucleotide sequence using specific adapter that can be identified and utilized to separate the sequences during post-sequencing analysis.

16.4.6 Fast-Cycling PCR

Fast-Cycling PCR has high speed of amplification due to use of modified polymerases and/or affinity increasing buffers. This PCR protocol can significantly reduce the time of amplification cycles and aid in rapid disease diagnosis in clinical settings.

16.4.7 RNase H-Dependent PCR

RNase H are sequence non-specific endonucleases that cleave DNA/RNA hybrids. These enzymes are used with RNA blocked primers, which can only conduct amplification after successful RNase H activity. As commonly used RNase H are active at higher temperature, this method provides an alternative to the Hot start PCR.

16.4.8 Use of Different Primer Setups in PCR

16.4.8.1 Multiplex PCR

Multiplex PCR allows amplification of multiple target sequences in a DNA strand using multiple primers at different temperature ranges in a single PCR run. The primers used in the protocol are designed in a manner that all the primers have the same annealing temperature. This causes simultaneous attachment of the sequence-specific primers, and a temperature-mediated DNA polymerase is used for the reaction. The protocol is employed wherein simultaneous detection of multiple sequences is required. The protocol is highly benefitting as it provides simultaneous detection and is economic and less tedious. Application of the technique includes genotyping, STR analysis, detection of mutations, indels, polymorphism, and study and detection of disease-causing pathogens and other organisms.(Elnifro et al. 2000)

16.4.9 Allele-Specific PCR

The technique is employed in laboratories for detection of SNPs detection. As the name suggests, the method is efficient in differentiating the alleles present in the sample DNA. The newly modified method in the technique uses two forward primers each containing mismatch of the allele to be amplified. The reverse primer used is common. The two forward primers used are of varied lengths ensuring their separation and identification when visualized on the agarose gel post-PCR. Gaudet et al. (2009) The technique is applicable in detecting single nucleotide variations, blood group genotype determination, etc. (Fig. 16.3).

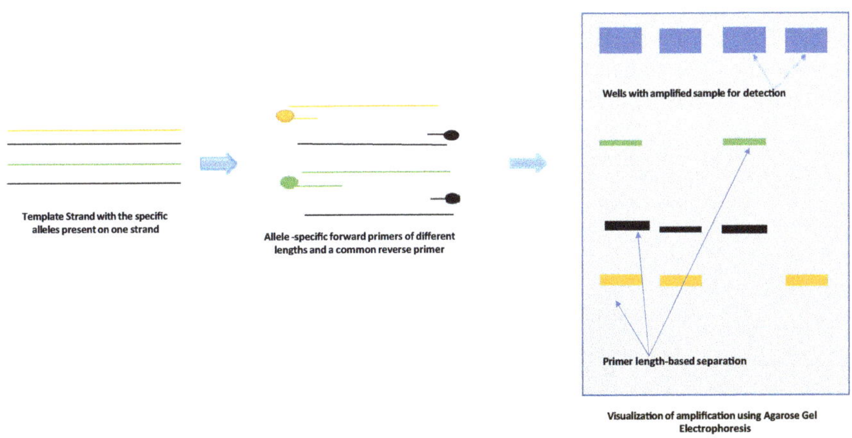

Fig. 16.3 Allele-specific PCR

16.4.10 Asymmetric PCR

In this version of the PCR protocol, there are two primers used, forward and reverse like the regular PCR protocol with the modification being that one of the primers is added in excess amounts. This causes higher amplification of one of the DNA strands as compared to the other. The PCR product generated at the end of the reaction is in fraction as compared to the total product as the total product is inclusive of the Single Stranded DNA (ssDNA) formed due to the primer added in excess. This technique is employed when preferential amplification is expected, and the preferential amplification can be detected using agarose gel electrophoresis or probe-based methods (Fig. 16.4).

A modification to this protocol is the Linear After the Exponential (LATE) PCR wherein the limiting primer uses a higher temperature, and at the start of the PCR reaction, the efficiencies of both primers are equal thereby providing a good reaction efficiency, and upon the complete depletion of the limiting primer, the excess synthesis of the ssDNA begins and carries on for the remaining cycles of the run. Applications of asymmetric PCR include sequencing, mutagenesis studies, and hybridization studies.

16.4.11 Intersequence-Specific (ISS) PCR

As the name suggests, the modified PCR protocol employs use of repetitive sequences present within the desired DNA strands. Primers complementary to these repeating sequences are designed and then used for amplification of the DNA strand. The primers designed are short usually between 26 and 25 bp long ensuring high sensitivity of amplification. This causes the DNA strand to produce a unique fingerprint product as the number of the repetitive sequences is unique for each DNA sequence. The varied applications of the protocol include genomic fingerprinting, phylogenetic analysis, species-specific testing, genetic mapping, assessment of genetic diversity, and gene tagging.

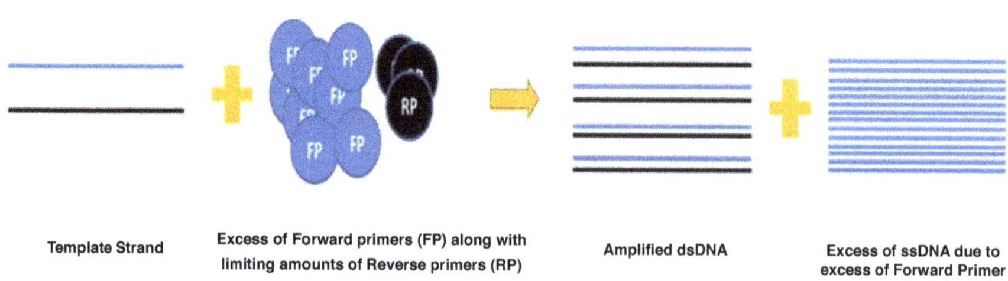

Fig. 16.4 Asymmetric PCR

16.4.12 Mini-Primer PCR

This PCR protocol allows use of primers of short lengths (9–10 nucleotides) by using an engineered thermostable polymerase enzyme specifically designed to carry out the reaction. This allows specific binding of the primer to the regions of the DNA strand of highly conserved nature. This protocol is specifically designed to target 16s rRNA sequences undetectable by using standard primers. Other applications of Mini-primer PCR include targeting smaller primer binding regions and amplification of highly conserved DNA sequences such as 16 (or eukaryotic 18S rRNA genes).

16.4.13 Nested PCR

The protocol is highly sequence-specific. Prior knowledge of the complete sequence of interest to be amplified is essential before employing this method. The method uses two different primers for amplification which are used in tandem. Two PCR runs are required for complete amplification of the product. The first run is identical to the standard PCR protocol wherein the region of interest is amplified. In the second run, PCR primers specific to a secondary region of the amplified product are used ensuring the amplification of only the targeted region. This technique is highly specific and provides high-quality results and can be employed in cases of low initial DNA concentration. Application of this technique includes genetic testing, phylogenetic studies, migration studies, and detection of different pathogens.

16.4.14 RAPD: Rapid Amplified Polymorphic DNA Analysis

RAPD uses around 10 bp long random primers that bind and amplify non-specific sites from the template of interest. This protocol allows molecular exploration of organisms by producing unique band patterns on agarose gel distinguishing between individuals, strains, species, and other higher taxonomic class.

16.4.15 RACE: Rapid Amplification of cDNA Ends

The RACE protocol completes the unknown end of cDNA by using primers binding at small known regions within mRNA. If the protocol is amplifying 3′ end of the mRNA, then it is called 3′-RACE, else it is known as 5′-RACE. Regardless of its type, the complete protocol may amplify the full-length cDNA. As this method uses an anchored sequence within mRNA to complete its end, it is also known as anchored PCR.

16.4.16 Differential Display PCR

Before the development of modern techniques to study differential gene expression between tissues or cells such as RNA-seq, microarrays, and qRT-PCR, the differential display PCR was used to separately amplify mRNA load from the samples in comparison and difference in their agarose gel band patterns was observed to detect differential gene expression. This method is a cheap alternative to selectively study a small number of tissue or cell-specific mRNA expressions.

16.4.17 Unidirectional Genome Walking

In this PCR method, partial sequence information is used to design a primer that can randomly amplify the genome in one direction. It enables charting of unexplored gene and genome region. Because the method can be utilized to complete the linear or circular chromosome, it is also called chromosome walking.

16.4.18 Use of Dyes and Chemicals in PCR

16.4.18.1 Real-Time PCR (q-PCR)

It is the qualitative method for estimating the PCR product formed in real time. The protocol uses the standard PCR techniques with fluorescent dyes that aid in detection and quantification of the PCR product. The process is carried out by adding a fluorescent probe (reporter) along with a quencher into the DNA sequence during the denaturation step. The binding of the reporter and the quencher keeps the reporter from emitting fluorescence. As the target sequence begins to amplify, the bond between the reporter and the quencher breaks causing the fluorescent probe to emit fluorescence. This emitted fluorescence is detected by the detector, and based on the intensity of fluorescence, a real-time graph is plotted of the amplification products. The protocol has received high importance during the Covid-19 pandemic for its ability to detect the SARs-CoV virus. Other application of the technique includes detection of SNPs, genotyping, oncological detection, testing of bacterial and viral loads, and GMO detection. Kubista et al. (2006)

16.4.19 High-Resolution Melt (HRM) PCR

The protocol of HRM PCR is remarkably like q-PCR with the modification being the increased sensitivity of the protocol towards detection of SNPs. This protocol follows steps like the q-PCR wherein dyes are used as probes for detection of the amplification product. Herein, the dyes used are highly saturated and can detect PCR products with very minor temperature variation of 0.01–0.02 °C. Highly concentrated dyes such as LC Green PLUS, Eva Green, etc. are used as opposed to SYBR Green used in q-PCR reactions. Detections of SNPs are through their variation in melting behaviour due to change in their nucleotide composition. Analysis of the PCR product is carried out by studying the amplification and melting curves. To obtain high-quality results, standardization is essential in this protocol. Application of this protocol includes detection of indels, SNPs, and food safety analysis. (Druml & Cichna-Markl 2014)

16.4.20 Methylation-Specific PCR (MSP)

Sodium bisulfite treatment converts unmethylated cytosines to uracil while leaving methylated cytosines in the DNA sequence. To find the methylated cytosine using PCR in known sequence, two sets of primers are designed from which one set binds to unchanged methylated cytosine and another set binds to the uracil representing the unmethylated cytosine. Successful amplification from specific primer set will confirm the cytosine methylation status in the target sequence. It is crucial to conduct a stringent PCR for this method because of non-specific primer template binding resulting from uracil presence.

16.4.21 Advanced PCR Protocols and Systems

16.4.21.1 Digital PCR (d-PCR)

The most advanced generation of the PCR technique, dPCR allows amplification of rare targets. The technique can also be used in case the sample consists of high amounts of inhibitors. The principle of d-PCR is based on q-PCR workflow. Kuypers and Jerome (2017) The reagents and amplification occur in the fashion as that of q-PCR. The detection method is binary based on Poisson Distribution (1,0). The protocol has a high mutation detection rate with mutations less than or equal to 0.001%; the difference in this technique is that d-PCR uses nanoplates in place of the 96 well plates employed for q-PCR reactions.

The protocol for dPCR includes loading master mix into the nanoplates followed by sealing the upper end of the plate using the rubber seal. There are probes present in the thermal cycle that push the samples into the nano wells after which the nano wells are sealed from the bottom separating each well. Each nano partition then undergoes thermal cycles and nanoparticles with the template strand amplify. The templates that amplify emit fluorescence based on the concentration of the amplified template. Precaution to be taken in case of protocol execution is that the initial concentration of DNA template taken for amplification should be low as higher concentrations cause errors in analysis. DNA samples that can be used as template include gDNA, Formalin fixed paraffin embedded DNA, and circulating free DNA.

Applications of the technique include mutation detection, SNP genotyping, assessing Copy Number Variations, Microbiome analysis, and Wastewater RNA determination. Along with all the modifications in the technique, the technique throughout the recent years has miniaturized to make it easily portable and compact. These PCR models available are either Droplet-based systems, Chip-based systems, and Hybrid-systems. Development of these compact models has led to less use of reagents thereby saving experimentation costs and have increased sensitivity and thus increasing the accuracy of the protocol. Also, the time required for carrying out amplification is reduced with the development of the techniques.

16.4.22 Droplet-Based System

Herein, the sample added into the system is in the form of a droplet containing the sample. This droplet is enclosed within a system of mineral oil or an immiscible liquid. For performing the protocol, the sample is first split to form droplets which are then subjected to DNA extraction and purification followed by encircling the dots with the immiscible liquid. These droplets are then subjected to PCR and the amplified product is detected by fluorescence. This technique is beneficial as the sample requirement is in nanolitres and can be specifically used in case of single cell studies or in sequencing. Example of this system is emulsion PCR used by Ion torrent NGS platform where primer coated Ion sphere particles are used to amplify the template DNA in a water-in-oil emulsion droplet before sequencing.

16.4.23 Chip-Based System

Chip-based systems used a chip as a base embedded with microwells or microfluidic channels wherein amplification takes place. These chips are constructed from materials such as silicon, glass, or other polymeric entities. The reaction mixture is added into the chips and the amplification process is carried out within the chips itself. Well-known example of such chip-based amplification system is the flow cell used to conduct NGS by illumina.

16.4.24 Hybrid System

In this system, the reaction mixture placed on a stationary surfaced is covered by an immiscible liquid. The area within the liquid acts as a closed system for the reaction to occur. The necessary temperature ranges for amplification are provided to the system externally by heating and cooling apparatus. The amplified product is visualized using fluorescence-based detection system.

16.4.25 Colony PCR

Colony PCR is heuristic approach carried out directly with addition of intact cells into the reaction mixture to bypassing the separate nucleic acids extraction protocol. (Cavanaugh & Bathrick 2018) This method physically lyses the cells during denaturation step and releases nucleic acids for amplification via standard PCR protocol. Usually, this method is conducted with an isolated colony of bacteria to amplify either 16S rRNA gene or gene present on a plasmid and hence known as colony PCR.

Similar shortcut approach can be applied to other types of cells if it does not inhibit or contaminate the PCR protocol.

16.4.26 Single-Cell PCR

Current benchtop molecular techniques use the total nucleic acids extracted from a mix population of cells which can overshadow the unique features of individual cells or cell type. This limitation can be rectified using advanced cell sorting tools such as flow cytometry which can separate individual cells based on physical appearance, dye tagging, and surface markers. After enrichment of specific cells, PCR with high sensitivity is applied to even amplify a low abundant nucleic acid sequence from the single cell-derived template.

16.4.27 In Situ PCR

This method is used to study the localization of cell-specific nucleotide sequences and markers. Tissue or cells of interests are carefully processed and fixed on a slide or a chip before treating with suitable radiolabelled, fluorescently labelled, or biotin-labelled nucleic acid primers that can release amplification signals as a result of PCR protocol. The intensity of these signals depends upon the pre-processing and fixing of the sample. In situ PCR can be used to assess the disease and development progression in its native state.

16.5 Primer Designing

An ideal primer should strongly and selectively bind with the target region in template only at a specific annealing temperature to promote successful PCR protocol without any non-specific amplification. Therefore, designing of primers requires consideration of template sequence information, primer GC content, primer length, and primers' complimentary and melting temperature difference with those of other primers used in the same reaction. Usually, length of primer depends upon the type of PCR it is designed for such as standard PCR primers are 15–30 bps long, whereas multiplex PCR uses primers in range of 21–30 bps to avoid any inter primer similarities leading towards formation of primer-dimers. Common GC content of standard primers is between 30 and 60% with circumventing presence of multiple C or G in a row at 3′ end. It is crucial to have a strong complementary between 3′ end of primer and the template at a unique sequence region for higher selectivity. The maximum temperature difference between the forward and reverse primer strands should be within 5 °C. Design and selection of best set of primers can be achieved by use of various in silico tools such as Primer3. Elsalam (2003), Dieffenbach et al. (1993), Green and Sambrook (2019), Koressaar and Remm (2007), Nimbkar and D, Bhatt V. (2022). Predesigned commonly used primer sets can also be procured from a manufacturer.

Above-mentioned designing parameters are followed when nucleotide sequence of the template is well-known where all forward and all reverse primers are identical and complementary to the targeted sequence in the template without any cross binding. However, degenerate primers are designed when only amino acid sequence information is available. Degenerate primers are mixture of primers having base change in various positions to offset the degeneracy of genetic code resulting while nucleotide sequence translation to amino acid sequence. Even with mixture of variable primers, only few sets will be complementary with the template and the rest might have similarities with other primers in the mixture, hence amplification conditions should be optimized to minimize the non-specific amplifications with careful designing of these primers. Degenerate primers should avoid any degeneracy in 3′ end to improve template binding efficiency. Degree of degeneracy in these primers can be increased by incorporating selective mismatches toward 5′ end to cover more combination that may successfully amplify the unidentified target from the template. It is advisable to keep degeneracy bellow fourfold at any given position in a primer for better results. Once the desired primers are designed, they should be checked against template sequence of interest using in-Silico PCR before their procurement to verify feasibility of successful amplification and detection of potential non-specific binding sites in the genome or template of interest. This computational step will make the trial-and-error iteration of your primer standardization time- and cost-effective Nybo (2013).

16.6 PCR Optimization

Successful amplification and reproducibility of a PCR workflow depends upon many factors including features of primers, initial amount of template, type of DNA polymerase used, nature of reaction additives, and parameters used for the reaction condition. Results of a PCR protocol are improved by modifying one or more of these parameters such as the denaturation duration is decided based on the GC content of the sample as higher GC content is directly proportional to the time required for denaturation, but longer denaturation steps can also damage the template. However, presence of chemicals such as glycerol and DMSO can lower the denaturation time. For annealing, the temperature range is lower than denaturation and the formula for calculating annealing temperature or Tm or melting temperature is: $4(G + C) + 2(A + T)$. Time required for annealing can vary based upon presence of modified nucleotides and additives within the reaction mixture. Extension efficiency is also known as processivity, and processivity of up to 60 bases/s is considered efficient in case of Taq polymerase enzyme. The extension time required is dependent on the polymerase incorporated and the length of the strand to be constructed. As each PCR protocol is unique based on the components used in reaction mixture and the objective of the protocol, case-specific optimization is required to obtain a robust result.

16.6.1 Optimizing Primer Concentration and Annealing Temperature Using Gradient PCR

First step of PCR optimization is to identify optimal concentration and annealing temperature of designed primer by trying range of primer concentration across range of annealing temperatures. This can be achieved with use of modern gradient temperature-enabled PCRs where the machine creates either horizontal temperature gradient covering 12 different temperature and 8 reaction tubes per temperature or vertical temperature gradient covering 12 reaction tubes per 8 individual temperature setups. Running a single gradient PCR protocol allows assessment of 96 reaction condition from which most optimal condition can be selected for efficient and reproducible future PCR protocols.

16.6.2 Improving Amplification Yield by Modifying PCR Cycles

Second step of PCR optimization is finding an efficient cycling iteration to produce strong amplification without any non-specific contaminations. As single PCR cycle doubles the initial template concentration, the number of cycles is selected according to the initial concentration of template used in the reaction. In case of normal standard PCR protocol, ~30 cycles are run which produce approximately $(10)^{10}$ copies of the initial strands. In case of lower initial template concentration, cycles must be increased by 5–10 iterations to potentially produce detectable level of amplification. Increasing PCR cycles can also improve the amplification signal of low abundance nucleotide sequence from the sample.

16.6.3 Improving Amplification Sensitivity Using Touch Down PCR

In touch down PCR protocol, the annealing temperature of the primers is set in a temperature range wherein the temperature reduces with each cycle of the reaction until the actual annealing temperature of the primer is achieved. This initial high annealing temperature reduces non-specific primer annealing within the DNA sequence. The protocol is used in cases wherein the DNA sequence contains high GC content. Till date, no limitations of the technique have been reported and it is suggested to be employed on regular basis as the standard PCR protocol due to its increased sensitivity. (Korbie and Mattick 2008)

Earlier mentioned "Hot-start PCR" and "RNase H-dependent PCR" can also improve the amplification sensitivity by controlling polymerase activity using weak inhibitor molecules and extra enzymes. However, the Touch Down PCR is simpler and easy to operate as it does not require any extra molecular aid and improve the sensitivity by gradual annealing temperature change.

16.6.4 Optimizing Long length PCR Amplification

Usual PCR reaction amplifies 3–4 kb long amplicons, but longer products are difficult to produce because of secondary structures in template, suboptimal PCR cycling condition, problematic primer binding, and depurination damage. These challenges can be overcome by optimizing cycling condition, buffer additives, and proofreading activity. Depurination damage during long-range amplification results in smear band pattern and can be prevented or minimized by shorting the denaturation time (~10 s) and lowering the extension temperature (68 °C). GC-rich region in template can manifest hairpin loops which can halt the amplification process for long-range PCR. It can be resolved by adding denaturing agents such as DMSO which can change the melting behaviour of DNA opens secondary structures at lower temperature. Use of standard Taq polymerase increases the probability of error in long products and leads to false positive mutations in the products. This issue can be solved by addition of proofreading polymerases in the reaction and improves the efficiency of long-range PCR.

16.7 Conclusion

PCR is one of the benchmark technologies used by forensic investigators in genetic testing. Based on the DNA samples collected from the scene of crime, investigators extract the DNA to determine its source using reference samples of possible suspects. In cases, wherein the concentration of the extracted DNA is low, amplification of the extracted DNA provides the investigator with necessary concentration for analysis. PCR is also a quintessential protocol that can be carried out when working with samples of Mass Disaster Victims. PCR-based paternity testing can also be performed in disputed paternity cases. Along with this, migration pattern analysis, species determination, and phylogenetic analysis can be carried out. Other applications of PCR protocol involve q-PCR-based amplification for biotechnological, biomedical, microbiological, and clinical studies.

Multiple Choice Questions

1. q-PCR is the generation of the PCR technique
   (A) 1st
   (B) Next generation
   (C) 3rd
   (D) 2nd
2. Are q-PCR and RT-PCR terms interchangeable

(A) No
(B) Yes
(C) Inadequate information
(D) None of the above

3. Within the d-PCR workflow, wells are used
   (A) micro
   (B) PCR
   (C) 96
   (D) nano
4. In asymmetric PCR, there is excess formation of
   (A) gDNA
   (B) cDNA
   (C) ssDNA
   (D) Template strand
5. Co-factor essential for extension in PCR is
   (A) Mg^{+2}
   (B) Ca^{+2}
   (C) Na^{+}
   (D) None of these

Answers

1. (D)
2. (A)
3. (D)
4. (C)
5. (A)

References

Cavanaugh SE, Bathrick AS (2018) Direct PCR amplification of forensic touch and other challenging DNA samples: a review. Forensic Sci Int Genet 32:40–49

Dieffenbach CW, Lowe TMJ, Dveksler GS (1993) Genome Res 23:566–573

Druml B, Cichna-Markl M (2014) High resolution melting (HRM) analysis of DNA – its role and potential in food analysis. Food Chem 158:245–254

Eckert KA, Kunkel TA (1991) DNA polymerase fidelity and the polymerase chain reaction. Genome Res 1:17–24

Elnifro EM, Ashshi AM, Cooper RJ, Klapper PE (2000) Multiplex PCR: optimization and application in diagnostic virology. Clin Microbiol Rev 13:559–570

Elsalam KAA (2003) Bioinformatic tools and guideline for PCR primer design. Afr J Biotechnol 2: 91–95

Gaudet M, Fara A-G, Beritognolo I, Sabatti M (2009) Allele-specific PCR in SNP genotyping. In: Komar AA (ed) Single nucleotide polymorphisms: methods and protocols. Humana Press, Totowa, pp 415–424

Green MR, Sambrook J (2019) Polymerase chain reaction. Cold Spring Harb Protoc 2019:436–456

Joshi M, Deshpande JD (2010) Polymerase chain reaction: methods, principles. Int J Biomed Res 5: 81–97

Korbie DJ, Mattick JS (2008) Touchdown PCR for increased specificity and sensitivity in PCR amplification. Nat Protoc 3:1452–1456

Koressaar T, Remm M (2007) Enhancements and modifications of primer design program Primer3. Bioinformatics 23:1289–1291

Kubista M, Andrade JM, Bengtsson M, Forootan A, Jonák J, Lind K et al (2006) The real-time polymerase chain reaction. Mol Asp Med 27:95–125

Kuypers J, Jerome KR (2017) Applications of digital PCR for clinical microbiology. J Clin Microbiol 55:1621–1628

Nimbkar PH, D. Bhatt V. (2022) A review on touch DNA collection, extraction, amplification, analysis and determination of phenotype. Forensic Sci Int 336:111352

Nybo K (2013) Primer design. Biotechniques 54:249–250

Rahman MT, Uddin MS, Sultana R, Moue A, Setu M (2013) Polymerase chain reaction (PCR): a short review. Anwer Khan Mod Med Coll J 4:30–36

Roux KH (2009) Optimization and troubleshooting in PCR. Cold Spring Harb Protoc 4:66

DNA Electrophoresis

17

Aditi Mishra, Pratiksha H. Nimbkar, and Tanya Chauhan

Abstract

Electrophoretic techniques have come a long way since their discovery and show continuous potential for advancement. The basic principle of the technique is the separation of either DNA, RNA, or proteins based upon their charge and size. Concerning DNA electrophoresis, numerous techniques are currently available for one to choose from based on the size and type of the DNA sample to be separated. Today's techniques not only offer separation, but also aid in identification of the DNA sequences, thereby reducing the time required for analysis. Due to electrophoresis being a very versatile technique, it has applications not only in the field of research, but also in various allied fields such as clinical studies, biological, pathological and toxicological laboratories, genotoxicity studies, food safety surveillance, ecological studies, phylogenetic studies, infection control and outbreak investigations as well as forensic investigations to name a few. Conscious and continuous efforts are being put in improving the existing methods by reducing the time required for analysis, making them compact for field work and creating more economical and eco-friendly protocols that can create a sustainable future with minimal waste of the reagents.

A. Mishra (✉)
Kristu Jayanti College, Bangalore, India

P. H. Nimbkar
School of Applied Sciences and Technology, Gujarat Technological University, Ahmedabad, Gujarat, India

T. Chauhan
Department of Forensic Science, National Forensic Science University Delhi Campus (LNJN NICFS), New Delhi, Delhi, India

A. Puri et al. (eds.), *Fundamentals of Forensic Biology*,
https://doi.org/10.1007/978-981-99-3161-3_17

Keywords

DNA electrophoresis · Separation · Application · Identification

17.1 Definition of Terms

Deoxyribonucleic acid (DNA)	The genetic material present inside a cell
Buffer	An aqueous solution that either contains a weak acid or a conjugate base or vice-versa that resists a slight change in pH upon addition of a strong acid or a strong base
Bioanalytes	A biological sample under analysis
Magnetophoresis	A process combining magnetism and electrophoresis
Transillumination	Process of illuminating light through a sample

17.2 Principle

Ferdinand in 1807 was the first to notice that clay particles dispersed in the water remained in a steady state and exhibited no movement. However, upon application of an electric field across the water sample, peculiar movement of clay particles is observed. When the phenomenon was studied, it was understood that the movement was due to the presence of minute charges on the surface of the clay particles. When subjected to an external electric field, the clay particles travelled in the direction opposite to the charge they carried. This observation prompted the development of the modern-day electrophoretic methods. Electrophoresis can be defined as the movement of charged particles to the opposite side of their respective charges when placed in an electric field. The evolution of electrophoresis is a fascinating story. After the observations of Ferdinand, for almost a century, there was very little development in the area. During the 1900s, Tiselius, published his doctoral thesis titled, "The moving-boundary method of studying the electrophoresis of proteins". For his contributions, he was awarded a Nobel Prize in Chemistry in 1948. Tiselius Apparatus Vesterberg (1989) for performing electrophoresis developed around the 1930s and was reinvented around the 1960s to form U-shaped tubes with minimized internal diameter wherein the separation could take place in liquid form. Cellulose acetate gel in electrophoresis was first proposed by Joachin Kohn. Cellulose acetate gels have remained consistently in use since their discovery and are a routine practice for clinical tests such as the determination of sickle cell anaemia (Serjeant et al., 1974). Smithies' innovative experiments inspired by the use of starch gels as a separation medium in electrophoresis. The use of gels provided greater resolution and hence was worked upon for further advancements. In addition to the use of starch, many compounds were tested for their properties of gelling and separation. Acrylamide gels were the first to be developed in 1959 by Orstein, Raymond and

Winstraub with their counterpart, the Polyacrylamide gels by Orstein and Davis and Raymond and Winstraub, presently being utilised for protein separation. Svensson in 1961 developed the isoelectric focusing apparatus through the use of a pH gradient. His principles are applied even in today's 2-D Electrophoresis protocols where the proteins are first separated using 1-D Electrophoresis based on their charge and then via 2-D electrophoresis based on size. The addition of sodium dodecyl sulphate (SDS) by Weber and Osborn to the polyacrylamide gel electrophoresis was made in 1969. The use of two different polyacrylamide gels as a modification to the existing protocols was proposed by Laemmli. Disc electrophoresis, developed by Davis and Orstein in 1964, used discontinuous buffers of different compositions for protein resolution using PAG. Along with acrylamide and polyacrylamide gels, agar and agarose were experimented with to evaluate their potential as gelling media. Both agar and agarose are provided with larger porosity as compared to the acrylamide and polyacrylamide gels. However, as compared to agar, agarose provided more stable conditions for DNA separation. The first recorded use of agarose as a support medium for electrophoresis is around 1950s. This method with time has evolved to accommodate the various needs concerning DNA separation and detection. This book chapter strives to provide a comprehensive summary of the different electrophoretic methods that have developed with their applicability and feasibility.

The applicability of electrophoresis concerning DNA has primarily been to separate and visualise DNA fragments. The protocol has been modified in numerous ways to suit specific needs. Use of DNA electrophoresis in purification, analysis of PCR products, characterisation of enzymes involved in DNA reaction, studying DNA replication intermediates, monitoring genotoxicity, apoptosis, and regeneration of cells, determination of sequencing products, and analysis of cellular biochemical reactions has made the technique a quotidian practice. The basic principle for any DNA electrophoresis protocol is the separation of the different DNA fragments based upon the electric field applied and their rate of migration within a specific matrix. DNA—a negatively charged moiety, when placed in an electric field tends to travel from the anode towards the cathode. When we discuss the kinetics within the gel matrix, it is observed that the separation is not only based upon the size, but is also affected by the gel viscosity (resistance), pH of the buffer, temperature of the system, run time, the orientation of the DNA, and the sample volume.

17.3 DNA Separation Methods and Instrumentation

DNA separation involves the extraction of the DNA from the nuclear membrane into the eluant. The presence of the DNA needs to be confirmed and this is where the electrophoretic techniques come into play. Based on the type of the DNA sample, the length of the DNA strands, and the accuracy requirements, there are a plethora of techniques available for selection with the most basic and economical one being the Agarose Gel Electrophoresis. Along with AGE, its modifications (Pulse Field Gel Electrophoresis, Comet Assay, etc.,) Heintz and Gong (2019) and other electrophoretic techniques such as Capillary Electrophoresis, Microfluidic Chip

Electrophoresis, to name a few, are available for selection. The various techniques along with their instrumentation are mentioned in brief as follows:

17.3.1 Agarose Gel Electrophoresis (AGE)

Customarily, the agarose gel electrophoresis (AGE) protocol is a routine in laboratories for DNA fragment detection. The AGE setup includes a buffer tank, gel casting tray, well comb, an electric power supply, and electrodes. A basic AGE protocol adopted worldwide comprises an agarose solution of the desired concentration (usually 0.8–1.2%) prepared by mixing agarose powder in a buffer and the mixture is heated using a microwave oven for 1–1.5 min in bursts of 30 s. Next, the solution is allowed to cool down before the addition of a visualisation dye. Once the dye is added, the gel is poured into a casting tray and allowed to solidify. On partial solidification, the well comb is inserted which is responsible for the formation of sample loading wells and is removed just prior to sample loading. The cast gel is then removed from the casting tray and placed onto the platform of the buffer tank containing the running buffer to the marked level. A note of caution, while adding the buffer into the tank one should be precautious of not spilling the buffer on the electrodes and to completely immerse the gel within the buffer. This setup is also known as a submarine electrophoresis setup. The previously extracted DNA sample is mixed with a tracking dye and loaded into the wells of the gel. The setup is then supplied with 90 V and allowed to run for 30 min till the tracking dye reaches half of the gel length. The electric supply is then cut off and the gel is removed from the buffer tank and placed in the gel documentation system (Fig. 17.1).

Different laboratories around the world have their own standardised protocols for performing AGE. Once a gel is cast, it can be used numerous times until all the wells present in the gel have been utilised. When the number of DNA samples is less than the number of wells in the gel, the gel can be stored and reused. Herein, the question as to the changes within the gel structure arises when stored for a longer time period. To understand the occurrence of any structural changes within the gel, a study using Nuclear Magnetic Resonance (NMR) as a non-invasive technique was conducted by Descallar and Matsukawa (2020). The results of the study found a linear relationship between the gel storage time period and the tightening of the mesh structure within the gel. The tightening structure forms close aggregates which tend to get more closely bound with time. This altering gel structure has the potential to affect the electrophoretic mobility of the DNA sample under analysis. Reuse of gels is one of the ways by which laboratories try to create more environmentally safe protocols for conducting experiments. Another environmentally friendly working laboratory protocol for reusing gels has been devised by which the used agarose gels can be reused without causing discrepancies in the results. A freeze and thaw method devised Sasagawa (2018) to reduce the harmful effects of using EtBr on the environment involves freezing of utilised gels in a paper or plastic container and freezing it overnight at −20 °C. The next day, the gel is thawed and throughout the thawing process, the container with the gel is kept inverted to separate the water content and

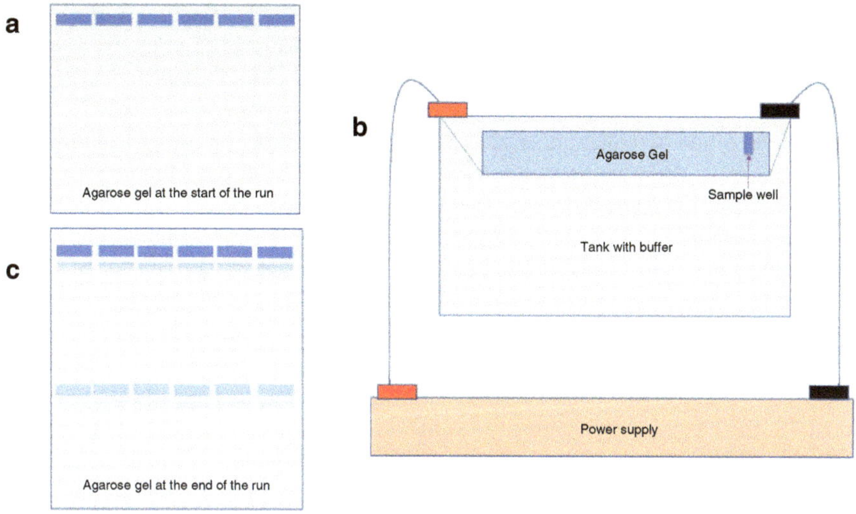

Fig. 17.1 Diagrammatic representation of AGE setup. (**a**) Agarose gel with sample wells and loaded with the sample plus dye; (**b**) submarine electrophoresis setup wherein the gel is completely immersed in the tank buffer and voltage is applied; (**c**) The run of the sample can be tracked by visualising the run of the dye

buffer. First the Buffer along with the dye is eluted. This concentrate can then be discarded through proper channels. The gel once frozen changes its physical structure from gel-like to flake-like resulting in the formation of agar flakes. These flakes can be diluted again by adding 100 mL of water and freezing the gel overnight. The same steps can be repeated a total of 5 times till the gel becomes completely devoid of the nucleic acid and dyes. The flakes then retrieved need to be evaluated for the concentration of agar present within them. Based on the concentration of agar, these flakes can be diluted using a buffer of necessary volume and then heated to get an agarose solution of the desired concentration. The regular protocol of electrophoresis can be performed hereafter with the newly prepared processed gels.

17.3.2 Pulse Field Gel Electrophoresis (PFGE)

AGE can efficiently determine DNA fragments up to a length of 50 kb. For visualisation of DNA fragments larger than 50 kb and up to 10 Mb, the pulse field gel electrophoresis (PFGE) protocol was developed in 1984. Based on the size of DNA fragments, PFGE can be efficiently used to differentiate between bacterial genomes. Considered to be a gold standard in bacterial DNA typing, the technique is used to separate large DNA fragments using low agarose concentration. The method can be practiced for food safety surveillance, vector insertion, phylogenetic studies, infection control, and outbreak investigations (Lopez-Canova et al. 2019).

Out of the different variations of PFGE available, namely, Rotating Gel Electrophoresis (RGE), Transverse Alternating Field Electrophoresis (TAFE), Field Inversion Gel Electrophoresis (FIGE), and Contoured clamped Homogeneous Electric Field (CHEF) gel electrophoresis, CHEF is the most widely used method. RGE utilises one of the two alternating electric supplies at a time. PFGE by Rotating Gel Electrophoresis (RGE) is capable of distinguishing between apoptotic Double-Stranded DNA breaks (DSBs) and DNA damage-induced DSBs. The applicability is appreciated in the fields of toxicology, biology, oncology, immunology, and environmental science wherein the determination of cause is necessary. In TAFE, the alternating electric field is applied transversely to the electrophoresis tank. The angles in the application of the electric field are 115° at the top and 165° at the bottom of the tank. This results in the fine separation of the DNA bands in the second two thirds of the gel and the formation of DNA blurs in the last one third of the gel. In FIGE, the electric field is applied at 180° and the electric charges are altered by inversing the polarity of the applied system. The two methods result in the separation of the DNA bands, but the necessary resolution is not attained. The limitations of the above-mentioned variations are overcome by using CHEF wherein the angle between the alternating electric fields is 120°, which provides a good resolution of the DNA bands. The run time for PFGE is high, wherein a CHEF setup can require 14–16 h of run time for effective separation and visualisation of the DNA strands. To reduce the run time and obtain fragmentation of desired resolution as the large-scale instruments, miniature versions of the PGFE are designed. The miniature versions of PFGE available include TAFE and CHEF PFGE. The instrumentation of the variants remains the same and is just scaled down to accommodate the specifications in smaller dimensions. The most widely and routinely used PFGE variant in the miniature and regular scale is the CHEF gel electrophoresis.

The workflow of the regular PFGE consists of sample cells embedded in agarose solution which are solidified into small gel blocks. These gel blocks are termed plugs. The cells are then subjected to lysis and the DNA extracted from these cells is present within the plugs. Furthermore, the DNA is treated with restriction enzymes to create DNA fragments. The plugs are loaded into the wells of the agarose gel cast for performing AGE. A pulsating charge is applied at a 120° angle between the two charge points to the gel in the PFGE setup. The pulsating charge causes the DNA fragments to travel in a zig-zag travel pattern of the DNA within the gel as opposed to the linear pattern resulting in an increased retention time of the fragments in the gel. Hitherto, the retention time of DNA fragments is directly proportional to their separation within the matrix. Hence, the results obtained are more precise and possess greater resolution. The results of AGE and PFGE are visualised using a gel documentation system (Figs. 17.2 and 17.3).

17.3.3 Supramolecular Gel Electrophoresis (SUGE)

The limitation associated with AGE and PFGE is that both methods use agarose as a medium of gelling and hence are capable of separating DNA fragments. To

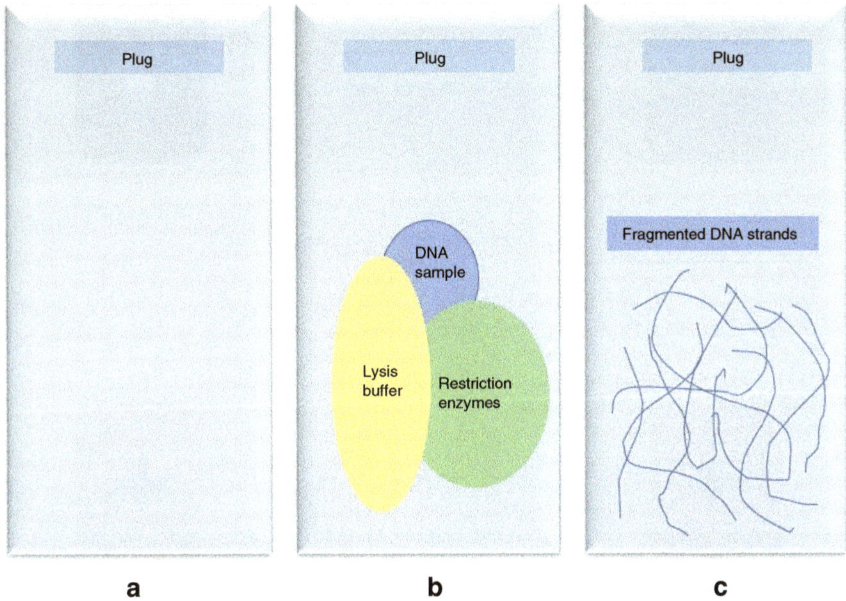

Fig. 17.2 Process of PFGE sample preparation in electrophoresis plug. (**a**) The electrophoresis plug; (**b**) DNA extraction by addition of lysis buffer and treatment of extracted DNA by using restriction enzymes; (**c**) DNA fragments within the electrophoresis plug to be loaded into the PFGE gel

overcome this limitation, a novel alternative to the use of PFGE is devised Yamanaka (2018), i.e. the use of supramolecular gels for performing electrophoresis. Supramolecular gels can be defined as compounds that form highly flexible non-covalent bonds between them, thus behaving like a gel (Chivers & Smith, 2019). There are various compounds used as supramolecular gels. One of them being is composed of C3-symmetric tris-urea, TGS solution having 25 mM Tris, 195 mM glycine, and 3.5 mM SDS (Fig. 17.4).

17.3.4 Comet Assay (CA) or Single Cell Gel Electrophoresis (SCGE)

Analogous to AGE, comet assay or single cell gel electrophoresis (SCGE) resembles the size of the DNA fragments that it can analyse. Employed in ecological biomonitoring and genotoxicity testing, they can determine the presence of denatured DNA and can monitor the effects of genotoxicity within a single cell. DSBs and SSBs can be studied along with cell apoptosis and regeneration. In case of DSBs, visualisation of the damaged DNA is feasible in neutral conditions, whereas for the investigation of SSBs within the DNA, alkaline conditions are necessary for performing SCGE.

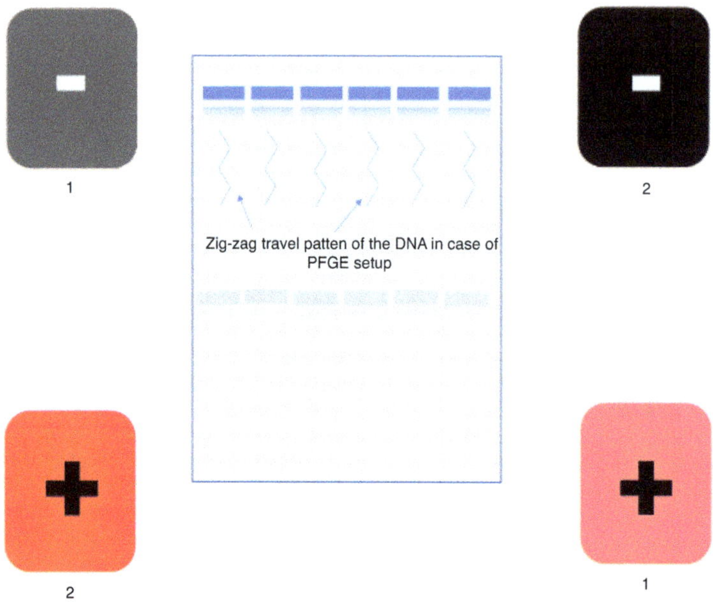

Fig. 17.3 Movement of DNA sample (zig-zag pattern) within the PFGE setup (on application of pulsating charges diagonally making a 120° angle (1 and 2 indicate the sets of pulsating electric charge power supply))

A generalised comet assay protocol entails coating a glass slide with 0.5% w/v low melting agarose gel. The sample of interest (10 μL) along with 0.5% w/v low melting agarose gel (90 μL) is taken and placed onto the slide. Subsequently, the slide is placed in a submarine electrophoresis apparatus and subjected to an alkali electrophoresis buffer for 20 min before the application of the current. The step ensures that any damaged DNA (DSBs or SSBs) present within the cell (if any) gets unwound and can be visualised post the electric field application. The voltage of 0.7V/cm is applied for 20 min. The temperature requirement for the setup is <10 °C. Upon electrophoresis, the slides are placed in a neutralisation buffer for 5 min, dehydrated using 99.6% ethanol and allowed to air dry. Slides are stained using highly diluted SYBR Gold dye and the results are viewed using an image analyser system. Intact DNA with no breaks in between or the undamaged DNA is observed as circular fluorescent spots. However, in the case of damaged DNA due to apoptosis or necrosis leading to damage to the cell's nuclear membrane, the damaged DNA of lengths different from the intact DNA is present outside the nuclear membrane. This damaged DNA flows through the gel matrix based on the varied fragment length giving it a comet-like appearance with a bright fluorescent head and the lengths of the fragments in decreasing order appearing as the comet tail. A cell undergone complete apoptosis/necrosis is observed as a burst entity (Fig. 17.5).

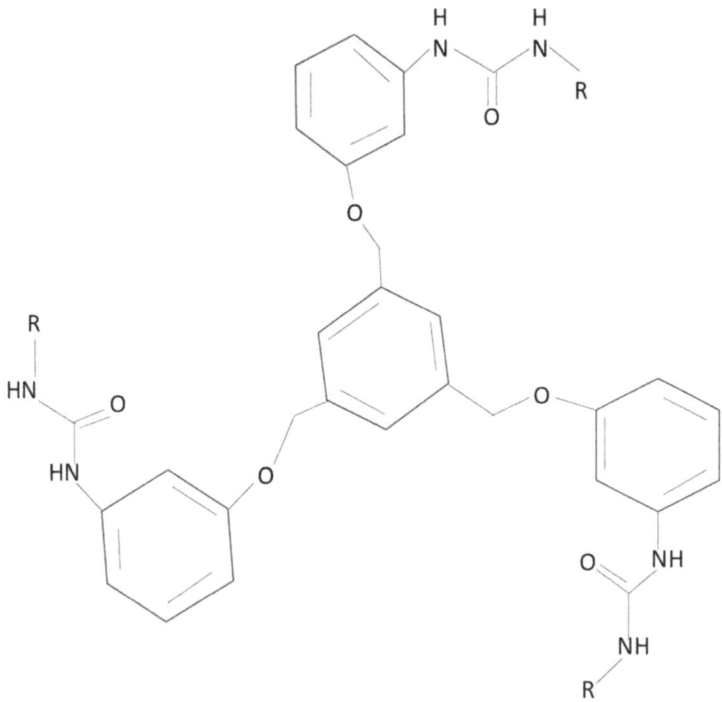

Fig. 17.4 Chemical structure of amphiphilic C3-symmetric

To study programmed cell death (DNA fractionation) and protein fractionation simultaneously, a novel high-throughput mechanism termed Single-cell Electrophoresis-Based Viability And Protein assay (SEVAP) can be utilised. Through the assay, events with respect to the fractionation of the DNA during apoptosis can be visualised along with the cleaved DNA-repair proteins in the same setup. Apoptosis in mammalian cells resulting from the DNA-repair proteins occurring apoptotic mechanisms are fragmented to lengths approximating 50 kbps. Polyacrylamide is used as the gelling medium in the SEVAP assay as it allows only fragmented DNA and proteins to travel within the matrix, keeping the intact DNA as a single entity. The setup is an open microfluidic device on which the results are obtained in a fashion similar to the ones observed in comet assay. In the case of intact DNA, there is no fragmentation of the DNA or the repair proteins observed and only one spot corresponding to the intact DNA observed is identified.

17.3.5 Capillary Electrophoresis (CE)

Developed in 1983, capillary electrophoresis is widely adopted for the separation of DNA, RNA, and proteins. Capillary electrophoresis revolutionised the slab gel

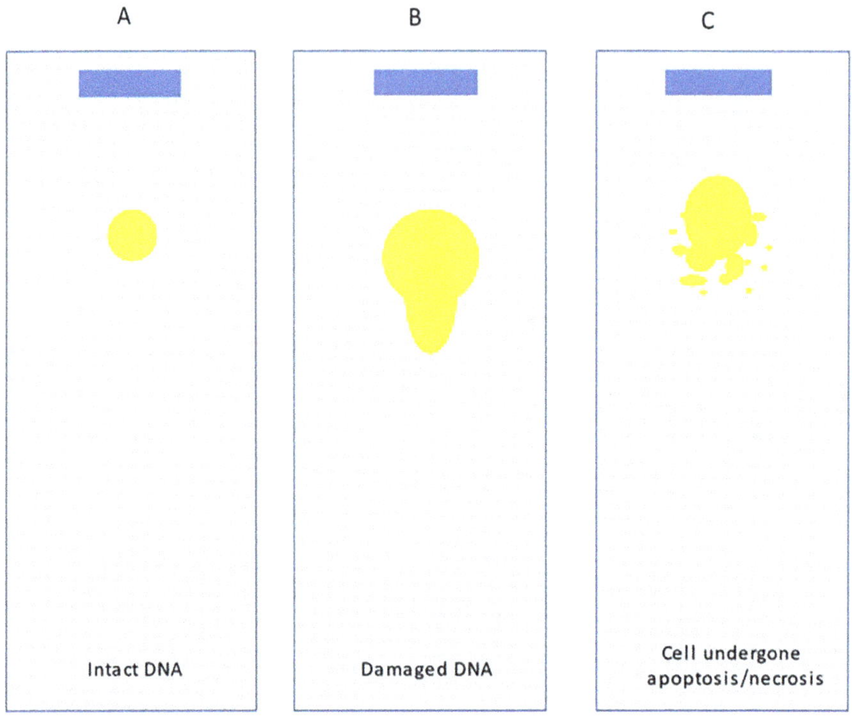

Fig. 17.5 Results observed for Comet assay. (**a**) In case of Intact DNA; (**b**) In case of damaged DNA; (**c**) In case of cell undergone apoptosis or necrosis

electrophoresis technique by bringing automation to the process and providing higher resolution. It is capable of producing high-quality resolution from minute sample quantities. It also provides a more detailed analysis of the sample with applications in various industries such as biopharmaceuticals, clinical analysis, genetic mapping, and forensic analysis. Apart from nuclear DNA, CE can also be utilised effectively for mtDNA analysis in disaster victim identification, maternity testing, and historical reconstruction. The methods of CE used for DNA analysis are either free-zone or gel-facilitated sieving. Free-zone CE has limitations as it can separate DNA-based only on charge and not size. This limitation is overcome through the development of gel-facilitated sieving wherein the pore size of the gel incorporated within the capillaries aids in separating the DNA based on size. For effective separation of the DNA fragments and opposed to agarose or polyacrylamide gels as in traditional electrophoresis, CE utilises soluble polymers. Three mechanisms for DNA separation are observed in CE: free-solution, Ogston sieving, and Reptation. The Ogston mechanism of DNA separation assumes DNA to behave rigidly and the separation of DNA fragments is based on their mobility within the gel as rigid blocks. However, the reptation mechanism assumes that the DNA fragments are flexible and hence they travel through the gel matrix in a flowy manner.

When discussed about the general idea of the instrumentation involved in CE, the setup closely resembles the model of a Gas Chromatography instrument consisting of six parts: the injector, the capillary, the voltage system, two buffer reservoirs, a detector, and an output device (Oorschot and Ballantyne 2013).

The basic CE protocol involves DNA extraction followed by PCR and pre-treatment of the DNA with barcodes and detectors that fluoresce when excited. This pre-treated DNA sample is then loaded into the capillary via the injector. The injector port allows the blend of samples to enter into the capillary tube which has a diameter of less than 100 μm and a length of up to 80 cm. The injection port operates through differences in charge upon applying an electric field when the injection is in contact with the DNA sample. The PCR products inserted into the capillaries if not purified before injecting have the possibility of containing unbound dNTPs and chloride ions in the loading sample. These additions are negatively charged and hence will be facilely transported through the injector into the capillary column. This causes less volume of the required DNA sample to be loaded into the capillary column, thereby interfering with the final results. Subsequently, the sample enters the capillary tubes coated with soluble polymers such as polybrene, PolyE-323, etc.; the sample volume injected in total affects the results in CE. Through a study conducted by Nakazumi and Hara (2018), it was observed that, at the same of the DNA concentration, for sample injection volumes of 10 μL and 20 μL, the sample injection volume of 10 μL provided better results as compared to an injection volume of 20 μL in terms of migration time of DNA and its size, motility, and size and resolution length and size. The factors influencing separation in capillary columns include voltage applied, run time, the buffer used, and the type of the capillary column used.

17.3.6 Microfluidic Chip Electrophoresis (MCE)

The instrumental setup for CE used today has scaled down proportionately leading to the development of Microfluidic Chip Electrophoresis (MCE). It is the miniature version of the traditional CE and can be easily used by researchers as a portable version performing the same tasks as that of CE at an accelerated rate. MCE provides instantaneous results with exponentially low requirements of reagents. The applicability of MCE for biomolecule analysis includes analysis of body fluids and other bioanalytes. A typical MCE setup is ‘T’-shaped, composed of either silicon, glass, or quartz. The latest single-use chips have shifted to using paper and toner for providing users with more economic options. Within the chip, one end of the ‘T’ acts as a sample loading zone, whereas the other end acts as a waste collection area. The stem of the ‘T’ is where the separation of the sample takes place upon application of the electric current. After separation, a detector is added at the end for converting the data to a comprehensible format.

Continuous modification of the existing methods has driven the history of electrophoresis. Despite that, the process of performing electrophoresis has paradoxically evolved and simultaneously retrogressed. Current separation methods involve highly sophisticated instrumentation utilising minute reagent quantities. Yet, in the case of the vehicle used for separation, the time has circled back to the use of paper. The newer ways of electrophoresis embody fibre-based designs to offer features such as portability and handy instrumentation.

The use of fibre-based microchip analytical devices is a branch explored wherein the researchers have tested materials such as resin, wax, and paper as carrier vehicles for DNA separation. A customised microfluidic fibre-based device for DNA separation (μPADs) caters to the research requirements. The setup embodies a wax paper heat-fixed microchannel structure on paper. The ends of the setup are coated with conductive ink to facilitate the flow of current within the system.

Lastly, the coated areas are supplied with voltage via alligator clamps. The paper strips after the run can be eluted in a DNA elution buffer and can then be verified and visualised using agarose gel electrophoresis and can be used for further amplification (Heinsohn et al., 2022).

17.3.7 Magnetophoresis

A modification to the existing MCE suggested by Schneider et al. (2021) incorporates the use of Solid Phase Reversible Immobilisation (SPRI) magnetic beads to remove any impurities within the DNA samples. The modification significantly reduces the number of pipetting steps required in the library preparation for CE or MCE. The steps of DNA fragmentation, end repair, adenylation, and adapter ligation in library preparation are automated via this process, reducing mechanical errors and enabling the researchers to get more accurate results for analysis. The procedure encompasses magnetic properties as well as electrophoresis and hence can prove beneficial. The whole mechanism is incorporated within the microfluidic chip setup wherein the SPRI beads along with the purified DNA travel in the direction of the negative charge applied to the system and the unbound DNA is left on the positive side of the applied electric field. The positive applied electric field is near the sample loading well, ensuring that only the negatively charged DNA with the magnetic beads travels to the other end of the setup. This process aids in DNA purification as only the purified DNA bound to the magnetic beads is obtained when the run is complete. The process mentioned has several potentials for incorporation into DNA sequencing, nucleic acid sizing, and genotyping. Statistically, there is no significant difference in the library preparation done by both methods, i.e. through the use of microchip and manual off-chip method; there is no statistical difference between the values obtained.

The demand for single-use DNA separation kits is growing as they offer faster analysis, are more economic, provide greater convenience to use in the field, and have a compact design with the microfluidic chip electrophoresis pioneering in creating portable devices with low reagent requirements and high-throughput values encouraging future ventures into the subject matter.

17.3.8 Ideal Filter Capillary Electrophoresis (IFCE)

Ideal Filter Capillary Electrophoresis (IFCE) uses DNA aptamers for selective binding and differentiating between binders and non-binders. The setup used is that of CE with the difference between the binders and non-binders being that they travel in opposite directions within the column, leading to efficient separation of the two. Using the method, the separation efficiency of the protein-DNA complexes from the unbound DNA in IFCE is greater than regular CE. The separation occurs due to specific Electroosmotic Flow (EOF) of the running buffer within the capillary tube by altering either the buffer's pH orionic strength. The DNA bound with the proteins has slower electrophoretic mobility as compared to unbound DNA and hence will experience a greater drag force within the gel matrix. This principle acts as a backbone for determining the range of EOF mobility range. For determining the pH range and ionic strength, it has been shown to depict a linear relationship with EOF when experimented with using Tris HCl as a running buffer. Hence, based on the two statements, one can predict the necessary range of EOF required for the separation of the protein-DNA complex and the unbound DNA using IFCE.

17.3.9 Denaturing Gradient Gel Electrophoresis (DGGE)

For analysis of PCR products containing DNA fragments of similar lengths, regular methods of gel electrophoresis cannot efficiently result in DNA band separation and the bands are often observed as an aggregate band representing all the amplified reads. To overcome this limitation, Denaturing Gradient Gel Electrophoresis (DGGE) can be utilised. The technique uses a gel of varying gradients causing a variable porosity vertically throughout the gel. Two gels are prepared of different densities (high denaturing and low denaturing) and are injected simultaneously in between the glass plates. The setup of DGGE is vertical and requires pre-heating the running buffer up to 55 °C before inserting the gel rack. As mentioned previously, gels of two densities (or concentrations) are prepared and then loaded into two separate syringes. Glass plates previously cleaned with detergent and wiped dry are taken and attached tightly to the gel rack. The gel rack is then clamped to the stand with a sponge bottom to hold it firmly in position while the gel is being poured into the setup. The syringes are then attached to the wheel mechanism and the Y tubing is attached with the two arms each attached to a syringe and the stem of the tube being attached to a needle from which the gel is to be poured in between the plates. The wheel is rotated causing the syringes to release the gel starting with the high-density gel and then slowly fusing it with the low-density gel. Once the gel is poured in between the plates, it is allowed to solidify for an hour and the well comb is inserted within the gel to create wells. Upon solidification, the well comb is removed and the gel rack is dismantled from the sponge bottom. It is then attached to the gel rack to be inserted into the buffer tank. The gel rack is inserted into the buffer tank with the high-density gradient gel at the bottom and the low-density gradient gel at the top.

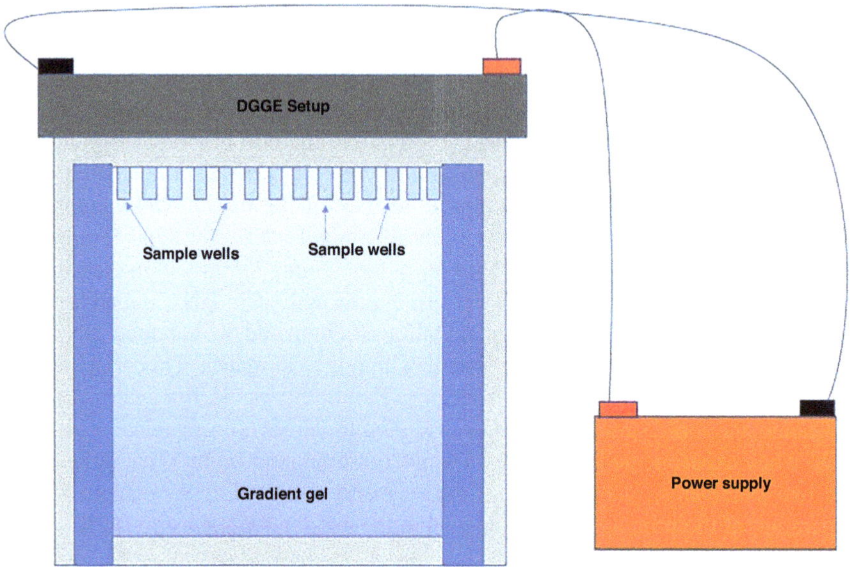

Fig. 17.6 DGGE setup

The tank is then completely filled with the buffer and the samples are loaded within the wells (Fig. 17.6).

17.4 DNA Detection Methods

DNA detection methods provide visual confirmation of the presence of the DNA as well as the presence of the separated DNA strands in case of determination of different DNA fragment lengths using suitable techniques. The three main techniques for DNA visualisation in post-electrophoresis are use of gel documentation system, fluorescence microscopy, and use of fluorescent detectors using CCD.

Specifically, selected dyes are added within the gel (visualisation dye) as well as when the DNA samples are loaded (loading/tracking dyes). Examples of visualisation dyes include Ethidium bromide (EtBr), SYBR Green I, and Thiazole Orange (TO), (O'Neil et al. 2019) whereas bromophenol blue, Orange G, and xylene cyanol FF are the widely used tracking dyes. Visualisation of the DNA bands following electrophoresis is achieved using transilluminating agents such as UV light/blue light. The documentation apparatus is a boxed structure fitted with a tray, transilluminating light source, and camera for imaging. Within the documentation system, options are available for the selection of the visualisation dyes, and once selected, the transilluminator illuminates the gel with UV light and the image with the DNA bands is recorded and can be visualised on the screen. A note of caution while using the gel documentation system, it is crucial to close the gel documentation tray before the start of imaging as the UV light can cause serious health hazards if exposed to it.

The mechanism of DNA band visualisation is via nucleic acid binding dye. The binding of the dye is directly proportional to the amount of DNA present within the extracted sample. The higher the DNA concentration, the greater is the dye binding and the more prominent is its visualisation in the captured image. EtBr is the most commonly used nucleic acid binding dye. The use of EtBr is preferred due to its ease of availability and economic aspects. Nevertheless, it has a few hazardous effects on human health as well as the environment. EtBr is a mutagen and has specific disposal requirements. As a more environmentally friendly and less harmful alternative to EtBr, thiazole orange (TO) can be used. It is less mutagenic than EtBr. Visualisation of DNA bands in the case of TO is possible even in blue light which is advantageous as it causes less damage to the DNA as compared to EtBr.

For visualisation of the DNA in Comet Assay, SYBR Green I dye is used, and post-electrophoresis, the DNA is viewed at 10× using a fluorescent microscope. However, in the case of apoptotic cells, there are several smear-like patterns observed linearly corresponding to the DNA and protein fragments, respectively.

Specialised detectors are used in case of CE and MCE due to their complex instrumentation as compared to the regular AGE setup. At the terminal end of the capillary column, the detector (widely used—fluorescence detector) reads the bases based on the colour of fluorescence emitted and records the data. The sample then gets eluted into the output device. The recorded data are then converted to comprehensible format and further analysed for various purposes such as STR analysis, SNP profiling, or genome sequencing. Detectors in CE and MCE primarily are UV/Visible detectors or fluorescence detectors. The fluorescence detectors offer more sensitivity. Laser-Induced Fluorescence (LIF) is employed in both CE and MCE as it offers sensitivity to 10^{-10}–10^{-12} M. LIF has other advantages such as MS CE and MCE both offer parallel analysis of multiple samples and hence their usage has become coherent in forensic analysis.

17.5 Conclusion

The evolution of any technique is a continuous process and can be witnessed with the changing design and complexity. For electrophoresis, the design has come a long way incorporating many variations throughout the way. The modern electrophoresis setups still pay heed to the initial designs by incorporating its simplicity, but have evolved in a manner to provide high-resolution results necessary to keep up with current requirements. The DNA electrophoretic techniques used today not only confirm the presence of the DNA, but also allow the researchers to determine its size, sequence, and the nature of the separated DNA. These additions have made the use of electrophoretic techniques a regular in many different analyses. Further evolution of these techniques will primarily revolve around making the techniques more economical, more compact, and never the least eco-friendly.

Multiple Choice Questions

1. The fragment size that can be analysed using PFGE is
   (A) Between 50–70 Mb
   (B) Between 50 kb to 10 Mb
   (C) >50kb
   (D) None of these
2. Results in Comet Assay for damaged DNA with DSBs/SSBs are observed as
   (A) A comet-shaped appearance
   (B) Burst entity
   (C) A circular fluorescent spot
   (D) All of the above
3. The instrumentation setup of Capillary electrophoresis resembles that of a
   (A) HPTLC
   (B) UV-Vis Spectrophotometer
   (C) ICP-MS
   (D) GC
4. Detectors in CE and MCE are usually
   (A) Fluorescence and UV/Vis detectors
   (B) Gel documentation system
   (C) UV transilluminator
   (D) Fluorescence microscope
5. A more environmentally friendly alternative to EtBr is
   (A) SYBR green I
   (B) Bromophenol blue
   (C) Thiazole orange
   (D) FastBlue B

Answers

1. (B)
2. (A)
3. (D)
4. (A)
5. (C)

References

Descallar FBA, Matsukawa S (2020) Change of network structure in agarose gels by aging during storage studied by NMR and electrophoresis. Carbohydr Polym 245:116497. https://doi.org/10.1016/j.carbpol.2020.116497

Heintz N, Gong S (2019) Examination of bacterial artificial chromosome (BAC) DNA quality and quantity by pulsed-field gel electrophoresis. In: Cold spring harbor protocols. Cold Spring Harbor Laboratory Press, New York, pp 746–747. https://doi.org/10.1101/pdb.prot098020

Lopez-Canovas L, Martinez Benitez MB, Herrera Isidron JA, Flores Soto E (2019) Pulsed field gel electrophoresis: past, present, and future. In: Analytical biochemistry, vol 573. Academic, New York, pp 17–29. https://doi.org/10.1016/j.ab.2019.02.020

Nakazumi T, Hara Y (2018) Influence of the sample volume upon injection on capillary gel electrophoresis for the separation of small DNA fragments. Int J Chem Eng Appl 9(1):16–19. https://doi.org/10.18178/ijcea.2018.9.1.692

O'Neil CS, Beach JL, Gruber TD (2019) DNA electrophoresis using thiazole orange instead of ethidium bromide or alternative dyes. JoVE 145:59341. https://doi.org/10.3791/59341

Ou X, Chen P, Huang X, Li S, Liu BF (2020) Microfluidic chip electrophoresis for biochemical analysis. J Sep Sci 43(1):258–270. https://doi.org/10.1002/jssc.201900758

Sasagawa N (2018) A freeze-and-thaw method to reuse agarose gels for DNA electrophoresis. Biosci Trends 12(6):627–629. https://doi.org/10.5582/bst.2018.01267

Schneider L, Fraser M, Tripathi A (2021) Integrated magneto-electrophoresis microfluidic chip purification on library preparation device for preimplantation genetic testing for aneuploidy detection. RSC Adv 11(24):14459–14474. https://doi.org/10.1039/d1ra01732b

van Oorschot RAH, Ballantyne KN (2013) Capillary electrophoresis in forensic biology. In: Encyclopedia of forensic sciences, 2nd edn. Elsevier, Amsterdam, pp 560–566. https://doi.org/10.1016/B978-0-12-382165-2.00242-7

Vesterberg O (1989) History of electrophoretic methods. J Chromatogr 480:3–19

Yamanaka M (2018) Supramolecular gel electrophoresis. Polym J 50(8):627–635. https://doi.org/10.1038/s41428-018-0033-y

Short Tandem Repeats Profiling

18

Tanya Chauhan, Shreya Arora, Rutwik Shedge, and Astha

Abstract

DNA is the blueprint of life, found in almost every living organism, and working as a set of instructions governing every aspect of the functioning and composition of each individual. While most of the DNA shared by organisms of a species is identical, certain locations of DNA exhibit variations in sequences. The analysis of these locations or loci, with variations in DNA sequences, serves as the basic principle behind individual identification through DNA profiling. One commonly practiced method of individual identification using DNA profiling of mapping variations in the length of repeated sequences is the Short Tandem Repeats Profiling. Short Tandem Repeats (STR) have played a significant role in advancing the applications of DNA science in forensic individual identification. The discovery of Short Tandem Repeat (STR) DNA and its application in forensic science initiated the implementation of the DNA profiling process to create databases of DNA at a national level (Gill et al., Forensic Sci Int Genet 18: 100–117, 2015).

T. Chauhan (✉)
Department of Forensic Science, National Forensic Science University Delhi Campus (LNJN NICFS), New Delhi, Delhi, India

S. Arora
CTM IRTE, Faridabad, Haryana, India

R. Shedge
Department of Forensic Science, NFSU Tripura, Agartala, India

Astha
LNJN National Institute of Criminology and Forensic Science, Delhi, India

A. Puri et al. (eds.), *Fundamentals of Forensic Biology*,
https://doi.org/10.1007/978-981-99-3161-3_18

Keywords

DNA · STR · STR genotyping · PCR amplification · CODIS

18.1 Definitions

STR: Short Tandem Repeats (STR), also known as Simple Sequence Repeats (SSR), are short-length sequences of DNA repeated multiple times within a genome.

STR genotyping: Process of identification of number of short tandem repeats present at any particular locus in an individual's DNA.

PCR: Polymerase chain reaction is a process of amplification of a specific DNA sequence.

CODIS: Combined DNA Index System is a DNA database of the Federal Bureau of Investigation (FBI) of the Government of the United States of America.

18.2 STR Markers

Short Tandem Repeats, commonly called STR or microsatellites, are short-length sequences of DNA repeated multiple times within a genome. These variations are found at different locations within the non-coding region of the genome and constitute approximately 3% of the whole human genome. The short tandem repeats are one of the most polymorphic loci in the human genome (Willems et al. 2014) with the number of repeats varying in each individual. Short tandem repeats (STRs) are sequences of DNA where two to seven nucleotides long core repeat units are tandemly repeated about a half dozen to several dozen times (Butler 2005). The most commonly used STR loci in forensic science are 100–500 base pairs in length. The repeats are present between flanking regions, a sequence of DNA that is non-variable or constant. These constant regions play an important role in the identification of these sequences by primers used in PCR amplification during STR profiling.

There are various advantages of using STR loci as compared to VNTR loci which include better amplification of loci and even reduction of preferential amplification. Also, the profiling of mixed samples and degraded DNA samples will be carried out with improved profiles. The resolution of electrophoretic separation, as well as result analysis, is much better in STR.

The number of nucleotides in a single repeat unit is called the repeat unit length. STR nucleotides can range from being two to six in number (Fig. 18.1). In dinucleotide repeats, only two nucleotides repeat over and over again. Trinucleotides have three nucleotides, while tetranucleotides have four and so on in the core repeat unit length. The tetrameric repeats constitute approximately 9% of the total STRs in the human genome. Though the pentameric and hexameric repeats are highly polymorphic, they are utilized in forensic applications in less number as they are less abundant in the human genome as compared with the other set of repeats.

Fig. 18.1 Types of repeat unit length in the human genome

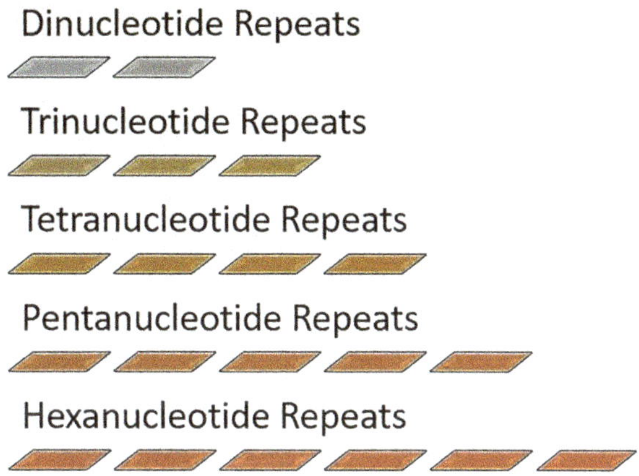

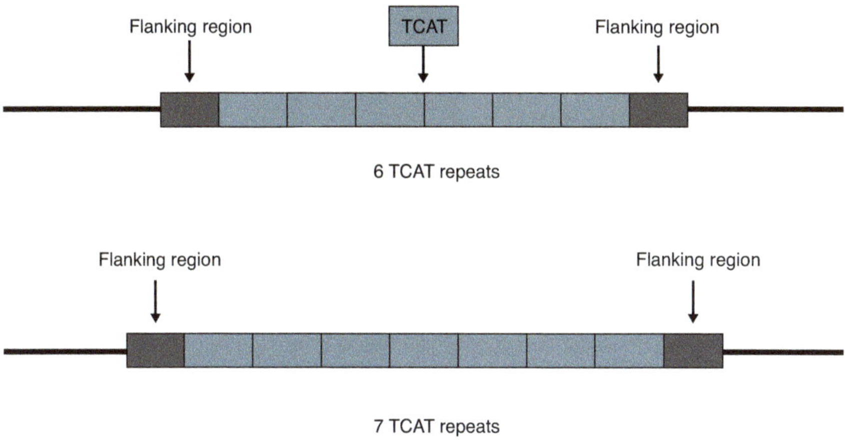

Fig. 18.2 Short Tandem Repeats (STRs)

The STR has multiple variants or alleles, represented by the number of repeats of the DNA sequence present. Individuals get a set of chromosomes from each of their parents, and depending on the sequence they inherit, an individual can be homozygous or heterozygous for an STR allele. Homozygous have a set of alleles of equal lengths for loci, while the length differs in heterozygous conditions. STR profiling techniques use specific markers to aid in individual identification through a Polymerase Chain Reaction- (PCR) based DNA profiling method (Fig. 18.2). The STR profiling system analyses multiple loci to provide a high power of discrimination to the result. A commonly used STR system is the 13 tetrameric short tandem repeat (STR) loci system used by CODIS or the Combined DNA Index System database. The FBI selected this set of markers in November 1997 (Butler 2005). It has been used for determining parentage, forensic identification, and for medical diagnostics.

18.3 STR Genotyping

STR analysis involves DNA extraction from evidence, quantitation of DNA, subjecting extracted DNA samples to a multiplex polymerase chain reaction (PCR), separation of alleles and sizing, typing of STR loci, and interpretation on profile interpretation (Fig. 18.3). In some cases, the statistical significance of the match is also reported (Evett et al. 1996).

The genotyping process begins with subjecting the DNA extracted from evidence to a sequence of methods to measure the allele repeats at specific STR loci. It involves multiplex PCR amplification, capillary electrophoresis, and genotyping. Polymerase chain reaction yields products from a minimum quantity of biological evidence in form of different nucleotide lengths based on the number of short tandem repeats. STR profiling uses Multiplex PCR where multiple STR loci are amplified simultaneously using different coloured dyes and different length DNA sequences.

This method allows for performance analysis with a small account of DNA material; 1 ng or less (Butler 2005).

18.3.1 Sex Identification

The ability to identify whether a sample is obtained from a male or female source is very useful as well as important especially in case of sexual assault cases, where it is very important to distinguish between the accused and the victim. Similarly, sex identification is beneficial in cases of mass disaster as well as missing person investigations. There are various markers now to distinguish between a male and female sample, but the most commonly used marker is Amelogenin.

Amelogenin is a gene that codes for the proteins that are present in tooth enamel. Forensic laboratories have a set of primers to detect this locus in any biological sample collected. The PCR amplification of this generally results in 106 bp and 112 bp DNA lengths from the X and Y chromosomes, respectively. Hence, in the case of female, one peak of 112 bp marker is observed, while in the case of male two peaks of 106 and 112 bp are observed in an electropherogram due to their homozygous and heterozygous condition, respectively.

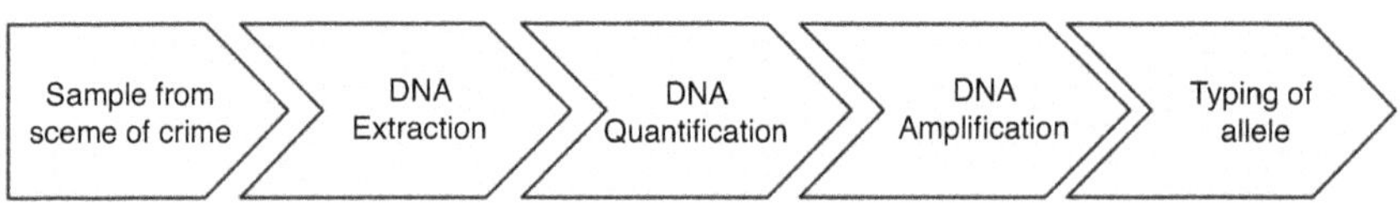

Fig. 18.3 Schematic representation explaining the stages of DNA genotyping

18.4 STR Kits

The pieces of evidence encountered at the scene of a crime are commonly degraded due to environmental exposure and decay. STR analysis works with short-sized base pair DNA sequences, which favours the possibility of getting information from degraded DNA obtained from such evidentiary material. STR genotyping is also performed using commercially available kits with a standardized mix of components such as primers, DNA polymerase, enzyme buffers, and Deoxynucleoside triphosphate (dNTP) for uniform and accurate results along with allelic ladders for genotyping. There are a number of kits available commercially for single or multiplex PCR amplification of STR markers for DNA fingerprinting. Commercial kits also help standardize the procedures for performing analysis. For example, in a 13 loci STR system, the likelihood that any two individuals, except identical twins, have the same DNA profile can be as high as 1 in 1 billion or greater. The adoption of 13 STR markers started in the USA which eventually led to the development of STR markers multiplexes. This technology evolved in the late 1990s for being more sensitive and rapid extents of STR alleles. Also, STR markers can be amplified simultaneously which makes it more favourable than other techniques.

18.4.1 CODIS

One of the most prominent application of STR typing for forensic purposes is CODIS. CODIS or the Combined DNA Index System by the FBI (Federal Bureau of Investigation) was developed with the aim to provide assistance to law enforcement agencies in the investigation of cases where DNA evidence can provide potential information to identify those involved in the crime. The program uses STR-based loci for identification and the criminal justice DNA databases. The program comprises DNA profiles of the alleles at one or two of the 20 CODIS Core Loci, collected from the participating forensic laboratories across the states. The 20 CODIS loci used are CSF1PO, FGA, THO1, TPOX, VWA, D3S1358, D5S818, D7S820, D8S1179, D13S317, D16S539, D18S51, D21S11, D1S1656, D2S441, D2S1338, D10S1248, D12S391, D19S433, and D22S1045 (Budowle et al. 1999). The chromosomal location and repeat motifs of each of the aforementioned markers are shown in Table 18.1.

The use of CODIS for investigation begins with the collection of DNA samples from the scene, suspect, or victim of a crime. The DNA extracted from such a sample is treated as an unknown profile that is then searched against the DNA database. Any match is confirmed by the laboratory, and based on the result, further investigation is carried out.

Table 18.1 The 13 STR loci involved in STR profiling

STR loci	Chromosomal Location	Repeat Motif
CSF1PO	5q33.1	AGAT
FGA	4q31.3	CTTT/TTCC
THO1	11p15.5	TCAT
TPOX	2p25.3	AATG
VWA	12p13.31	TCTA/TCTG
D3S1358	3p21.31	TCTA/TCTG
D5S818	5q23.2	AGAT
D7S820	7q21.11	GATA
D8S1179	8q24.13	TCTA/TCTG
D13S317	13q31.1	TATC
D16S539	16q24.1	GATA
D18S51	18q21.33	AGAA
D21S11	21q21.1	TCTA, TCTG
D1S1656	1q42	TAGA
D2S441	2p14	TCTA/TCAA
D2S1338	2q35	TGCC/TTCC
D10S1248	10q26.3	GGAA
D12S391	12p13.2	AGAT/AGAC
D19S433	19q12	AAGG/TAGG
D22S1045	22q12.3	ATT

18.5 Challenges of Genotyping for Forensic Samples

STR genotyping involves subjecting extracted DNA samples to a polymerase chain reaction (PCR) and automated sequencers. The genotyping process begins with subjecting the DNA extracted from evidence to a sequence of methods to identify the allele repeats at specific STR loci. It involves multiplex PCR amplification, capillary electrophoresis, and genotyping. Polymerase chain reaction yields products in the form of different nucleotide lengths based on the number of short tandem repeats.

In forensic science, the samples found at the scene of a crime can be highly variable according to the environment and the method of collection and preservation adopted. Hence, the end of highly sensitive and reproducible techniques arises. The following are the challenges faced by the experts in terms of samples.

18.5.1 Degraded Samples

The biological evidence with high environmental exposure including high humidity and temperature can lead to DNA degradation and eventually lead to DNA fragmentation. The more the degradation, the more the fragmentation. This is one of the major reasons that in forensic science a shift was observed from VNTRs to STRs.

Since after fragmentation the larger alleles tend to break into smaller strands, the chances of STRs amplicons which is 100–500 bp in length remain safe. This also leads to incomplete or failure in the generation of genetic profiles. To overcome this, issue amplification was introduced after extraction so that the alleles can be multiplied multiple folds for better detection.

Wiegand and Kleiber (2001) demonstrated that when the DNA is highly degraded, it indicates a very less amount of DNA present. Here, the primers can be redesigned to a closer region of STR core repeat as compared to the other longer amplicons. They are known as mini-STRs. The major disadvantage of this mini-STR is that very few loci can be simultaneously amplified as the size aspect is considered. The chances of having an allele dropout also increase.

18.5.2 PCR Inhibition

The other challenge in profiling an individual from the sample collected at a crime scene is that PCR amplification process is affected by the inhibitors present either inside the sample or collected while collecting the sample. In outdoor cases, body fluids are present on soil, sand, or even dyed cloth pieces. These inhibitors hinder the activity of cell lysis in the initial steps of DNA extraction, can also capture and degrade the nucleic acid extracted, and even inhibit the DNA polymerase enzyme during PCR (Rådström et al. 2008). These inhibitors result in partial profiling in most cases as they fail to amplify the large STR loci.

There are various strategies proposed to overcome these inhibitions which include dilution of the sample extracted which eventually dilutes the inhibitors also (Rådström et al. 2008), and the addition of bovine serum albumin (BSA) or betaine. Also, the use of sodium hydroxide treatment has neutralized the inhibitors on various occasions including the use of filters specifically made for this only (Moreira 1998).

18.5.3 Low Copy Number DNA Testing

DNA analysis in forensic science requires the testing of very fewer amounts of DNA in a sample collected which could be as less as 100 pg. This type of DNA is collected from fingerprints and specifically touch DNA. But these low levels of DNA extracted are further amplified using a PCR machine to increase the number as well as the sensitivity of the analysis.

18.5.4 Mixed DNA Profiles

In forensic science, we encounter various types of cases including sexual assault cases in which the evidence recovered from the victim contains a mixed sample of both accused and the victim. Here the profile generated is of both the contributors.

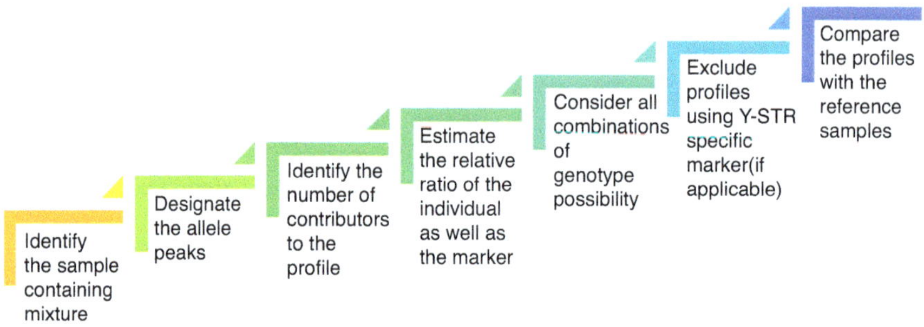

Fig. 18.4 Schematic steps for the interpretation of mixed genetic profiles (Clayton et al. 1998)

The steps involved in interpretation of mixed STR profiles are shown in Fig. 18.4. The greater the number of contributors, more difficult the interpretation of STR profiles become. One commonly encountered forensic case with this issue is a sexual assault case (Laberke et al. 2012).

18.6 Factors Affecting Genotyping Results

There are a number of factors that may affect the accuracy of genotypic profiles.

18.6.1 Mutation and "Off-Ladder Allele"

The selection of STR loci in forensics is also based on the mutation rate of that region. Hence, loci with low mutation rates are preferred especially for the identification of individuals in mass disasters and paternity cases. Rare alleles are also encountered in the human population which may differ from common or wild types. However, in some cases mutation occurs, which eventually affects the profiling results. And if due to the sequence variations observed there is a difference between the consensus alleles already present as a reference, they can be referred to as 'off-ladder allele' (Cupples et al. 2009).

18.6.1.1 Mutation at STR Core Repeat Regions

This mutation occurs in the core region of STR loci. Mutation can further occur in two types of cells in the body. Germ cell mutation is where mutation occurs in the germ line of the individual and will be inherited by the offspring. This type of mutation helps in the characterization of loci in a population study. However, somatic mutation means mutation occurs in the somatic cells of the individual and will be inherited by the offspring of the individual.

Also, the sequence variation due to insertion, deletion, or substitution is compared and is called ad microvariants.

18.6.1.2 Gene and Chromosome Duplication

A duplication-bearing mutation within the STR loci core repeat region can affect the number of tandem repeat units. In wild conditions, there are two alleles inherited from each parent. But in special conditions either three alleles or three chromosomes (trisomy) often affect the STR profile associated with the concerned locus or chromosome, while the unaffected do not hinder in profiling (Moreira 1998).

18.6.1.3 Point Mutations

These types of mutations include a change of one single nucleotide due to deletion/insertion or substitution. They can be a part of flanking region or the core repeat region on STR loci. If the nucleotide mutation occurs in the flanking region, then the primer binding region is affected which eventually causes non-specific binding of primer that reduces the efficiency of binding and/or amplification will be failed. The complete failure of the gene amplification is called a silent allele or null allele (Clayton et al. 2004).

18.6.2 Amplification Artifacts

18.6.2.1 Stutter Products

In this, a minor allele peak is observed in an electropherogram which is known as a stutter peak (Fig. 18.5). The repeat units are either shorter or longer than the parent allele. They are typically less than 15% of the genuine allele peak height. This phenomenon is observed infrequently in pentameric and hexameric repeat unit loci as compared to the other types.

It is also believed that stuttering is observed due to the slippage of polymerase occurring during amplification. The stutter peaks longer than the parent allele that

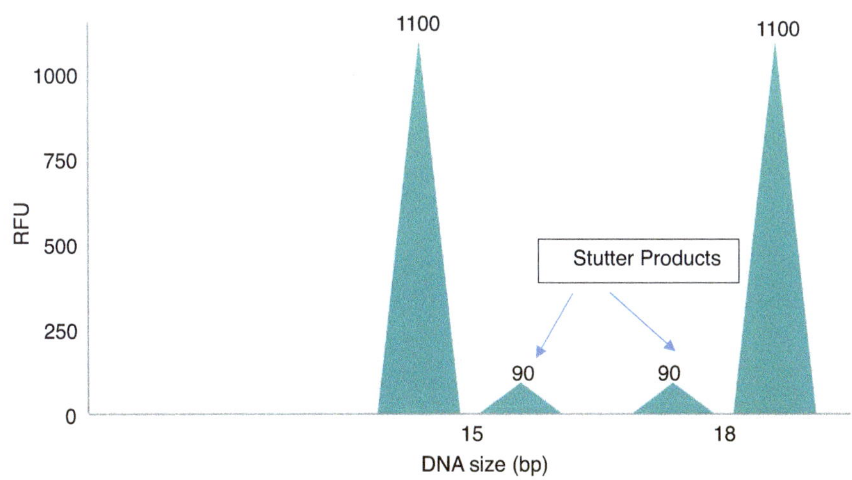

Fig. 18.5 Schematic representation of electropherogram having stutter peaks

are very rare. Also, stutter products have an impact on the electropherogram; if they are almost of same size as an actual allele, then it becomes challenging to determine whether a small peak is a real peak or a stutter product (Walsh et al. 1996).

18.6.2.2 Non-Template Adenylation

During amplification, sometimes DNA polymerase adds an extra nucleotide which is usually adenosine to the 3′ end of the amplicon. This phenomenon is generally known as non-template adenylation. However, the adenylation is dependent on the sequence of the template strand. This results in the addition of an extra nucleotide group (+A) which signifies the amplified product towards the electropherogram and without the extra nucleotide (−A) which signifies the parent allele of the desired loci. These peaks are observed generally in the mixture during partial non-template addition. Also, sharper peaks increase the likelihood that the system's genotyping further calls for accuracy (Gomes et al. 2007).

18.6.2.3 Heterozygote Imbalance

It is when one of the alleles has a much greater peak area or amplitude than the other allele in heterozygous conditions on the same locus (Fig. 18.6). It is said that this type of condition is observed when the sample extracted contains unequal proportions of DNA template at the heterozygous loci. This may be due to preferential amplification which eventually interferes with the interpretation of the samples provided and results in heterozygous imbalance.

18.6.2.4 Allelic Dropout

Allelic dropout is the condition in which one of the alleles of a heterozygous individual fails to get detected and leads to an assumption that the individual is

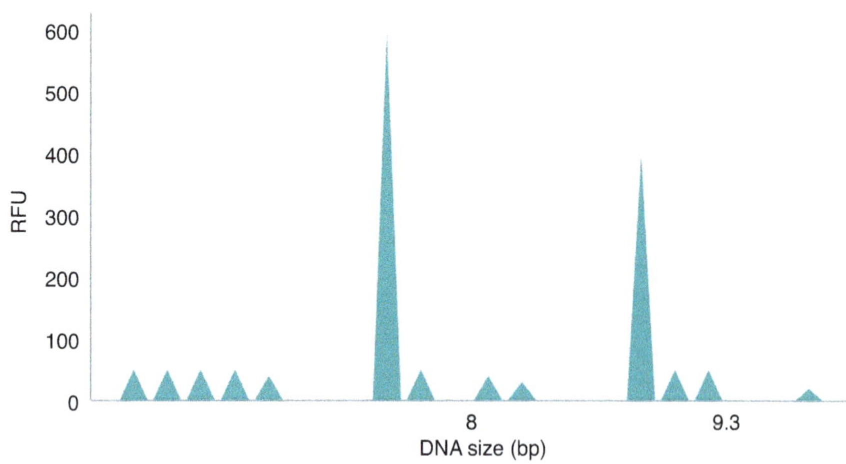

Fig. 18.6 Schematic diagram of heterozygote imbalance. Here the intensity signal of one allele is quite great than the other allele on the same locus

homozygous for that particular locus. Though very limited causes of this condition have been identified, one of them is an extreme situation of preferential amplification which also leads to heterozygote imbalance. Mutations can cause allelic dropout as well.

18.6.3 Electrophoretic Artifacts

18.6.3.1 Pull-Up Peaks

This occurs when a minor peak of an electropherogram in one colour is pulled up from a major peak of an allele in another colour. The colours are of overlapping spectra and can cause inaccuracy of the genetic profiling. This happens when a sample is overloaded or the calibration is not done for the matrix file.

18.6.3.2 Spikes

They are sharp peaks with almost similar signal intensities present in all the colour panels of an electropherogram. They are caused because of an air bubble or urea crystals trapped in the electrophoretic platform of the electropherogram. These are non-reproducible in nature and sometimes even the voltage spikes contribute to it. The electrophoretic artifacts can be confirmed by repeating the process of electrophoresis and the electropherogram can be analysed accordingly.

18.6.3.3 Contamination

Genetic profiling is one of the most sensitive techniques for individualization. The laboratory-standardized protocols are followed so to increase the extraction and quantification rate of DNA. But the problem appears when proper care regarding amplification or detection is not done. Contamination can occur in any stage of profiling, but the most likely result is exclusion or inconclusive result (Diegoli 2015).

To avoid the interpretation of contaminated results, the negative control samples hold a lot of importance. The negative control basically involves the blank which tests the tubes involved including the reagents used. There are various chances of contamination starting from the scene of the crime from where the sample is being collected to the final profiling step. Hence, in each step collection and testing of the control sample are necessary (Tucker et al. 2012).

18.7 Data Interpretations

There are several guidelines for the interpretation of STR profiles which are proposed by the Scientific Working Group on DNA Analysis Methods (SWGDAM) and also other scientific bodies involved in deciphering the profiles. Generally, a conclusion can be given as follows:

1. Inclusion (100% match).

In this, the STR loci peaks are compared between the profiles generated that match exactly. This infers that the sample collected from the crime scene or the victim has the same origin as that of sample collected by the accused.

2. Inconclusive Result:
   The data provided or deduced from the results obtained don't support any inclusion or exclusion. When the analyst cannot obtain any conclusion due to insufficient information available.
3. Exclusion Result:
   The genotypes of two or more samples in comparison are different. In other words, they are from different origin as the markers between the samples do not overlap.

18.8 Statistical Evaluation of DNA Typing Results

The allelic distribution in an individual follows the Mendel law of inheritance. The gametes are formed during the process known as meiosis and the haploid set of chromosomes is received from one parent and the next set is received from the other parent. A diploid individual is composed of 22 pairs of chromosomes and 2 sex chromosomes (XX in females and XY in males).

To calculate the proportion of alleles at a given locus, the Hardy-Weinberg Principle can help in estimating the heterozygosity (Ladd et al. 2001).

Also, to determine whether the genotypes of a population obey the Hardy-Weinberg principle, which is eventually used for database construction using the chi-square formula (Gazi et al. 2010).

$$\chi^2 = \sum \frac{(O_i - E_i)^2}{E_i}$$

Where: O_i = observed genotype frequency; E_i = expected genotype frequency; n = total number of genotypes.

The allele frequency of one of the CODIS STR Loci is shown in Table 18.2.

18.8.1 Probability of Match

This is basically the discriminating power of the genetic locus that can be estimated using population match probability (P_m). They define it as the possibility of matching a genotype between any two randomly chosen individuals. The lower the value of $P_{m,}$ it is less likely to get a match between two randomly chosen individuals.

Table 18.2 The allelic frequency of one of the alleles of 13 CODIS marker. (Source: (Budowle et al. 1999))

Allele	Allelic Frequency (%)		
VWA	African American	Caucasian	Hispanic
	$N = 180$	$N = 196$	$N = 203$
11	0.278	0.000	0.246
13	0.556	0.510	0.000
14	6.667	10.204	6.158
15	23.611	11.224	7.635
16	26.944	20.153	35.961
17	18.333	26.276	22.167
18	13.611	22.194	19.458
19	7.222	8.418	7.143
20	2.778	1.020	1.232
21	0.000	0.000	0.000

$$P_{\mathrm{m}} = \sum_i \left(p^2\right)^2 + \sum_j (2pq)^2$$

Where: P and q = the frequencies of two different alleles; Pm = can also be used to compare the discriminating powers of different loci.

It is also stated to evaluate the strength of DNA profiling results and their possibility of matching between the two random individual profiles. The chances to have a matching profile from a suspect to the crime scene sample can be because of the following reasons:

(a) The samples from both the crime scene and suspect belong to the same individual and have a common origin.
(b) The suspect happens to have the same genetic profile as that of the sample found at the scene of the crime. The significance of the match can be calculated using statistical calculations to determine how rare it is to have the same DNA profile of 2 non-related samples.

This analysis is done by considering the alleles involved with the heterozygosity and multiplying the heterozygosity of each allele to determine the interpretation (Vanderheyden et al. 2007).

18.9 Preparation of Reports

After DNA extraction and profile generation, a report is prepared stating whether the profiles between the samples collected from the accused match with the sample collected from the crime scene or victim. In this, the STR loci using which were selected to generate the profile are matched. The electropherogram generated should overlap in both cases. Also, proper statistical analysis is involved and calculation is done to estimate the probability of a match in the respective case (Riccardi et al. 2009).

18.10 Other Usage of STR Typing

In this particular chapter, we have seen the use of STR typing in forensic science. This particular area of study not only helps in forensic-related problems, but has a wide area of research.

18.10.1 Cell Lines

Human cell lines are used for various research studies and their authentication is equally important. When these cell lines are the major issue faced by the research group in the cross-contamination of cell lines with other cell lines, STR typing enables the discovery of such conditions as well as analysing any tri/tetra allelic traits and patterns due to imbalance in the cell cultures (Lucy et al. 2007).

18.10.2 Observing Transplants

STR typing can help in generating the profile of the transplant recipient as well as the donor to match the grafting profiles and diagnose the graft failure and/or relapse probability. Generally, the transplants are bone marrow or blood stem cell transplants where the rejection rate from the body is quite high.

18.10.3 Detection of Cancerous Tumours

The loss of heterozygosity (LOH) is also one of the methods to monitor the genetic mutations including deletions which eventually causes tumours and many other types of cancer. This causes severe imbalance at loci and can be amplified using PCR. This also helps in the prior detection of individuals who are more prone to cancer than other normal individuals in terms of genetic makeup. During this, using a tissue sample nine STR loci are studied to check the severe imbalance in the peaks of the designated markers.

18.10.4 Mapping of Genetic Diseases

The genetic scans comprising the disease genetic mapping are performed for around 400 STR markers covering the whole human genome at an approximate distance of 5–10 centimorgan (cM). The study of the allele frequency between a normal and diseased population helps in the estimation and association of genetic diseases with the proposed markers. The markers are further correlated with the help of linkage analysis. A set of 5 STR loci has already been identified for this type of study (Willems et al. 2014).

Multiple Choice Questions

1. Which locus can be used for determining both male and female gender in DNA fingerprinting?
   (a) DYS 19
   (b) DYS 393
   (c) Amelogenin
   (d) Y-plex ladder.
   Answers: (c)
2. The genetic markers which are seen close together on the same chromosome exhibit.
   (a) Genetic linkage
   (b) Homozygosity
   (c) Genetic concordance
   (d) Independent segregation.
   Answers: (a)
3. What is the length of Amelogenin gene marker in X - Chromosome?
   (a) 120 bp
   (b) 112 bp
   (c) 108 bp
   (d) (D)106 bp.
   Answers: (d)
4. D13S317 and TPOX are basically:
   (a) STR markers.
   (b) Regions of mt DNA.
   (c) Types of viruses.
   (d) Restriction enzymes.
   Answers: (a)
5. A computerized database that allows to obtain the information related to an individual's DNA profile is:
   (a) PCR.
   (b) CODIS.
   (c) AFIS.
   (d) PDQ.
   Answers: (b)

References

Budowle B, Moretti TR, Baumstark AL, Defenbaugh DA, Keys KM (1999) Population data on the thirteen CODIS core short tandem repeat loci in African Americans, US Caucasians, Hispanics, Bahamians, Jamaicans, and Trinidadians. J Forensic Sci 44(6):1277–1286

Butler JM (2005) Forensic DNA typing: biology, technology, and genetics of STR markers. Elsevier, Amsterdam

Clayton TM, Whitaker JP, Sparkes R, Gill P (1998) Analysis and interpretation of mixed forensic stains using DNA STR profiling. Forensic Sci Int 91(1):55–70

Clayton TM, Hill SM, Denton LA, Watson SK, Urquhart AJ (2004) Primer binding site mutations affecting the typing of STR loci contained within the AMPFlSTR SGM Plus kit. Forensic Sci Int 139(2–3):255–259

Cupples CM, Champagne JR, Lewis KE, Cruz TD (2009) STR profiles from DNA samples with "undetected" or low Quantifiler™ results. J Forensic Sci 54(1):103–107

Diegoli TM (2015) Forensic typing of short tandem repeat markers on the X and Y chromosomes. Forensic Sci Int Genet 18:140–151

Evett IW, Gill PD, Scrange JK, Weir BS (1996) Establishing the robustness of short-tandem-repeat statistics for forensic applications. Am J Hum Genet 58(2):398

Gazi NN, Eaaswarkhanth M, Ferdous A, Qadri SS, Singh L, Thangaraj K (2010) Genetic polymorphisms of 15 autosomal STR loci in three isolated tribal populations of Bangladesh. Forensic Sci Int Genet 4(4):265–266

Gomes AV, Mauricio-da-Silva L, Raposo G, Vieira JRC, dos Santos Silva R (2007) 13 STR loci frequencies in the population from Paraíba, Northeast Brazil. Forensic Sci Int 173(2–3): 231–234

Laberke PJ, Grossenbacher R, Hausmann R, Balitzki B (2012) Method to predict the chance of developing a male profile out of mixtures of male and female DNA. Int J Legal Med 126(1): 157–160

Ladd C, Lee HC, Yang N, Bieber FR (2001) Interpretation of complex forensic DNA mixtures. Croat Med J 42(3):244–246

Lucy D, Curran JM, Pirie AA, Gill P (2007) The probability of achieving full allelic representation for LCN-STR profiling of haploid cells. Sci Justice 47(4):168–171

Moreira D (1998) Efficient removal of PCR inhibitors using agarose-embedded DNA preparations. Nucleic Acids Res 26(13):3309–3310

Rådström P, Löfström C, Lövenklev M, Knutsson R, Wolffs P (2008) Strategies for overcoming PCR inhibition. Cold Spring Harb Protoc 2008(3):pdb-top20

Riccardi LN, Melean G, Rada A, Tirado N, Genuardi M, Ricci U (2009) Genetic profiling of Bolivian population using 15 STR markers of forensic importance. Legal Med 11(3):149–151

Tucker VC, Hopwood AJ, Sprecher CJ, McLaren RS, Rabbach DR, Ensenberger MG et al (2012) Developmental validation of the PowerPlex® ESX 16 and PowerPlex® ESX 17 systems. Forensic Sci Int Genet 6(1):124–131

Vanderheyden N, Mai A, Gilissen A, Cassiman JJ, Decorte R (2007) Identification and sequence analysis of discordant phenotypes between AmpFlSTR SGM plus™ and PowerPlex® 16. Int J Legal Med 121(4):297–301

Walsh PS, Fildes NJ, Reynolds R (1996) Sequence analysis and characterization of stutter products at the tetranucleotide repeat locus vWA. Nucleic Acids Res 24(14):2807–2812

Wiegand P, Kleiber M (2001) Less is more – length reduction of STR amplicons using redesigned primers. Int J Legal Med 114:285–287

Willems T, Gymrek M, Highnam G (2014) 1000 genomes project consortium Mittelman D., Erlich Y. *the landscape of human STR variation*. Genome Res 24:1894–1904

Sex Chromosome Haplotyping 19

Monisha Samuel and Rutwik Shedge

Abstract

Chromosomes X and Y have been true homologues and have evolved differently over the years. The Y chromosome has shrunk to about 60 Mb in size as a result of deletions throughout time. Still there is a lot of sequence homology with the X chromosome. The Y chromosome is made up primarily of heterochromatin and has the fewest genes of any chromosome. When it comes to sexual development, the genes on the Y chromosome are crucial (sex-determining region on the Y gene, SRY, which only determines male sex) (Skaletsky et al. Nature 423:825–837, 2003). Two pseudoautosomal areas on the Y chromosome can potentially recombine with the X chromosome during spermatogenesis. These regions are located at both ends of the chromosome. Ninety-five percent of the Y-chromosome contains functional genes and transcription-inert heterochromatin region which contributes to formation of the non-recombining area. This is only found in males and is passed down intact from father to son and is abundant in micro- and minisatellite DNA as well as polymorphic repetitive elements. The Y chromosome's non-recombining region contains short tandem repeat (STR) loci, which are inherited as a block of linked haplotypes. The identification of unknown people, the determination of a person's paternity, the detection of the male DNA profile in mixtures and azoospermic people, and the confirmation of amelogenin-deficient males are all forensic applications that greatly benefit from Y-STR haplotyping (Quintana-Murci and Fellous J Biomed Biotechnol 1:18, 2001). The use of X-chromosomal markers is also numerous in forensic contexts. More than 30 STRs have been recognized as forensic markers. Joint typing of STRs that are extremely closely connected produces stable haplotypes that can be

M. Samuel (✉) · R. Shedge
Department of Forensic Science, National Forensic Sciences University, Tripura Campus, Agartala, Tripura, India

A. Puri et al. (eds.), *Fundamentals of Forensic Biology*,
https://doi.org/10.1007/978-981-99-3161-3_19

utilized to determine the relationship between distant relatives and have been very beneficial in kindship testing (Kauppi et al. Ann N Y Acad Sci 1267:18, 2012).

Keywords

Y-STR · Y-SNP · Haplotype · PAR · Sex typing · X Chromosome · Amelogenin · SRY · Forensic science

19.1 Definitions

Haplotype: A set of genetic determinants located on a single chromosome.

SNP (Single nucleotide polymorphism): In genetics, a single-nucleotide polymorphism is a germline substitution of a single nucleotide at a specific position in the genome.

Sex-typing: the stereotypical categorization of people, or their appearance or behaviour, according to conventional perceptions of what is typical of each sex.

Polymorphism: Polymorphism, as related to genomics, refers to the presence of two or more variant forms of a specific DNA sequence that can occur among different individuals or populations.

19.2 Y Chromosome Haplotyping

19.2.1 Y Chromosome Genome in Humans

Genetic traits on the human Y chromosome serve as a lineage marker for males as they directly transmit a single haplotype from father to son [R]. Along with 22 pairs of autosomal chromosomes, the Y chromosomes are one of a pair, with the X chromosome, of the human sex chromosomes that are found in the nucleus of human cells. Although there are possibilities of variances in the chromosome numbers which are very rare and other specific mutations that can alter the phenotype, individuals with an X and Y chromosome are typically phenotypically masculine, and those with two X chromosomes are female (de Knijff 2022). The Y chromosome in males encodes a large number of genes necessary for male-specific processes, such as sex determination and spermatogenesis. The human Y chromosome is made up of 50–60 genes and has a base pair count of around 59 million which represent around 2%–3% of a haploid genome. The pseudoautosomal region (PAR) and the male-specific Y region are two very distinct divisions of the Y chromosome (Quintana-Murci and Fellous 2001).

19.2.2 Pseudoautosomal Region

Two pseudoautosomal regions (PARs) are present in the terminal ends human Y chromosomes, one in the end of short arm (Yp) and other in the long arm (Yq) which are separated by centromeres (Figure). The genes PAR1 and PAR2, which both can recombine with regions on the X chromosome, are enriched in the lightly packed chromatin with the remaining sections part of male-specific region (MSY) and are non-recombing in nature. PAR1 is 2.6 Mb in size and within the PAR1, 24 genes have been known, whereas four genes are currently known to exist within the 320 kb of the PAR2 (Quintana-Murci and Fellous 2001).

The correct separation of the X and Y chromosomes during meiosis depends on the PARs. Sex chromosomal aneuploidies are caused by the improper segregation of the X and Y chromosomes and have clinical effects like infertility and the Klinefelter syndrome. Thus, all chromosomes must identify their homologous partners and complete recombination and pairing in order for meiosis to be successful. The pseudoautosomal regions, and very specifically PAR1, are the regions where Y chromosome pairs and exchanges genetic material with the pseudoautosomal region of the X chromosome during male meiosis. Compared to this, PAR2 is substantially shorter and exhibits lower pairing frequency than PAR1, therefore deletion of PAR2 results in less severe phenotypes (Kauppi et al. 2012).

19.2.3 Male-Specific Y Region

The male-specific region of the Y chromosome, the MSY which was earlier called the nonrecombining Y (NRY) region, separates the sexes and encompasses 95% of the chromosome's length. There are approximately 23 megabases (Mb) of euchromatic DNA sequences in the MSY, with 8 Mb on the short arm (Yp) and 14.5 Mb on the long arm (Yq). The MSY is an assortment of heterochromatic sequences and three classes of euchromatic sequences which are X-transposed, X-degenerate, and ampliconic. These classes comprise the entire 156 known transcription units. This further includes 78 protein-coding genes that collectively encode 27 distinct proteins. The X-transposed sequences parade 99% identity to the X chromosome and thus when choosing Y chromosome-specific markers for forensic DNA profiling, areas having similarity to the X chromosome should be avoided. The X-degenerate sequences are remains of ancient autosomes from which the modern X and Y chromosomes evolved (Skaletsky et al. 2003).

19.2.4 Polymorphic Sequences

The Y chromosome consists of many polymorphic sequences along with many variable number of tandem repeats (VNTR), single nucleotide polymorphisms (SNPs) and mobile elements. Some of these sequences are extremely important in forensic investigation. Most importantly, SNPs and mobile elements have far lower discrimination power than Y chromosome short tandem repeats (Y-STRs). Due to the high-throughput analysis and robust discrimination strength, Y-STRs are most frequently employed for Y chromosomal DNA testing.

19.2.4.1 Y-STRs

Y-chromosome STRs and single nucleotide polymorphisms on the Y chromosome (Y-SNPs) are two very distinctly different types of Y-chromosome polymorphisms that can be used separately or in combination for routine forensic or population genetic DNA research. In the Y chromosomal genome, around 400 STR sites have been found. Using information from the sequencing of the human genome, the precise positions of these loci have been successively mapped. About 60% of the 400 known Y-STR loci are found on the long arm of the chromosome, 22% are on the short arm, and a small number are found in the centromeric region. In the earlier days, only a small subset of Y-STRs could be genotyped well by PCR. The tri-nucleotide repeat locus DYS392 and the tetra-nucleotide repeat loci DYS19, DYS385a and b, DYS389-I, DYS389-II, DYS390, DYS391, and DYS393, made up this core collection of Y-STRs. All of the major Y-STR multiplex kits contain this set of nine Y-STR-loci known as the European minimal haplotype locus set /minimal haplotype loci as recommended by the International Y-STR User Group for forensic applications. Three additional tetra-nucleotide loci—DYS437, DYS438, and DYS439, have also been included in all of the Y-STR kits as approved by Scientific Working Group for DNA Analysis Methods (SWGDAM) in 2003. Currently, new multiplex systems are being developed using a large number of newly discovered Y-STR loci.

There are now commercially accessible kits with additional Y-STR loci that have been approved for forensic usage.

The majority of Y-STR sets, however, have poor capacity to distinguish between related males who share similar patrilineage. Therefore, patrilineal relatives of the suspect cannot be excluded by existing forensic Y-STR profiling. Ballantyne et. al. initially introduced rapidly modifying Y-STR (RM-Y- STRs) in Y-STR genotyping and sequencing in the year 2012. RM Y-STR have mutation rates which are above 10^{-2} and this is significantly greater than the average Y-STR mutation rates. These RM Y-STRs provide far better probability of being able to distinguish closely related male relatives by paternity, which is undoubtedly a significant benefit in some forensic instances (de Knijff 2022).

The Y-STR loci, DYS385 and DYS389, are multilocal (MLL) in nature. The STR that is present at several sites on the Y chromosome as a result of duplication is referred to as MLL Y-STRs. About 50 of these MLL Y-STRs have been found thus far. The Y-STR loci DYS464 occurs at least four times in the highly palindromic region close to the centre of the long arm of the Y-chromosome. The use of highly polymorphic markers is helpful in forensic casework applications because the amount of typable DNA material may be constrained in order to reduce the number of markers required to distinguish unrelated individuals (Butler and Schoske 2004) (Table 19.1).

Table 19.1 Common Y-STR Loci Covered by some Commercial Kits like AmpFlSTR® Yfiler® PCR Amplification Kit (Applied Biosystems) & PowerPlex® Y23 System (Promega)

SL. No.	Core loci	Additional loci
1.	DYS19	DYS437
2.	DYS385a/b	DYS448
3.	DYS389 I	DYS456
4.	DYS389 II	DYS458
5.	DYS390	DYS481
6.	DYS391	DYS533
7.	DYS392	DYS549
8.	DYS393	DYS570
9.	DYS438	DYS576
10.	DYS439	DYS635
11.		DYS643
12.		Y-GATA-H4

19.2.4.2 Y-SNPs

A significant class of biallelic markers on the Y chromosome includes insertion/deletions (indels) and single nucleotide polymorphisms (SNPs). Due to their lower mutation rate than STRs (10^{-8} vs. 10^{-3} mutations per generation), these markers are also referred to as unique event polymorphisms (UEPs). STRs, which can contain a dozen or more alleles (or allelic combinations in the case of multi-copy Y-STRs), provide more information per marker than SNPs, which only have two alleles. The history of Y-SNP screening and application has been substantially more nuanced. Six alternative haplogroup nomenclatures and a wide range of screening methodologies were employed by various research groups when employing Y-SNP polymorphisms for population genetic goals. Fortunately, the mtDNA example was used to resolve the discrepancy in Babylonian nomenclature, leading to the terminology we still use today (de Knijff 2022). With the advent of whole genome sequencing and datasets widely available, large scale screening methods for Y-SNP have begun developing rapidly (Karafet et al. 2008; Tillmar et al. 2021).

19.3 X Chromosome Haplotyping

The X chromosome has a special structure that has evolved to create gender-specific genetic distinctions that are not shared by its counterpart, the Y chromosome, or the autosomes. The pseudoautosomal regions, PAR1 and PAR2, are the only places where the X and Y chromosomes can recombine in male individuals. As a result, in males the X chromosome is (nearly) fully passed down to female offspring. However, females have two copies of the X chromosome, which recombines along the entire chromosome during female meiosis and is passed on to both female and male offspring. These transmission of traits makes the X chromosome an indispensable genetic tool for population genetic studies as well as forensic investigations. The potential uses that result from its distinctive features are what primarily underpin the use of X chromosomal polymorphisms in population genetics and in human identification. X chromosome markers may provide crucial information in a variety of study fields, either independently or in addition to information provided by the autosomes, Y chromosome, or mtDNA markers (Gomes et al. 2020). It must be emphasized that identity testing utilising X-STRs in specific situations, namely in cases of (complicated) kinship testing, may be the only method available to solve some issues (Chen et al. 2020).

The exclusion strength of autosomal STRs is significantly diminished in paternity cases involving close blood relatives as putative fathers, and X-STRs may be preferable to autosomal markers. For instance, if two putative fathers are father and son, then they would not share any X-chromosomal alleles identical by descent, thus making X-STR analysis significant (Hering et al. 2010).

Markers do not segregate independently if they are close to each other and thus the phenomenon of linkage comes to play. The alleles of the two loci are inherited together as a haplotype, when two loci are very close to each other. Thus, it is beneficial to use the recombination fraction (or recombination frequency), which is the proportion of recombinants produced by chromosomal crossing between two loci during meiosis among all the progeny, to estimate how closely two distinct loci are related. Commercially available kits separate the ChrX into the linkage groups like 1-4 at Xp22.2, Xq12, Xq26, and Xq28, which are a collection of unlinked STRs and produce independent genotype information (Gomes et al. 2020).

19.4 Gender Identification

In forensic investigations specially involving sexual assault or missing persons, sex determination might be very crucial role. Genetic typing of biological traces is required, and the classification of the sex is significant in such cases. Sex determination using genetic typing of fragments specific to the amelogenin gene is one of the most common methods. Amelogenin is the protein that is most prevalent in secretory stage of enamel. Defect in *AMELX* causes the X-linked amelogenesis imperfecta. The human X chromosome houses the AMELX gene (Xp22.2) and the human Y chromosome has the AMELY gene (Yp11.2), which are homologous genes that make up the AMEL locus. Despite being a homologous pair, the genes' size and sequence vary. The AMELY gene's coding sequence is 3272 bp long, while the AMELX gene's coding sequence is 2872 bp long. From a single main RNA transcript, alternative splicing produces several amelogenin isoforms. The amelogenin gene, which has X and Y chromosome variants (AMELX and AMELY, respectively), can be used to determine the sex of unidentified human samples. Comparing AMELX's intron 1 to AMELY's, there is a 6-base-pair loss. The polymerase chain reaction (PCR) and gel electrophoresis can be used to identify this difference. If both the AMELX and AMELY variants of the gene are present (i.e. the sample is from a male), then two bands of DNA at 555 bps and 371 bps are resolved, whereas one band of DNA at 555 bps is observed if the AMELX version alone is present (i.e. the sample is from a female) (Morrill et al. 2008).

This method of sex determination is not 100% accurate, however, due to AMELY variance among people and populations. PCR amplification may not function if mutations occur in AMELY intron 1 areas that are frequently employed as primer annealing sites. An amplicon with the same length as AMELX is produced by inserting a 6 bp sequence into intron 1 of AMELY. Some males may completely lose the gene for AMELY. In each of these situations, just one band is visible during the gel electrophoresis of the PCR product, leading to the sample being mistakenly identified as female. Problems could occur due to amelogenin-specific fragments being falsely detected (or not detected at all), chimerism (bone marrow transplants), micro chimerism (pregnant women carrying male foetuses), and potential gender inconsistencies between the biological gender and the (forensically relevant) legal gender on personal identity documents.

To be able to overcome this issue, other gene like SRY is co-amplified with AMEL which aids in the detection of AMELY-null samples. Sex-determining region Y (SRY) gene located on the Yp11.31 of the Y chromosome is a very useful marker. The sex-determination pathway is assumed to be steered towards male development by SRY (Morikawaa et al. 2011). The ability to combine the SRY gene assay for gender determination into other STR analyses using commercial kits that also contain AMEL has been shown in many studies and has been validated as well. The PCR product of this unique marker is 96 bp long and makes it possible to determine gender when typing extremely damaged forensic evidence (Dash et al. 2020).

A multiplex amplification of the SRY, STS (steroid sulphatase), and amelogenin gene areas and their homologous sequences has also been used by Morikawa et al. (Morikawaa et al. 2011) as a method for determining sex. The male DNA component of mixed samples with a male: female ratio as low as 1:10 may be detected using this approach, which has a detection limit of 63 pg of genomic DNA. The STS gene for steroid sulfatase (an enzyme) is responsible for converting sulfated steroid precursors into physiologically active steroids like estrogens and androgens. The STS gene is found at Xp22.31, which is the distal end of the short arm of the X chromosome. Ten exons and nine introns make up the 146 kb long STS gene. The STSP1 pseudogene is a 100 kb long STS gene located on the long arm of the Y chromosome. Despite certain sequence homologies between STSP1 and the STS gene, no functional genes are encoded by the pseudogene. *STS* and *STSP1* alleles are both used for sex typing.

Additionally, the pentanucleotide microsatellite DXYS156, which can be used to determine sex, maps to the pseudoautosomal region of both sex chromosomes. DXYS156 is a multi-allelic STR with geographic-specific allelic distribution that aids in identifying a person's sex as well as their likely geographical place of origin (Mukherjee et al. 2013).

Multiple Choice Questions

1. Gene that aids in detection of AMELY-null samples are.
   (a) SRY.
   (b) DXYS156.
   (c) Both.
   (d) DYS19.
   Answer: (c)
2. The alleles of the two loci are inherited together as a haplotype, when two loci are very close to each other. This phenomenon is known as.
   (a) Sex typing.
   (b) Linkage.
   (c) Polymorphism.
   (d) Genetic variation.
   Answer: (b)
3. If both the AMELX and AMELY variants of the gene are present, then two bands of DNA at 555 bps and 371 bps are resolved using electrophoresis of the PCR products indicating the sample belongs to the gender:
   (a) Female.
   (b) Male.
   (c) Both.
   (d) Unidentified.
   Answer: (b)
4. The sex-determination pathway is assumed to be steered towards male development by which gene?
   (a) DXYS156.
   (b) Both.
   (c) DYS19.
   (d) SRY.
   Answer: (d)
5. Higher probability of being able to distinguish closely related male relatives by paternity can be achieved using the following:
   (a) Y-STR.
   (b) RM Y-STR.
   (c) X-STR.
   (d) SRY.
   Answer: RM Y-STR (Randomly mutating Y-STR)
6. Recombination is more localized to which region of the Y chromosomes?
   (a) PAR 1.
   (b) PAR 2.
   (c) Both.
   (d) Male specific Y-region.
   Answer: (c)
7. X-STRs are more preferably utilized in which of the following cases?

(a) Sex typing.
(b) Kinship testing.
(c) Paternity.
(d) None.
Answer: (b)

8. Identify the Y-STR which are multilocal in nature.
(a) DYS385.
(b) DYS389.
(c) Both the above.
(d) DXYS156.
Answer: (c)

References

Butler JM, Schoske R (2004) Forensic value of the multicopy Y-STR marker DYS464, vol 1261, pp 278–280

Chen M, Ren H, Liu Z et al (2020) Genetic polymorphisms and mutation rates of 16 X-STRs in a Han Chinese population of Beijing and application examples in second-degree kinship cases. Int J Legal Med 134:163–168. https://doi.org/10.1007/s00414-019-02047-8

Dash HR, Rawat N, Das S (2020) Alternatives to amelogenin markers for sex determination in humans and their forensic relevance. Mol Biol Rep 47:2347–2360. https://doi.org/10.1007/s11033-020-05268-y

de Knijff P (2022) On the forensic use of Y-chromosome polymorphisms. Genes (Basel) 13:898. https://doi.org/10.3390/genes13050898

Gomes I, Pinto N, Antão-Sousa S et al (2020) Twenty years later: a comprehensive review of the X chromosome use in forensic genetics. Front Genet 11:926. https://doi.org/10.3389/fgene.2020.00926

Hering S, Edelmann J, Augustin C et al (2010) X chromosomal recombination—a family study analysing 39 STR markers in German three-generation pedigrees. Int J Legal Med 124:483–491. https://doi.org/10.1007/s00414-009-0387-y

Karafet TM, Mendez FL, Meilerman MB et al (2008) New binary polymorphisms reshape and increase resolution of the human Y chromosomal haplogroup tree. Genome Res 18:830–838. https://doi.org/10.1101/GR.7172008

Kauppi L, Jasin M, Keeney S (2012) The tricky path to recombining X and Y chromosomes in meiosis. Ann N Y Acad Sci 1267:18. https://doi.org/10.1111/J.1749-6632.2012.06593.X

Morikawaa T, Yamamoto Y, Miyaish S (2011) A new method for sex determination based on detection of SRY, STS and amelogenin gene regions with simultaneous amplification of their homologous sequences by a multiplex PCR. Acta Med Okayama 65:113–122. https://doi.org/10.18926/AMO/45270

Morrill BH, Rickords LF, Schafstall HJ (2008) Sequence length polymorphisms within primate amelogenin and amelogenin-like genes: usefulness in sex determination. Am J Primatol 70:976–985. https://doi.org/10.1002/ajp.20590

Mukerjee S, Mukherjee M, Ghosh T et al (2013) Differential pattern of genetic variability at the DXYS156 locus on homologous regions of X and Y chromosomes in Indian population and its forensic implications. Int J Legal Med 127:1–6. https://doi.org/10.1007/S00414-011-0646-6

Quintana-Murci L, Fellous M (2001) The human Y chromosome: the biological role of a "functional wasteland". J Biomed Biotechnol 1:18. https://doi.org/10.1155/S1110724301000080

Skaletsky H, Kuroda-Kawaguchl T, Minx PJ et al (2003) The male-specific region of the human Y chromosome is a mosaic of discrete sequence classes. Nature 423:825–837. https://doi.org/10.1038/NATURE01722

Tillmar A, Sturk-Andreaggi K, Daniels-Higginbotham J et al (2021) The FORCE panel: an all-in-one SNP marker set for confirming investigative genetic genealogy leads and for general forensic applications. Genes (Basel) 12:1968. https://doi.org/10.3390/GENES12121968

Mitochondrial DNA Profiling

20

Pratiksha H. Nimbkar, Aditi Mishra, Ulhas Gondhali, Sarthak Misra, and Tanya Chauhan

Abstract

Mitochondrial DNA (mtDNA) is a versatile tool in understanding the clinical and non-clinical aspects attributed by the nucleotide sequences. The presence of the nucleotide sequences and their constant mutations within provide unique phenotypic characters to the bearer and are a probable cause of numerous disorders related to the functional output of the mitochondrion. The study and understanding of these nucleotide sequences have clinical and forensic applications. Clinical applications include the prognosis of mitochondrial diseases (mitochondriopathies) and undertaking preventive or delaying measures, whereas forensic applications of mtDNA are very essential in the case of missing persons or the case of Disaster Victim Identification (DVI). Apart from these, mtDNA can also give an idea of the maternal lineage of the victim or suspect. In cases of disputed maternity, mtDNA can provide information concerning the heredity of the disputed child. These benefits of the mtDNA make it a valuable asset for

P. H. Nimbkar
School of Applied Sciences and Technology, Gujarat Technological University, Ahmedabad, Gujarat, India

A. Mishra
Kristu Jayanti College, Bangalore, India

U. Gondhali
Jindal Institute of Behavioural Sciences, O.P. Jindal Global University, Sonipat, India

S. Misra (✉)
Indira Gandhi Institute of Medical Sciences, Patna, India

T. Chauhan
Department of Forensic Science, National Forensic Science University Delhi (LNJN NICFS), New Delhi, Delhi, India

A. Puri et al. (eds.), *Fundamentals of Forensic Biology*,
https://doi.org/10.1007/978-981-99-3161-3_20

research and further study to understand the rate and occurrence of the mutations and their effects on the phenotypic characters of individuals.

Keywords

Mitochondrial DNA (mtDNA) · Phenotypic characters · Mitochondriopathies · Forensic applications

20.1 Definition of Terms

Disaster Victim Identification (DVI)	Process of recovery and identification of the people who have died during events of multiple deaths such as disasters
Ubiquitination	The process of attachment of ubiquitin to a protein for its degradation
Biopsy	Process of removal of a tissue sample for testing

20.2 Human Mitochondrial Genome

Known as the powerhouse of the cell, mitochondria are special due to their unique property of energy generation not acquired by any other organelle within a cell. There can be up to 1000 mitochondria present within a single cell with each of the mitochondrion having 2-10 copies of the mtDNA (Amorim et al. 2019; Ingman and Gyllensten 2001). They are double membrane rod-shaped structures. It performs various functions such as calcium signalling, haem synthesis, and steroid synthesis and aids in apoptosis. Mitochondrial DNA (mtDNA) is a double-stranded circular moiety present within the mitochondria. First isolated and identified in 1963 by Margit Nass and Sylvan Nass, the complete sequence of the mtDNA was first published in 1981 by Cambridge Reference Sequence (CRS). Similar to nuclear DNA, mtDNA has most of its sequence that is common among the various mitochondria within the cell. This property of homogeneous sequences is termed homoplasmy. Vice versa, the presence of the mutant regions with the wild-type sequences is termed heteroplasmy. The understanding of the occurrence of heteroplasmy within the mtDNA is crucial in studying mitochondrial diseases.

The mitochondrion over the years has accommodated itself within the eukaryotic cell. Many theories emerge as to the evolution of the mitochondrion. The most widely known theory is the theory of the endosymbiotic origin of mitochondria. According to the theory, the mitochondrion was believed to be an aerobic prokaryote that was engulfed by a eukaryotic cell. The aerobic prokaryote then formed a symbiotic relationship with the eukaryote. This led to the development and evolution of the current mitochondrion. When we trace back the ancestry of the mitochondrion, the closest relative from which it can be said to have evolved is the α-proteobacteria. The presence of these organelles today is representative of those

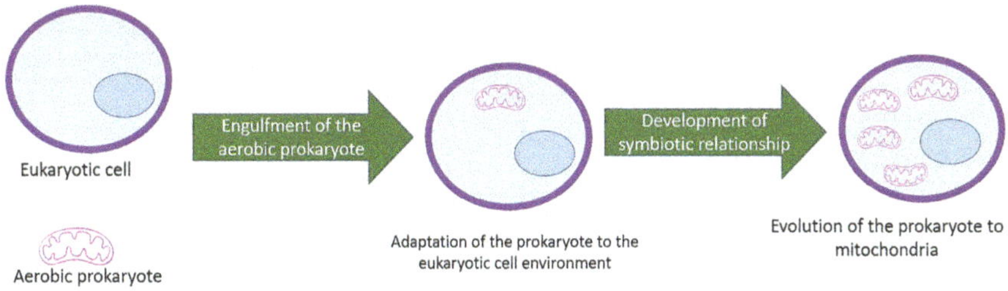

Fig. 20.1 Diagrammatic representation of the theory of the endosymbiotic origin of mitochondria

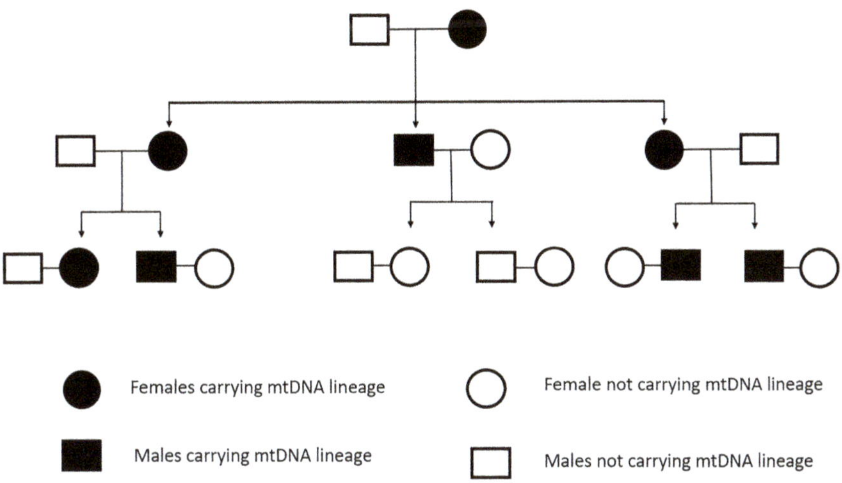

Fig. 20.2 Pedigree diagram depicting the transfer of mtDNA through generations. mtDNA inherited by males is not passed on to the next generation due to its loss during embryogenesis

ancient symbiotic relationships. The mitochondrion is similar to an ancient bacterial cell having its DNA and possessing the ability to generate ATP through oxidative phosphorylation (Fig. 20.1).

mtDNA has been a subject of interest for researchers due to its nature of conservation through generations. It is representative of the maternal lineage and ancestry. During fertilization, the mitochondria within the sperm cell have a very low copy number of the mtDNA as compared to an ovum and the transferred mitochondria are somehow lost during embryogenesis. Another probable reason for only maternal mtDNA inheritance is the ubiquitination of the paternal mitochondria. The mtDNA continuously undergoes mutations. This property of the mtDNA along with its maternal conservation through generations makes it ideal for studying migratory patterns. The mutations result in highly specific nucleotide sequences evolved over generations within populations residing in a relative geographic region and having a common maternal ancestor, providing these individuals with distinct genetic fingerprints (Fig. 20.2).

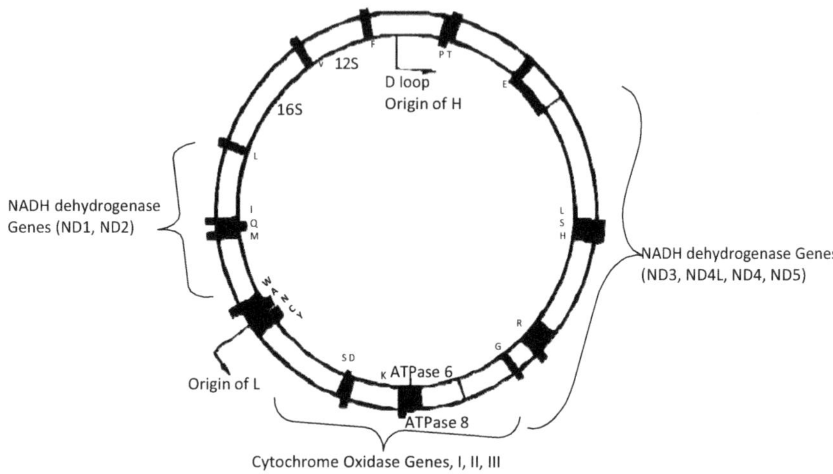

Fig. 20.3 Structure of the mtDNA (Alphabets correspond to the amino acid codes for the 22 mitochondrial tRNAs), rRNAs are marked as 12S and 16S, respectively, the origin of the Heavy (H) and Light (L) strands are marked along with major genes such as CO, ATPase, and ND

Since the copy number of human mtDNA within cells is relatively very high as compared to nuclear DNA and has a small size of 16569 bps (Anderson et al. 1981), it can easily be studied. When we specifically talk about the human mitochondrial genome, the complete structure has been previously studied and recorded. It is a double-stranded structure consisting of an outer heavy (H) strand and an inner light (L) strand. The H strand is heavy due to the higher concentration of guanine residues as compared to the higher concentration of cytosine residues in the L strand. Furthermore, the H strand consists of the majority of the genes for two rRNAs, 14 tRNAs, and 12 polypeptides. The L strand has genes for eight tRNAs and only one polypeptide (Taanman 1999). The mtDNA is primarily composed of coding regions and is devoid of introns except for one regulatory region. This makes the mtDNA very systematized as compared to nuclear DNA. A major site controlling the mtDNA expression is found as a short three-stranded structure that contains complementary base pairs for the L-strand. This region is termed the Displacement loop or the D-loop. The leading strand origin of replication and major promoters for transcription are present on the D-loop along with the genes for Phenylalanine and Proline tRNAs (Fig. 20.3).

Applications of mtDNA studies include clinical studies, ancestry tracing, maternal phylogenetic studies, and forensic science. Clinical studies involve the determination of mtDNA regions hosting sequences for diseases such as cancer, neurogenerative disorders, atrophies, etc. mtDNA holds great importance in forensic science. mtDNA has allowed investigators to construct a partial profile of suspected perpetrators. It can be used as evidence in cases of disputed maternity. mtDNA becomes a great tool for identification in mass disaster victim identification cases due to its presence in samples such as hair, teeth, and bone which degrade at very slow rates.

20.3 mtDNA Polymorphic Regions

As discussed earlier, understanding the reasons for the occurrence of heteroplasmy is essential for various reasons. Heteroplasmy can be easily defined as the occurrence of more than one type of detectable mtDNA type within an individual (Luo et al. 2018). Since each cell consists of numerous mitochondria, probability of one tissue being heteroplasmic and the other tissue exhibiting homoplasmy increases. The possibility of occurrence of the said case is very low, but can be considered as an exception. Heteroplasmy can be found within the mtDNA naturally (if the maternal lineage consists of heterozygous alleles). However, other reasons for heteroplasmy can include the accidental transfer of the mtDNA from the paternal lineage into the ovum during fertilization (Sinha et al. 2020). A few recent studies have supported the hypothesis, but in the majority of the cases, as mentioned previously, the mitochondria acquired from the sperm cell get destroyed during embryogenesis, conserving only the maternal lineage. Transfer of the paternal mtDNA can prove beneficial as it allows recombination of the mtDNA, leading to the strengthening of the genetic lineage through the introduction of new nucleotide sequences. Advantages of heteroplasmy include removal of deleterious mutations present within the mtDNA and reduction of the effect of "mother's curse" (a hypothesis stating that strict conservation of the mtDNA can have little or no effect on females, but can prove to be harmful to males). Heteroplasmy can be divided into two classes, wherein class one relates to the length polymorphism within the mtDNA and the second class deals with the point mutations (SNPs) within the mtDNA sequence. Heteroplasmy is significant in determining migration patterns. Mutant determination is based upon whether the number of mutant individuals is more than the control samples, the severity of the mutations, and their positive expression within the mutant population expressing the presence of mutant genes. The presence and absence of the level of dominance of specific alleles within the mtDNA are indicative of specific geographical locations. The presence of heteroplasmy affects various mitochondrial diseases. Mitochondria primarily deal with the energy requirements of the cell and hence diseases associated with mtDNA are generally observed to affect body organs and systems with high energy requirements such as the central nervous system or the heart. A brief explanation of the said phenomena can be as follows:

1. In the case of people residing in high-altitude areas such as Tibet, the 3394C allele is present dominantly as compared to the people residing in lower altitude regions (Stewart and Chinnery 2021). The presence of this allele thus becomes indicative of the geographic-specific mutations as well as if an individual from the region travels to a different region of the world; the presence of the gene can help in determining the migratory pattern of the individual.
2. The level of mutation can affect the intensity of the disorders associated. For example, individuals possessing heteroplasmic mutations (A/G) for the mtDNA tRNA coding for Leucine exhibit different disorders based on the concentration of G present. In the case of 20–30% mutation (3243G), individuals are diabetic

patients, whereas a mutation of 50–90% results in individuals having neuromuscular disease and a 100% mutation in the mtDNA leads to lethal perinatal disease (Wallace 2015).

3. Specific sequence variations within populations create sub-populations having an almost identical mtDNA sequence, but with a variation in the common region. These sub-populations can be termed as halogroups of the original population. These subgroups harbour nucleotide sequences that are not necessarily heteroplasmic, but the resultant sequence variations can lead to the groups having associations with specific mitochondrial disorders. For example, within the European continent, due to migrations, 10 major halogroups have been found to co-exist within the populations. These halogroups have, however, shown to harbour sequences for disorders such as Leber's hereditary optic neuropathy (LHON). The disorder results in gradual loss of vision as a result of increasing age (Chinnery and Hudson 2013).
4. Similarly, as the disorders associated with mtDNA are more related to the metabolically active body systems, changes in the mtDNA sequences can result in age-related neurogenerative disorders such as Parkinson's disease, Alzheimer's disease, and other age-related disorders affecting the central nervous system, heart, and muscular tissues.
5. Maintenance of mtDNA is looked upon by genes such as OPA1 and MFN2. These genes not only act as regulators for the nucleotide sequences within the mtDNA, but polymorphism due to deletions can result in optic atrophy along with other disorders visible as phenotypic characters such as deafness and neuromuscular disease. Polymorph of MFN2 is associated with Charcot-Marie-Tooth disease (CMT2A) and hereditary motor and sensory neuropathy.

20.4 mtDNA Testing

Within the mtDNA, along with the high number of mutations in the hypervariable regions H I and H II, mutations also occur outside these regions. These mutations can give rise to SNPs within the mtDNA sequences, making the nucleotide sequences unique for the particular maternal lineage. Studying the nucleotide sequences within the mtDNA can either concern with understanding the various genetic aspects of inheritance of diseases and other phenotypic characters or the second approach for studying mtDNA can be concerning analysing and comparing mtDNA sequences of reference and test samples for identification from a forensic perspective. mtDNA presents itself as a suitable candidate for identification as it has a high copy number among the cells; it can be extracted from older samples such as hair, teeth, and bones with low nDNA concentration.

Mitochondriopathies or diseases inherited due to the mtDNA mutations are mutations in the nDNA affecting the normal mitochondrion function. Mitochondriopathies can also result in a reduction in the number of mitochondria within the cell, damage in the replication of the mitochondria, and changes in the oxidation potential within the mitochondrial membranes. These diseases as previously stated are found to affect the oxygen utilization and energy production processes (oxidative phosphorylation—OXPHOS).

Clinical diagnostics for mtDNA testing involve liquid biopsies which offer a less invasive diagnosis. The liquid biopsies concerned with testing body fluids such as blood, urine, cerebrospinal fluid (CSF), etc. biomarkers are present within the body fluids that are indicative of the presence of mitochondriopathies. For example, Fibroblast growth factor 21 (FGF21), growth/differentiation factor 15(GDF-15), cell-free circulating mtDNA, and ROS-sensitive miRNA-9/9* are widely used biomarkers used in mtDNA testing. For determination of Alzheimer's, biomarkers within the blood and CSF such as Apolipoprotein E4 (ApoE), lipofuscin-like pigments, Cytochrome oxidase (COX) activity, mitochondrial aconitase levels, and reduced antioxidant levels among others give a clear indication as to the presence or possible future development of Alzheimer's within individuals. Similarly, there are specific biomarkers for the detection of other mitochondriopathies such as Parkinson's disease or Carcinogenesis. These biomarkers have facilitated the easy and early diagnosis of various mitochondriopathies providing patients with the option of starting an early treatment to delay the effects of the disorders. When discussing the forensic aspect of mtDNA testing, more sophisticated and advanced testing protocols are adopted. Since the analysis is based on nucleotide sequence matching, analysis is usually performed by using PCR-based methods, Sanger sequencing, or by using the Next Generation Sequencing Technologies (NGS). PCR methods capable of detecting hetroplasmy can also detect SNP's presence in the mtDNA sequences. These highly sensitive protocols are highly focused on the detection of either specific regions of nucleotide sequences as in PCR-RFLP or determining changes within single base pairs using single-stranded conformation polymorphism (SSCP). With the use of qPCR, one can even assess the level of heterogenicity within the samples. The only drawback of using PCR-based techniques is the prior requirement of a reference sequence and in the absence of a reference sequence, the PCR techniques cannot fulfil the requirements. Hence, forensic laboratories prefer using sequencing techniques for analysis. Most laboratories globally prefer using Sanger sequencing for mtDNA matching. Slowly, the newer platforms of NGS such as Ion torrent and Illumina Sequencing have started to be used on a routine basis in forensic laboratories. mtDNA analysis, similar to nDNA analysis, is reference and query sequence matching, where one of the samples received from a potential crime scene or suspect is matched with a reference sample for checking the matching percentage. The only difference between mtDNA testing and nDNA testing is that mtDNA testing is limited to class characteristics (Parsons and Coble 2001) identification and not individualization as two individuals of the same maternal lineage can have matching mtDNA sequences.

20.5 Forensic mtDNA Testing

Forensic analysis in DNA is based completely upon comparative analysis between a test and reference sample to determine their match percentage. In case of mtDNA analysis, it provides wider opportunities for researchers and law enforcement agencies with a more confident say as to the partial origin of the samples specifically restricted to their maternal lineage. It is one of the preferred methods of testing as it allows good quality analysis in case of very small sample volume as well in case of samples that have been degraded or have small amounts of contamination.

There have been numerous protocols for mtDNA analysis, but since the development of Next Generation Sequencing (NGS) technology, it has become more efficient, flexible, and robust due to the ease of entire genome mapping of the mitochondrial DNA. NGS has allowed researchers as well as forensic investigators to analyse highly compromised and degraded DNA samples such as in cases of Disaster Victim Identification (DVA) or multiple DNA donors on a crime scene. For analysis of disaster victims, short read sequencing can prove more accurate in determining the samples as high amounts of mtDNA are available in small sample volumes. This allows the investigators to study more broadly the whole genome structure as well as the associated SNPs within them. This assists in separation of mixed samples where there is more than one contributor involved for one DNA sample pool. Majority of the mutations within the mtDNA are present in the hypervariable regions of the genome. These regions are focused upon in forensic investigations as they provide reliable information as to individual identification due to the presence of unique SNP and STR regions. Forensic analysis of mtDNA can be conducted using numerous techniques such as Mass spectrometry, MALDI-TOF (Nahar Sultana 2018), Whole Genome Sequencing (WGS), Amplicon sequencing, along with another few. These methods provide accurate analysis for testing and building of mtDNA databases.

A technical challenge associated with mtDNA analysis is hetroplasmy. As priorly mentioned, hetroplasmy is one of the defining features of mtDNA as it allows a single cell to have different genetic compositions within the mitochondrion within it. This creates bias in mtDNA testing as a single donor can possess mutations within the samples of different tissues making it difficult to identify the donor based on the available test and reference sample.

Additional assistance in forensic mtDNA testing is the presence of reference mtDNA databases that can reduce the time required for sequence comparison and matching. As CODIS, for mtDNA, databases include EDNAP Mitochondrial DNA Population Database (EMPOP) and Mitomap. EMPOP is used as a reference database for comparing query sequences. The advantage of EMPOP is that it provides high-quality results and hence acts as a quality control tool for reference matching. Mitomap mtDNA database provides various selections to the user based on the nature of the search to be made. Designed in 1996, the database has data that can be categorized based on user requirements such as geographic locations and disease-specific variants. Both the mentioned databases allow ease of sequence comparison and analysis and are fit for use by geneticists, forensic scientists, and doctors interested in understanding inheritance patterns (Amorim et al. 2019).

Concerning mtDNA or nDNA testing, apart from the numerous benefits that they provide, there are a few legal and ethical aspects to their analysis. Whether nDNA or mtDNA analysis, the results of the analysis are bound to reveal not just the requested response, but also other aspects of the nucleotide sequence revealing any current or future progressive disorders that the individual has without his knowledge or any disorders that can progress in the future. Hence, it comes under the boundaries of ethics for the analyst to take action according to the policies set by the said institution. Therefore, to avoid legal and ethical dilemmas, the analysis of the nucleotide sequences is being kept limited to the non-coding regions of the mtDNA that have not yet been identified with diseases or peculiar phenotypic characters.

Apart from this, mtDNA can be utilized in wildlife forensics. In cases of wildlife crimes, the samples retrieved for analysis are very difficult to identify due to absence of phenotypic species identification features such as scales, hair, etc. In such cases when species identification is essential to establish commission of crime, mtDNA analysis can offer accurate species level identification. Previous literature has mention of wildlife crime case samples being identified up to species level using mtDNA gene markers like Cytb and COI (Kumar et al. 2019). Preference in analysis is given for mtDNA in such cases as its rates of degradation and mutation are inversely proportional and hence provide stable conditions for analysis even with long passages of time since the commission of the crime.

Food adulteration can be detected using mtDNA sequences in food products such as frozen meat and fish, exotic food products, and processed animal products. This provides a wide scope for researchers to explore the different applications of mtDNA apart from the clinical and forensic applications.

Thus, the application of mtDNA is not only restricted to humans, but can also be applied to animals for species determination. However, for plants, mtDNA is not considered as a major key player in DNA analysis due to slow evolution and very low number of mutations within its genome sequence.

20.6 Conclusion

The studies with respect to mtDNA have shown the potential that it has in understanding various aspects of the organisms. The evolution of mtDNA explains the evolutionary symbiotic relationships existing for millions of years between organisms. Today, mtDNA analysis is not only capable of providing information concerning inheritance and migration patterns, but can also predict a clear picture of numerous mitochondriopathies that exist or can affect a person in their later stages of life. The prognosis of possible disorders can help patients by getting early treatment, thereby increasing their lifespan of individuals. From a forensic perspective, the mtDNA analysis takes investigators one step closer to the identification of suspects or missing persons by providing partial genealogical details in context to the case at hand. This reduces the stretch of search that the investigators need to undertake, thus saving valuable time. With this, the horizons of mtDNA testing are continuously expanding as it continues to prove to be a key player in DNA analysis. The field of mtDNA analysis is being evolving and is now widely used not only for clinical, but also for forensic investigations and food adulteration monitoring purposes.

Multiple Choice Questions

1. The size of the mtDNA is.
   (a) 14,899 bps.
   (b) 22,000 bps.
   (c) 16,569 bps.
   (d) 15,255 bps.
   Answer: (c)
2. Can a father carrying mtDNA lineage pass it on to his future generation?
   (a) No.
   (b) Yes.
   (c) Inadequate Information.
   (d) None of the above.
   Answer: (a)
3. A non-invasive method for the detection of mitochondriopathies is.
   (a) Regular Biopsy.
   (b) Surgery.
   (c) Both a and b.
   (d) Liquid Biopsy.
   Answer: (d)
4. Mitomap and EMPOP are.
   (a) Mitochondrial DNA Databases.
   (b) Ballistic Databases.
   (c) Fingerprint Databases.
   (d) Bacterial Genome Databases.
   Answer: (a)
5. For ethical and legal reasons, mtDNA analysis is restricted to.
   (a) Coding regions.
   (b) Complete mtDNA.
   (c) Non-coding regions.
   (d) Nuclear DNA.
   Answer: (c)

References

Amorim A, Fernandes T, Taveira N (2019) Mitochondrial DNA in human identification: a review. PeerJ 7:e7314. https://doi.org/10.7717/peerj.7314

Anderson S, Bankier A et al (1981) Sequence and organization of the human mitochondrial genome. Nature 290(9):457–465

Chevigny N, Schatz-Daas D, Lotfi F, Gualberto JM (2020) DNA repair and the stability of the plant mitochondrial genome. Int J Mol Sci 21(1):328. https://doi.org/10.3390/ijms21010328

Chinnery PF, Hudson G (2013) Mitochondrial genetics. Br Med Bull 106(1):135–159. https://doi.org/10.1093/bmb/ldt017

Ingman M, Gyllensten U (2001) Analysis of the complete human mtDNA genome: methodology and inferences for human evolution. J Hered 92:454; https://academic.oup.com/jhered/article/92/6/454/2187179

Kumar V, Chandra K, Kundu S, Tyagi K, Amin Laskar B, Singha D, Chakraborty R, Pakrashi A (2019) Utility of mitochondrial DNA in wildlife forensic science: reliable identification of confiscated materials from eastern India. Mitochondrial DNA B Resour 4(1):583–588. https://doi.org/10.1080/23802359.2018.1561216

Luo S, Valencia CA, Zhang J, Lee NC, Slone J, Gui B, Wang X, Li Z, Dell S, Brown J, Chen SM, Chien YH, Hwu WL, Fan PC, Wong LJ, Atwal PS, Huang T (2018) Biparental inheritance of mitochondrial DNA in humans. Proc Natl Acad Sci 115(51):13039–13044. https://doi.org/10.1073/pnas.1810946115

Nahar Sultana GN (2018) Medicolegal aspects and examination techniques in exhumation cases with alleged human rights violation. J Forensic Sci Crim Invest 9(1):1. https://doi.org/10.19080/jfsci.2018.09.555755

Parsons TJ, Coble MD (2001) Increasing the forensic discrimination of mitochondrial DNA testing through analysis of the entire mitochondrial DNA Genome. Croat Med J 42(3):304; www.cmj.hr

Sinha M, Rana M, Kushwaha P (2020) Applications of mitochondrial DNA in forensic science. In: Forensic DNA typing: principles, applications and advancements. Springer, Singapore, pp 329–343. https://doi.org/10.1007/978-981-15-6655-4_16

Stewart JB, Chinnery PF (2021) Extreme heterogeneity of human mitochondrial DNA 1 from organelles to populations. Nat Rev Genet 22:106

Taanman J-W (1999) The mitochondrial genome: structure, transcription, translation and replication. Biochim Biophys Acta 1410(2):103

Wallace DC (2015) Mitochondrial DNA variation in human radiation and disease. Cell 163(1):33–38. https://doi.org/10.1016/j.cell.2015.08.067

Single Nucleotide Polymorphism as Evolutionary Evidence of Individuality

21

Sarthak Misra, Parth Sharma, Aditi Mishra, Ulhas Gondhali, and Tanya Chauhan

Abstract

Since the development of DNA fingerprinting by Sir Alec Jeffery, the technique has always had a special relevance to forensic science. With the new emerging technologies, DNA fingerprinting has been performed through detection of specific DNA sequences within reference and query samples by techniques such as RFLP analysis and SSCP analysis to name a few. Recent advancement into determination of individuality includes the detection and analysis of Single Nucleotide Polymorphs (SNPs) within the samples. These analyses have proven significance due to their uniqueness within the genetic sequences by acting as biological markers. SNP detection protocols focus on highlighting the presence of the sequence modifications by using electrophoretic techniques, probes, primers, and high-throughput methods such as Sanger sequencing and NGS. The high-throughput techniques allow simultaneous multi-sample analysis through sequence by synthesis. With respect to individualization, these techniques have

S. Misra
Indira Gandhi Institute of Medical Sciences, Patna, India

P. Sharma
School of Applied Sciences and Technology, Gujarat Technological University, Ahmedabad, Gujarat, India

A. Mishra (✉)
Kristu Jayanti College, Bangalore, India

U. Gondhali
Jindal Institute of Behavioural Sciences, O.P. Jindal Global University, Sonipat, India

T. Chauhan
Department of Forensic Science, National Forensic Science University Delhi Campus (LNJN NICFS), New Delhi, Delhi, India

A. Puri et al. (eds.), *Fundamentals of Forensic Biology*,
https://doi.org/10.1007/978-981-99-3161-3_21

been adopted worldwide on regular basis for forensic investigation analysis of recent and cold cases.

Keywords

Single Nucleotide Polymorphs (SNPs) · Individualization · Polymorphic characters · Evolutionary relations · Migratory patterns

21.1 Definition of Terms

Deoxyribonucleic Acid (DNA)	Molecule coding for genetic information of an organism
Next-generation sequencing (NGS)	Nucleotide sequencing technique that can execute sequencing of multiple nucleotide sequences parallelly
Polymerase chain reaction (PCR)	Molecular technique that selectively amplifies targeted region of nucleotide sequence from the sample
Chemical cleavage of mismatch (CCM)	Use of chemical agent to cut the nucleotide sequence at the position of mismatch
Electrophoresis	Movement of charged particles due to electric field

21.2 Basic Characteristics

Traditionally, organisms have been classified based on their differentiating physical traits and adaptations, as these divergences in morphological features provided evidence for the evolution of a group or species. However, identifying an individual organism from a group becomes confusing because of similarity in morphological traits. Therefore, reliance on the molecular remains of evolution on Deoxyribonucleic acid (DNA) sequence level would be more accurate to spot an individual from the group.

Adaptation at molecular level occurs through various exposure events producing mutations in the DNA sequences and changing it gradually towards acquisition of novel traits in the organism Darwin et al. (1859), Barreiro et al. (2008). Such mutations limited to change of a nucleotide in DNA sequence are called single nucleotide variations (SNVs). However, SNVs fixed in germline with the potential to transmit across generations are considered as single nucleotide polymorphisms (SNPs) Brookes (1999). Hence, distribution of SNPs in individuals can be utilized for their identification and assessment of their ancestry. Moreover, SNPs may also influence the fitness of an organism by producing functional transfiguration or inefficient translation causing changes in resulting protein sequences Arshad et al. (2018), Robert & Pelletier (2018). Therefore, only nonlethal SNPs may be conserved through long-term evolution Morin et al. (2004).

In eukaryotic organisms such as Humans, DNA sequences are present in the nucleus and mitochondria of a cell Bogenhagen & Clayton (1974). Nuclear DNA (nDNA) sequences are inherited from both parents equally, whereas mitochondrial DNA (mtDNA) is maternally inherited. For that reason, nDNA can be used to identify both paternal and maternal-related ancestry, while mtDNA can only resolve maternal side ancestry. Furthermore, mtDNA collects mutations faster than nDNA in the vertebrates, accordingly mtDNA markers are widely used to infer evolutionary relations in animals Allio et al. (2017). Combination of DNA markers from both sources can be analysed to improve confidence of relatedness and even track migration patterns of the ancestors Gunter et al. (1994). Due to all of above reasons, DNA marker to assess diversity should be selected based on objectives of the study.

In this book chapter, we are discussing experiment drafting to detect SNP signatures as individuality marker by covering gel electrophoresis analysis, PCR amplification analysis, Next Generation Sequencing (NGS) analysis, and development of regional level high-throughput techniques.

21.3 Strategies for SNPs Profiling and Its Applications

SNPs are nucleotide substitutions that do not always result in phenotype such as genetic diseases or fitness improving traits Fiatal & Adany (2018). Therefore, to identify individuals using SNPs, genetic approach is preferred. Extraction of genetic materials from the sample requires to be standardized based on the sample type and its downstream applications. For the SNP-based individualization, a query SNP distribution pattern is superimposed on the reference SNP patterns to check their similarity and validate the source identity.

21.3.1 Reference Sequence Generation and Regional Database Creation Are Essential to Use SNP Profiles to Identify Individuals from the Population

SNPs are always searched against a relative sequence, and this known sequence is called a reference sequence. Therefore, development and availability of robust reference database are required for accurate SNP-based identifications Marth et al. (1999). For the development of reference sequence database, planned location-based sampling and sequencing is important for the development of high-throughput SNP profiling analysis.

Event-based reference sequences can also be generated and used to profile the SNPs distribution in the sequence isolated from a query sample. For example, sequence generated from a crime scene can be considered as a reference sequence and compared with a query sequence obtained from suspects to identify the individual leaving the DNA at the crime scene. However, to conduct such SNP-based analysis, quality sampling and extraction of nucleic acid from the evidence are technically challenging as the biological material present at the event is usually scarce and contaminated. As the whole genome from such limited biological sample is difficult, gene marker-based reference sequence generation is suggested and may result in cost and time efficiency compared to other type of SNP profile analysis and database building. Moreover, reference sequences can be supplemented with reference specific metadata such as disease condition, unique phenotypes, geographical location, migration data, and ancestry information.

21.3.2 Isolation of Query Sequences from the Target Samples to Compare SNP Profiles with Reference Sequences Is the Key to Individualization of the Sample

Once the reference sequence is available for the comparison, generation of a query sequence from target sample brings its own challenges such as ethical constrains with respect to sample collection and the biological source selection. As sample can be human or non-human, their availability or accessibility may yield low nucleic acid concentration and create contamination-based biases. For example, if sample source went through blood transfusion, SNP profile generated from their blood sample may not match with their SNP profile from other body cells. Hence, such false negative biasness can be avoided by selecting robust sampling source. Another challenge is to recover individual sequence from degrading mixed samples collected from environmental sources. Apart from these technicalities, especially when working with non-human IUCN-listed wildlife samples, nucleic acid sequences are unavailable from invasive methods due to their conservation status. To bypass such ethical hurdles, non-invasive nucleic acid extraction from scats and shed exoskeleton is advisable but usually results in limited yield.

21.3.3 Gene Marker-Based SNP Profiling Approaches for Identification Are Effective and Efficient Compared to de Novo Whole Genome Sequencing

As gene marker-based approaches selectively amplify a part of the genome using PCR, it is further sequenced for the identification and sequence compression. Whereas whole genome sequencing technology sequences random fragments from the genome and compiles them to identify nucleotide variations, this approach is tedious and impractical for the high-throughput applications Hillier et al. (2008). Therefore, selection of gene marker-based approaches is more suitable for large-scale and efficient screening of nucleotide variations Cargill et al. (1999).

With respect to forensic investigations, phenotypic gene marker tends to be very useful as they allow to proximate physical characters from nucleotide variations with high confidence. For such analysis, multi-gene marker panel such as HIrisPlex-S is available in the market with various phenotypic gene combinations that can predict the morphological features of the individual Sham et al. (2002). Genes that are used in the above panel are mentioned in Table 21.1. Presence of these genes is correlated with its database having 9466 SNP references for eye colour, 1878 SNP references

Table 21.1 Phenotypic gene combination of HIrisPlex-S panel system (modified from https://hirisplex.erasmusmc.nl/ & Breslina et al. 2019)

Gene marker	SNP variation reference	Chromosome Number	Nucleotide Position	Reference Allele	Substitut ed. Allele	Amplicon Number
TYR	rs1042602	11	88,911,696	C	A	TYR Amplicon
BNC2	rs10756819	9	16,858,084	G	A	BNC2 Amplicon
MC1R	rs1110400	16	89,986,130	T	C	BNC2 Amplicon
TYR	rs1126809	16	89,017,961	G	A	MC1R Amplicon 5
HERC2	rs1129038	15	28,356,859	G	A	HERC2 Amplicon 4
MC1R	rs11547464	16	89,986,091	G	A	MC1R Amplicon 2
IRF4	rs12203592	6	396,321	C	T	IRF4 Amplicon
OCA2	rs12441727	15	28,271,775	G	A	OCA2 Amplicon 5
KITLG	rs12821256	12	89,328,335	T	C	KITLG Amplicon
LOC1053 70,627	rs12896399	14	92,773,663	G	T	SLC24A4 Amplicon
HERC2	rs12913832	15	28,365,618	A	G	HERC2 Amplicon 1
TYR	rs1393350	11	89,011,046	G	A	TYR Amplicon
SLC24A5	rs1426654	15	48,426,484	A	G	SLC24A5 Amplicon
OCA2	rs1470608	15	28,288,121	G	T	OCA2 Amplicon 3

(continued)

Table 21.1 (continued)

Gene marker	SNP variation reference	Chromosome Number	Nucleotide Position	Reference Allele	Substitut ed. Allele	Amplicon Number
OCA2	rs1545397	15	28,187,772	A	T	OCA2 Amplicon 4
HERC2	rs1667394	15	28,530,182	C	T	HERC2 Amplicon 5
SLC45A2	rs16891982	5	33,951,693	C	G	SLC45A2 Amplicon 2
SLC24A4	rs17128291	14	92,882,826	A	G	SLC24A4 Amplicon
OCA2	rs1800407	15	28,230,318	C	T	OCA2 Amplicon 1
OCA2	rs1800414	15	28,197,037	T	C	OCA2 Amplicon 2
MC1R	rs1805005	16	89,985,844	G	T	OCA2 Amplicon 2
MC1R	rs1805006	16	89,985,918	C	A	OCA2 Amplicon 2
MC1R	rs1805007	16	89,986,117	C	T	OCA2 Amplicon 2
MC1R	rs1805008	16	89,986,144	C	T	OCA2 Amplicon 2
TUBB3	rs1805009	16	89,986,546	G	C	MC1R Amplicon 4
MC1R	rs201326893	16	89,986,122	C	A	MC1R Amplicon 4
MC1R	rs2228479	16	89,985,940	G	A	MC1R Amplicon 3
HERC2	rs2238289	15	28,453,215	A	G	HERC2 Amplicon 2

PIGU	rs2378249	20	33,218,090	G	A	PIGU Amplicon
SLC24A4	rs2402130	14	92,801,203	G	A	SLC24A4 Amplicon
SLC45A2	rs28777	5	33,958,959	C	A	SLC45A2 Amplicon 1
ANKRD1 1	rs3114908	16	89,383,725	T	C	ANKRD11 Amplicon
MC1R	rs3212355	16	89,984,378	C	T	MC1R Amplicon 6
LOC1053 74,875	rs4959270	6	457,748	C	A	EXOC2 Amplicon
RALY	rs6059655	20	32,665,748	A	G	RALY Amplicon
ASIP	rs6119471	20	32,785,212	C	G	ASIP Amplicon
HERC2	rs6497292	15	28,496,195	A	G	HERC2 Amplicon 3
TYRP1	rs683	9	12,709,305	C	A	TYRP1 Amplicon
DEF8	rs8051733	16	90,024,206	A	G	DEF8 Amplicon
MC1R	rs885479	16	89,986,154	G	A	MC1R Amplicon 1

for hair colour, 854 SNP references for hair shades, and 1423 SNP references for skin colour to predict the phenotype of sample based on sequence variations.

Apart from above-mentioned phenotype-specific gene marker panel, other independent gene markers are also used for objective-oriented studies or investigations. HLA typing is one of such popular single gene marker-based approaches to identify an individual Huchard et al. (2006). Another common target genomic region for such investigation is microsatellite markers which are highly unstable and accumulate high numbers of mutations, hence can be used for individualization. Moreover, if the target individual is a carrier of genetic disease, then disease-specific gene markers can also be used for the same purpose, but resolution of such gene marker may be of low confidence as frequency of disease-causing mutations might be highly similar in the population Glazier et al. (2002), Buetow et al. (1999), Liu et al. (1998).

21.4 SNP Techniques

21.4.1 Assessment of Mismatches and Fragmentation Patterns to Find SNPs in the Target DNA Sequence

Variation at single nucleotide results in development of mismatch in the DNA molecule and their presence can be identified with chemical modification. Moreover, their positions can also be identified using chemical cleavage followed by electrophoresis. Usually, target sequence is amplified using polymerase chain reaction (PCR) prior to SNP detection techniques.

21.4.1.1 Oxidative Modifications of DNA Mismatches Can Be Detected Via Differential Optical Absorbance

Treatment of potassium permanganate (pink in colour) oxidizes mismatches in hetroduplex DNA and generates hypomanganate diester (yellow in colour) with release of manganese dioxide. Formation of hypomanganate diester can be detected by the absorbance profile at 420 nm wavelength using standard UV/vis spectrophotometer. Presence of absorbance change due to above oxidation reaction confirms the presence of single or multiple SNPs in the target sequence. As potassium permanganate modifies mismatched thymine nucleotide, further sequence validation is suggested to identify other types of SNPs Hayatsu et al. (1991).

21.4.1.2 Chemical Cleavage of Mismatch (CCM) Can Fragment DNA at the Mismatch Site and Its Position Can Be Identified Using Electrophoresis

Chemical cleavage of mismatch (CCM) uses two-step process to first tag mismatches in the target sequence, and second cleave it at the position of the tag. Chemical modification as a tag is carried out by hydroxylamine, and potassium permanganate for cytosine, and thymine mismatches, respectively, before cleavage by the piperidine at the same sites. Lastly, cleaved fragments are separated on electrophoresis gel to identify the position of the SNPs in the target DNA sequence Cotton et al. (1988).

21.4.1.3 SNPs in the Restriction Sites of Endonucleases Can Be Detected Using the PCR–Restriction Fragment Length Polymorphism (RFLP)

Another convenient and cost-effective method to detect SNPs is the use of selective endonucleases for the identification of unique presence or absence of restriction sites resulting from nucleotide variations. Target sequence is searched for the reference distribution of the restriction sites to select correct endonucleases using tools such as SNP cutter. Next, target DNA sequence is amplified using PCR and digested with the selected endonucleases. Lastly, resulting DNA fragments patterns are compared between reference sequence and target sequence to identify the SNP distribution Ota et al. (2007).

21.4.2 Detection and Screening of SNPs in Target Sequence Using Sequence-Dependent Denaturation Profile

Sequence-specific denaturation property of DNA can be used to separate nucleotide variation-bearing sequences from the reference sequences, as denatured DNA molecules move slowly than intact DNA molecules in the gel electrophoresis. In this chapter we will discuss different techniques to utilize DNA's physical properties such as size, electrical charge, and denaturation rate for the detection of SNPs.

21.4.2.1 Denaturing Gradient Gel Electrophoresis (DGGE) Can Distinguish Between Different DNA Denaturation Rates Resulting from the Presence of SNPs Using Chemical Gradient

Denaturation property of DNA sequence depends upon its GC content and can be utilized to separate DNA sequences with varying melting characteristics using DGGE (Fig. 21.1) Myers et al. (1987), Tully et al. (2000). Higher resolution of DNA separation results from partial denaturation of the sequence. Therefore, DGGE usually function better with higher density of nucleotide variation, whereas addition of GC-clamp (40 bp GC rich sequence) may improve the separation of sequences with single nucleotide variation Sheffield et al. (1989), Fischer & Lerman (1983), Theophilus et al. (1989). *In silico*-simulated standardization of DGGE with or without GC-clamp to detect SNPs is advisable using BIO-RAD's DNA melting profile analysis software called WinMelt for windows, and MacMelt for mac-based computers. To obtain higher resolution via the GC-clamp requires prior knowledge of sequence for the primer designing. These primers with GC-clamp are used to amplify targeted sequence using polymerase chain reaction (PCR) before running on DGGE.

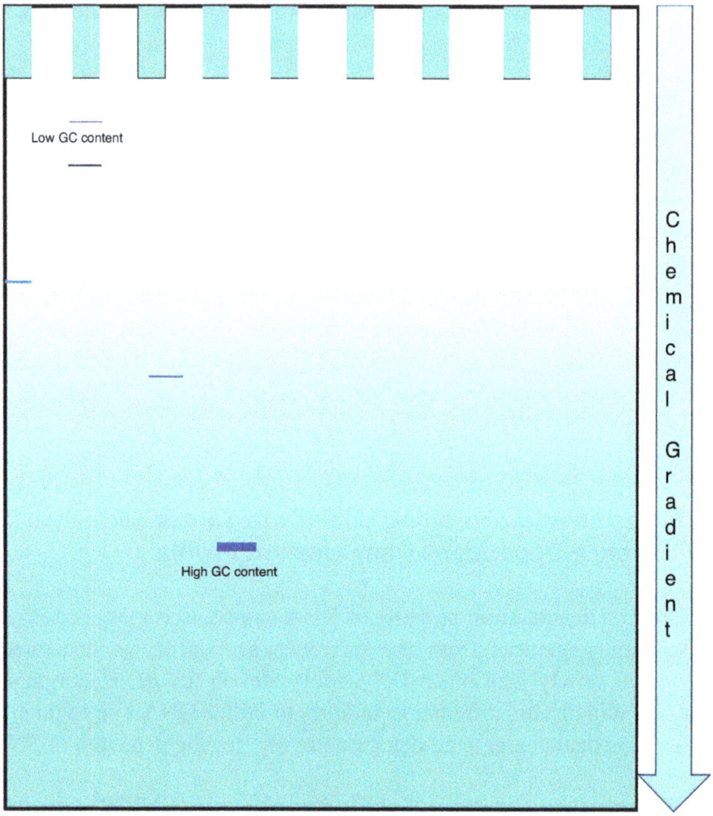

Fig. 21.1 Acrylamide gel created with denaturing agents such as urea and formamide

21.4.2.2 Temporal Temperature Gradient Electrophoresis (TTGE) Utilizes Differential Denaturation Property of SNP Bearing Sequence Via Gradual Temperature Increase for the Detection

Next electrophoretic technique to detect sequences with nucleotide variation and varying denaturation property is TTGE. This method can be used to discover SNPs in nDNA and mtDNA. Because this protocol doesn't use chemical denaturation gradient, it has higher reproducibility than DGGE, specifically for mtDNA Chen et al. (1999). As this method can also separate heteroplasmic mtDNA, it has been considered suitable method to detect mutations in mtDNA Ito et al. (2001), Wong et al. (2004). Target sequence must be amplified before loading on TTGE, and use of GC-clamp can further improve resolution of separation as it prevents complete denaturation of the sequence (Fig. 21.2). TTGE standardization can be simulated using WinMelt/MacMelt as described before to reduce the experimental cost.

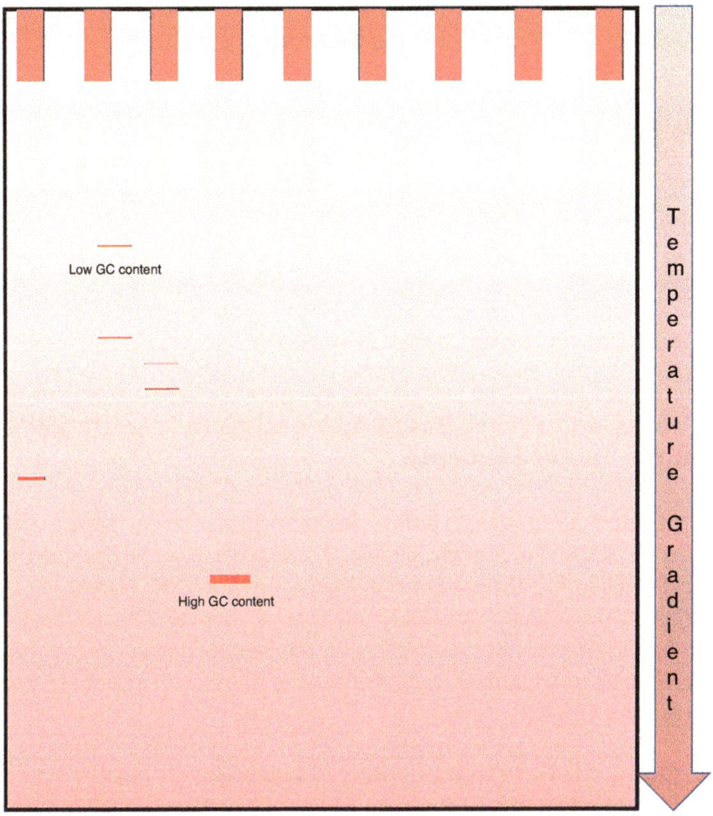

Fig. 21.2 Electrophoresis setup that gradually increase the temperature

21.4.2.3 Zn^{2+}–Cyclen-Based Charge Reduction and Conformation Change at Thymine Sites in DNA Can Be Used to Identify SNPs Using Polyacrylamide Gel Electrophoresis (PAGE) System

Zn^{2+}–1,4,7,10-tetraazacyclododecane (Zn^{2+}–cyclen) have property to reversibly bind with thymine (dT$^-$) bases in DNA sequence and change its conformation and charge at ph 8.0. Binding of Zn^{2+}–cyclen with dT breaks the hydrogen bond between Adenine and Thymine and reduce the melting temperature of the DNA sequence Kikuta et al. (1999), Kukita et al. (2002), Kinoshita et al. (2009), Kinoshita-Kukita (2002), Shionoya & Shirot (1993). Use of these Zn^{2+}–cyclen with PAGE and heteroduplex PCR amplification separates DNA sequences in maximum of four bands which may be further sequenced to identify the SNPs.

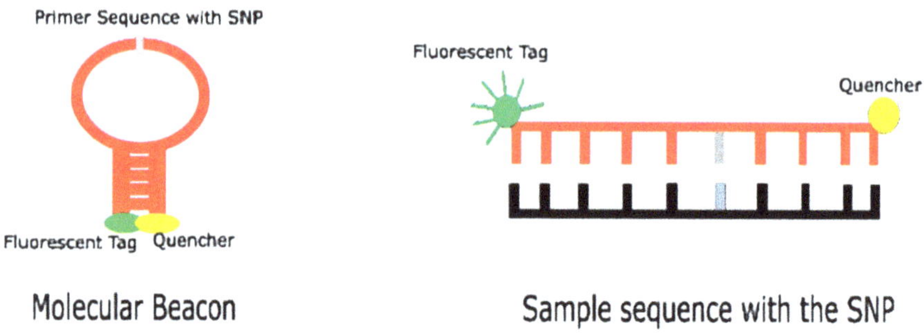

Fig. 21.3 Fluorescence Quenching Exhibited by Probes/Primers

21.4.3 Probes and Primer-Based Approaches to Identify SNPs in the Target Sequence

Based on the reference sequence, probes and primers can be designed to detect and amplify specific sequence from the sample. These probes can be designed with dye or fluorescent tags to flag the presence of sequences with SNPs Gao et al. (2019). In case of primers, it can have chemical modification that can change the migration pattern of the DNA molecules with SNPs on the electrophoresis gel or it can have specific nucleotide substitutions to selectively amplify with matching nucleotide variation (Fig. 21.3).

21.4.3.1 Single Strand Conformation Polymorphism (SSCP) Can Be Detected by Amplification of SNPs Containing DNA Sequences Prior to Their Fluorescent Tagging and Separation by Capillary Gel Electrophoresis

Precise identification of SNPs and indel mutations can be verified by post-labelling PCR product. Post-labelling reduces PCR error of added 3’-end labelled nucleotides. SNP positions can be detected by running the sample through automated capillary electrophoresis setup van Oorschot & Ballantyne (2013). In case of pooled DNA restricted to equal parts DNA contributions, results can be obtained by mapping allele frequencies in pooled and reference samples Kukita et al. (1997), Kukita et al. (2002), Higasa et al. (2002), Orita et al. (1989a, b).

21.4.3.2 Formation of Hybridized Probes Using Excimers for Detection of SNPs in Bacterial Strain

Excimers, herein Pyrene derivatives, emit fluorescence and act as probes for detection of single nucleotide variations. Fluorescence is emitted in the range of 500–-510 nm Paris et al. (1998), Yamana et al. (2002), Schü et al. (2018). The technique can also be utilized for studying genetic mutations.

21.4.3.3 Molecular Beacons in Association with TAG SNPs for Detection of Single Nucleotide Variation

Molecular beacons are highly selective stem-loop structures working as probes for SNP detection Lohrer & Tangen (2000). Ends of the stem are bound with a fluorescent tag and a quencher (Fig. 21.3) Varona & Anderson (2019). The apparatus helps in determination of the specific nucleotide substitution sites. Upon adhesion to the site of substitution, fluorescence is emitted confirming presence of polymorphism Tyagi & Kramer (1996). Nucleotide substitutions are less variable for populations in close contact. These substitutions can be termed as TAG SNPs. They provide a picture of the genetic diversity within a region. High-throughput techniques allow detection of several Tag SNPs simultaneously.

21.4.3.4 5′ -Nuclease Allelic Discrimination Probe-Based Assay for SNP Detection Through Use of Forward and Reverse Primers

The TaqMan method for SNP detection is probe-based. Fluorescently labelled probes are incorporated within the forward and reverse primers targeting the test sequence during the PCR reaction. Opposed to molecular beacons, they are a string with at 5' end and a quencher at the 3'end. Detection of homozygosity or heterozygosity is based on the intensity of the colour of fluorescence emitted. Presence of SNP is verified by unbinding of primer sequences due to nucleotide polymorphism.

21.4.4 Sequence Analysis to Accurately Detect SNPs in the Target Sequence

Technical advancement of nucleotide sequencing allowed to detect new SNPs accurately and efficiently. Development of modern panel-based amplification of multiple phenotype-specific gene markers and advanced bioinformatics tools with rich database are identifying individuals by accurately generating high-resolution phenotypic features and deciphering long lost migration histories Tillmar et al. (2021), Dario et al. 2011.

21.4.4.1 Sanger Sequencing Is Considered a "Gold Standard" to Accurately Detect SNPs in Targeted and Limited Numbers of Sequences

Frederick Sanger used dideoxynucleotides substituting normal deoxynucleoties which can inhibit the DNA replication and produce truncated DNA fragments from the site of incorporation Sanger et al. (1977). Length of these fragments suggests the position of respective nucleotide in the DNA sequence and assembling of all the possible fragments provide complete sequence of the target DNA molecule. Later on, above-mentioned technique was automated by Applied Biosystems, Inc. (ABI) with addition of fluorescent nucleotides (Inazuka et al. (1996)) and used to sequence the first human genome which was published in 2003 Lander et al. (2001). Moreover, this Sanger sequencing has been known as "gold standard" method to

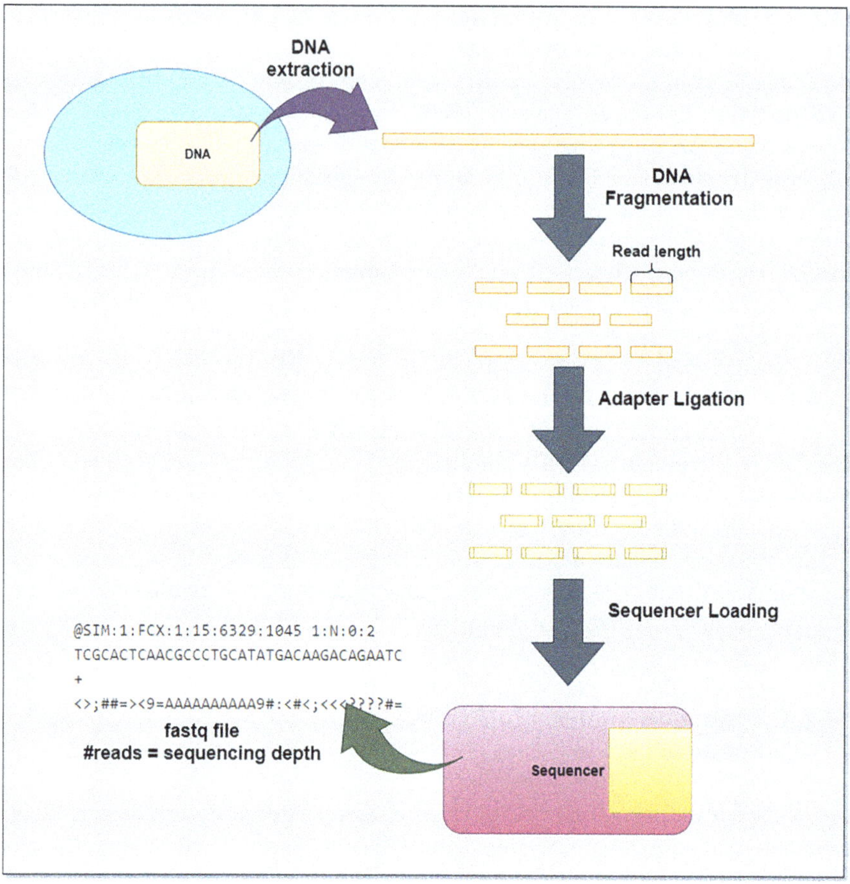

Fig. 21.4 NGS work flow

study SNPs. Limitation of Sanger sequencing is that it is low throughput, as it can only sequence 96-384 sequences per run.

Sanger sequencer generates nucleotide sequence in fasta format and its quality assessment can be done from its chromatogram. This sequence represents an individual sample and can be compared with reference sequence representing the comparison point to identify SNPs. For the comparison, query sequence is aligned on the reference sequence using sequence alignment tools such as MEGA, JalView, and SeqMan Pro to name a few.

21.4.4.2 Next-Generation Sequencing (NGS) Is High-Throughput Technique to Identify SNPs in longer and Many Numbers of Sequences

There are several platforms developed as the advancement of Sanger sequencing and all of these platforms uses their respective nucleotide detection method to sequence

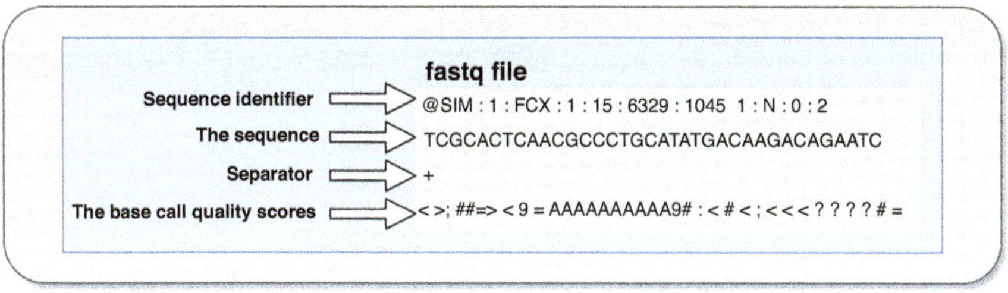

Fig. 21.5 Organization of raw NGS fastq file format

targeted sample (Fig. 21.4) Margulies et al. (2005). Illumina is widely used NGS platform which can sequence more than 150 bp long DNA fragments and considerably high sequencing depth with respectively low error rates. It uses sequencing by synthesis (SBS) technology to detect nucleotide sequence. Apart from illumina, other platforms also exist such as Ion torrent, nanopore, PacBio, etc. with their respective advantages and limitations. Selection of NGS platform and the sequencing workflow should be according to the objective of the study Ratan et al. (2013). For instance, if project expects to sequence complete gene sequence longer than 1 kb than long read sequencers such as nanopore, or PacBio should be selected, whereas if sequencing speed and the cost is the limiting factor then Ion torrent platform should be preferred. Next, shotgun sequencing can be used to explore an unknown nucleotide sequence to establish a reference data to compare with re-sequenced nucleotide sequences from the population either via shotgun or amplicon sequencing to identify SNPs. These newly produced reference sequences can be collected in public databases such as HapMap, dbSNP Entrez, and National Center for Biotechnology Information (NCBI) for easy distribution and access by larger scientific community Kruglyak & Nickerson (2001), International HapMap Consortium (2003), Sachidanandam et al. (2001).

Raw reads in fastq format produced from NGS platform contain sequence with quality scores (Fig. 21.5), which can be further used for SNP detection by process called the variance calling. In variance calling, raw reads coming from the query sample are aligned with the reference sequence using reference mapping tools such as bowtie2 to generate alignment file called Sequence Alignment Map (SAM) or Binary Alignment Map (BAM) before identifying SNP distribution by generating Variance Call File (VCF).

21.5 Use of Cases for Human SNP Profiling

SNPs provide information of individualization and their geographical residence. They are results of evolution and migratory patterns. Studying SNPs provide information as to the geographical abundance of a haplotype within a population Belmont et al. (2005), Oldoni et al. (2019). Migratory patterns can be established with certainty for human cases in presence of reference nucleotide sequences Bas Yavaser et al. (2021). Majority of SNP profiles are generated in cases of disputes as to confirm the absence or presence of the polymorphic characters for individualization Bardan (2019).

21.5.1 April M. Tinsley Case

April Tinsley (8 years old), Fort Wayne, Indiana, was the victim of sexual assault and murder in April, 1988. The body was found few days after the incident. DNA evidence of suspected perpetrator was obtained from the clothing of the victim. After a few days of discovery, writing on a nearby barn confessing the murder by the perpetrator was found with another threat to kill. Law enforcement agencies working on the case were unable to identify potential suspects. After many rounds of questioning of potential suspects based on suspect lists of identical crimes, no match was found. Within this time, more letters of confession and threat from the perpetrator were received by the law enforcement agencies along with evidence containing semen sample of the perpetrator which when compared with the sample found on the victim's clothing provided with an accurate match. The case remained unsolved for a long time period as there was no suspect that matched with the profile generated by the Law enforcement agencies. In the mid-2018, the case was reopened as the lead genealogist CeCe Moore of a reputed lab found ancestral data proving relevant to the case. Phenotypic profiles of probable suspects were created based on the reference SNP profile of DNA sample collected by the Law enforcement agencies during investigation. From the suspect list, based upon probability and life expectancy, two matches living in close vicinity from the scene of crime were detected. Surveillance was carried out for each suspect and their DNA samples were collected from trash samples and were matched with the reference standard. One sample (John Miller) was found to be matching and the suspect was called upon for questioning. Upon questioning, he admitted to the crimes committed against April Tinsley and was found guilty and sentenced to 80 years in prison for his crimes Armentrout (2019).

21.5.2 Population Mixture and Human Migration of China

Ancient genomes can be sampled from the mortal remains of humans found from different geographical locations which further can be compared with modern genomes from related geographical location to track down the migratory path of humans through spatiotemporal landscape. Robust database- (>629433 SNP sites) backed ADMIXTOOLS is applied to infer migratory patterns on the data generated from similar experiment setup described above. In this study, authors have deciphered the hazy genetic history of Yellow and West Liao River-associated human society. They suggested the probable migration of Yellow River-associated humans to modern day southern and southeast Chinese population following the movement of humans to yellow river from West Liao River. They also discussed the possible need for such migration to be likeness towards farming and accessibility to developed economies Ning et al. (2020).

21.6 Conclusion

SNPs act as unique tags in identification of individuals. Evolution and adaptation of the different species within varied environments provide them with their unique identification tags. Analysis of these polymorphic characters via the different techniques gives an exact overview of their genetic composition. The data generated from the analysis are raw and upon processing have numerous clinical and non-clinical applications.

Currently, human identification and individualization through the use of SNPs has gained importance in the field of forensic science. Forensic DNA Phenotyping (FDP) based on Externally Visible Characteristics (EVC) has opened new horizons for law enforcement agencies allowing them to generate possible suspect profiles based on the trace amounts of DNA evidence obtained from the scene of crime. Single nucleotide variations have enabled the investigating agencies to narrow down suspect list considerably as the SNPs provide adequate data for phylogenetic analysis.

Further, research within the subject has great potential not only with respect to criminalistics, but also within areas of clinical researches like oncological and migratory studies, etc. Generation of SNPs databases and construction of SNP Tags will offer a helping hand to researchers and law enforcement agencies looking forward for research and analysis.

Multiple Choice Questions

1. What is responsible for accumulation of SNPs in nucleotide sequence?
   (a) Transcription.
   (b) Mutation.
   (c) Translation.
   (d) Post-translation modification.
   Answers: (b)
2. SNP produces.
   (a) Mismatch.
   (b) Difference in melting temperature.
   (c) Sequence variation.
   (d) All of the above.
   Answers: (d)
3. SNP may result in.
   (a) Phenotypic variation.
   (b) Disease condition.
   (c) Both of above.
   (d) None of above.

Answers: (c)

4. Who solved the cold case of April M. Tinsley using SNP profiling.
   (a) Sir Francis Galton.
   (b) CeCe Moore.
   (c) Sir Alec Jeffery.
   (d) Dr. Edmond Locard.
   Answers: (b)
5. Which tool can be used to detect migration pattern from SNP profile.
   (a) ADMIX tools.
   (b) Alignment tools.
   (c) Phylogenetic tools.
   (d) All of above.
   Answers: (a)

References

Allio R, Donega S, Galtier N, Nabholz B (2017) Large variation in the ratio of mitochondrial to nuclear mutation rate across animals: implications for genetic diversity and the use of mitochondrial DNA as a molecular marker. Mol Biol Evol 34:2762–2772

Armentrout S. Using SNP genotyping to solve crimes: the cold case of April Tinsley warning: this presentation contains graphic sexual content. 2019

Arshad M, Bhatti A, John P (2018) Identification and in silico analysis of functional SNPs of human TAGAP protein: a comprehensive study. PLoS One 13:1–13

Bardan F. New forensic DNA profiling techniques for human identification. 2019

Barreiro LB, Laval G, Quach H, Patin E, Quintana-Murci L (2008) Natural selection has driven population differentiation in modern humans. Nat Genet 40:340–345

Bas Yavaser G, Hulya Yukseloglu E, Cavus Yonar F, Erkan I (2021) Assessment of 13 single nucleotide polymorphisms loci for identification in forensic sciences for Turkish population. Int J Biol Chem 14:56–63

Belmont JW, Boudreau A, Leal SM, Hardenbol P, Pasternak S, Wheeler DA et al (2005) A haplotype map of the human genome. Nature 437:1299–1320

Bogenhagen D, Clayton DA (1974) The number of mitochondrial deoxyribonucleic acid genomes in mouse L and human HeLa cells. Quantitative isolation of mitochondrial deoxyribonucleic acid. J Biol Chem 249:7991–7995

Breslin K, Wills B, Ralf A, Ventayol Garcia M, Kukla-Bartoszek M, Pospiech E et al (2019) HIrisPlex-S system for eye, hair, and skin color prediction from DNA: massively parallel sequencing solutions for two common forensically used platforms. Forensic Sci Int Genet 43: 102152

Brookes AJ (1999) The essence of SNPs. Gene 234:177–186

Buetow KH, Edmonson MN, Cassidy AB (1999) Reliable identification of large numbers of candidate SNPs from public EST data. Nat Genet 21:323–325

Cargill M, Altshuler D, Ireland J, Sklar P, Ardlie K, Patil N et al (1999) Characterization of single-nucleotide polymorphisms in coding regions of human genes [published erratum appears in Nat Genet 1999 Nov;23(3):373]. Nat Genet 22:231–238

Chen TJ, Boles RG, Wong LJ (1999) Detection of mitochondrial DNA mutations by temporal temperature gradient gel electrophoresis. Clin Chem 45:1162–1167

Cotton RG, Rodrigues NR, Campbell RD (1988) Reactivity of cytosine and thymine in single-base-pair mismatches with hydroxylamine and osmium tetroxide and its application to the study of mutations. Proc Natl Acad Sci U S A 85:4397–4401

Dario P, Ribeiro T, Dias D, Corte-Real F, Geada H (2011) Complex casework using single nucleotide polymorphisms. Forensic Sci Int Genet Suppl Ser 3:379–380

Darwin C (1859) On the origin of species by means of natural selection, or preservation of favoured races in the struggle for life. John Murray, London; 1809-1882

Fiatal S, Ádány R (2018) Application of single-nucleotide polymorphism-related risk estimates in identification of increased genetic susceptibility to cardiovascular diseases: a literature review. Front Public Heal 5:5

Fischer SG, Lerman LS (1983) DNA fragments differing by single base-pair substitutions are separated in denaturing gradient gels: correspondence with melting theory. Proc Natl Acad Sci U S A 80:1579–1583

Gao Y, He Z, He X, Zhang H, Weng J, Yang X et al (2019) Dual-color emissive AIEgen for specific and label-free double-stranded DNA recognition and single-nucleotide polymorphisms detection. J Am Chem Soc 141:20097–20106

Glazier AM, Nadeau JH, Aitman TJ (2002) Genetics: finding genes that underline complex traits. Science 298(80):2345–2349

Gunter LE, Kochert G, Giannasi DE (1994) Phylogenetic relationships of the Juglandaceae. Plant Syst Evol 192:11–29

Hayatsu H, Atsumi G, Nawamura T, Kanamitsu S, Negishi K, Maeda M (1991) Permanganate oxidation of nucleic acid components: a reinvestigation. Nucleic Acids Symp Ser 25:77–78

Higasa K, Kukita Y, Baba S, Hayashi K (2002) Software for machine-independent quantitative interpretation of SSCP in capillary array electrophoresis (QUISCA). BioTechniques 33:1342–1348

Hillier LDW, Marth GT, Quinlan AR, Dooling D, Fewell G, Barnett D et al (2008) Whole-genome sequencing and variant discovery in C. elegans. Nat Methods 5:183–188

Huchard E, Cowlishaw G, Raymond M, Weill M, Knapp LA (2006) Molecular study of Mhc-DRB in wild chacma baboons reveals high variability and evidence for trans-species inheritance. Immunogenetics 58:805–816

Inazuka M, Tahira T, Hayashi K (1996) One-tube post-PCR fluorescent labeling of DNA fragments. Genome Res 6:551–557

International HapMap Consortium. International HapMap consortium. The International HapMap Project Nature 2003; 426: 789–796

Ito M, Tran Le S, Chaudhari D, Higashimoto T, Maslim A, Boles RG (2001) Screening for mitochondrial DNA heteroplasmy in children at risk for mitochondrial disease. Mitochondrion 1:269–278

Kikuta E, Murata M, Katsube N, Koike T, Kimura E (1999) Novel recognition of thymine base in double-stranded DNA by zinc(II)- macrocyclic tetraamine complexes appended with aromatic groups. J Am Chem Soc 121:5426–5436

Kikuta E, Aoki S, Kimura E (2002) New potent agents binding to a poly(dT) sequence in double-stranded DNA: Bis(Zn2+−cyclen) and tris(Zn2+−cyclen) complexes. J Biol Inorg Chem 7: 473–482

Kinoshita E, Kinoshita E, Koike T. Chapter 10 Zn (II)– Cyclen polyacrylamide gel electrophoresis for SNP. 2009

Kinoshita-Kikuta E, Kinoshita E, Koike T (2002) Erratum: a novel procedure for simple and efficient genotyping of single nucleotide polymorphisms by using the Zn2+−cyclen complex (nucleic acids research (2002) vol. 30 (e126)). Nucleic Acids Res 30:5593

Kruglyak L, Nickerson DA (2001) Variation is the spice of life. Nat Genet 27:234–236

Kukita Y, Tahira T, Sommer SS, Hayashi K (1997) SSCP analysis of long DNA fragments in low pH gel. Hum Mutat 10:400–407

Kukita Y, Higasa K, Baba S, Nakamura M, Manago S, Suzuki A et al (2002) A single-strand conformation polymorphism method for the large-scale analysis of mutations/polymorphisms using capillary electrophoresis. Electrophoresis 23:2259–2266

Lander ES, Linton LM, Birren B, Nusbaum C, Zody MC, Baldwin J et al (2001) Erratum: initial sequencing and analysis of the human genome: international human genome sequencing consortium (nature (2001) 409 (860-921)). Nature 412:565–566

Liu W, Smith DI, Rechtzigel KJ, Thibodeau SN, James CD (1998) Denaturing high performance liquid chromatography (DHPLC) used in the detection of germline and somatic mutations. Nucleic Acids Res 26:1396–1400

Lohrer HD, Tangen U (2000) Investigations into the molecular effects of single nucleotide polymorphism. Pathobiology 68:283–290

Margulies M, Egholm M, Altman WE, Attiya S, Bader JS, Bemben LA et al (2005) Genome sequencing in microfabricated high-density picolitre reactors. Nature 437:376–380

Marth GT, Korf I, Yandell MD, Yeh RT, Gu Z, Zakeri H et al (1999) A general approach to single-nucleotide polymorphism discovery. Nat Genet 23:452–456

Morin PA, Luikart G, Wayne RK (2004) SNPs in ecology, evolution and conservation. Trends Ecol Evol 19:208–216

Myers RM, Maniatis T, Lerman LS (1987) Detection and localization of single base changes by denaturing gradient gel electrophoresis. Methods Enzymol 155:501–527

Ning C, Li T, Wang K, Zhang F, Li T, Wu X et al (2020) Ancient genomes from Northern China suggest links between subsistence changes and human migration. Nat Commun 11:1–9

Oldoni F, Kidd KK, Podini D (2019) Microhaplotypes in forensic genetics. Forensic Sci Int Genet 38:54–69

Orita M, Iwahana H, Kanazawa H, Hayashi K, Sekiya T (1989a) Detection of polymorphisms of human DNA by gel electrophoresis as single-strand conformation polymorphisms. Proc Natl Acad Sci U S A 86:2766–2770

Orita M, Suzuki Y, Sekiya T, Hayashi K (1989b) Rapid and sensitive detection of point mutations and DNA polymorphisms using the polymerase chain reaction. Genomics 5:874–879

Ota M, Fukushima H, Kulski JK, Inoko H (2007) Single nucleotide polymorphism detection by polymerase chain reaction-restriction fragment length polymorphism. Nat Protoc 2:2857–2864

Paris PL, Langenhan JM, Kool ET (1998) Probing DNA sequences in solution with a monomer-excimer fluorescence color change. Nucleic Acids Res 26:3789–3793

Ratan A, Miller W, Guillory J, Stinson J, Seshagiri S, Schuster SC (2013) Comparison of sequencing platforms for single nucleotide variant calls in a human sample. PLoS One 8:1–10

Robert F, Pelletier J (2018) Exploring the impact of single-nucleotide polymorphisms on translation. Front Genet 9:1–11

Sachidanandam R, Weissman D, Schmidt SC, Kakol JM, Stein LD, Marth G et al (2001) A map of human genome sequence variation containing 1.42 million single nucleotide polymorphisms. Nature 409:928–933

Sanger F, Nicklen S, Coulson AR (1977) DNA sequencing with chain-terminating inhibitors. Proc Natl Acad Sci U S A 74:5463–5467

Schürch AC, Arredondo-Alonso S, Willems RJL, Goering RV (2018) Whole genome sequencing options for bacterial strain typing and epidemiologic analysis based on single nucleotide polymorphism versus gene-by-gene–based approaches. Clin Microbiol Infect 24:350–354

Sham P, Bader JS, Craig I, O'Donovan M, Owen M (2002) DNA pooling: a tool for large-scale association studies. Nat Rev Genet 3:862–871

Sheffield VC, Cox DR, Lerman LS, Myers RM (1989) Attachment of a 40-base-pair G + C-rich sequence (GC-clamp) to genomic DNA fragments by the polymerase chain reaction results in improved detection of single-base changes. Proc Natl Acad Sci U S A 86:232–236

Shionoya M, Shirot M (1993) A New Ternary Zinc(II)Complex with [12]aneN4 (= 1, 4, 7, 10-tetraazacyclododecane) and AZT (= 3'-azido-3'-deoxythymidine). Highly selective recognition of thymidine and its related nucleosides by a zinc (II) macrocyclic tetraamine complex with novel complementary associations. J Am Chem Soc 4:6730–6737

Theophilus BDM, Latham T, Grabowski GA, Smith FI (1989) Comparison of RNase a, a chemical cleavage and GC-clamped denaturing gradient gel electrophoresis for the detection of mutations in exon 9 of the human acid β-glucosidase gene. Nucleic Acids Res 17:7707–7722

Tillmar A, Sturk-Andreaggi K, Daniels-Higginbotham J, Thomas JT, Marshall C (2021) The FORCE panel: an all-in-one SNP marker set for confirming investigative genetic genealogy leads and for general forensic applications. Genes (Basel) 12:12

Tully LA, Parsons TJ, Steighner RJ, Holland MM, Marino MA, Prenger VL (2000) A sensitive denaturing gradient-gel electrophoresis assay reveals a high frequency of heteroplasmy in hypervariable region 1 of the human mtDNA control region. Am J Hum Genet 67:432–443

Tyagi S, Kramer FR (1996) Molecular beacon probes that fluoresce on hybridiztion. Nature 14: 303–308

van Oorschot RAH, Ballantyne KN (2013) Capillary electrophoresis in forensic biology. In: Encyclopedia of forensic sciences: second edition, 2nd edn. Elsevier Ltd, London

Varona M, Anderson JL (2019) Visual detection of single-nucleotide polymorphisms using molecular beacon loop-mediated isothermal amplification with centrifuge-free DNA extraction. Anal Chem 91:6991–6995

Wong L-JC, Chen T-J, Tan D-J (2004) Detection of mitochondrial DNA mutations using temporal temperature gradient gel electrophoresis. Electrophoresis 25:2602–2610

Yamana K, Iwai T, Ohtani Y, Sato S, Nakamura M, Nakano H (2002) Bis-pyrene-labeled oligonucleotides: sequence specificity of excimer and monomer fluorescence changes upon hybridization with DNA. Bioconjug Chem 13:1266–1273

Insect DNA Testing

22

Moumita Sinha, Arjun Rao Isukapatla, Prashant Kumar, Paromita Banerjee, and Neelam Ahirwar

Abstract

Insects are one of the most diverse and abundant groups of animals on Earth. There are over a million known species of insects. Insects can be used as forensic evidence in crime scene investigations. They can help determine the time and location of death and the cause of death. Additionally, insect DNA can be used to establish the time of death and the location of a crime scene. This could help investigators narrow down the timeline of a crime or determine if a murder occurred in a specific area. Insects may also be used to detect traces of drugs or poisons at a crime scene. Insects could also be used in cases of mass disasters, to help identify victims and determine the cause of death. Insects may not be the most efficient or reliable way to obtain forensic evidence, but they could provide valuable insight and help investigators solve crimes.

Keywords

Insects · Forensic · Identification · DNA · Postmortem

M. Sinha
Department of Forensic Science, Kalinga University, Raipur, Chhattisgarh, India

Department of Cardiology, AIIMS, Rishikesh, India
e-mail: moumita.sinha@kalingauniversity.ac.in

A. R. Isukapatla
Department of Cardiology, AIIMS, Rishikesh, India

Faculty of Forensic Science, School of Life Sciences, Christ (Deemed to be University), Bengaluru, Karnataka, India
e-mail: arjun.rao@christuniversity.in

P. Kumar · P. Banerjee (✉) · N. Ahirwar
Department of Cardiology, AIIMS, Rishikesh, India
e-mail: paromitabanerjee08@gmail.com

A. Puri et al. (eds.), *Fundamentals of Forensic Biology*,
https://doi.org/10.1007/978-981-99-3161-3_22

22.1 Introduction

Insects are any small, terrestrial arthropod (an invertebrate with an exoskeleton and segmented body) that has six legs, three body sections (head, thorax, and abdomen), and one pair of antennae. Insects can be found in almost every habitat on Earth, from tropical rainforests to the coldest polar regions. There are over a million known species of insects, making them the most diverse and abundant group of animals on the planet.

Insects are one of the most diverse and abundant groups of animals on Earth. They are found in nearly every habitat and can adapt to a wide variety of environments. There are over a million known species of insects, and scientists estimate that there may be as many as 30 million species that have yet to be discovered. Insects can be divided into several types, including beetles, flies, bees and wasps, ants, butterflies and moths, true bugs, grasshoppers, and crickets (Janzen 1976; May 1988).

Insects can be used as forensic evidence for detectives and scientists as they can provide important information about a person's time and place of death. Insects can be found at a crime scene and can help determine the victim's time of death. For example, the presence of certain species of flies or beetles can indicate the time of year or even the time of day of a death. Insects can also provide information about the location of a crime scene, as different species of insects are found in different climates and regions. The presence of insects can also indicate whether a body has been moved, as different insect species will be attracted to different sites. Using insects as evidence in a crime scene, investigation can help to provide valuable clues and aid in the investigation process. They can help in crime scene investigations and help to provide evidence that can be used in court. Insects are often used to determine the time and location of death, as well as the cause of death. They can also be used to determine the type of weapon used and the presence of drugs or poisons (Amendt et al. 2004). Insects can also identify the presence of other materials such as fibers, hair, and blood, which can help link suspects with a crime scene. In addition, the presence of insects can provide valuable insight into the environment at the time of the crime, which can be used to help reconstruct the events leading up to the crime.

Forensic entomology is the study of insects and their relationship to legal determines the victim's time of death leads to provide evidence in criminal cases by studying the insects associated with a crime scene. This evidence can help determine the time and place of death and reconstruct how a crime took place. Forensic entomologists can also provide information about the presence of illicit drugs or poisons in a victim's body (LeBlanc and Logan 2010).

22.2 Insects of Forensic Importance

22.2.1 Necrophagous Insects

22.2.1.1 Blow Flies

These insects are often the first to arrive at a corpse and are important for determining the time of death.

Blow flies are a type of fly that is typically known for their metallic blue or green color. They are found on all continents except Antarctica and are especially prevalent in the summer months. Blow flies are usually found near garbage or decaying animal matter, as they are attracted to the smell of it. They lay their eggs in or near the decaying material, and their larvae feed on it. The larvae, also known as maggots, can be used in forensic investigations to estimate the time of death of an animal or person (Gary and Lance 2002).

22.2.2 Common Species of Blow Flies in Forensic Cases

The three most common species of blow flies used in forensic investigations are *Calliphora vicina*, *Calliphora vomitoria*, and *Lucilia sericata*. These species of blow flies typically lay their eggs on decaying material, such as carrion or corpses, making them ideal for forensic investigations. They can provide information about the time of death and other aspects of the crime scene. The eggs are easily identifiable and can be used to identify the species of blowfly present. They can also be used to determine the postmortem interval, or how long ago the death occurred (Fig. 22.1).

22.2.2.1 Carpet Beetles

These insects have been known to infest homes and are often used as evidence in cases of arson and insurance fraud.

Carpet beetles are small insects that feed on carpets, fabrics, and other materials found in homes. They can cause significant damage to furniture and clothing and are

Fig. 22.1 Blow Flies

considered a nuisance pest. Carpet beetles are round or oval in shape and vary in size from 1/16 to 1/8 of an inch. They are usually black or brown in color, but can vary from species to species. Some carpet beetles have colorful markings, such as iridescent scales. Carpet beetles lay their eggs in dark, hidden areas, often near fabrics or carpets. The larvae are typically yellow or white and covered in tiny hairs. They feed on fabrics, carpets, and other materials and can cause significant damage if left unchecked. Carpet beetles can be difficult to control, so it is important to identify the source of the infestation and remove it. Chemical treatments may be necessary in some cases.

22.2.3 Common Species of Carpet Beetles in Forensic Cases

The most common species of carpet beetles in forensic cases are the varied carpet beetle (*Anthrenus verbasci*) and the black carpet beetle (*Attagenus unicolor*). The varied carpet beetle feeds on animal products such as wool, fur, feathers, and animal skins. The black carpet beetle feeds on plant material such as grains, cereals, and other starchy items. Both of these species can infest homes, furniture, and carpets and can be difficult to eradicate. In forensic cases, these species can be important evidence for investigators as they can provide clues about the environment in which the incident occurred (Fig. 22.2).

22.2.3.1 Flesh Flies

These insects are often used to determine the age of a corpse and can also be used to help determine the cause of death.

Flesh flies are one of the few species of flies that can reproduce without mating. Flesh flies are particularly attracted to rotting meat and are often found near slaughterhouses, butcher shops, and other places where there is a lot of decaying organic matter. Flesh flies are important pollinators of many plants and flowers. Flesh flies are not known to transmit any human disease, although they can spread

Fig. 22.2 Carpet Beetles

the larvae of some parasites. Flesh flies are one of the few species that have been observed to have parental care, in which the female guards her eggs until they hatch. Flesh flies are also an important food source for many animals, including birds and amphibians. Flesh flies are often used as model organisms for studying animal behavior.

22.2.4 Common Species of Flesh Flies in Forensic Cases

Sarcophaga carnaria: This species is the largest and most common flesh fly in forensic cases. It is mostly found in areas of high human activity, such as refuse dumps, carcasses, and occasionally in homes.

Sarcophaga haemorrhoidalis: This species is typically found near human corpses and is common in forensic cases.

Wohlfahrtia magnifica: This species is most commonly found near cadavers and is also a common species in forensic cases.

Calliphora vomitoria: This species is found near dead animals, garbage, and occasionally in homes. It is also a common species in forensic cases.

Calliphora vicina: This species is most commonly found in urban areas, in garbage, and occasionally in homes. It is also a common species in forensic cases (Fig. 22.3).

Fig. 22.3 Flesh Flies

22.2.5 Hematophagous Insects

22.2.5.1 Termites

These insects are important for determining whether a structure has been damaged by water or other external forces.

Termites are small, winged insects that are found in most parts of the world. They feed on wood and other materials that contain cellulose, such as paper and cardboard. Termites are considered pests because they can cause extensive damage to buildings and furniture. They can also cause significant health issues since they can spread diseases and allergens. To help prevent termite infestations, homeowners should seal off any potential entry points, such as cracks in the foundation, and inspect their property regularly for signs of termites (Gary and Lance 2002).

22.2.6 Common Species of Termites in Forensics Cases

- Eastern Subterranean Termites: These are the most common termites found in forensic cases. They live in the soil and build mud tubes to travel to and from their food sources.
- Formosan Subterranean Termites: Found primarily in the Southeastern United States, this species is especially destructive, as they can cause significant damage to homes and other structures.
- Drywood Termites: These termites are found in coastal areas and are able to survive in dry wood.
- Dampwood Termites: These termites are usually found in damp wood and can cause significant damage.
- Conehead Termites: These species is found mostly in Florida, and they can cause damage to structures and trees (Fig. 22.4).

22.2.6.1 Sand Flies

Sand flies can be found in forensic cases, as they are attracted to human blood and decomposing bodies. They can be found in areas where bodies have been dumped, and they can also be found on corpses. Sand flies are also used to help determine time of death, as the larvae can be used to study the rate of decomposition of a body.

Fig. 22.4 Termites

Fig. 22.5 Sand Flies

Additionally, sand flies can help identify the location of a crime scene since they feed and breed in specific areas (Fig. 22.5).

22.2.6.2 Deer Flies

Deer flies have been used in forensic cases to help identify the presence of persons at the scene of a crime. The presence of deer fly larvae in the clothing of a victim or suspect can provide valuable evidence in criminal investigations. Deer fly larvae are microscopic and can remain on clothing material for weeks or months after contact with a person. For example, in cases involving homicides, investigators have used deer fly larvae as evidence to determine when a victim was last alive or to show the movement of a suspect. Deer fly larvae can also be used to trace the movement of a suspect or victim in a particular area, as well as to provide evidence of the presence of a particular individual at the scene of a crime (Gary and Lance 2002) (Fig. 22.6).

22.2.7 Common Methods Used in Forensic Entomology

22.2.7.1 Temperature Analysis

Temperature is one of the most important environmental factors influencing entomological evidence in forensic cases. Temperature can affect the rate of insect development, adult longevity, and insect species composition. Temperature can also be used to estimate a body's postmortem interval (PMI).

22.2.7.2 Insect Taxonomy

Taxonomic identification of insect specimens is an important part of forensic entomology. Specimens can be identified to the family, genus, or species level.

22.2.7.3 Insect Colonization

The presence and abundance of insect species in a given area can be used to determine the PMI of a body. Different species show up at different times after death and can indicate the age of a corpse.

Fig. 22.6 Deer Fly

22.2.7.4 Maggot Morphology

The morphology of maggots can also be used to determine the PMI. Maggots of different species can have different shapes and sizes and can be used to estimate the age of a corpse.

22.2.7.5 Forensic Light Sources

Forensic light sources are used to detect evidence that may be invisible to the naked eye, such as insects or insect parts. These light sources can range from simple handheld devices to complex laboratory equipment.

22.2.8 Insect DNA for Human Identification in Forensic Cases

The use of insect DNA for human identification in forensic cases is a relatively new and emerging field. Scientists have recently been able to use insect DNA to distinguish human remains, linking them to specific individuals. This can be done by analyzing the DNA of an insect that has fed on the remains. Insects such as blowflies, flesh flies, and other carrion-feeding insects are able to ingest human DNA and store it in their gut for a period of time. By extracting and analyzing the insect DNA, scientists can match it to the remains and thus identify the deceased. Insect DNA identification methods are particularly useful for identifying remains that have been exposed to the elements for a long period of time and are too decomposed or damaged for other methods of identification such as fingerprints or dental records to be used. Insect DNA can also be used to identify unidentified human remains, helping to provide closure to families and law enforcement. Insect DNA identification is still an emerging field and is currently limited in its accuracy and reliability, as the insect DNA must be isolated from other sources of contamination and DNA from the surrounding environment. As such, it is not yet a reliable method for human identification and should be used in conjunction with other methods of identification (Wells et al. 2008).

22.2.9 DNA-Based Approach Used for Insect DNA Testing in Human Identification

One of the most common applications for insect evidence gathered during a death investigation is calculating the postmortem interval (PMI). When based on the age of a maggot (fly larva), this estimate can indicate the time since death. Because the age of the larva is only relevant to PMI if all of the maggot's development and feeding occurred on the victim. This assumption is usually justified because the larva was collected directly from or near a corpse that showed other signs of decomposition. However, such a direct and certain association cannot always be made, and during the course of any casework, a number of situations encountered in which an alternative method of associating a maggot with a corpse would have been useful.

22.2.9.1 Polymerase Chain Reaction (PCR)

This method involves amplifying a specific section of DNA to produce multiple copies of the target sequence, which can then be used for further analysis.

Polymerase chain reaction (PCR) is a powerful technique used to amplify small amounts of DNA. It is a widely used tool in forensic science, as it can be used to amplify even trace amounts of insect DNA found at a crime scene. This enables investigators to narrow down the possible suspects by determining the species of insect present. PCR can also be used to determine the age of an insect, as well as its sex. In addition, PCR can be used to determine whether the insect was exposed to a particular environmental factor, such as a chemical or pesticide. By amplifying the insect DNA, it is possible to obtain enough material for further analysis, such as DNA profiling.

22.2.9.2 Mitochondrial DNA Analysis in Insects and Humans

Insect mitochondrial DNA analysis can be used in forensic cases to help identify and differentiate between different species of insects. By analyzing the mitochondrial DNA of insects, researchers can determine the species, subspecies, and even age of the insect. This can be used to help identify the insect species present at a crime scene, allowing investigators to gain valuable insight into the crime or incident. Additionally, insect mitochondrial DNA can be used to compare insect species present at different crime scenes, helping to determine if a particular species of insect was present at both locations. This can be used to help establish links between different crimes, or to reconstruct the events that took place during a criminal act (Wells et al. 2001).

In forensic cases, mitochondrial DNA analysis of insect gut can be used to identify individuals or even species of insects. This method can help to determine the origin of a particular insect, or determine the source of contamination in a particular area. It can also be used to identify human remains that have been exposed to insects, such as in cases involving outdoor crime scenes.

Analysis of mitochondrial DNA from insect gut can be used to identify individuals or species of insects and to determine the age of an insect. This type of analysis involves extracting DNA from the gut of an insect and then using polymerase chain reaction (PCR) to amplify the mitochondrial DNA for further analysis. Once the amplified DNA is sequenced, the sequences can be compared to known sequences from other insects, or to a reference database. This can help to identify the species of insect, as well as the geographic origin of the insect.

In addition to identifying the species of insect, analysis of mitochondrial DNA from insect gut can also be used to identify human remains that have been exposed to insects. This type of analysis involves extracting DNA from the gut of an insect and then using PCR to amplify the mitochondrial DNA. The amplified DNA is then compared to a reference database of human mitochondrial DNA sequences. If the sequences match, then the insect can be used to identify a particular individual or species (Samerjai et al. 2019).

Overall, analysis of mitochondrial DNA from insect gut can be a useful tool in forensic cases. It can help to identify individuals or species of insects, as well as human remains that have been exposed to insects.

22.2.9.3 Y-Chromosome Analysis

This method involves analyzing the DNA in the Y chromosome, which is passed down from father to son. Human Y-DNA analysis from insect gut in forensic cases is a relatively new method of DNA analysis that is being developed to help solve crimes. This method involves extracting DNA from the gut of insects that have fed on a human body, such as flies and beetles. The insects ingest human DNA as they feed, and this can be used to identify the individual from whom the insect is fed. This method has been used to identify victims of mass disasters, such as plane crashes, as well as to identify an individual in a criminal case. While this method of analysis is still in the early stages of development, it has the potential to provide an additional source of information for forensic investigations (Luise et al. 2008).

22.2.9.4 Human STR Analysis from Insect Gut Materials

Short Tandem Repeats (STRs) are a type of genetic marker used to identify individuals based on their DNA. STR analysis of human DNA from maggots fed on decomposing bodies can be used to help identify the deceased individual by comparing the STR profile of the maggot-derived DNA to the STR profile of the potential victim. The STR analysis involves looking at the number of repeated sequences of nucleotides present in the DNA, as well as the length of the repeats. This information can be used to determine whether the maggot-derived DNA is a match for the victim's DNA and can be used to help identify the deceased individual (Njau et al. 2016).

22.2.9.5 DNA Sequencing

This method involves analyzing the entire sequence of DNA in an organism, which can then be used to identify the species. DNA sequencing techniques can be used to analyze the gut DNA of insects. This involves extracting the DNA from the gut tissue of the insect, amplifying the DNA using PCR and then sequencing the amplified DNA. This can be used to identify the presence of specific genes or sequences within the insect gut, as well as to identify particular species present in the gut. This can help to understand the diet of the insect and to determine the composition of the gut microbiota.

Also, DNA sequencing techniques are increasingly being used in insect gut DNA analysis in forensics. Insects are often present at crime scenes and can provide valuable evidence. By looking at the DNA in the gut of an insect, it is possible to determine if the insect was in contact with a particular person, animal, or object. This can help in identifying suspects and establishing links between different individuals and items. DNA sequencing techniques are used to identify the insect species, as well as to determine the presence of specific genetic markers. This can be used to identify individuals and link them to a particular crime scene. Additionally, DNA sequencing can be used to analyze the insect gut microbiome, which can provide important clues about the environment and circumstances in which the crime took place.

22.3 Conclusion

In forensic cases, insects can be used as vectors of DNA to aid in the identification of victims or perpetrators. Insects are often found in and around crime scenes, as they are attracted to the organic material present. Forensic entomologists can use these insects to extract and analyze DNA from them such as mitochondrial DNA, which is often used in forensic cases. The extracted DNA can then be compared to samples from potential victims or suspects to aid in the identification process.

Further, insects can carry DNA from human remains to a laboratory, where the DNA can be extracted and analyzed. This could help investigators quickly identify a person's identity or other important information about them. DNA can be used to match individuals to a crime scene or to confirm a suspect's identity. Additionally, insect DNA can be used to establish the time of death and the location of a crime scene. This could help investigators narrow down the timeline of a crime or determine if a murder occurred in a specific area. Insects may also be used to detect traces of drugs or poisons at a crime scene. Insects could also be used in cases of mass disasters, to help identify victims and determine the cause of death. Insects may not be the most efficient or reliable way to obtain forensic evidence, but they could provide valuable insight and help investigators solve crimes.

References

Amendt J, Krettek R, Zehner R (2004) Forensic entomology. Naturwissenschaften 91:51–65

Gary M, Lance D (2002) Medical and veterinary entomology. Academic Press, London, pp 1–13; ISBN 9780125104517

Janzen D (1976) Why are there so many species of insects? Proceedings of XV International Congress of Entomology, Washington, DC, p 8494

LeBlanc HN, Logan JG (2010) Exploiting insect olfaction in forensic entomology. In: Amendt J, Goff ML, Campobasso CP, Grassberger M (eds) Current concepts in forensic entomology. Springer, Dordrecht, pp 205–221

Luise ED, Magni P, Staiti N, Spitaleri S, Romano C (2008) Genotyping of human nuclear DNA recovered from the gut of fly larvae. Forensic Sci Int Genet Suppl Ser 1(1):591–592

May RM (1988) How many species are there on earth? Science 241:441–1449

Njau GD, Muge EK, Kinyanjui PW, COA O, Sophie M (2016) STR analysis of human DNA from maggots fed on decomposing bodies: assessment of the time period for successful analysis. Egyptian J For Sci 6(3):261–269

Samerjai C, Sukontason KL, Sontigun N, Sukontason K, Klong-Klaew T, Chareonviriyaphap T, Kurahashi H, Klimpel S, Kochmann J, Saeung A, Somboon P, Wannasan A (2019) Mitochondrial DNA-based identification of forensically important flesh flies (Diptera: Sarcophagidae) in Thailand. Insects 11(1):2

Wells JD, Introna F Jr, Di Vella G, Campobasso CP, Hayes J, Sperling FAH (2001) Human and insect mitochondrial DNA analysis from maggots. J Forensic Sci 46(3):685–687

Wells JD, Stevens JR (2008) Application of DNA-based methods in forensic entomology. Annu Rev Entomol 53:103–120

Animal DNA Testing

23

Tilak Ram Chandrakar and Ajay Biswas

Abstract

Animal DNA can play a key role in poaching, illegal trafficking, and crimes involving animal products. DNA from nonhuman sources can be more prone to DNA damage compared to human DNA. Animal DNA is often used in the field of wildlife crime for their detection and identification. DNA profiling can be used to identify animals. The Charge switch Technique and prep GEM Technique can be used on insect specimens that have been preserved. The Puregene method, DNAzol method, and DNeasy method can all be used to obtain DNA extract from newly collected insects. Using the QiaQuick PCR purification kit, DNA could be extracted from Copal. It is possible to analyze the DNA of Xylophagus insects using the CTAB and modified CTAB-PVP methods. The extraction of insect DNA within 24 h can now be accomplished using a new high molecular-weight DNA extraction technique.

Keywords

Animal · DNA · Poaching · Wildlife · Crime · Nonhuman

T. R. Chandrakar (✉) · A. Biswas
Department of Forensic Science, Medi-Caps University, Indore, MP, India
e-mail: tilak.ram@medicaps.ac.in; ajay.biswas@medicaps.ac.in

A. Puri et al. (eds.), *Fundamentals of Forensic Biology*,
https://doi.org/10.1007/978-981-99-3161-3_23

23.1 Introduction

23.1.1 DNA Characteristics

Deoxyribonucleic acid (DNA) is often regarded as the preeminent biomolecule present within cellular structures, and it create a compacted form within the chromosomes of many living entities. Every eukaryote has two chromosomes, which are linear DNA molecules (Pray 2008). The DNA for these chromosomes is typically compacted into a denser form rather than existing frequently in a loose state. Nucleosomes are bead-like structures made up of certain DNA-binding proteins called histones. Because the DNA is kept coiled around these nucleosomes, the length of the DNA string as a whole is reduced. The "beads-on-a-string" configuration is then further wrapped into a configuration with a width of about 30 nm, which is appropriately referred to as the 30 nm fiber. This is how DNA is created, and it lasts for the bulk of the lifespan of the cell (Travers and Muskhelishvili 2015). DNA is also present in other nonhuman sources which included the Plants, Bacteria's, Fungi, Virus and many more. To decide how to best proceed with its examination, the unique properties of DNA derived from nonhuman sources must be taken into consideration. When compared to DNA from a variety of other sources, DNA from one source can need a completely different extraction or analysis technique from others merely based on the diverse properties of nucleic acid present in a specific organism. The source of DNA could be intra-organism having the genome ranges of 10–1 lakh Mb in length in which the nuclear DNA analysis approach must be different for the Mitochondrial DNA (mtDNA) to Nuclear DNA (nDNA) (Taanman and Kroon 2019). Thus, several reports revel on properties of circular mtDNA and say it is more prone to DNA damage compared to nDNA. The mtDNA molecules lack the histone protection that sustained the DNA repair and are situated near the electron transport chain (ETC), which continuously produces oxidizing by products known as reactive oxygen species (ROS). As a result, in reaction to DNA-damaging chemicals, the mutation rate of mtDNA has been found to be up to 15-fold higher than that of nDNA (Claus et al. 2011).

23.1.2 Animals DNA in Wildlife Crime

Wildlife crime is the unlawful seizing, trading, exploiting, possessing, or killing of animals or plants in violation of local, state, federal, or international laws. Animals play a crucial role in many people's lives; therefore, it makes sense that their DNA may have a big impact on the forensics field. Animals or animal products can play a key role in a range of situations, including poaching, illegal trafficking, and crimes involving animal products. Any DNA that could be extracted from the animal or product in these cases would probably be able to considerably aid the investigation because the animal or animal product is either the primary victim of the crime or is particularly crucial to whatever that crime is. However, in other circumstances, animals could still be involved in a situation that is largely unrelated. Animal products, like hair or fur, can be used to connect a person to a place because of the material's special qualities (Butler 2023).

23.1.3 Poaching

Animal DNA is obviously often used in the field of wildlife crime for their detection and identification. The crime related to poaching can be overcome while utilizing the tool named as DNA profiling. Several cases are reported by various government and nongovernment organizations about the cruelty against the wild animals for several purposes around the world. The broadly affected animals are Tigers (*Panthera tigris*), Elephants (*Elephantidae*), Rhinos (*Rhinocerotidae*), Bears (*Ursidae*), Pangolins (*Pholidota*), Dear Species (*Cervidae*) etc. (Sahajpal et al. 2021). The product obtained from the animal sources are been used as medicines that is encouraged by Chinese traditional medicine, which solicitated the poachers' activities. The DNA Profiling tool plays a crucial role to identify and minimize the poaching issue of animals and can improve the conviction rate.

23.1.4 The Blocking of Illegal Trade Around the World Using Molecular Techniques

Illegal trading or trafficking of animals is widely increased around the world. Poachers targeted the easiest route to transport the animal remains in border areas and by air routes. The case reported by Rodionov et al. (2021) explained the poaching of western tur *Capra caucasica*, an endangered goat antelope native to the Caucasus Mountains. The police authorities seized the samples and send it for their analysis. The samples were analyzed using the genome-wide SNP genotyping using DNA chips for species assignment and using the 14 STR (microsatellite) genetic markers for confirming their identity. The avian species, specially the Macau's parrot bird, is threatened and illegally transported from Europeans countries to others. Their eggs also found to be transported that is closely monitored for suspected illegal trade (Coghlan et al. 2011). Another tool named as Mitochondrial DNA (mtDNA) testing specially using the Cyt-B gene and Cytochrome C-Oxidase genes helps to identify the origin of samples. The species of smuggled samples can now be determined using sequencing technologies, allowing to fix the identity for subsequent identification. The new rapid identification technology enables more prompt enforcement of laws for prohibiting the trade in wildlife as well as criminal prosecution of individuals responsible in poaching.

23.1.5 Fake Animal Product and Linking the Products with Suspected

There are numerous case studies cited by certain agencies that show the fabrication of animal products for financial gain; a report submitted by a Chinese police official looking into 57 occurrences of alleged unlawful ivory trading that occurred since 2011. Only 513 of the 1714 items actually turned out to be synthetic, despite 27 of them attempting to hide samples of real ivory by mixing them with imitation ivory (Zhou 2014). According to Carvalho et al. (2015), the seafood industry also alters the authentic seafood and sells fraudulent food, particularly fish, in numerous restaurants. There have been numerous instances of fake animal products reported throughout the world, but they can simply be prevented by using a promising approach like the DNA barcoding technique. The linking evidence connects the samples with the offenders, specially if suspect directly involves with animal species. The sample that may be recovered are the shredded hair, fur, skin, and dander as part of their natural growth cycles. The analysis of said samples can provide a lead to investigate the wildlife crime in an appropriate direction.

23.1.6 Plants DNA and Its Forensic Application

Although plant DNA fingerprinting for forensic purposes is underdeveloped, it is an essential and rapidly developing field that could provide valuable information. The plant DNA is utilized in the cases about the illegal deforestation and harvesting of important and protected plant species. In terms of an Indian continent, the plant species are protected under the wildlife protection act 1972 and their unlawful transportation violets the norms and becomes punishable offence (Sharma and raturi 2018). The identification of botanical material to a specific species may help with an investigation and give significant evidence, depending on the crime. Such species-assignment calls for the construction of a database of species-specific sequences at one or more genetic loci, frequently on the chloroplast genome, as well as a barcoding technique. As an alternative, botanical material may be matched to a specific plant, clone, or population using DNA fingerprinting if the source species is identified. Multilocus genotypes at highly variable loci, such as DNA microsatellites, also known as simple sequence repeats (SSRs), short tandem repeats (STRs), or variable number tandem repeats (VNTRs), could be used to do this. The DNA database of sequence (DNA Barcodes) helps to build up the records of a threatened and protected plant species that may be matched if any incidents or crime comes in a light. Plant architecture and morphology can also be utilized to identify the species from botanical evidence like stem, roots, leaves, or seeds. However, identification only based on physical traits might be ambiguous or imprecise, which requires a knowledgeable botanist having the huge experience in taxonomy information, and the evidence must be largely undamaged. The minute evidence makes the identification procedure more robust and difficult due to their shape, leaf fragments, seeds, or pollen. The trace evidence like this may mislead the evidence also if investigator is not aware with their properties and features. These evidences are sometimes also found with the human or animal-related cases such as in poisoning condition where a digested form of seed or leaves is also found into their stomach that also makes the identification procedure more difficult. However, DNA may be recovered from even the smallest plant pieces, and sequence information gained from this material may be used to potentially use the evidence to clearly identify the plant species and can be used to differentiate between the wild protected species to non-protected and helps to provide a significant result to minimize the crime related with the plant species.

23.1.7 Evidence Connections with the Suspect and Harmed Species

The value of plant samples as linking evidence in criminal cases has increased as a result of advancements in research into plant DNA markers. Plant fragments including pollen, seeds, and leaves may be discovered in close proximity to a suspect or misplaced pieces of evidence. The suspect or the evidence can then be connected to that specific crime scene via plant remains and DNA fingerprinting, which links the plant material to the crime scene (Caccianiga et al. 2021).

23.1.8 Microbial DNA Testing; Prokaryotes and Their Genome

The genome of prokaryotes is made up of a single double-stranded DNA molecule that is shaped like a loop or circle. A nucleoid is the area of the cell that houses this genetic material. Additionally, some prokaryotes include tiny DNA loops called plasmids that are not necessary for typical growth. These plasmids can be transferred between bacteria, and occasionally the recipient bacteria receive advantageous additional genes that they can add to their chromosomal DNA. One characteristic that frequently spreads inside a bacterial colony via plasmid exchange is antibiotic resistance. In comparison to prokaryotes, the genome of eukaryotes is made up of several double-stranded linear DNA molecules organized into chromosomes. The number of chromosomes in the cell nuclei varies depending on the kind of eukaryote such as Human gametes (sperm or eggs); each has 23 paired chromosomes; some animals have 37 paired chromosomes which vary according to each species. The packaging of the prokaryotic genome is the primary difference which makes them different from other organisms. The basal form of most prokaryotes is a single circle, but by a process called supercoiling, the size is reduced. By changing the number of turns in the double helix molecule, torsional stress is introduced to the entire molecule. The molecule then turns and winds around itself to release this stress as much as it can. Finally, this totally twisted molecule is joined to a nucleoid made of proteins that concentrates the genome that controls the prokaryotic transcription (Dorman and Dorman 2016). The bacterial genome's content is distinct from the eukaryotic genome that only 11% of the entire genome in prokaryotes is noncoding, and their genomes are quite small in comparison to many multicellular and many unicellular eukaryotes. In this queue the virus acts as an infectious microorganism that is made up of a protein-coated segment of nucleic acid (either DNA or RNA). A virus can't multiply by itself; it has to infect cells in order to utilize the host cell's components to make copies of itself. The forensic investigation leads to reveal the geographical identity of a prokaryotes organism and open up their sources for justice purposes.

23.1.8.1 Viruses

Viruses are the tiniest microorganism found on earth. Virus is not widely considered to be alive since they are found inside the host for their propagation of their lineage through parasitism and are unable to survive outside of the host cells. Virus is generally composed of the proteins and nucleic acids where few viruses also have a lipid coating that contains glycoproteins. A virus's DNA is held in a symmetrical protein capsid, which makes up the virus's structure. The region of the genome that codes for these constituent proteins can be quite minimal due to the symmetric and repetitive nature of the capsid. The double- or single-stranded DNA or RNA makes up the nucleic acid inside the capsid. Most of the time, the host cell's resources are used to replicate this genome. The only contribution of the virus is the transfer of the genome; all other requirements for energy, resources, and activities are met by parasitizing the host cell. Additionally, viruses come in a wide variety of sizes and shapes. Viruses can have a diameter of up to 400 nm and can take on a variety of forms. Viral capsids specifically come in three distinct shapes: icosahedral, linear, and complicated (Chaitanya 2019). Viruses can be divided into monopartite and multipartite types as well. While multipartite viral genomes are divided into numerous independent molecules, monopartite viral genomes have the entire genes inside one single molecule.

23.1.8.2 Bacteria

Bacteria, viruses, and fungi are naturally present in the human body and can be found in places including the vagina, skin, and gastrointestinal system. Due to their astounding ubiquity, bacteria can be surprisingly valuable in forensic investigations. In every living thing bacterium contains and plays an essential part of life. Some bacteria are pathogenic, and as a result, they are frequently exploited as destructive agents in bioterrorism. Still other bacteria are selective enough in their environment to behave as distinguishing characteristics of particular organic molecules. Last but not the least, the enormous diversity among bacterial groups enables distinct communities of these organisms to be detected and potentially serve as fingerprint-like indicators for a particular person. The earlier molecular methods that were used for classifying the bacteria relied on factors including GC content, plasmid profiling, and adaptability to genetic modification. Presently, nucleotide sequencing and hybridization are two essential molecular applications that are widely used in the detection and identification of bacteria. Methodologies used for the detection and identification of microbes that are based on hybridization include the loop-mediated isothermal amplification (LAMP), Southern, PCR, real-time PCR, microarray, and universal tagging methods (Barghouthi 2011).

23.1.8.3 Fungi

The fungi (Mycota) are multicellular organisms having the mycelial structure that is composed with the thin filaments or hyphae (2–10 m in diameter). The hyphae found may be unbranched or branched, septate or nonseptate, and are frequently multinucleate in shapes. Numerous fungi pose a threat to human interests. They have the potential to harm humans either directly or indirectly through their toxins, such as mycotoxins and mushroom poisoning. They may potentially spread illnesses to both plants and animals. Fungus can be very useful sometimes in crime investigation or during postmortem if pores are found and can make a link between suspect and victim with a particular geographic area. Currently, the phenol chloroform isoamyl alcohol (PCI) technique, the spin-column procedure using a Qiagen DNA extraction kit, and the spin-column method using a Norgen DNA extraction kit are used to isolate DNA from fungi (Kumar and Mugunthan 2018).

23.1.9 Insect DNA Identification Techniques

Insects are essential to the health of the forest environment because they consume wood and help the food chain. In a general sense, the term "insect" usually refers to well-known pests or disease vectors, such as bedbugs, houseflies, clothes moths, beetles, mosquitoes, fleas, horseflies, etc. A lot of insects are helpful to humans because they pollinate plants, make useful compounds, kill harmful insects, scavenge trash, and feed other animals. The genome is an insect's complete set of DNA. In insects, the nuclear genomes are made up of chromosomes, which contain DNA and proteins. Within an insect, the nuclear genome contributes the most genetic data. However, the cytoplasmic organelles known as mitochondria are also parts of the genome. The DNA of several small insects with short storage times in 95% ethanol has been discovered to be more successfully isolated by a method that is easier, more dependable, cost-effective, and requires less equipment and chemicals; like nuclear gene and cytochrome oxidase I gene (Microsatellites). Salting out and Chelex Techniques are the other methods employed for isolating insect DNA. The Chargeswitch Technique and prepGEM Technique can be used on insect specimens that have been preserved. The Puregene method, DNAzol method, and DNeasy method can all be used to obtain DNA extract from newly collected insects. Using the QiaQuick PCR purification kit, DNA could be extracted from Copal. It is possible to analyze the DNA of Xylophagus insects using the CTAB and modified CTAB-PVP methods. The extraction of insect DNA within 24 h can now be accomplished using a new high molecular weight DNA extraction technique (Asghar et al. 2015).

References

Asghar U, Malik M, Anwar F, Javed A, Raza A (2015) DNA extraction from insects by using different techniques: a review. Adv Entomol 3:132–138. https://doi.org/10.4236/ae.2015.34016

Barghouthi SA (2011) A universal method for the identification of bacteria based on general PCR primers. Indian J Microbiol 51(4):430–444. https://doi.org/10.1007/s12088-011-0122-5

Butler J (2023) Recent advances in forensic biology and forensic DNA typing: INTERPOL review 2019–2022. Forensic Sci Int Synerg 6:100311. https://doi.org/10.1016/j.fsisyn.2022.100311

Caccianiga M, Caccia G, Mazzarelli D et al (2021) Common and much less common scenarios in which botany is crucial for forensic pathologist and anthropologists: a series of eight case studies. Int J Legal Med 135:1067–1077. https://doi.org/10.1007/s00414-020-02456-0

Carvalho DC, Palhares RM, Drummond MG, Frigo TB (2015) DNA barcoding identification of commercialized seafood in south Brazil: A governmental regulatory forensic program. Food Control 50:784–788. https://doi.org/10.1016/j.foodcont.2014.10.025

Chaitanya KV (2019) Structure and organization of virus genomes. In: Genome and genomics. Springer Singapore, pp 1–30. https://doi.org/10.1007/978-981-15-0702-1_1

Claus D, Maiken LM, Keshav KS, Lene JR (2011) The importance of mitochondrial DNA in aging and cancer. J Aging Res 2011:407536. https://doi.org/10.4061/2011/407536

Coghlan ML, White NE, Parkinson L, Haile J, Spencer PBS, Bunce M (2011) Egg forensics: An appraisal of DNA sequencing to assist in species identification of illegally smuggled eggs. Forensic Science International: Genetics 6(2):268–273. https://doi.org/10.1016/j.fsigen.2011.06.006

Dorman CJ, Dorman MJ (2016) DNA supercoiling is a fundamental regulatory principle in the control of bacterial gene expression. Biophys Rev 8(3):209–220. https://doi.org/10.1007/s12551-016-0205-y

Kumar M, Mugunthan M (2018) Evaluation of three DNA extraction methods from fungal cultures. Med J Armed Forces India 74(4):333–336. https://doi.org/10.1016/j.mjafi.2017.07.009

Pray L (2008) Discovery of DNA structure and function: Watson and Crick. Natur Educ 1(1):100

Rodionov A, Deniskova T, Dotsev A, Volkova V, Petrov S, Kharzinova V, Koshkina O, Abdelmanova A, Solovieva A, Shakhin A et al (2021) Combination of multiple microsatellite analysis and genome-wide SNP genotyping helps to solve wildlife crime: a case study of poaching of a Caucasian tur (*Capra caucasica*) in Russian Mountain National Park. Animals 11:3416. https://doi.org/10.3390/ani11123416

Sahajpal V, Sudhanshu M, Deepika B (2021) Forensic Analysis in Wildlife Crime Cases: Microscopy. DNA Profiling and Isotope Analysis. https://doi.org/10.5772/intechopen.98252

Sharma R, Raturi A (2018) Wildlife PROTECTON in India: a LEGAL analysis. JCRT 6(2):1

Taanman J-W, Kroon A (2019) Mitochondrial DNA: structure, genetics, replication and defects. https://doi.org/10.1016/B978-0-12-811752-1.00005-5

Travers A, Muskhelishvili G (2015) DNA structure and function. FEBS J 282:2279. https://doi.org/10.1111/febs.13307

Zhou ZM (2014) Synthetic ivory fails to stop illegal trade. Nature 507:40. https://doi.org/10.1038/507040a

Microbial DNA Testing

24

Paromita Banerjee and Prashant Kumar

Abstract

Microbiology is the study of microorganisms, which are tiny organisms that are typically invisible to the naked eye. These include bacteria, viruses, fungi, protozoa, and algae. It is a diverse and interdisciplinary field that encompasses many different sub-disciplines, including bacteriology, virology, mycology, parasitology, and phycology.

Keywords

Skull · Scavenger · Excavation dimorphic

1. **Introduction:** In this section, we have provided an overview of the importance of microbial DNA testing, its applications, and the various techniques used to perform the tests.
2. **Sample collection and preparation**: Here, we have described the methods used to collect and prepare samples for microbial DNA testing, such as swabbing, plating, and centrifugation. We have also discussed the importance of proper sample handling and storage to ensure the accuracy of the test results.
3. **DNA extraction**: This section covers the various techniques used to extract DNA from microbial samples, such as boiling, lysis, and chemical methods. We also

P. Banerjee (✉)
Department of Cardiology, AIIMS, Rishikesh, India

P. Kumar
Department of Bioinformatics, Kalinga University, Raipur, Chhattisgarh, India
e-mail: kumar.prashant@kalingauniversity.ac.in

A. Puri et al. (eds.), *Fundamentals of Forensic Biology*,
https://doi.org/10.1007/978-981-99-3161-3_24

discussed the pros and cons of each technique and the factors that influence the efficiency of DNA extraction.

4. **PCR-based methods**: We provided an overview of PCR-based methods, such as quantitative (qPCR) and multiplex PCR, and their applications in microbial DNA testing. This section also covers the advantages and disadvantages of PCR-based methods and the parameters that affect their performance.
5. **DNA sequencing**: This section describes the various DNA sequencing techniques, such as Sanger sequencing and next-generation sequencing (NGS), and their applications in microbial DNA testing. We also discussed the advantages and disadvantages of different sequencing techniques and the factors influencing sequencing accuracy.
6. **Data analysis**: Different software and bioinformatics tools have been demonstrated to analyse the data generated by microbial DNA testing. We also discussed the importance of data validation and quality control in ensuring the accuracy of the results.
7. **Applications:** In this section, we described the various applications of microbial DNA testing, such as pathogen detection, microbiome analysis, and genetic diversity studies. The potential of microbial DNA testing in medicine, agriculture, and biotechnology has also been discussed.
8. **Conclusion:** Summarise the key points discussed in the chapter and provide an overview of the current state of microbial DNA testing and future directions for research in this field.

Microbiology is the study of microorganisms, which are tiny organisms that are typically invisible to the naked eye. These include bacteria, viruses, fungi, protozoa, and algae. It is a diverse and interdisciplinary field that encompasses many different sub-disciplines, including bacteriology, virology, mycology, parasitology, and phycology.

Some of the key areas of research in microbiology include:

- Taxonomy and systematics: The classification and identification of microorganisms, including the development of new techniques for identifying microorganisms and understanding their evolutionary relationships.
- Microbial genetics and genomics: The study of the genetic makeup of microorganisms and how they evolve, including the use of DNA sequencing and other molecular techniques to study microorganisms.
- Microbial ecology: The study of how microorganisms interact with each other and with their environment, including the impact of microorganisms on the Earth's ecosystem.
- Microbial physiology: The study of the role of microorganisms in human health and disease, including the development of new diagnostic methods and therapies for treating infectious diseases.
- Biotechnology: The application of microbial physiology in the production of useful products such as antibiotics, enzymes, and vaccines.

- Industrial microbiology: The application of microbiology in industrial processes such as fermentation, waste treatment, and pollution control.

Overall, it plays a vital role in understanding the biology of microorganisms and how they interact with the world around us. It also has many practical applications in fields such as medicine, agriculture, and industry.

In today's world, Recombinant DNA technology is a method for combining genetic material from different sources to create new combinations of genes. This technology can be used to manipulate the DNA of microbes, such as bacteria and yeast, to produce useful products such as proteins, enzymes, and antibiotics. It can also be used to create genetically modified organisms (GMOs) with specific characteristics, such as resistance to disease or pests. This technology has a wide range of applications in medicine, agriculture, and biotechnology. All of these aspects come into light while we deal with microbial DNA testing. Because, in no way or the other when researchers have to reach in a definite conclusion, they have to undergo through a process of microbial testing. In the upcoming pages, we will get to learn a lot regarding this topic of Microbial DNA testing.

24.1 Introduction

Microbial DNA testing is crucial because it identifies and characterises microorganisms, providing insight into the causes of infections and other diseases. Additionally, microbial DNA testing can track the spread of infectious diseases, monitor the effectiveness of treatment, and determine the presence of antibiotic resistance. It can also be used in environmental monitoring, food safety and quality control, and biotechnology applications. Overall, it is a powerful tool that can help us understand and combat microorganisms that can cause harm to human health and the environment.

This has a wide range of applications, including:

1. Medical and clinical applications: Microbial DNA testing can be used to identify the cause of infections, such as bacterial, viral, and fungal infections, and to monitor the effectiveness of treatment. It can also be used to detect antibiotic resistance in microorganisms.
2. Environmental monitoring: Used to monitor the presence of microorganisms in water, soil, and air, which can help to identify sources of pollution or contamination.
3. Food safety and quality control: Detects harmful microorganisms in food and ensure that food products meet safety and quality standards.
4. Biotechnology: Identifies and characterises microorganisms used in the production of fermented foods and beverages, as well as in the production of enzymes, antibiotics, and other bioproducts.
5. Forensic applications: Identifies microorganisms in forensic samples, such as blood or skin cells, which can help solve crimes.

6. Research and discovery: Can even identify and study new microorganisms and understand the diversity and evolution of microorganisms in different environments.

Overall, microbial DNA testing plays a vital role in many areas of science and technology, and it will likely continue to be used in new and innovative ways.

Several techniques can be used to perform microbial DNA testing, including:

1. Polymerase chain reaction (PCR): This widely used technique can amplify small amounts of DNA. It can be used to detect and identify specific microorganisms in a sample.
2. DNA sequencing: This technique can be used to determine the complete genetic sequence of a microorganism. It can be used to identify microorganisms and to study their genetic diversity and evolution.
3. Microarray: This technique uses a chip with a large number of DNA probes to detect and identify multiple microorganisms in a sample at the same time.
4. Fluorescence in situ hybridisation (FISH): This technique uses fluorescently labelled DNA probes to detect and identify specific microorganisms in a sample.
5. Mass spectrometry: This technique can identify microorganisms by analysing the unique mass-to-charge ratios of their DNA or other biomolecules.
6. Next-generation sequencing (NGS): This technique can be used to sequence the entire genome of a microorganism, allowing for high-throughput, rapid, and cost-effective identification and characterisation of microorganisms.

The choice of technique will depend on the specific application, the type of microorganism being tested, and the available resources and expertise.

24.2 Sample Collection and Preparation

Several methods can be used to collect and prepare samples for microbial DNA testing, including:

1. Swabbing: This method involves using a swab to collect a sample from a surface or an area of infection. The swab is then placed in a transport medium to preserve the microorganisms for testing.
2. Biopsy: This method involves collecting a small tissue sample from an area of infection or disease.
3. Blood culture: This method involves collecting a sample of blood, which is then placed in a particular culture medium to grow any microorganisms present.
4. Faecal or urine sample.
5. Water or soil sample.

All of these samples along with the tissue sample is then placed in a transport medium to preserve the microorganisms for testing.

Once the sample is collected, it may need to be prepared before it can be tested. This may include steps such as:

- Homogenising the sample to break up any clumps of microorganisms.
- Extracting the DNA from the microorganisms.
- Purifying the DNA to remove any contaminants that could interfere with the test results.

The specific method of sample preparation will depend on the type of microorganism being tested and the type of DNA testing that will be performed.

Proper sample handling and storage are crucial for ensuring the accuracy of test results for microbial DNA testing. Improper handling and storage can lead to contamination, degradation, or loss of the microorganisms in the sample, which can result in inaccurate or unreliable test results.

Here are a few examples of how proper sample handling and storage can ensure the accuracy of test results:

1. Sample collection: Samples should be collected using sterile techniques to avoid contamination. Any swabs or other materials used to collect the sample should be sterile, and the sample should be stored in a transport medium that will preserve the microorganisms.
2. Sample transport: Samples should be transported to the lab as soon as possible after collection to minimise the risk of contamination or loss of microorganisms. They should also be stored at the appropriate temperature to preserve the microorganisms.
3. Sample preparation: Sample preparation should be done under sterile conditions to avoid contamination. The DNA should be extracted and purified as soon as possible after the sample is collected to minimise the risk of degradation.
4. Sample storage: Once the sample is prepared, it should be stored at the appropriate temperature to minimise the risk of degradation. Long-term storage should be done in a -80 °C freezer.
5. Sample testing: The sample should be tested as soon as possible after collection to minimise the risk of degradation or loss of microorganisms.

Overall, proper sample handling and storage are essential for ensuring the accuracy of test results for microbial DNA testing. It is important to follow standard operating procedures (SOPs) to minimise the risk of contamination, degradation, or loss of microorganisms in the sample.

24.3 DNA Extraction

Several techniques can be used to extract DNA from microbial samples for microbial DNA testing, including:

1. Phenol–chloroform extraction: This widely used technique combines phenol and chloroform to separate DNA from other cellular components. The DNA is then recovered from the aqueous layer after centrifugation.
2. Chelex extraction: This technique uses chelex beads, which contain a chelating agent, to remove positively charged ions and other contaminants that can interfere with DNA extraction.
3. Organic extraction: This technique uses organic solvents, such as ethanol or isopropanol, to separate the DNA from other cellular components. The DNA is then recovered by precipitation.
4. Silica-based extraction: This technique uses silica-based beads to bind the DNA, which is then washed and eluted from the beads.
5. Magnetic bead-based extraction: This technique uses magnetic beads coated with silica or other materials to bind the DNA, which is then washed and eluted from the beads using a magnetic field.
6. Commercial kits: There are also a variety of commercial DNA extraction kits available that can be used to extract DNA from microbial samples. These kits typically combine the abovementioned methods; some are tailored to specific microorganisms or samples.

The choice of technique will depend on the specific application, the type of microorganism being tested, and the available resources and expertise. It is essential to follow the protocols and standard operating procedures (SOPs) to ensure the quality and integrity of the extracted DNA.

Each technique used to extract DNA from microbial samples for microbial DNA testing has advantages and disadvantages. The choice of technique will depend on the specific application, the type of microorganism being tested, and the available resources and expertise.

1. Phenol–chloroform extraction: This widely used technique is relatively simple and inexpensive. However, it can be time-consuming and may not be as efficient at removing contaminants as other methods.
2. Chelex extraction: This technique is relatively quick and easy to perform and is suitable for removing positively charged ions and other contaminants. However, it may not be as efficient at extracting DNA from some types of microorganisms as other methods.
3. Organic extraction: This technique is inexpensive and can extract DNA from various microorganisms. However, it may not be as efficient at removing contaminants as other methods.
4. Silica-based extraction: This technique is suitable for extracting DNA from microorganisms that are difficult to lyse, and it is relatively quick and easy to perform. However, it may not be as efficient at removing contaminants as other methods.
5. Magnetic bead-based extraction: This technique is relatively quick and easy to perform and is suitable for extracting DNA from various microorganisms. However, it can be more expensive than other methods.

6. Commercial kits: These kits are pre-packaged and often tailored to specific types of microorganisms or samples, making them convenient to use. However, they can be more expensive than other methods and may provide less flexibility regarding sample type or downstream applications.

Factors that can influence the efficiency of DNA extraction include the type of microorganism being tested, the amount and quality of the sample, the presence of inhibitors or contaminants, and the conditions used during the extraction process. The pH, temperature, and incubation time can also affect extraction efficiency. It is essential to follow the protocols and SOPs and to optimise the conditions of extraction to ensure the quality and integrity of the extracted DNA.

24.4 PCR-Based Methods

PCR-based methods, such as quantitative (qPCR) and multiplex PCR, are widely used in microbial DNA testing.

1. Quantitative PCR (qPCR): Widely used technique in microbial DNA testing. It is a sensitive and specific method that can be used to detect and quantify small amounts of DNA from microorganisms. In qPCR, specific primers are designed to bind to specific regions of the target DNA. Then, the polymerase chain reaction (PCR) process is used to amplify the DNA. The amount of amplified DNA is measured using fluorescent dyes, which allows for quantifying the target DNA. QPCR can detect and quantify specific microorganisms in a sample, such as bacteria, viruses, and fungi. This makes it useful for the diagnosis of infections, as well as for monitoring the effectiveness of treatment. qPCR can also be used to track the spread of infectious diseases by detecting and quantifying the DNA of specific microorganisms in a sample. QPCR can detect and quantify specific microorganisms in various samples, including blood, urine, swab, and other clinical and environmental samples such as water and soil. Overall, qPCR is a powerful tool for microbial DNA testing that allows for the sensitive and specific detection and quantification of microorganisms in a sample. It is widely used in medical and clinical applications, environmental monitoring, food safety and quality control, and biotechnology. qPCR is a sensitive and specific method that can be used to detect and quantify small amounts of DNA. It involves using primers that bind to specific regions of the target DNA and then amplifying the DNA using the polymerase chain reaction (PCR) process. The amount of amplified DNA is then measured using fluorescent dyes. qPCR can be used to detect and quantify specific microorganisms in a sample, as well as to monitor the effectiveness of treatment or track the spread of infectious diseases.
2. Multiplex PCR: A variation of PCR that allows for the simultaneous amplification of multiple target DNA sequences. This technique is also widely used in microbial DNA testing, as it detects multiple microorganisms in a single reaction. Multiplex PCR is based on the same principle but uses multiple primers, each

specific to a different target DNA sequence. Different sets of primers and fluorescent dyes make detecting multiple microorganisms in a single reaction possible. This can be useful for diagnosing infections, where multiple microorganisms may be present, or for detecting outbreaks of infectious diseases, where multiple microorganisms are responsible for spreading the disease. Multiplex PCR can detect and identify multiple microorganisms in various samples, including blood, urine, swab, and other clinical and environmental samples such as water and soil. One of the main advantages of multiplex PCR is its ability to detect multiple microorganisms in a single reaction, which can be more efficient and cost-effective than running multiple individual PCR reactions. However, multiplex PCR can be more complex to design and optimise than traditional PCR. The number of fluorescent dyes available limits the number of targets amplified in a single reaction. Overall, multiplex PCR is a powerful tool for microbial DNA testing that allows for the simultaneous detection of multiple microorganisms in a single reaction. It can be used in various medical and clinical applications, environmental monitoring, food safety and quality control, and biotechnology. Multiplex PCR is a variation that allows for the simultaneous amplification of multiple target DNA sequences. It involves using multiple primers, each specific to a different target DNA sequence. This allows for the detection of multiple microorganisms in a single reaction. Multiplex PCR can be used to detect and identify multiple microorganisms in a sample at the same time, which can be helpful in the diagnosis of infections or for the detection of outbreaks of infectious diseases.

Both qPCR and multiplex PCR are sensitive and specific methods that can detect and identify microorganisms in a sample. They can be used in various medical and clinical applications, environmental monitoring, food safety and quality control, and biotechnology. The choice of method will depend on the specific application, the type of microorganism being tested, and the available resources and expertise.

24.5 DNA Sequencing

DNA sequencing determines the order of nucleotides in a DNA molecule. DNA sequencing can be used in microbial DNA testing to identify and study microorganisms.

1. Sanger sequencing: Sanger sequencing, also known as dideoxy sequencing, is a chain termination method that uses specific nucleotides called dideoxynucleotides (ddNTPs) to stop the synthesis of new DNA strands. This method generates multiple DNA fragments of different lengths, which can be separated by size using gel electrophoresis. The sequence of the DNA can then be determined by reading the order of the nucleotides in the fragments. Sanger sequencing is considered a standard gold method for DNA sequencing, but it is relatively slow and costly compared to newer methods.

2. Next-generation sequencing (NGS): Next-generation sequencing (NGS) is a newer method that allows for high-throughput, rapid, and cost-effective sequencing of DNA. NGS technologies include Illumina, PacBio, and Nanopore. These methods can sequence billions of DNA molecules simultaneously and can be used to sequence the entire genome of a microorganism. NGS is a powerful tool for microbial DNA testing, as it allows for the identification and characterisation of microorganisms and the study of their genetic diversity and evolution.
3. Whole Genome Sequencing (WGS): WGS is a method of DNA sequencing that aims to sequence the entire genome of an organism. This method allows for the identification of all genetic information, including the presence of virulence factors, antibiotic resistance genes, and other genetic markers important for understanding the biology of the microorganism.

The choice of DNA sequencing technique will depend on the specific application, the type of microorganism being tested, and the available resources and expertise. Sanger sequencing is considered a standard gold method for DNA sequencing, but it is relatively slow and costly. Next-generation sequencing (NGS) is a newer method that allows for high-throughput, rapid, and cost-effective sequencing of DNA. Whole genome sequencing (WGS) is the most comprehensive method for identifying all genetic information.

24.6 Data Analysis

Several factors can influence the accuracy of DNA sequencing, including:

1. Quality of the DNA sample: The quality of the DNA sample can affect the sequencing accuracy. High-quality DNA samples free from contaminants and inhibitors will yield more accurate sequencing results than poor-quality samples.
2. Sample preparation: Sample preparation is a critical step affecting sequencing accuracy. Proper extraction, purification, and quantification of the DNA sample are essential to ensure the quality and integrity of the DNA.
3. Sequencing technology: The sequencing technology used can also influence sequencing accuracy. Different sequencing technologies have different error rates; some are more accurate than others.
4. Sequencing conditions: The conditions used during sequencing, such as the temperature, pH, and reagent concentrations, can affect sequencing accuracy. Optimal conditions should be determined for each sequencing technology and followed to ensure sequencing accuracy.
5. Data analysis: The data analysis methods can also influence sequencing accuracy. The choice of software and the parameters used can affect the accuracy of the sequencing results.
6. Quality control: Quality control measures, such as repeat sequencing or reference standards, can also be used to check the accuracy of sequencing.

7. Mapping to reference genome: In the case of genome sequencing, the accuracy of the results is also influenced by the quality of the reference genome used for mapping.
8. Coverage: Sequencing coverage can also have an impact on the accuracy of the sequencing results. Higher coverage means a higher chance of detecting rare variants or mutations.

Overall, the accuracy of DNA sequencing can be influenced by various factors, including the quality of the DNA sample, the sequencing technology used, the sequencing conditions, data analysis methods, and quality control measures. By taking these factors into account and following standard operating procedures (SOPs), the accuracy of DNA sequencing can be improved, and the results can be more reliable.

24.7 Applications

Microbial DNA testing has a wide range of applications, including:

1. Medical and clinical applications: Microbial DNA testing can diagnose and monitor infectious diseases, such as bacterial and viral infections. It can also be used to monitor the effectiveness of treatment, track the spread of infections, and detect outbreaks of infectious diseases.
2. Environmental monitoring: It monitors the presence and abundance of microorganisms in the environment, such as in water and soil. This can be useful for monitoring the quality of water resources, detecting pollution, and assessing the impact of human activities on the environment.
3. Food safety and quality control: This technique can detect and identify microorganisms in food and agricultural products. This can be used to ensure food products' safety and quality and track the source of foodborne illness.
4. Biotechnology: Identifies and studies microorganisms that have potential applications in biotechnology, such as biofuels and other chemicals.
5. Pharmaceuticals: Microbial DNA testing can identify and study microorganisms used to produce drugs and other pharmaceuticals.
6. Forensics: Used in forensic investigations to identify and track the source of microorganisms in forensic samples.
7. Biomedical research: Microbial DNA testing can be used in biomedical research to study genetics and biology of microorganisms and their interactions with other organisms and their environment.

Overall, microbial DNA testing is a powerful tool that can be used in many applications, including medical and clinical applications, environmental monitoring, food safety and quality control, and biotechnology.

Microbial DNA testing can be used for various applications, including pathogen detection, microbiome analysis, and genetic diversity studies.

1. Pathogen detection: Microbial DNA testing can identify specific pathogens, such as bacteria, viruses, and fungi. This is commonly used in medical and clinical applications to diagnose and monitor infectious diseases, as well as to track the spread of infections and detect outbreaks of infectious diseases.
2. Microbiome analysis: Microbial DNA testing can study the diversity and composition of microorganisms in a specific environment or sample, such as the human gut microbiome. This can be used to understand the role of the microbiome in health and disease, as well as to develop new therapies and diagnostic tools.
3. Genetic diversity studies: Microbial DNA testing can study the genetic diversity of microorganisms, such as bacteria and fungi. This can be used to understand the evolution and spread of microorganisms and identify genetic markers that are important for understanding the biology of the microorganism.

Overall, microbial DNA testing is a powerful tool that can be used for various applications, including pathogen detection, microbiome analysis, and genetic diversity studies. It allows for the sensitive and specific detection and identification of microorganisms and the study of their genetics and biology.

Microbial DNA testing has the potential to play a significant role in the field of medicine. Some of the ways that it can be used include:

1. Diagnosis of infectious diseases: Microbial DNA testing can detect and identify specific pathogens, such as bacteria, viruses, and fungi. This can be used to diagnose infectious diseases and monitor the effectiveness of treatment.
2. Tracking the spread of infections: Microbial DNA testing can be used to track the spread of infectious diseases by detecting and identifying specific microorganisms in a sample. This can be used to identify outbreaks of infectious diseases and to monitor their progression.
3. Monitoring antibiotic resistance: Microbial DNA testing can detect and identify antibiotic resistance genes in microorganisms. This can be used to monitor the spread of antibiotic resistance and to guide the selection of antibiotics for treatment.
4. Personalised medicine: Microbial DNA testing can be used to study the diversity and composition of the microbiome, which can be used to understand the role of the microbiome in health and disease. This can be used to develop new therapies and diagnostic tools and to guide personalised medicine.
5. Biomedical research: Microbial DNA testing can be used to study genetics and biology of microorganisms and their interactions with other organisms and their environment. This can be used to understand the underlying causes of infectious diseases and to develop new therapies and diagnostic tools.

Overall, microbial DNA testing has the potential to play a significant role in the field of medicine by allowing for the sensitive and specific detection and identification of microorganisms, as well as the study of their genetics and biology. This can be used to diagnose and monitor infectious diseases, track the spread of infections, and develop new therapies and diagnostic tools.

Microbial DNA testing has the potential to play a significant role in the field of agriculture. Some of the ways that it can be used include:

1. Soil and plant health: Microbial DNA testing can be used to study the diversity and composition of microorganisms in soil and on plants. This can be used to understand the role of the microbiome in soil health and plant growth and to develop new strategies for improving crop yields and reducing chemical fertilisers and pesticides.
2. Food safety and quality control: Microbial DNA testing can detect and identify microorganisms in food and agricultural products. This can be used to ensure food products' safety and quality and track the source of foodborne illness.
3. Biocontrol and biopesticides: Microbial DNA testing can identify and study microorganisms with potential applications in biocontrol and biopesticides. This can be used to develop new strategies for controlling crops' pests and diseases without using chemical pesticides.
4. Production of biofuels and other chemicals: Microbial DNA testing can identify and study microorganisms that have potential applications in producing biofuels and other chemicals. This can be used to develop new strategies for producing biofuels and other chemicals from renewable sources.

Overall, microbial DNA testing can play a significant role in agriculture by allowing for studying the diversity and composition of microorganisms in the soil, plants, and food products. This can be used to understand the role of the microbiome in soil health, plant growth, and food safety and to develop new strategies for improving crop yields, controlling pests, and producing biofuels and other chemicals from renewable sources.

Microbial DNA testing has the potential to play a significant role in the field of biotechnology. Some of the ways that it can be used include:

1. Microbial strain identification and characterisation: Microbial DNA testing can identify and study microorganisms that have potential applications in biotechnology. This can be used to identify and characterise novel microorganisms and study the genetics and biology of known microorganisms.
2. Metagenomics: Microbial DNA testing can be used to study the diversity and composition of microbial communities in different environments, such as soil, water, and air. This can be used to identify novel microorganisms and genes that have potential applications in biotechnology.
3. Bioprocess development: Microbial DNA testing can identify and study microorganisms that have potential applications in producing biofuels, chemicals, and other bioproducts. This can be used to develop new strategies for producing bioproducts from renewable sources and to improve the efficiency and cost-effectiveness of bioprocesses.
4. Bioremediation: Microbial DNA testing can identify and study microorganisms that have potential applications in bioremediation, using microorganisms to degrade or remove environmental pollutants.

5. Synthetic biology: Microbial DNA testing can identify and study microorganisms that have potential applications in synthetic biology. This can be used to develop new strategies for engineering microorganisms to produce bioproducts, biofuels, and other chemicals.

Overall, microbial DNA testing can play a significant role in biotechnology by allowing for the identification and characterisation of microorganisms that have potential applications in biotechnology. This can be used to study the genetics and biology of microorganisms and develop new strategies for producing bioproducts, biofuels, and other chemicals from renewable sources, improving bioprocess.

Microbial DNA testing has the potential to revolutionise the field of biotechnology by providing a powerful tool for identifying and characterising microorganisms. This technology can be used to identify specific microbes in a sample, determine the genetic makeup of a microbe, and track changes in the genetic makeup of a microbe over time. In addition, microbial DNA testing can be used to identify the presence of pathogens and to monitor the effectiveness of treatments for infectious diseases. Overall, microbial DNA testing is a highly versatile and valuable tool for advancing our understanding of microorganisms and their roles in various biological systems.

24.8 Conclusion

Microbial DNA testing is a method used to identify and quantify the presence of microorganisms in a sample. This can include bacteria, viruses, fungi, and parasites. The test typically involves extracting DNA from the sample, amplifying specific DNA regions using polymerase chain reaction (PCR), and then detecting and identifying the microorganisms present using gel electrophoresis, sequencing, or array-based methods.

Microbial DNA testing can be used for a variety of applications, including:

- Diagnosis of infectious diseases: Microbial DNA testing can quickly and accurately diagnose various infectious diseases, including bacterial and viral infections.
- Environmental monitoring: Microbial DNA testing can be used to monitor the presence of microorganisms in the environment, such as in water or air samples.
- Food safety: Microbial DNA testing can be used to detect the presence of harmful microorganisms in food, such as E. coli and Salmonella.
- Pharmaceuticals and medical devices: Microbial DNA testing can be used to test for the presence of microorganisms in pharmaceuticals and medical devices, such as implants, to ensure their safety and purity.
- Microbial forensics: Microbial DNA testing can be used to identify the source of a microbe or a disease outbreak by comparing the DNA of microbes from different sources.

Overall, microbial DNA testing is a powerful tool for identifying and quantifying the presence of microorganisms in a wide range of samples and provides a more rapid and accurate diagnosis of infectious diseases than traditional culture-based methods.

Microbial DNA testing is a rapidly growing field that involves identifying and characterising microorganisms using DNA-based methods. This can include techniques such as polymerase chain reaction (PCR) and DNA sequencing. These methods have been used in various applications, including medical diagnostics, food safety testing, and environmental monitoring. In the medical field, microbial DNA testing is used to identify and track infectious diseases, including tuberculosis, gonorrhoea, and some forms of cancer. In the food industry, it is used to detect harmful bacteria and other microorganisms in food products. In the environmental field, it is used to monitor and track the presence of microorganisms in water, soil, and air. Overall, microbial DNA testing is a powerful tool helping to improve our understanding of the microbial world and its impact on human health and the environment.

There are several potential areas for future research in microbial DNA testing. One area of interest is the development of more rapid and accurate diagnostic methods for identifying and characterising microorganisms. This could include using portable, point-of-care devices for testing and developing new DNA sequencing technologies that can provide more detailed information about the genetic makeup of microorganisms.

Another area of research is microbial DNA testing for tracking and understanding the spread of infectious diseases. This could include developing new methods for identifying and tracking outbreaks of infectious diseases in real time, as well as using microbial DNA testing to understand the evolution and transmission of pathogens.

In addition, research in metagenomics and metatranscriptomics is expected to grow and be widely used. Metagenomics allows the study of the genetic diversity of microorganisms in a specific environment, while metatranscriptomics allows the study of the functional diversity of microbial communities.

Another direction of research is microbial DNA testing in precision medicine, which would allow for the personalised diagnosis and treatment of patients based on their unique microbial profiles.

Finally, research on applying artificial intelligence (AI) and machine learning (ML) in microbial DNA testing is expected to grow. These techniques will be used to analyse large datasets, improve diagnostic accuracy, and speed up identifying and characterising microorganisms.

Overall, the future of microbial DNA testing is expected to be characterised by developing more sophisticated and efficient methods for identifying and characterising microorganisms and applying these methods to a wide range of fields, including medicine, agriculture, and environmental monitoring.

DNA Phenotyping

25

Astha, Tanya Chauhan, Shreya Arora, and Rutwik Shedge

Abstract

The genesis of DNA biology has been a fundamental point in advancing the knowledge of living beings and their functioning. DNA is the blueprint of life and determines the functional and physical traits of an individual. In recent times, DNA has emerged as an indispensable part of research and applications around the world. Scholars around the world have used the knowledge from DNA to solve mysteries of evolution, heredity, health, and identity. DNA research has found an important role in the field of human identification. DNA-based identification is based on the fact that each individual has about 0.01% of DNA unique to them which can be analysed for identification purposes. The major benefit of DNA-based identification has been to the field of forensic science. Forensic Science employs a range of scientific theories and techniques in the aid of justice. DNA fingerprinting found an important and indispensable role in the field of forensic science (Gill et al. 1985). The role of DNA analysis in the advancement of the criminal justice system is commendable with its applications aiding in faster and more accurate interpretation of evidence, identification and prosecution of criminals in legal cases, and helping resolve disputed paternity cases. DNA-based techniques are used in many aspects of forensic investigation, the

Astha
LNJN National Institute of Criminology and Forensic Science, Delhi, India

T. Chauhan (✉)
Department of Forensic Science, National Forensic Science University Delhi Campus (LNJN NICFS), New Delhi, Delhi, India

S. Arora
CTM IRTE, Faridabad, Haryana, India

R. Shedge
Depatment of Forensic Science, NFSU, Agartala, Tripura, India

A. Puri et al. (eds.), *Fundamentals of Forensic Biology*,
https://doi.org/10.1007/978-981-99-3161-3_25

science began with incorporating the technique of DNA fingerprinting by Sir Alec Jeffrey for individual identification and later expanded to paternity determination and ancestry determination and in recent research involving tissue identification and DNA-based phenotyping.

Keywords

Forensic DNA phenotyping · EVC · Single nucleotide polymorphism · SNaPshot technology

25.1 Definitions

Forensic DNA Phenotyping: Prediction of appearance of an unknown donor of a biological sample through their DNA.

Externally Visible Characteristics (EVC): Features such as the eye colour, skin colour, hair colour, etc., that define the appearance of a human being.

Human Genome Project: Project aimed at sequencing the entire human DNA.

SNaPshot technology: Phenotypic model that uses information from SNP analysis of a DNA sample to predict ancestry and physical traits.

25.2 Introduction to DNA

Humans have 23 pairs of chromosomes, out of which 22 pairs are called autosomal chromosomes and the rest are sex chromosomes that determine the sex of the individual. The sex chromosomes are X and Y, an individual with XX chromosome will be biologically female and one with XY will be biologically male. The chromosomes are the basis of inheritance in humans; they are inherited equally from both parent, which means that each individual's half set of chromosomes is inherited from each of their parents. The chromosomes exist in pair (except XY) and so does the genes present in them. The alternate forms of genes are referred as alleles and each allele is present on either of the two homologous chromosomes. The alleles define the genotype or genetic composition of the individual, which establish the phenotype or physical composition of the individual. The complete genotype of an individual constitutes their genetic profile.

DNA are packaged into Deoxyribonucleic Acid (DNA) that is responsible for heredity in individuals. Deoxyribonucleic Acid (DNA) is a type of nucleic acid, composed of monomeric units called nucleotides. The nucleotides that are the major constituent of DNA, they are composed of a nitrogenous base, a pentose sugar (Deoxyribose), and a phosphate. The different nitrogenous bases include Adenine, Thymine, Guanine, and Cytosine, represented by A, T, G, and C, respectively. They are classified into Purines and Pyrimidines. The Purine bases of the DNA are Adenine and Guanine, whereas the Pyrimidine bases are Cytosine and Thymine. Out of the four bases, two of them, adenine and guanine, are purines, which are double ring molecules, while the other two, cytosine and thymine, are pyrimidines, which are single ring molecules. The nitrogenous bases exhibit complementary base pairing characteristics, where adenosine molecules bind with thymine molecules through a double hydrogen bond and the guanine molecules bind with cytosine molecules through a triple hydrogen bond. The nitrogenous bases exist and arrange themselves in a DNA molecule in accordance of the Chargaff's rules, proposed by Erwin Chargaff in the 1940s. They state that the number of guanine molecules is equal to those of the cytosine molecules and similarly the numbers of adenine and thymine molecules are same in a double-stranded DNA molecule. Thus, that the ratio of Adenine to Thymine and Guanine to Cytosine molecules is 1 in a double-stranded DNA molecule. These rules helped in establishing the double helix structure of the DNA. The sequence in which these bases occur governs the genetic characteristics attained by an individual.

DNA stores the information for the sustaining and functioning of human body. DNA exists in almost all the cells of the human body, the exception being the red blood cells or the erythrocytes, which are enucleated cells. DNA is present in the nuclei as well as the mitochondrion of a cell. It is distinguished as nuclear and mitochondrial DNA, respectively. The DNA is a complex molecule with an intricate structural organization and a wide range of roles to play in making life possible.

DNA was first isolated by Friedrich Miescher in 1869 from the pus cells found on discarded surgical bandages. He named the substance 'Nuclein'. Later, in 1889, 'nuclein' was renamed as 'nucleic acid' by Richard Altmann, who was one of Miescher's pupils. The role of DNA in inheritance was given by Frederick Griffith, through the findings of his experiment conducted on the virulent and non-virulent strains of pneumonia. He called DNA 'The transformation molecule'. The transformation molecule was finally confirmed to be the deoxyribonucleic acid (DNA) by the experimental findings of Oswald Avery. Phoebus Levene in 1929 discovered the constituents of a DNA molecule; they were four nitrogenous bases, sugar, and a phosphate molecule. He also gave idea about the order in which these molecules were arranged. In the 1940s, Erwin Chargaff gave the Chargaff's rule to explain the correlation among the quantity of the four bases of DNA. The fact that DNA has a helix shape was given by Rosalind Franklin and Maurice Wilkins. The famous and widely accepted double helical structure of DNA was given by James Watson and Francis Crick in 1953 which expanded the knowledge about the stability, interactions and working of the DNA molecule.

25.3 What a DNA Can Reveal

The DNA material in chromosomes is composed of 'coding' and 'non-coding' regions. A gene generally ranges from a few thousands to ten thousands base pairs. Eukaryotic DNA have a lot of repeated DNA sequence. These sequences vary in size and are known by the length of the core repeat unit which can contain several hundred to several thousand base pairs in the core repeat region. These regions are known as satellite DNA. The core repeat unit for approximately medium

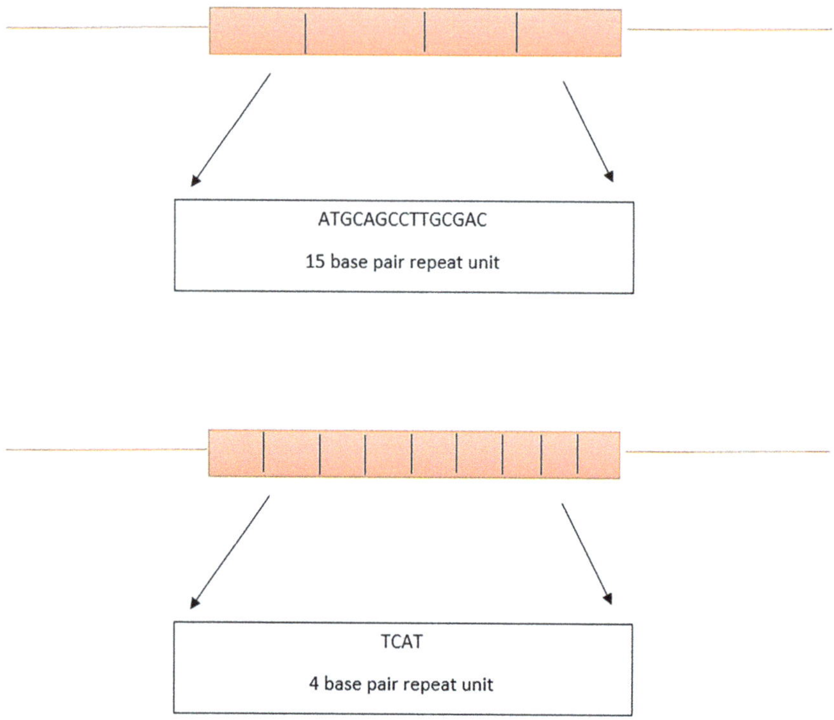

Fig. 25.1 Schematic representation of VNTR and STR DNA markers. The region of DNA surrounding a VNTR and STR is known as flanking region which helps us in designing the primers for the foresaid DNA marker

length repeats is known as Variable Number Tandem Repeat (VNTR). It ranges for approximately 10–100 base pairs in length. Variable Number Tandem Repeat also known as minisatellite. The forensic DNA marker D1S80 is actually a mini-satellite with 16 base pairs in a repeating unit (Jeffreys 2005).

The DNA regions where the repeat units vary for 2-base pairs in length are called microsatellites, or short tandem repeat (STR). These STRs are now a preferred DNA marker since they are easy to amplify and the variability in repeats in individuals is quite high which makes it more effective for forensic biology to function. They account for approximately 3–4% of the total human genome and occur on average after every 10,000 nucleotides. The schematic representations of VNTR and STR markers are shown in Fig. 25.1, and the differences between them are depicted in Table 25.1.

A single base variation in sequence between individuals at a particular location in the human genome is referred to as Single Nucleotide Polymorphism (SNP). They are quite abundant in the human genome and are also being used for studying the genetic diseases. There are millions of SNPs in each individual and can be further

Table 25.1 Difference between VNTR and STR demonstrating their use and feasibility in forensic DNA analysis. Also, STR markers are much more compatible for amplification as compared to VNTR markers

Variable Number Tandem Repeat (VNTR)	Short Tandem Repeat (STR)
Also known as minisatellites	Also known as microsatellites
It is a type of tandem repeat in which sequence of 10–60 base pairs are repeated a variable no. of times at a particular locus	It is a type of tandem repeat in which sequence of 2–6 base pairs are repeated a variable no. of times at a particular locus
It may consist of 10–1500 repeats in a unit	It may consist of 5–200 repeats in a unit
It forms an array of 0.5–1.6 kb	It forms an array of 10–1000 base pairs

Table 25.2 Comparison between STR and SNP markers. Though the frequency of SNP markers in human genome is quite high but the polymorphism is better in STRs than SNPs. Also, the detection is much convenient in case of STR marker analysis

Short Tandem Repeat (STR)	Single Nucleotide Polymorphism (SNPs)
Their frequency in human genome is approximately 1 in 15 kb	Their frequency in human genome is approximately 1 in 1 kb
They are highly informative regarding the informatics of genetic characteristics	They are 25–30% low informative as compared to STR regarding the informatics of genetic characteristics
There are generally more than 5 alleles per marker	There are generally 2 alleles per marker
They can be detected using gel/capillary-based electrophoresis	They can only be detected using sequence analysis

used in forensics for individualization. Restriction Fragment Length Polymorphism is the technique in which sequence detectable by variation in the length of fragments produced by specific restriction enzymes, Variable Number of Tandem Repeats. The chromosomes exist in pair (except sex chromosomes in case of XY) and so the genes as well as the genetic markers exist in pair, each of which is inherited from each of their parent and hence can be used to trace and identify blood relations. The differences between STR and SNP markers are shown in Table 25.2.

25.3.1 Externally Visible Characteristics Influenced by DNA

Pigmentation governs the externally visible characteristics in individuals. It is controlled by the quantity and type of melanin produced by melanosomes in melanocytes. Melanin exists in two forms eumelanin and pheomelanin and the ratio of their quantity determines the colouration to the eye, iris, and hair. High eumelanin content contributes to darker hair, eye, and skin colour, whereas high pheomelanin content leads to blonde or red hair and lighter skin tone. The melanocortin 1 receptor influences the pigmentation process and is controlled by the MC1R gene. The melanocortin 1 receptor, when activated, promotes the

Table 25.3 Genetic markers for externally visible characteristics (Marano and Fridman 2019)

EVCs	Nearest genes/chromosome region associated
Eye color	*HERC2, IRF4, LOC105370627, OCA2, SLC45A2, TYR*
Hair color	*EXOC2, HERC2, IRF4, KITLG, MC1R, OCA2, SLC24A4, SLC45A2, TYR*
Skin color	*ANKRD11, ASIP, BNC2, DEF8, HERC2, IRF4, KITLG, MC1R, OCA2, PIGU, RALY, SLC24A4, SLC45A2, TYR, TYRP1, SLC24A5*
Height	*ACAN, DNM3, EFEMP1, FBXW11*, GH region, *GHSR, GPR126, HHIP, HMGA1, HMGA1, IHH, LCORL, MICA, NOG, NPR3, PML, PPIF, SDR16C5, SOCS2*
Facial features	*ADAMTS2, ASPH, C5orf50, COL11A1, COL17A1, CTNND2, DNMT3B, EVC2, FBN1, FGFR1, FGFR2, GDF5, LRP6, PAX3, POLR1D, PRDM16, RAI1, RELN, ROR2, SATB2, SEMA3E, SLC35D1, TP63, UFD1L, WNT3*
Baldness	*AR/EDA2R, EBF1, HDAC9, TARDBP*, 20 p11

Abbreviation: EVCs, externally visible characteristics.

synthesis of eumelanin; an inactivated or blocked receptor promotes pheomelanin synthesis.

Externally visible characteristics (EVCs) such as eye colour, skin colour, hair colour, height, male pattern baldness, height, and facial features can be predicted from DNA markers. EVCs are outcome of distinct phenotype expressed in individuals as an outcome of insertion and/or deletion and single nucleotide polymorphism in the DNA (Marano and Fridman 2019). MC1R gene polymorphism produces the array of pigmentation variation that exists in the population. Other pigmentation-governing genes are ASIP, MATP, SLC24A5, TYR, TYRP1, and OCA2 (Tully 2007). Similarly, variants of the following were attributed to different phenotypic characters such as SLC24A4 for eye and hair colour (Sulem et al. 2007), KITLG (Sulem et al. 2007), ASIP (red colour) (Sulem et al. 2008), TPCN2 (Sulem et al. 2008) for hair colour, TYR (Sulem et al. 2007), HERC2 (Kayser et al. 2008) for colour of eye, ASIP (Sulem et al. 2008), and TYR and 6p25.3 (Sulem et al. 2007) for the presence of freckles. Some of the EVCs and their associated genes/chromosomes are shown in Table 25.3.

There have been population-specific studies to identify genes responsible for different phenotypes in the population. Genes for pigmentation of hair like HCL3 (Eiberg and Mohr 1996) for brown coloured hairs and BEY2 and GEY for brown and green eye colour in the population of Denmark (Eiberg and Mohr 1996) and the polymorphism in SLC24A5, TYR, and SLC45A2 for skin colour in the South Asian population (Stokowski et al. 2008). Male pattern baldness is another externally visible character with significant value for identification. The trait can be predicted through DNA-based phenotyping of AR, EDA2R, EBF1, TARDBP, and HDAC9 genes (Marcińska et al. 2015). The pattern of hairs whether straight or curled can be predicted through markers such as FGFR2 and EDAR genes for straight and thick hairs in the Asian population (Fujimoto et al. 2007, 2009) and TCHH gene for straight hair in European population (Medland et al. 2009). There are also recognized markers for height estimation identified as IHH, HHIP, PTCH1, EFEMP1, ADAMTSL3, ACAN, CDK6, HMGA2, and DLEU7 (Weedon et al. 2008) among others.

Forensic DNA phenotyping systems are single nucleotide polymorphism-based multiplex assay capable of simultaneous analysis of multiple markers. They developed as the outcome of identification of genetic marker for phenotypic characters and allowed prediction of different visible traits. IrisPlex DNA test system predicted the blue and brown eye colour from DNA samples using 6 SNP-based markers (Walsh et al. 2010). HIris-Plex was an advanced version of phenotype prediction models, and predicted 24 eye colour and hair colour from DNA samples (Walsh et al. 2013). HIris-Plex-S DNA test system added the application of predicting skin colour to the previous HIris-Plex system. The HIris-Plex-S DNA system worked with novel assay of 41 SNP markers and can help predict 17 phenotypes of skin colour (Chaitanya et al. 2018).

In the current times, advancements in technology have allowed for affordable sequencing of genome and identify single nucleotide polymorphisms within it. This trend has allowed for public and private platforms to perform genome-wide analysis for a multitude of individuals and organism for diverse purposes. Parabon labs in the United States studied the genomic data of subjects with known phenotypes to build a phenotype prediction model. The model was used to deduce the physical appearance of an individual based on the genomic data. The Snapshot DNA Phenotyping System is another phenotypic model that uses information from SNP analysis of a DNA sample to predict ancestry and physical traits, such as skin colour, hair colour, eye colour, freckling, and even face morphology.

Indian SNP data was organized by Centre for Cellular and Molecular Biology. Their study was based on the knowledge that India has a diverse human genetic pool with 4635 anthropologically well-defined populations and maintaining endogamy for the thousands of years. Indian populations are the admixture of 2 ancestral populations; Ancestral North Indians (ANI) with genetic affinities with Middle Easterns, West-Eurasians, and Europeans and the Ancestral South Indians (ASI), which is not related to any group outside Indian subcontinent. They performed genome-wide association study (GWAS) with existing reference panels, which include African, European, and East Asian HapMap population samples; South-Asians of 1000 genome project; Indian population samples (Indo-Europeans and Dravidians); and combined HapMap and Indian population samples.

Human genetic variation studies in Indian populations were performed by The Centre for DNA Fingerprinting and Diagnostics (CDFD). They studied genetic diversity among the Indian populations and utilize them for improving the current human DNA profiling technology for forensic human identification (HID). They designed a panel comprising of 70 SNPs on the autosomes with desired features for HID application for Indian populations and also on genotype–phenotype correlation studies on the association of DNA-based markers with characteristics like skin colour, body-mass index (BMI), and height.

The technique has also found its application in fields of archaeology and anthropology where the phenotyping analyses of ancient remains have generated information about the appearances, ancestry, and behavioural characteristics of early populations. Such was a case of the famous 'Cheddar man' fossil recovered from a cave in England whose physical appearance and possible lactose intolerance was deduced using DNA-based phenotyping.

25.3.2 Characterization of Cell-Lines and Monitoring Transplants

Human cell line authentication is carried which enables the discovery of cross-contamination between the cell lines and can also be served as universal standard. These types of analysis are called cell culture forensics.

To monitor the donor cells after the transplant helps in checking the success rate of the transplantation. This is done by checking the STR profile of the graft. In various cases, chimerism is observed where a mixed profile is generated.

25.3.3 Mapping Genetic Diseases and Detection of Cancer Tumours

A genetic scan is performed to detect a disease and mapping is done using approximately 400 STR markers that covers the whole human genome. They study the association of the marker with the respective disease in contrast with the patient's population. Through linkage analysis, a correlation can be further analysed between the gene marker and the diseased gene of interest. For example, a genetic disease known as Meckel–Gruber syndrome is reported to be located near D8S1179 marker.

As we study the genetic maps, we can also know gene deletion and addition. There are many types of cancer that can be caused due to gene deletions, i.e. loss of heterozygosity. In this, approximately nine STR markers are used on the tissue extracted for cancer biopsy to confirm the cause of cancer in an individual.

25.3.4 Human Genome Project

The Human Genome Project that began in 1990 and went on till 2003 was an ambitious project, which caved way to numerous major scientific discoveries. The major goal of the project was to determine the sequence of nucleotides that make up the human DNA and eventually map the genes to further study the structural and functional importance. The project showed that among the entire human genome, only 2% of the human genome is the protein coding region and the rest 98% is the noncoding DNA region. The protein coding region as the name implies refers to the DNA that codes for the proteins. These regions have certain locations where the DNA sequence varies among individuals; this concept helped establish the science of individualization based on the DNA sequence. The human genome project holds the benefits in many fields including molecular biology, biotechnology, and bioinformatics and even to human evolution. It helps in understanding various diseases caused by bacteria, viruses, and even autoimmune disorders. The Human Genome

Project further revealed that 99.9% of the DNA is identical among individuals. The remaining fraction of DNA exhibited 'DNA Polymorphism'; this term was used to define the sites where differences in the DNA sequence of two individuals exist. The sequencing of the entire human genome helped to identify and understand the polymorphism of DNA sequences. This difference acts as genetic markers, and is the basis of identification and individualization through DNA Profiling or DNA fingerprinting based on DNA evidence. There are various kinds of polymorphism that exist in the human genome, such as the Single Nucleotide Polymorphism, Short Tandem Repeats, and Variable Number of Tandem Repeats (Collins et al. 1998).

25.4 Technology Used in DNA Phenotyping

In order to analyse the DNA extracted using various protocols, the DNA needs to be quantified and qualitied. There are various techniques which can be used for the analysis.

RFLP or Restriction Fragment Length Polymorphism was the first technique developed for identifying the variation in the genetic makeup of individuals. This technique was based on the variation of the DNA sequence at a specific location recognized by certain restriction enzymes. The restriction enzymes have unique characteristics of identifying target site in a DNA sequence and producing cuts at these sites and leading to fragmentation of the DNA molecule. These restriction enzymes exhibit high specificity for their target site and if there is variation in the DNA sequence, the restriction enzyme would not act at that site. This would lead to a variation in the fragment length produced by the enzyme. The restriction enzyme selected for the process have specific target sites and produces fragments of known length, and any variation in the resulting fragment indicates towards variation in the sequence. The variation of the resultant fragments is detected by performing electrophoresis or hybridization.

In order to distinguish various molecules from one to another, a step for separation is followed to pull the different sizes of DNA fragments apart. The separation is performed using the electrophoresis which is either conducted on slab gels or capillary environment. In slab gel electrophoresis, a slab consisting of solid matrix with a series of pores/holes inside a buffer solution through which the DNA molecules pass when the electricity is provided. The gels formed can be further of different types including Agarose Gels in which agarose is used to form a pore-like structure on a flattened plate. The pore size is approximately 200 nm. They are easily prepared using liquid agarose and Polyacrylamide Gels which have much smaller pore size than agarose. In this, gels are formed due to cross linking of acrylamide and bisacrylamide.

Capillary Electrophoresis is relatively a new outlook to this family. In this, the steps like injection, separation, and detection can be fully automated. Here, the movement of molecules are based on the size only but are detected because of fluorescence molecule attached. The modern-day instrument, Genetic analyser, is based on this technique only.

RFLP technique does not perform well with degraded samples or when the DNA extracted is in limited amount. To overcome this drawback, a better technique was introduced, i.e. Amplified Fragment Length Polymorphism. In this technique, the loci which is smaller than 1Kb was targeted using restriction enzymes amplified using Polymerase Chain Reaction (PCR). These amplified fragments are separated using Polyacrylamide Gel Electrophoresis (PAGE) and detected with the help of silver staining technique. The advances made in the science of DNA analysis has made the methods and the results more dependable and proficient and increased their validity as evidence in the court of law.

25.4.1 DNA Databases

The recent advances in technology have played an eminent role in establishing DNA databases, which are being used worldwide to store the genetic information of individuals by law enforcement agencies for faster identification of individuals in criminal investigation. In the year 1995, the UK established the one of the world's National databases for DNA matching and information termed NDNAD in which DNA data from England and Wales were uploaded and accessed. Though Northern Ireland and Scotland have their own database but they do contribute in NDNAD (Johnson et al. 2003). The major DNA database CODIS is maintained by FBI, USA, in 1998, where it stores the genetic information based on STR or short tandem repeats of an individual to aid in identification and comparison. They selected 13 STR loci that were selected for the data generation. There are various other countries including Austria, Germany, Netherlands, and various others that introduced their own databases by the end of 1998. Also, the criteria for the entry in database as well as removal criteria depends on Country's database rules and regulation (Francisco Corte-Real 2004).

CODIS has three hierarchical levels, the Local DNA Index System (LDIS) which is maintained by the crime laboratories functioning in police departments and local agencies (Fig. 25.2). All the data generated by LDIS is further transmitted and stored at State DNA Index System (SDIS) and National DNA Index System (NDIS) which is operated by the state laboratories. SDIS also acts as a bridge between LDIS and NDIS.

25.5 Forensic Applications of DNA

The uniqueness of the DNA sequence of an individual is the basis of the role DNA play in forensic investigation. No two individual exhibits the same DNA profile; recent scientific research has identified criteria of variation among the DNA sequence of identical twins. The variation in the DNA sequences of individuals has helped forensic scientists link or dismiss individuals as the source of given biological evidence in question. Since its origin, DNA evidences have been utilized

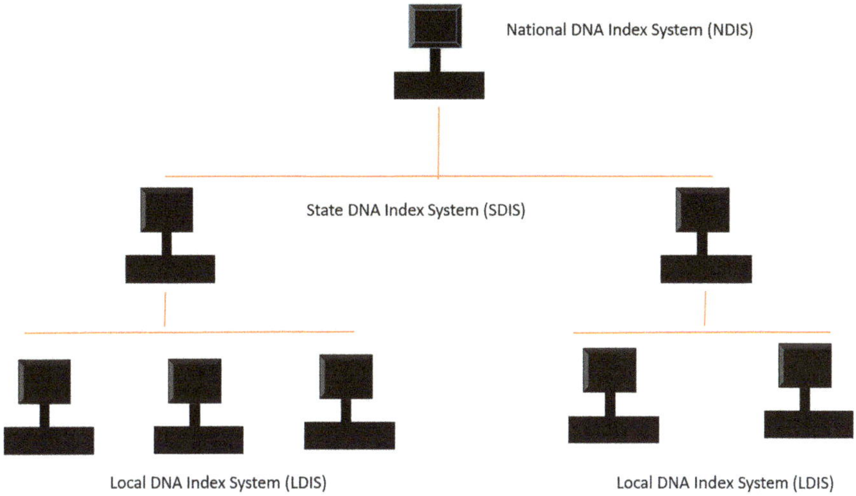

Fig. 25.2 Schematic diagram of hierarchical set-up of CODIS. It explains the maintaining body of this database while LDIS are the level which contribute to the database

to solve many old crimes as well as help establish the innocence of previously convicted individuals.

Forensic Analysis aims to help the investigation by revealing the facts based on the evidences. Identification and Individualization of evidences are two key concepts of forensic analysis. Edmund Locard gave the "Locard Transfer Theory" or "Locard's Exchange Principle" to explain the exchange or transfer of traces from one object to another during a contact between the two. This principle formed the basis of comparing evidences in forensic investigation. The evidences are analysed and compared with the reference samples by the forensic scientist to establish an association between them, concluding whether they are similar or originated from the same source or not. The science of DNA Analysis is a combination of various disciplines of science such as Molecular Biology, Genetics, Statistics, and Biochemistry. The science of genetics explains the concepts for use of DNA in paternity issues through heredity of traits by individuals from their biological parents. DNA analysis is successful because DNA is a comparatively more stable for longer durations of time, than many of the genetic markers analysed in forensic analysis. DNA Analysis is now an important entity in the criminal justice system, where it helps in the delivery of justice.

DNA evidence is one of the major biological evidence in crime investigation. The matching of DNA profiles is performed to establish the presence of a person at a scene of crime. They help link individuals to crimes or to establish their innocence. DNA Profiling is also used to establish relatedness, such as paternity, maternity, siblings, etc., among individuals in dispute owing to the fact that DNA sequence follow a defined inheritance pattern, i.e. both the parents equally contribute half of their chromosome to their progeny which on analysis helps establish the biological link between them. DNA also has role in determining bio-geographical ancestry using Ancestry informative markers (Phillips 2015) and single nucleotide polymorphism. Externally visible characteristics can be identified using prediction of phenotypic characteristics through established DNA markers.

DNA for forensic analysis can be obtained from a number of biological sources. DNA is present in almost all types of body cells, thus most of them serve as sources of DNA for analysis. Biological sources of DNA involve the biological fluids, such as blood, semen, saliva, etc., and the stains generated by these fluids. Biological materials such as Bone, Tooth, Nail, Tissue, etc., are also used to extract DNA. Many surfaces or objects remain in direct contact with the human body. Many objects or surfaces have the cells of the body transferred to them via the contact made with individuals. These cells may remain attached to these objects or surfaces, from where these cells can be utilized to extraction of DNA. Certain objects acquire the cells of an individual and when they come in contact with them, the cells adhere to them and thus these objects or surfaces serve as evidentiary material for the extraction of DNA. Many of these sources are commonly encountered at various scenes of crime, and help provide DNA-based evidence for investigation. One of the most common and frequently encountered biological evidence at the scene of crime is Blood. They are found at crime scenes ranging from homicide, assault, robbery, kidnapping, etc., and are also collected by doctors and medico-legal examiners from victims and suspects to serve as reference for analysis in investigation. In blood, the White Blood Cells or Leucocytes are the source of nuclear DNA, the red blood cells of blood lack nuclei, i.e. they are enucleated to facilitate their role in transport of oxygen. In forensic science, DNA plays a significant role in individualization. The development of DNA Profiling by Alec Jeffreys in 1984 began the incorporation of DNA profiling in forensic science. DNA fingerprinting helped solve immigration cases (Jeffreys et al. 1985) in its wake and found its first forensic application in the Enderby murder case, Leicestershire (Wong et al. 1987) where it helped to prove innocence of a previously accused individual and finding the true culprit. The method of DNA profiling advanced to using Short Tandem Repeats (STR) for profiling, used for the first time in 1990 with the war crime investigation for identification of skeletal remains of a war criminal (Jeffreys et al. 1992). Tetranucleotide Short tandem repeats developed as an improvement to the system giving better profiles for analysis (Edwards et al. 1991). Genetic profiles are conventionally generated from DNA isolated from sources such as old and new biological stains (Blood, Semen, Saliva, and Vaginal Secretions), Cigarette buds, Hair samples, Bones, etc. A significant development came in 1997 with the innovation of extraction of DNA from touched object began, allowing the extraction of DNA from numerous surfaces in contact with the body (van Oorschot and Jones 1997). The minute quantities of biological sample left at a crime scene are used to generate a DNA profile that can be compared to a database or to a reference sample from suspect or victim. The first DNA-based database was developed by Michael Howard for the United Kingdom in 1995 to store DNA profiles of criminals.

A DNA profile is representative of unique region in the non-coding part of an individual's genome such as short tandem repeats or STRs which are nucleotide repeats present at specific locations or loci and the numbers of repeats varies among individuals. Human identification based on the genetic profiling obtained from studying polymorphism is one of the standards accepted worldwide. STRs are a type of length polymorphism and approximately one million of them has been identified in the human genome (Marcińska et al. 2015). The profiling recruits about 13 of these STR markers and the profile provide the numbers of repeats at each of these locations for the tested sample. These STR loci are approved by the international scientific community for DNA profiling for forensic purposes. The markers are highly variable and the set number of tested loci provides a profile with a probability high enough to reduce the possibility of same profile for two individuals unless they are identical twins.

25.5.1 Future of Forensics: DNA-Based Phenotyping

"DNA fingerprinting" technique is commonly used by forensic scientists to identify individuals to help in the investigation process. The DNA profiling technique is a very popular and commonly practiced technique; however, despite its success rate, every year many crimes go unsolved due to lack of reference samples for comparison and identification and lack of other forensic evidences.

To overcome the limits of DNA profiling and to get better profiling, there was a need to explore more of the data that DNA contains about an individual. This led to the use of DNA-based phenotyping techniques for better profiling. DNA phenotyping involves the analysis of the DNA markers of an individual to predict their externally visible characteristics or phenotypic characteristics to generate a profile for identification. The analysis does not provide a certain result but predict the phenotypic characteristics of an individual to a degree of probabilistic likelihood (Samuel and Prainsack 2019). The analysis provides prediction of visible characteristics like eye colour, hair colour, hair type, skin tone, and facial characteristics (Walsh et al. 2017). The phenotype expressed by an individual depends on the genotype and the epigenetic influence on the genetic material of an individual. More than one gene can contribute to the expression of a phenotype and the role of epigenetic impact on the expression makes prediction a complex process. Many of the genes contributing to the visual characteristics of an individual have been identified by scientists. The knowledge of the genes or the set of genes contributing to the resulting phenotype are now a potential knowledge bank for forensic scientists for the future of individual identification. Prediction models based on the identified markers and their expression profiles are developed based on molecular, statistical, and computational analysis.

25.5.2 Case Studies

In 2001, a sexual assault case in USA was investigated using DNA phenotyping to predict the physical characteristics of the culprit in the absence of match with the database. The phenotypic analysis predicted the hair, eye, and skin colour of the culprit, which was later used for profiling in the neighbouring area. The search led to identification of an individual which was later confirmed through matching of STR profile of the biological evidence from the scene with the culprit. In 2017, an 8-year-old murder case in United states was resolved with arrest of the culprit through ancestry details provided by DNA phenotyping that contradicted the earlier profile the police assumed of the culprit.

A 41-year-old cold case got a lead through DNA phenotyping of a recovered skull in New Castle County, Delaware, by providing ancestry details and physical characteristics of the victim. A 1994 cold case of sexual assault and murder from Montgomery County, Maryland, was resolved using DNA-based phenotyping to apprehend the culprit in 2018. The culprit was wanted for similar offence in a similar older case as well.

25.5.3 Legal and Ethical Aspects

DNA profile for individual identification is based on STR profiling from the non-coding region of the DNA and this provides no personal information. However, the biggest concern for DNA phenotyping is the invasion of privacy due to the analysis outcome. There are arguments about the amount of information that become available to law enforcement officials if phenotyping is allowed. The collection and retention of samples for the use in forensic DNA has led to the quite rapid development and its associated law practices. Medical history among other personal information becomes susceptible to be extracted without consent if phenotyping is carried out in an unregulated way. Racial stereotyping also remains an issue at the international level where extradition of ancestry and phenotypic characteristics would add to the problem of racial profiling (Koops and Schellekens 2007). The discrimination concern was the reason Germany was reluctant in adopting DNA phenotyping in their investigation process. The issues include a surge for racial disparities. In France, a long debate regarding the "DNA fingerprints" was raised where they stated the major ethical issues with this technique to identify individuals as potential criminals and the risk of private investigation in cases like paternity or with insurance companies. This is the reason why in 1994 the French Parliament adopted one of strict laws and recommendation of usage of such databases is only permitted to accredited laboratories. This has reduced the difficulty in implementation of DNA fingerprinting for the criminal justice system (Machado and Silva 2019).

The concerns regarding DNA phenotyping are worth addressing while adopting the process in investigation process of law enforcement bodies. The process should aim at protecting the privacy and dignity of general public. Ensuring these criteria would help acceptance of the process in investigation and overcome the concern of utilization of the technique for targeting and discrimination towards individuals or groups. In France, a strict procedure is followed for all the genetic tests which comes under bioethics laws (1994) of the respective country in which the physician has to obtain consent from the individual before the test is actually carried out.

25.5.4 Limitations

DNA phenotyping does not provide a conclusive result as DNA profiling. In DNA phenotyping approaches, the analysis is affected by the complexity of fact that many physical traits are manifested as quantitative traits which are governed by a number of factors and coupled with variation arising due to environmental and lifestyle-based factors. The analysis involves complex mathematical models and statistical analysis to give results of high accuracy. The genomic data of the Indian population is limited and the genetic makeup of the diverse populations is very much unexplored. There is a need for extensive genomic data on various populations to generate a model with sufficient accuracy to provide dependable results. The currently available models have their data obtained from European populations and thus the results may not be appropriate for Indian population studies. The IrisPlex system, popular for phenotyping of eye colour from DNA samples of the European population, did not demonstrate the same accuracy in samples from Asian populations, suggesting that more studies should be performed on more distinct population samples.

Multiple Choice Questions

1. Genes that exist as an alternate expression at a particular locus are known as
   (a) Loci.
   (b) Antigen.
   (c) Allele.
   (d) Phenotype.
   Answer (c)
2. Which locus is used for determining male as well as female gender in DNA fingerprinting?
   (a) DYS 19.
   (b) DYS 393.
   (c) Amelogenin.
   (d) Y-plex ladder.
   Answer (c)
3. Short fragments of DNA that are labelled with radioactive tags are known as
   (a) Probes.
   (b) Primers.
   (c) STRs.
   (d) VNTRs.
   Answer (a)
4. VNTR is a type of
   (a) Length Polymorphism.
   (b) Sequence Polymorphism.
   (c) Both of the above.
   (d) None of the above.
   Answer (a)

5. The rate of migration of DNA within an agarose gel in the gel electrophoresis procedure is primarily based on
   (a) The volume of the DNA sample loaded.
   (b) The size of the DNA fragment.
   (c) The number of DNA fragments.
   (d) The negative charge of the DNA.
   Answer (b)
6. STR DNA analysis can help us in determining which of the following?
   (a) Age of an individual.
   (b) Race of an individual.
   (c) Sex of an individual.
   (d) Height of an individual.
   Answer (c)
7. RFLP determines which of the following?
   (a) Power of discrimination (PD)
   (b) Variation in the length of a defined fragment of DNA.
   (c) Hundreds of variations at each locus.
   (d) Age of bloodstain.
   (e) Choose the most appropriate answer from the options given below:
   (f) d, a, and c only.
   (g) b, d, and c only.
   (h) a, b and c only.
   (i) a, d, and b only.
   Answer (c)

References

Chaitanya L, Breslin K, Zuniga S, Wirken L, Pośpiech E, Kukla-Bartoszek M et al (2018) The HIrisPlex-S system for eye, hair and skin colour prediction from DNA: introduction and forensic developmental validation. Forensic Sci Int Genet 35:123. https://doi.org/10.1016/j.fsigen.2018.04.004

Collins FS, Patrinos A, Jordan E, Chakravarti A, Gesteland R, Walters L, Members of the DOE and NIH Planning Groups (1998) New goals for the US human genome project: 1998-2003. Science 282(5389):682–689

Corte-Real F (2004) Forensic DNA databases. Forensic Sci Int 146:S143–S144

Edwards A, Civitello A, Hammond HA, Caskey CT (1991) DNA typing and genetic mapping with trimeric and tetrameric tandem repeats. Am J Hum Genet 49(4):746–756. https://pubmed.ncbi.nlm.nih.gov/1897522

Eiberg H, Mohr J (1996) Assignment of genes coding for Brown eye colour (BEY2) and Brown hair colour (HCL3) on chromosome 15q. Eur J Hum Genet 4(4):237–241. https://doi.org/10.1159/000472205

Fujimoto A, Kimura R, Ohashi J, Omi K, Yuliwulandari R, Batubara L et al (2007) A scan for genctic determinants of human hair morphology: EDAR is associated with Asian hair thickness. Hum Mol Genet 17(6):835–843. https://doi.org/10.1093/hmg/ddm355

Fujimoto A, Nishida N, Kimura R, Miyagawa T, Yuliwulandari R, Batubara L et al (2009) FGFR2 is associated with hair thickness in Asian populations. J Hum Genet 54(8):461–465. https://doi.org/10.1038/jhg.2009.61

Gill P, Jeffreys AJ, Werrett DJ (1985) Forensic application of DNA ‘fingerprints’. Nature 318(6046):577–579. https://doi.org/10.1038/318577a0

Jeffreys A (2005) Genetic fingerprinting. Nat Med 11:1035–1039. https://doi.org/10.1038/nm1005-1035

Jeffreys AJ, Allen MJ, Hagelberg E, Sonnberg A (1992) Identification of the skeletal remains of josef mengele by DNA analysis. Forensic Sci Int 56(1):65–76. https://doi.org/10.1016/0379-0738(92)90148-P

Jeffreys AJ, Brookfield JFY, Semeonoff R (1985) Positive identification of an immigration test-case using human DNA fingerprints. Nature 317(6040):818–819. https://doi.org/10.1038/317818a0

Johnson P, Williams R, Martin P (2003) Genetics and forensics: making the national DNA database. Sci Stud (Helsinki, Finland) 16(2):22

Kayser M, Liu F, Janssens ACJW, Rivadeneira F, Lao O, van Duijn K et al (2008) Three genome-wide association studies and a linkage analysis identify HERC2 as a human iris color gene. Am J Hum Genet 82(2):411–423. https://doi.org/10.1016/j.ajhg.2007.10.003

Koops BJ, Schellekens M (2007) Forensic DNA phenotyping: regulatory issues. Colum Sci Tech L Rev 9:158

Machado H, Silva S (2019) What influences public views on forensic DNA testing in the criminal field? A scoping review of quantitative evidence. Hum Genomics 13(1):1–13

Marano L, Fridman C (2019) DNA phenotyping: current application in forensic science. Res Rep Forensic Med Sci 9:1–8. https://doi.org/10.2147/RRFMS.S164090

Marcińska M, Pośpiech E, Abidi S, Andersen JD, van den Berge M, Carracedo Á et al (2015) Evaluation of DNA variants associated with androgenetic alopecia and their potential to predict male pattern baldness. PLoS One 10(5):e0127852. https://doi.org/10.1371/journal.pone.0127852

Medland SE, Nyholt DR, Painter JN, McEvoy BP, McRae AF, Zhu G et al (2009) Common variants in the Trichohyalin gene are associated with straight hair in Europeans. Am J Hum Genet 85(5):750–755. https://doi.org/10.1016/j.ajhg.2009.10.009

Phillips C (2015) Forensic genetic analysis of bio-geographical ancestry. Forensic Sci Int Genet 18: 49. https://doi.org/10.1016/j.fsigen.2015.05.012

Samuel G, Prainsack B (2019) Forensic DNA phenotyping in Europe: views “on the ground” from those who have a professional stake in the technology. New Genet Soc 38(2):119–141. https://doi.org/10.1080/14636778.2018.1549984

Stokowski R, Pant P, Dadd T, Fereday A, Hinds D, Jarman C, Cox D (2008) A Genomewide association study of skin pigmentation in a south Asian population. Am J Hum Genet 81:1119–1132. https://doi.org/10.1086/522235

Sulem P, Gudbjartsson DF, Stacey SN, Helgason A, Rafnar T, Jakobsdottir M, Stefansson K (2008) Two newly identified genetic determinants of pigmentation in Europeans. Nat Genet 40(7): 835–837. https://doi.org/10.1038/ng.160

Sulem P, Gudbjartsson DF, Stacey SN, Helgason A, Rafnar T, Magnusson KP, Stefansson K (2007) Genetic determinants of hair, eye and skin pigmentation in Europeans. Nat Genet 39(12): 1443–1452. https://doi.org/10.1038/ng.2007.13

Tully G (2007) Genotype versus phenotype: human pigmentation. Forensic Sci Int Genet 1(2): 105–110. https://doi.org/10.1016/j.fsigen.2007.01.005

van Oorschot RAH, Jones MK (1997) DNA fingerprints from fingerprints. Nature 387(6635):767. https://doi.org/10.1038/42838

Walsh S, Chaitanya L, Breslin K, Muralidharan C, Bronikowska A, Pospiech E, Kayser M (2017) Global skin colour prediction from DNA. Hum Genet 136(7):847–863. https://doi.org/10.1007/s00439-017-1808-5

Walsh S, Lindenbergh A, Zuniga S, Sijen T, Knijff P, Kayser M, Ballantyne K (2010) Developmental validation of the IrisPlex system: determination of blue and brown iris colour for forensic intelligence. Forensic Sci Int Genet 5:464–471. https://doi.org/10.1016/j.fsigen.2010.09.008

Walsh S, Liu F, Wollstein A, Kovatsi L, Ralf A, Kosiniak-Kamysz A, Kayser M (2013) The HIrisPlex system for simultaneous prediction of hair and eye colour from DNA. Forensic Sci Int Genet 7(1):98–115. https://doi.org/10.1016/j.fsigen.2012.07.005

Weedon MN, Lango H, Lindgren CM, Wallace C, Evans DM, Mangino M, Consortium, C. G. E. M (2008) Genome-wide association analysis identifies 20 loci that influence adult height. Nat Genet 40(5):575–583. https://doi.org/10.1038/ng.121

Wong Z, Wilson V, Patel I, Povey S, Jeffreys AJ (1987) Characterization of a panel of highly variable minisatellites cloned from human DNA. Ann Hum Genet 51(4):269–288. https://doi.org/10.1111/j.1469-1809.1987.tb01062.x

Forensic Anthropology

26

Rutwik Shedge, Kam Salem Guite, Varsha Warrier, Tanuj Kanchan, and Kewal Krishan

Abstract

Forensic Anthropology is the application of the science of anatomy, physical anthropology, archaeology, and taphonomy to aid the law enforcement agencies in recovery, examination, and identification of skeletal remains. It involves determination of whether the material recovered at a scene of crime is osteological, determination of species of origin, minimum number of contributors of the skeletal remains, determination of ancestry of the individual, their sex, their age, their stature, and any individualistic features that can assist in positive identification of the remains.

Keywords

Forensic anthropology · Human identification · Osteobiography · Sex determination · Age estimation · Stature estimation · Facial reconstruction · Skull-photo superimposition

26.1 Definitions

Anthropology: Study of humans.

R. Shedge (✉)
School of Forensic Science, National Forensic Sciences University, Gandhinagar, Gujarat, India

K. S. Guite · V. Warrier · T. Kanchan
Department of Forensic Medicine and Toxicology, All India Institute of Medical Sciences, Jodhpur, India

K. Krishan
Department of Anthropology, Panjab University, Chandigarh, India

A. Puri et al. (eds.), *Fundamentals of Forensic Biology*,
https://doi.org/10.1007/978-981-99-3161-3_26

Forensic anthropology: Application of the science of physical anthropology, archaeology, human osteology, and anatomy to aid law enforcement agencies in human identification.

Osteobiography: Human identification by means of determining ethnicity, and sex, and estimating age, and stature.

Forensic facial reconstruction: Process of constructing the face of an unknown individual based on the underlying skeletal structure of their skull.

26.1.1 Introduction

The word 'anthropology' is a derivative of the new Latin word 'anthropologia' which meant 'the study of humanity', which in turn has its roots in the Greek word 'anthropos' which meant 'human being'. The science of Forensic Anthropology deals with the recovery, examination, and identification of human skeletal remains.

Forensic Anthropology is a multidisciplinary science that involves establishing the identity of an unknown individual (Krogman 1962). A forensic anthropologist needs to have thorough knowledge of a variety of subjects such as the human anatomy, osteology, physical anthropology, bioarchaeology, and taphonomy. Moreover, some basic expertise in zoology, entomology, and forensic medicine is necessary for excelling in this science.

A forensic anthropologist deals with generating the biological profile of an unknown individual, which also known as the person's osteobiography (Saul 1972). This involves finding out the individual's ancestry, sex, age at death, and their living stature. The general process of generating an individual's biological profile is detailed thoroughly in this chapter.

26.2 Locating and Recovering Skeletal Remains at a Scene of Incidence

Examination of skeletal remains begins with recovery of skeletal remains. Usually, skeletal remains buried in shallow graves are found accidently, or through the means of an eyewitness or an informant who can lead the police to the general area where the remains are buried (İşcan and Steyn 2013). In such cases, surface indicators of skeletal remains may be present, but they are of transient nature and can easily get destroyed or lost due to vagaries of nature and activity of scavengers. Multiple methods, from rudimentary observation to sophisticated techniques, exist for locating graves and skeletal remains (İşcan and Steyn 2013).

Usually, the surface changes of soil and its vegetation cover are assessed to find graves. When a perpetrator hides their victim's body by burial, they fill the grave of the victim with soil. Initially, the level of grave soil is equal with its surrounding. However, in the first few months after burial, there occurs a primary depression in the grave due to compaction of soil and decomposition of the victim's body. Since the largest volume to collapse during decomposition is the thorax, there may be observation of a secondary depression. The extent of both the depressions depends upon the type of soil, extent of water in the soil, depth of the grave, etc. (Krogman 1962).

Often, excess amount of soil may be dumped over the cadaver to hide it, and will result in the formation of a mound or pile of soil. There may be a colour difference in the resultant pile of soil and the surrounding soil due to intermixing of the soil from the grave and that on the surface. This colour change may be appreciable for a limited time.

Another indicator of the presence of a grave is the general vegetation cover of the land. Initially, when the body is buried, the surface of the grave may be completely devoid of any vegetation, or be covered with uprooted plants from the neighbouring areas (Adovasio and Dirkmaat 1996). As the time passes, newer plants will start growing from the grave. The first plants to appear will be weeds, and can often be distinguished from the plants present in the surrounding areas. In case of shallow grave, the growth of newer plants is prolific on the account of the nutrients present in the grave soil due to decomposition of the cadaver.

Scavenger activity can aid the law enforcement agencies in finding buried cadavers. Animals usually scavenge shallow graves, and the presence of seemingly random holes or disturbed soil beds can indicate the presence of a grave. Similarly, the animals could also disturb personal belongings or bones of the cadaver and bring them to the surface. Presence of these bone fragments or scraps of clothing can indicate the presence of a buried body (Haglund et al. 2001).

Apart from accidental recovery, gross observations, and standard crime scene search methods, sophisticated tools can also be used to find a grave. One of the most basic tools that a searcher may utilise is the metal detector. However, metal detectors are non-specific, and will find buried metal objects irrespective of whether they are linked to a cadaver or not. Due to advent of remote sensing technologies, scientists now have in their arsenal techniques such as ground penetrating radar, GPS, thermal scans, magnetometry, infrared and aerial photography, etc., that can be used for searching of graves (İşcan and Steyn 2013).

Once the site where the cadaver is identified, the extremely vital task of recovery of skeletal remains begins. As with crime scene investigation, securing the general area of the grave is of great importance. Barricading tape, ropes, wooden planks, etc., are utilised to cordon off a wide area around the grave. Only essential personnel are allowed to enter the barricaded area. Depending upon the landform of the area, extra precautions should be taken. For instance, if the grave is located near a flowing water body, additional searchers must be sent along the path of water flow to search for any skeletal evidences. In case the grave is near a slope, the searchers should look for skeletal evidences at the end of the slope, as bones such as skull may roll down the slope and get deposited elsewhere. During the recovery of skeletal remains, exhaustive and thorough documentation should be done. Clear and precise notes should be made, and everything should be photographed while upholding the highest standards. Mapping of evidence and sketching their location on the scene is also advised. The process of recovery should be conducted in a manner that preserves the integrity of the skeletal remains, and none of the osteological evidences should get damaged.

Vegetation should be cleared from the surface of the grave with gardening tools. The plants removed from the grave should be treated as botanical evidences and thus preserved as such. The entire surface of the grave and some of its vicinity should be marked with a grid pattern. This can be done using thin ropes. The process of excavation should involve careful removal of horizontal layers of soil matrix from each of the grid squares. The removal should follow the natural strata of the soil, and excavation layers should not exceed 15 cm. This can be done using a gardening trowel until the actual evidences start appearing. Upon discovering any evidence, prompt documentation by means of note making and photography should be done (Krogman 1962). Different coloured flags may be used to mark different types of evidences. Red flags may be used to show skeletal material, yellow flags may be used to show any plastic evidences, and blue flags may be used to show metal objects. Once sufficiently photographed, the surface of the evidence should be cleaned using a brush and photographed again. The entire process of cleaning, marking, and documenting is repeated for all the evidences. None of the evidences should be removed from the grave until the excavation process is completed. While taking photographs of the skeletal remains within the grave, an arrow pointing towards the cardinal direction north should be placed within the grave, along with a clearly visible scale. The soil removed from the grave during excavation should be sieved to find fragmented and small bones. Once the grave along with its skeletal remains is photographed, then the bones and other evidences are removed one by one. The bones are placed in a tarpaulin sheet spread near the grave. After completion of recovery, the tarpaulin sheet is wrapped up and moved to a vehicle for transport to the mortuary. In case the remains are only partially skeletonised, it may be necessary to remove the entire body in one piece. In this case, similar procedure as above is followed, with a tarpaulin sheet or a body bag is placed near the grave, and the entire body is transferred to it. All the evidences are sent to the mortuary for a detailed post-mortem examination and generation of a biological profile.

26.3 Developing Biological Profile from Skeletal Remains

Human identification is one of the most important tasks that a Forensic Anthropologist performs. Identification itself can be dubbed as either absolute identification or partial identification. Absolute identification, also known as ‘complete’ identification is when the identity of the deceased is established without any doubt or with certainty. Methods of absolute identification include dactylography and DNA profiling. Partial identification, also known as ‘incomplete’ identification is when the identity of an unknown individual cannot be established with absolute certainty, but can eliminate or narrow down a group of probable identities. Methods of partial identification include determination of ethnicity and sex, estimation of age and stature, ABO blood grouping, serum protein polymorphisms, enzyme typing, etc.

Case study: Skeletal remains of an individual were found near a river bank. Upon its examination by Forensic Medicine experts and Forensic Anthropologists, it was found that the skeleton belonged to a Caucasian female of age 16–18 years. Few of the teeth of the deceased were preserved by the Anthropologists for DNA profiling. The police used the information furnished by the anthropologists and compared it with the details of individuals who had gone missing in the general vicinity in the past few years. Police could thus 'eliminate' all the missing males, and non-Caucasian females younger than 16 years and older than 18 years from their search. Upon narrowing down to 2 possible victims, the DNA profile generated from the teeth of the cadaver was compared to the DNA profiles of the possible victims' parents, and 'absolute' identification of the deceased was done.

Human identification of the dead is of grave importance in both civil and criminal cases.

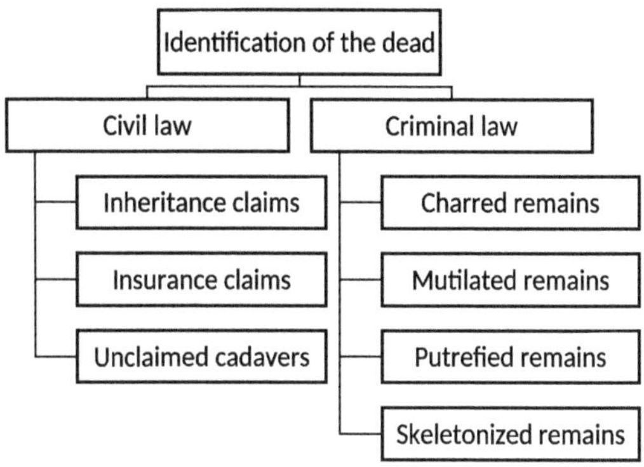

Identification of human remains is done through a process of generation of the cadaver's biological profile, also known as 'Osteobiography'. Osteobiography in simpler terms is the biography of bones, and deals with answering questions such as what ethnicity did the victim belong to, what their sex was, how old were they when they died, what was their living stature, and whether there were any unique, individualistic features in their skeleton, that can be used for absolute identification of the victim. Determination of sex, and estimation of age and stature are considered to be the 'Big 3' of human identification.

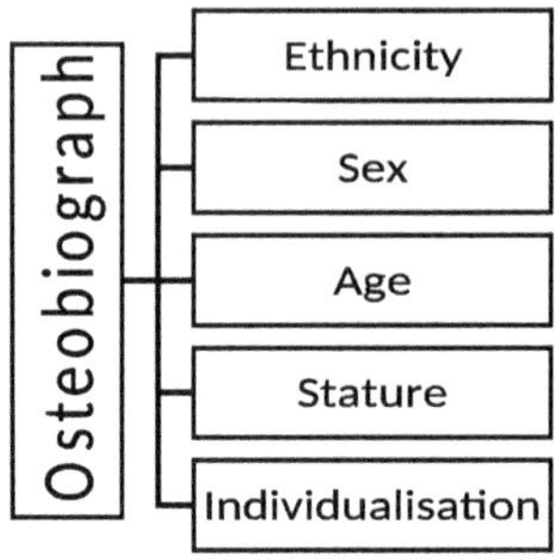

26.3.1 Determination of Ethnicity

Determination of ethnicity/ancestry of an individual (previously referred to as race) is an important part of the osteobiography of the individual. Broadly speaking, there are three major global ethnicities that a person can belong to: Caucasoids (European descent), Mongoloids (Asian descent), and Negroids (African descent). Apart from these, there exist multiple contemporary ethnic groups such as the Andamanoids, Australoids, Incans, etc. (Blue et al. 2021). Determination of ethnicity is based on individualistic features that can be appreciated externally on both recently dead individuals and skeletal remains.

In recently dead individuals, indication of their ethnicity can be given by their complexion, nature of hair, eyes, clothes, etc. (Table 26.1). The colour of skin of a Caucasoid may range from fair (European descent) to wheatish or brown (Indian descent) (Ubelaker 2006). Mongoloids have a yellowish complexion, while the complexion of Negroids ranges from brown to black. Mongoloids, Negroids, and Indian Caucasoids have eyes whose colour ranges from brown to black. European Caucasoids may exhibit green, grey, or blue eye colour. The hair of Caucasoids is usually wavy with a circular or oval cross section, and has even pigmentation. Mongoloids usually have coarse and straight hair with circular cross section and dark medulla, while Negroids have woolly or kinky hair which is densely pigmented and oval in cross section (Ubelaker 2018).

Table 26.1 Differences in complexion, eye colour and hair of the three major ethnicities

Feature	Caucasoids	Mongoloids	Negroids
Complexion	Fair (Europeans) to brown (Indians)	Yellowish	Brown to black
Eye colour	Brown, and black to blue, green, or grey (Europeans)	Brown, and black	Brown, and black
Hair	Wavy, evenly pigmented, circular/ oval cross section	Straight, coarse, solid and dark medulla, circular cross section	Woolly/kinky, densely pigmented, oval cross section

In putrefied bodies and skeletal remains, the aforementioned features may not give a clear indication of an individual's ethnicity, and hence other means are utilised. Broadly, methods of ancestry determination from skeletal remains can be classified into two groups: anthropometric methods and non-metric methods. Anthropometric methods involve measurement of certain skeletal features that are polymorphic in individuals of different ethnicities (Black and Ferguson 2011). On the other hand, non-metric methods rely upon observation rather than measurement of certain polymorphic features.

Anthropometric methods include calculation of different indices of the sull and long bones. One of the most used indices to determine the ancestry of an unknown individual is the cephalic index.

$$\text{Cephalic Index} = \frac{\text{Maximum width of the skull}}{\text{Maximum length of the skull}} \times 100$$

Negroids have a long and narrow skull, exhibit a cephalic index of less than 75, and are considered to be dolichocephalic (Greek origin: dolicho = long; cephalic = head) (Akinbami 2014). Mongoloids have a short and wide sull, exhibit a cephalic index of more than 80, and are considered to be brachycephalic (Greek origin: brachy = short; cephalic = head) (Ngeow and Aljunid 2009). Caucasoids have a round sull, exhibit a cephalic index that ranges between 75 and 80, and are considered to be mesocephalic or mesaticephalic (Buretic-Tomljanovic et al. 2007). Other indices that can be used to determine ethnicity are nasal index, brachial index, crural index, humerofemoral index, and intermembral index (Table 26.2).

Non-metric methods for determination of ethnicity involve observation of skeletal features such as shape of forehead, shape of orbits, shape of nasal aperture, inferior nasal margin, zygomatic projection, nasal spine projection, shape of palate, etc. The aforementioned features are described in detail in Table 26.3.

26.3.1.1 Determination of Ethnicity from Dentition

The three major ethnicities have a few unique dental features that can be used in generation of an unknown individual's biological profile. Caucasoid individuals may have an extra cusp on the mesiopalatal line of maxillary first molars. This additional cusp is called the Cusp of Carabelli, and is observed only in Caucasoids (Bhavyaa et al. 2021). Mongoloids may exhibit marked depressions in the lingual or palatal side of their incisors, giving the teeth a shovel-shaped appearance and hence the name – shovel-shaped incisors (Suzuki and Sakai 1964). While there is no unique dental feature in Negroids, they do show higher prevalence of supernumerary teeth (number of teeth greater than 32).

It is essential to understand that most forensic and biological anthropologists agree that there are no distinct human races, and humans cannot be classified as such based on their overt skeletal features. However, there exist a few similarities in the skeletal framework of individuals from the same geographical location or origin. While once upon a time the differences between individuals of two distant geographical locations may have been describable, centuries of population

Table 26.2 Indices to determine ancestry

Index	Caucasoid	Mongoloid	Negroid
Nasal index $= \frac{\text{nasal breadth}}{\text{nasal height}} \times 100$	Leptorhinae <47	Mesorhinae 47–51	Chamaerhinae >51
Brachial index $= \frac{\text{length of radius}}{\text{length of humerus}} \times 100$	<75	>80	75–80
Crural Index $= \frac{\text{Length of Tibia}}{\text{Length of Femur}} \times 100$	83.3	86.2	83.6
Humero – femoral index $= \frac{\text{Length of Humerus}}{\text{Length of Femur}} \times 100$	<72	>72	–
Intermembral index $= \frac{\text{Length of Humerus+Length of Radius}}{\text{Length of Femur+Length of Tibia}} \times 100$	>70	<70	–

Table 26.3 Ancestry determination using skeletal features

Feature	Caucasoid	Negroid	Mongoloid
Forehead	Elevated or raised	Small and short	Inclined
Orbital opening	Angular	Square/rectangular	Rounded/circular
Palate shape	Parabolic	Hyperbolic	Round
Nasal aperture	Narrow and long	Broad and square shaped	Rounded
Nasal spine projection	Prominent	Small/not prominent	Slightly prominent
Zygomatic projection	Intermediate	Small/not prominent	Prominent
Nasal sill	Sharp	Guttered	Intermediate

intermingling have diluted these polymorphic features (Krogman 1962). Hence, determination of race, ancestry, or ethnicity is an inexact science, and must be treated so by a Forensic Anthropologist while dealing with skeletal remains.

26.3.2 Determination of Sex

One of the important parts of developing an individual's biological profile is determination of their sex. The scientific basis for determination of sex of an unknown individual using their bones is the presence of certain anatomical and morphological areas that are dimorphic in males and females (Kanchan and Rastogi 2009). Wilton Marion Krogman estimated the accuracy of sex determination based upon what skeletal material is present (Krogman 1962). According to him, forensic anthropologists can determine the sex of an unknown individual with 100% accuracy if the entire skeleton is present. If only the pelvis and the skull are recovered, the accuracy of sex determination falls to 98%. If only the pelvis of the individual is recovered, anthropologists can determine the sex with 95% accuracy, and with 90% accuracy if only the skull is recovered. If only long bones of an individual are recovered, an anthropologist can determine whether the individual was a male or a female with 80% accuracy (Tables 26.4, 26.5, 26.6, 26.7 26.8 and 26.9), (Fig. 26.1, 26.2, 26.3, 26.4, 26.5, 26.6, and 26.7).

26.3.3 Estimation of Age

Estimation of biological age of an individual is a vital cog in the process of osteobiography. Age at death estimation is crucial in both civil and criminal law. Broadly, age of an individual can be estimated based on their skeletal maturity (Shedge et al. 2020, 2021), dental maturity (Kanchan et al. 2021), and in later ages, using morphological changes of their bones and teeth (Gustafson 1947).

26.3.3.1 Estimation of Age at Death Based on Skeletal Maturity

The bones of humans develop from a plethora of growth centres, also known as ossification centres (Cunningham et al. 2016). These ossification centres are not only

Table 26.4 Determination of sex using pelvis (Krogman 1962)

Sl. No.	Criterion	Males	Females
1	General appearance	Larger, thicker, tougher	Smaller, thinner, delicate
2	Muscle markings	More prominent	Less prominent
3	General shape	Deep funnel shaped	Flat bowl shaped
4	Iliac fossa	Deep	Shallow
5	Iliac crest	Curves are less marked	Curves are well marked
6	Preauricular sulcus	Not prominent (sometimes invisible)	Prominent and well defined
7	Acetabulum	Large, directed laterally	Smaller, directed anterolaterally
8	Obturator foramen	Large, oval	Small, triangular
9	Average width of greater sciatic notch	46.5 mm	51.5 mm
10	Sciatic notch index	145	166
11	Ischial spine	Inverted	Everted
12	Ischial tuberosity	Inverted	Everted
13	Chilotic line	Sacral part prominent	Pelvic part prominent
14	Chilotic line index	>100	<100
15	Body of pubis	Narrow, triangular	Wide, quadrangular
16	Symphysis pubis	Higher, narrower	Lower, wider
17	Pelvic brim	Heart shaped	Circular/elliptical
18	Pelvic outlet	Smaller	Larger
19	Subpubic angle	70–75^0, V-shaped	90–100^0, U-shaped
20	Coccyx	Inverted	Everted

Table 26.5 Determination of sex using sacrum (Krogman 1962)

Criterion	Males	Females
Size and shape	Longer and narrower	Shorter and wider
Curvature	Even	Uneven
Sacral promontory	Well-marked	Less marked
Body of first sacral vertebra	Larger	Smaller
Inner curvature	Uniformly curved	Abruptly curved at the last 2 segments
Sacro – Iliac articulation	Extend up to 21/2–3 segments	Extend up to 2–21/2 segments
Sacral index	105	115
Kimura's base-wing index (alar index)	65	80
Corpobasal index	45	40

Table 26.6 Determination of sex using skull (Krogman 1962)

Criterion	Males	Females
General appearance	Larger, more muscle markings	Smaller, less muscle markings
Architecture	Rugged	Smooth
Cranial capacity	1500–1550 cm^3	1350–1400 cm^3
Walls	Thicker	Thinner
Glabella	More prominent	Less prominent
Frontonasal junction	Distinct angulation	Smooth curve
Orbits: Size	Smaller	Larger
Orbits: Margins	Rounded	Sharp
Orbits: Shape	Square	Round
Supraorbital ridge	Prominent	Less prominent
Zygomatic arch	Prominent	Less prominent
Nasal aperture	Higher and narrower	Lower and wider
Parietal eminences	Small	Large
Occipital protuberance	Prominent	Less prominent
Occipital condyles	Larger	Smaller
Mastoid process	Large and blunt	Small and sharp
Digastric groove	Deeper	Shallower
Palate	Larger, broader, U-shaped	Smaller, narrower, parabolic
Foramen magnum	Larger and oblong	Smaller and rounder
Teeth	Larger	Smaller

Table 26.7 Determination of sex using mandible (Albalawi et al. 2019)

Criterion	Males	Females
General appearance	Larger, thicker	Smaller, thinner
Chin (symphysis menti)	Square or U-shaped	Pointed or V-shaped
Angle of body of ramus	Less obtuse (<125°)	More obtuse (>125°)
Appearance of angle	Everted	Inverted
Condyles	Larger	Smaller

Table 26.8 Determination of sex using sternum (Hunnargi et al. 2008)

Criterion	Males	Females
Sternal breadth	Larger	Smaller
Length of mesosternum	>95 mm	<80 mm
Ashley's rule	Total length > 149 mm	Total length < 149 mm
Hyrtl's law	Body >2X manubrium	Body <2X manubrium
Sternal index	46.2	54.3

Table 26.9 Determination of sex using femur (Purkait and Chandra 2004)

Criterion	Males	Females
General appearance	Heavier, stronger, with prominent muscle markings	Lighter, less prominent muscle markings
Head	Larger	Smaller
Vertical diameter of head	>46 mm	<42 mm
Angle of neck with shaft/ caputcollum-diaphyseal angle	125°	<125°
Bicondylar width	>78 mm	<72 mm
Angle of shaft with condyles	80°	75°

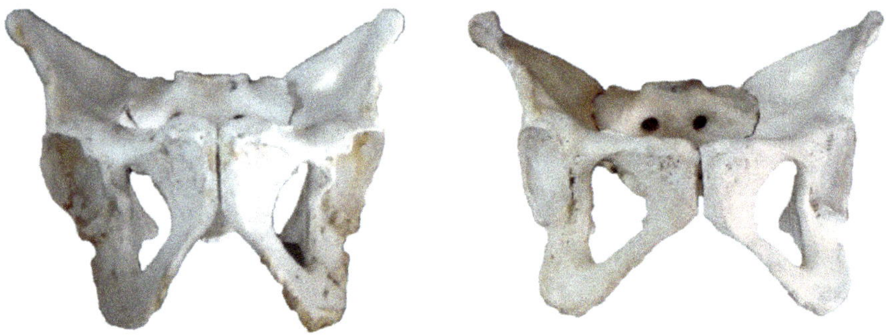

Fig. 26.1 Anterior view of the male and female pelvic bones

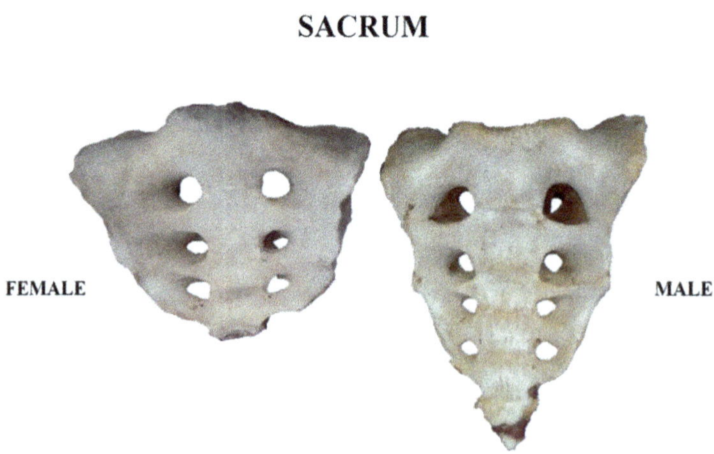

Fig. 26.2 Male and female sacra

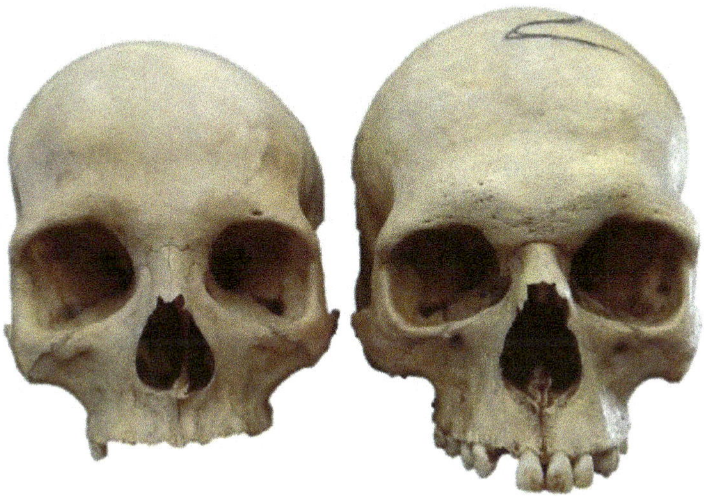

FEMALE MALE

Fig. 26.3 Human skull (normal frontalis)

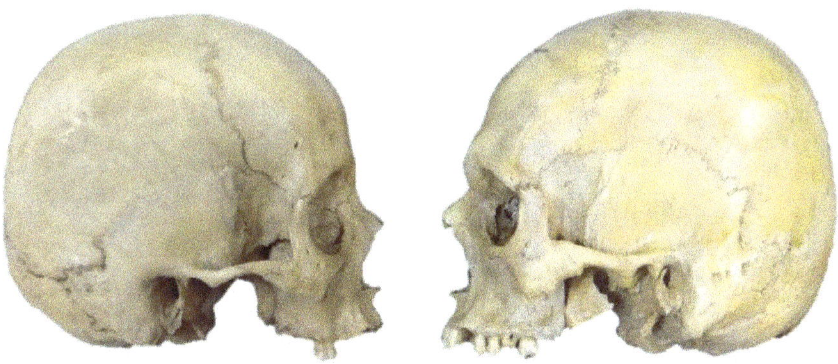

FEMALE MALE

Fig. 26.4 Human skull (normal lateralis)

involved in the formation of long bones, but also the short, flat, and irregular bones of the human skeleton. While it is a known fact that adult humans have 206 bones, a lesser-known fact is that the human foetus at the age of 11 weeks, has 806 ossification centres, and at birth it has about 450 ossification centres (Krogman 1962). Between 11th week to the final instance of epiphyseal union, 600 odd centres of bone growth coalesce in a fairly defined sequence.

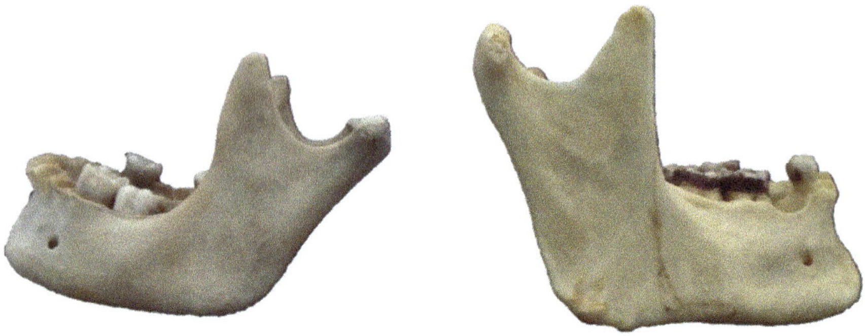

FEMALE MALE

Fig. 26.5 Human mandible (lateral view)

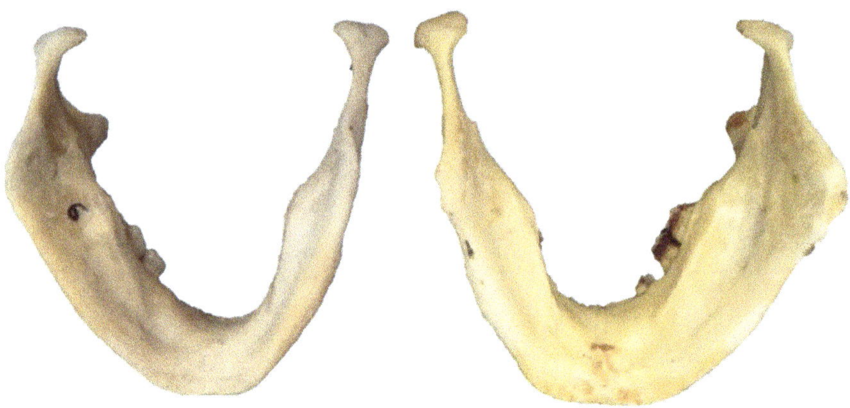

FEMALE MALE

Fig. 26.6 Human mandible (inferior view)

Prenatal Ossification

The age of a foetus can be determined based on the appearance of ossification centres during the intrauterine life (IUL) (Krogman 1962). A few of the important centres of ossification are outlined below in Table 26.10.

Postnatal Ossification

After birth, the primary centres of ossification grow and coalesce with secondary centres to form bones of the human body. The union of secondary ossification centres of the long bones (present in the epiphysis) with the primary ossification centre (diaphysis) is termed as epiphyseal fusion, and is mostly used in case of age estimation of skeletal remains (Krogman 1962). The appearance of ossification centres and their fusion for different bones are outlined in the following table (Table 26.11).

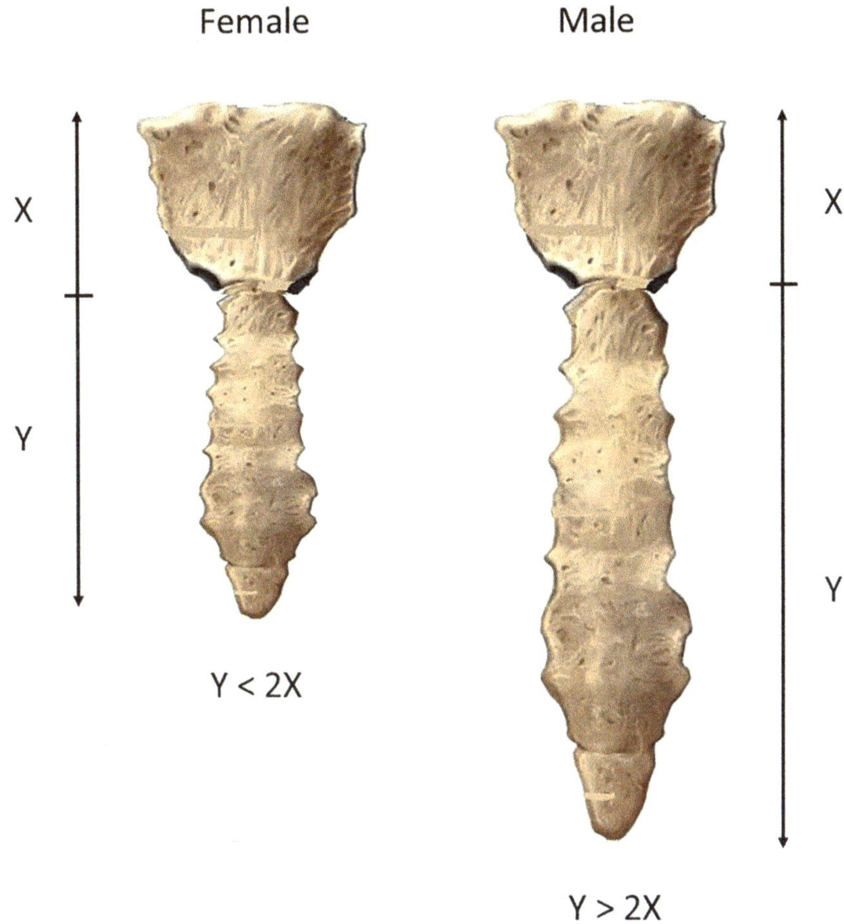

Fig. 26.7 Human sternum (illustrator to modify this image)

26.3.3.2 Estimation of Age Using Cranial Suture Closure

The skull of an infant is made up of 6 separate bones: the frontal bone, the occipital bone, a pair of parietal bones, and a pair of temporal bones. These bones are connected together by means of robust fibrous tissues known as sutures. As an individual grows older, these sutures start fusing, which is termed as closure or obliteration of sutures (Shedge and Kanchan 2019; Fan et al. 2020; Ruengdit et al. 2020). Different sutures of the human skull get obliterated at different time intervals, and this can allow forensic anthropologists to predict the age of death of an individual. The closure of cranial sutures begins endocranially first (5–7 years before ectocranial suture). The time ranges at which the different cranial sutures close are detailed in the Figure 26.8.

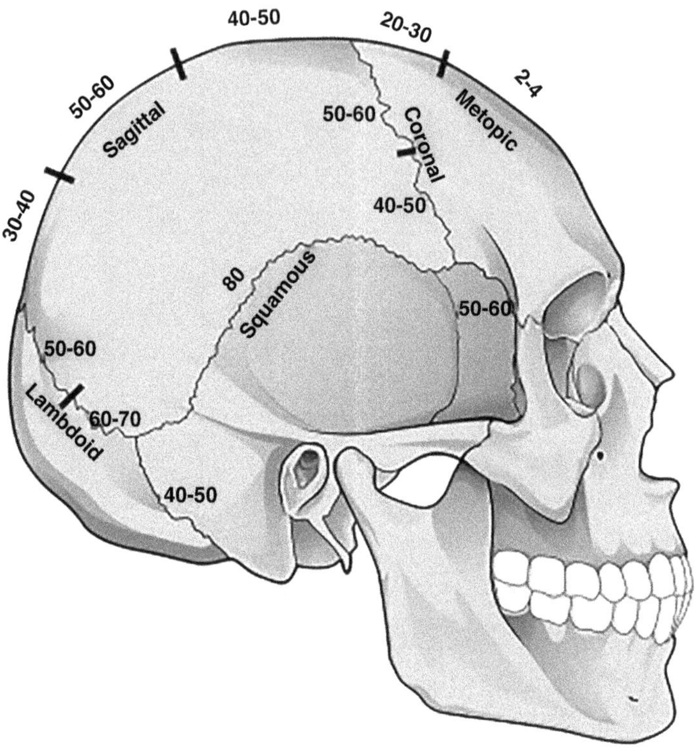

Fig. 26.8 Cranial sutures and their time of obliteration

Basi-occiput and basi-sphenoid sutures (not shown in the figure) get obliterated at the age of 18–20 years.

26.3.3.3 Estimation of Age Using the Pubic Symphysis

A symphyseal cartilage separates the left and the right pubic bones in the pelvic girdle. The surface of the pubic bones in this region undergoes certain morphological changes that can be used to estimate an individual's age (Warrier et al. 2021). the morphological changes of the pubic symphysis as well as the age at which they are observed, are detailed below (Table 26.12).

26.3.3.4 Estimation of Age Using the Mandible

As the age of an individual increases, their mandible starts exhibiting certain morphological changes (Aggrawal). These changes are detailed below (Table 26.13).

Table 26.10 Prenatal ossification centres and their time of appearance

Ossification centre	Appearance
Occipital bone	Weeks 8–10
Temporal bone	Weeks 7–8
Parietal bone	Weeks 7–8
Frontal bone	Weeks 6–7
Maxilla	Week 6
Mandible	Week 6
Clavicle	5–6 weeks of IUL
Centra of C4-S2	Month 3
Centra of C2–3 & S3–4	Month 4
Shaft of humerus	Week 6
Shaft of radius	Week 7
Shaft of ulna	Week 7
Shaft of femur	Week 7–8
Shaft of tibia	Week 7–8
Shaft of fibula	Week 8
Ilium	Month 2–3
Ischium	Month 4–5
Pubis	Month 6–8
Centra of C2–3 & S3–4	Month 4

26.3.3.5 Estimation of Age Using Dentition

Teeth are one of the most reliable features of the human body that can be used to estimate the age of an unknown individual. Teeth follow a predictable and fixed pattern of development and eruption that is used for estimation of age. Humans have two sets of dentitions. The first set of teeth that erupt is known as the primary dentition or the deciduous teeth. These teeth start erupting at the age of 6 months, and the entire set appears by the age of 2.5 years. The total primary dentition includes 20 teeth, with the dental formula being 2102 (2 incisors, 1 canine, 0 premolar, 2 molars). Falling off of deciduous teeth precedes eruption of secondary teeth. Secondary teeth, also known as permanent teeth, first erupt at the age of 6 years. The dental formula of secondary teeth is 2123 (2 incisors, 1 canine, 2 premolars, and 3 molars). The sequence and time or eruption of deciduous and permanent teeth are detailed below (Table 26.14 and 26.15).

The third molar is an anomalous tooth, which may erupt anywhere between 17 and 25 years, or may never erupt at all.

26.3.3.6 Gustafson's Method

After the age of 25 when all the teeth have erupted, one cannot estimate the age of an individual based on the presence of teeth. However, age can be based on the attritional changes of the teeth. One of the methods that utilises this is the Gustafson's method (Gustafson 1947). This method focuses on 6 changes in the teeth: degree of attrition (wear and tear of the occlusal surface of the tooth), degree of

Table 26.11 Postnatal centres of ossification and their times of appearance and fusion

Bone	Appearance	Fusion
Clavicle		
Medial end	11–19 years	18–25 years
Lateral end	Puberty	19–20 years
Sternum		
Manubrium	5 months	With S1: >60 years
Sternebra 1 (S1)	5–6 months	With S2: 25 years
Sternebra 2 (S2)	7–8 months	With S3: 20 years
Sternebra 3 (S3)	7–8 months	With S4: 15 years
Sternebra 4 (S4)	1 year	With xiphoid process: >40 years
Xiphoid process	3–6 years	With mesosternum: >40 years
Scapula		
Coracoid	Around 1 year	15–17 years
Sub coracoid	8–10 years	14–17 years
Acromion	14–16 years	18–25 years
Shoulder joint		
Head of humerus	1 year	All three unite at the age of 5–6 years to form the conjoint epiphysis of the shoulder
Greater tubercle	3 years	
Lesser tubercle	5 years	
Conjoint epiphysis	5–6 years	18–19 years
Elbow joint		
Capitulum	1 year	All three unite at the age of 14–16 years to form the conjoint epiphysis of the elbow
Trochlea	9–11 years	
Lateral epicondyle of humerus	11 years	
Conjoint epiphysis of elbow	14–16 years	16–17 years
Medial epicondyle of humerus	6–7 years	16–17 years
Head of radius	5 years	16–17 years
Upper end of ulna	9 years	16–17 years
Wrist joint		
Distal end of radius	2 years	17–18 years
Distal end of ulna	5–6 years	17–18 years
Base of first metacarpal	5–7 years	15–17 years
Pelvis		
Iliac crest	14 years	20–21 years
Triradiate cartilage	13 years	15 years
Ischiopubic ramus	–	7–8 years
Ischial tuberosity	16 years	20–21 years
Head of femur	1 year	17–18 years
Greater trochanter	4 years	17–18 years

(continued)

Table 26.11 (continued)

Bone	Appearance	Fusion
Lesser trochanter	14 years	17–18 years
Knee joint		
Lower end of femur	9th month of IUL	18–19 years
Upper end of tibia	10th month of IUL – At term	18–19 years
Upper end of fibula	4 years	18–19 years
Ankle joint		
Lower end of tibia	1–2 years	16–17 years
Lower end of fibula	1–2 years	16–17 years
Calcaneus	6–8 years	14–16 years

Table 26.12 Age-based pubic symphyseal changes

Age	Description
18–19 years	Surface of the pubic symphysis is uneven, and shows the presence of horizontal ridges which are separated by well-marked grooves
20–21 years	Grooves between the horizontal ridges start getting filled near the dorsal end
22–24 years	Dorsal margin gets more defined
25–40 years	Lower extremity starts becoming clearer, ridges start disappearing, bony/ossific nodule forms at upper extremity, outer margins get clearly defined
40–50 years	Surface becomes oval and smooth, margin has narrow beaded rim
50+ years	Symphyseal face becomes eroded, and shows erratic ossification. As the age increase, disfigurement increases too.

Table 26.13 Age estimation using the mandible

Feature	Infants	Adults	Elderly
Body	Shallow, thin	Thick and long	Shallow
Coronoid process	Higher than condyloid process	Lower than condyloid process	Higher than condyloid process, with the neck of condyloid process bent backwards
Sigmoid notch	Deep	Deeper	Shallow
Angle of ramus	140°	110°–120°	140°
Ramus	Short, oblique	Straight	Posteriorly inclined
Mental foramen	Midway between upper and lower margin and is directed forwards	Midway between upper and lower margin, under socket of second premolar	Near the alveolar margin.
Mental protuberance	Not well developed	Well developed	Moderately developed.

Table 26.14 Eruption of primary dentition in humans

Tooth	Eruption	Number of teeth
Central incisor (lower)	6–8 months	2
Central incisor (upper)	7–9 months	4
Lateral incisor (upper)	8–10 months	6
Lateral incisor (lower)	10–12 months	8
First molar	12–14 months	12
Canine	17–19 months	16
Second molar	20–30 months	20

Table 26.15 Eruption of secondary teeth in humans

Tooth	Eruption	Number of teeth
First molar	6–7 years	21–24
Central incisor	6–9 years	24
Lateral incisor	7–9 years	24
1st premolar	9–11 years	24
Second premolar	10–12 years	24
Canine	11–13 years	24
Second molar	12–14 years	25–28
Third molar	17–25 years	29–32

periodontosis (retraction of gums from the neck and neighbouring regions of the tooth), degree of secondary dentin deposition (deposition of dentin in the pulp cavity of the tooth), cementum apposition (deposition of newer layers of cementum around existing one), root transparency (increase in transparency of the roots), and root resorption (absorption of roots' apex into the gum). These 6 changes are scored from 0 to 3, with 0 indicating no change, and 3 indicating massive change. Root transparency is considered to be the best parameter among these 6 for estimation of age. Central incisors are found to be the best single indicator of age using the Gustafson's criteria. A combination of canines and first molars predicts the age using Gustafson's criteria better than the central incisor.

26.3.4 Estimation of Stature

Stature refers to an individual's height, and is one of the most crucial parts of human identification. On an average, an individual loses 0.6 mm of stature per year after the age of 30 years on account of degenerative changes of the body (Krogman 1962). Additionally, the stature of an individual is observed to be the maximum when lying flat on their back (2–3 cm more than while standing). After death, due to primary relaxation, the stature of an individual increases by 2–4 cm. However, once putrefaction sets in, the spinal column shortens and causes the overall stature to decrease by a cm.

26.3.4.1 Anatomical Method of Estimating Stature

Whenever the entire skeleton, or the cranium, vertebral column, hip bones, and at least one complete lower limb are recovered, stature of the individual can be estimated by the anatomical or the Fully method (Fully 1956). It involves arrangement of the skeleton/bones in an anatomical position and recording of following measurements:

1. Maximum height of the skull.
2. Distance between corpus of C2 and L5.
3. Length of anterior first sacral segment.
4. Physiological length of femur.
5. Maximum length of tibia.
6. Articulate height of talus and calcaneus.

As these measurements do not take into consideration the presence of soft tissues in between the bones, soft tissue correction factors are used. After all the measurements are recorded, soft tissue correction factors are added to the total, and the resultant value is attributed as stature of the individual.

26.3.4.2 Mathematical Methods of Estimating Stature

In cases where only a few bones such as the long bones are recovered, anatomical method of stature estimation cannot be applied. In such cases, forensic anthropologists use mathematical methods for stature estimation. These mathematical methods have a simple premise: the length of any long bone of the human body is proportional to the length of the body itself. Mathematical methods are of two types: use of multiplication factors to estimate stature, and use of population specific linear regression models to estimate stature.

Multiplication factors (MF) entail use of a number that can simply be multiplied to the corresponding bone length to get the stature of an individual (Aggrawal). For example, if the length of humerus of an unknown cadaver is 35 cm, and the multiplication factor for humerus is 5.3, the estimated stature of the individual is calculated to be $35 \times 5.3 = 185.5$ cm. the multiplication factors for different long bones of the human body are as follows:

Bone	Humerus	Radius	Ulna	Femur	Tibia/Fibula
MF	5.3	6.9	6.3	3.7	4.48

Linear regression models entail estimation of a person's stature based upon their bone length. Linear regression analysis is more accurate and has less error rate than multiplication factors. Linear regression models for any particular population group are first generated by conducting extensive research in that population group. Lengths of different long bones of individuals belonging to the particular population are recorded and regression analysis is conducted on the collected data. This leads to generation of a regression model which looks like: $Y = AX + B$, where Y is the stature of the person (independent variable), X is the length of the bone (dependent variable), A is the slope of the regression line, and B is the y-intercept.

26.3.4.3 Estimation of Stature Using Fragmented Bones

Often times, the skeletal remains recovered consist of broken pieces or fragments of long bones. Estimation of stature using such fragmented bones is vital, and involves use of linear regression analysis (İşcan and Steyn 2013). In such cases, regression analysis is done twice; first, the dependent variable is considered to be the bone fragment and length of the bone is estimated, and then the estimated value of the long bone is considered to be the dependent variable and stature of the individual is calculated.

26.3.5 Individualisation of the Skeletal Remains

Determination of race and sex, and estimation of age and stature are used in partial or incomplete identification of the skeletal remains. However, complete identification of an individual requires a few extra steps. The most commonly employed method involves isolation of DNA from either the bone marrow of long bones or from the dental pulp, followed by STR analysis (Li 2021). This allows forensic anthropologists to generate a DNA profile of the unknown individual, which may later on be compared with antemortem data of the suspected deceased, or their parents. Another method of individualisation is finding any surgical implants in the victim's bones, or presence of any dental work. Any surgical implants such as rods, bolts, replaced nee/hip may have serial number that can be retrospectively verified with a database to find out who the victim is. Even with dentition, evidences of dental work may be compared with antemortem records to find out the identity of the victim.

26.4 Restoration of Facial Details from the Skull

On a daily basis, humans are able to discriminate people they know from complete strangers using a variety of features such as their voice, touch, feel, etc. However, one of the biggest discriminating features, that humans rely upon the most, is the face. In case of skeletal remains, identification of the deceased can rarely be done based on identifying their facial features. However, since there is an undeniable relationship between the face and its underlying skull, Forensic Anthropologists have devised methods in which positive identification of a victim can be done based on reconstruction of their facial features, or by comparing photographs of the suspected victim's antemortem photo and the photo of the deceased's skull. These two major methods of identification, forensic facial reconstruction and skull-photo superimposition, are discussed in this chapter.

26.4.1 Forensic Facial Reconstruction

Forensic facial reconstruction, also known as forensic facial approximation is the process of constructing the face of an unknown individual based on the underlying skeletal structure of their skull. It involves careful layering and shaping of soft tissue over the skull to achieve an approximation of how the individual looed when alive. Forensic facial reconstruction is an amalgamation of art and science. Expertise in human anatomy There are three major approaches of forensic facial reconstruction:

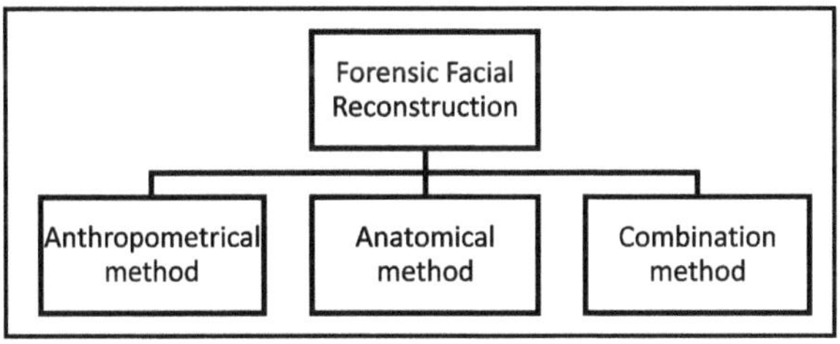

26.4.1.1 Anthropometrical/Tissue Depth/American Method

The anthropometrical method of forensic facial reconstruction involves use of tissue depth measurements of a particular population group and was developed by Krogman in 1946 (Gupta et al. 2015). Measurement of soft tissue thickness is recorded at 21 distinct anatomical landmarks, and these measurements are used in forensic case work. The tissue depth measurement is done by means of needles, X-rays, ultrasound, or MRI. The 21 landmarks include 10 midline points and 11 bilateral points as mentioned below.

Midline landmarks	Bilateral landmarks
Supraglabella	Frontal eminence
Glabella	Supraorbital
Nasion	Suborbital
End of nasals	Lateral orbit
Mid philtrum	Zygoma
Upper lip border	Inferior malar
Lower lip border	Supra M2
Lip-chin fold	Occlusal line
Mental tubercle	Sub M2
Gnathion	Gonion
	Supraglenoid

When working on an unknown skull, pegs are placed on the aforementioned 21 landmarks. These pegs are of varying lengths, with their length corresponding to the tissue thickness data acquired in previous studies. The spaces in between the pegs are covered with clay. The clay is moulded while taking into consideration the natural curvature and surface characteristics of the skull. Additional clay is moulded until the surface of the previously attached pegs becomes invisible. The finished skull is then photographed and compared with the antemortem photos of the suspected victim.

26.4.1.2 Anatomical/Russian Method

The anatomical method of forensic facial reconstruction involves layering of muscles, glands, and cartilages bit by bit on the unknown skull (Gupta et al. 2015). This method was developed by Mikhail Gerasimov in 1971, and requires an acute and thorough knowledge of the facial anatomy and biomechanics. All the different facial muscles, glands, and cartilages are shaped separately and then layered upon the bare skull. The different soft tissues that are constructed out of clay are: Temporalis muscle, masseter muscle, zygomaticus muscle, orbicularis oris, parotid gland, buccal fatty pad, and the fat pad of chin. The anatomical method is slower than the anthropometrical method, and is rarely used in the modern day and age.

26.4.1.3 Combination/British/Manchester Method

The combination method, as the name suggests, involves aspects of both the anatomical and the anthropometrical methods (Gupta et al. 2015). The skull is first placed in the Frankfurt-horizontal plane using a tubular Craniophore, and pegs are either attached to the different craniofacial landmarks, or are drilled into the skull. As in the anthropometrical method, each peg length represents the average tissue thickness/depth at the particular landmark. The different soft tissues mentioned in the anatomical method section are constructed out of clay, and moulded onto the skull at their points of origin or insertion.

26.4.1.4 Computerised 3D Forensic Facial Reconstruction

With the advent of 3D scanning technology in forensic science, the mode of operation with respect to facial reconstruction has shifted from manual to a computerised approach. These modern 3D scanners can scan the entirety of the skull in a matter of minutes, and using a computer, facial approximation of the unknown skull can be done by an individual with anthropological and computer modelling skills (Shedge and Kanchan 2021).

26.4.2 Skull Photo Superimposition

Skull photo superimposition is a human identification method that involves comparison of the skull of an individual with an antemortem photo of a missing individual (Ubelaker et al. 2019). The antemortem photos of the suspected victim can be readily acquired from the victim's family, but they will rarely be to the scale and in a full-frontal orientation. The challenge thus lies in manipulating the skull in such a manner, that any photograph taken of the skull will be in the same alignment and perspective as the antemortem photos, and can thus be superimposed with the antemortem photo to observe whether the skull belongs to the individual in the photo.

The entire process of the skull-photo superimposition begins with identifying measurable objects in the antemortem photo to establish scale. For example, if the antemortem photo that the family of the suspected victim gave to the investigators consists of the victim wearing a floral shirt, the family of the suspected victim may be asked to provide the same shirt to the investigators. The investigators can measure the dimensions of the flower patterns present on the shirt that are visible in the photo, and find out the magnification needed to enlarge the photo, so that the face of the suspected victim can be made life-sized. Once the extent of magnification required is found out, the original photo is magnified and printed as a positive transparency.

Simultaneously, the skull has to be oriented in the perspective identical to the suspected victim's face in the antemortem photograph. This orientation is achieved by means of a Craniophore attached to a pulse-driven motor. Once identical orientation is achieved, a photograph of the skull is taken and printed onto a positive transparency.

Both the positive transparencies are then placed above each other (superimposed) and it is observed that whether both of them align together or not. While comparing, the anthropologist conducting the entire process should be well versed with the positional relationship of the skull and the face. A few of such relationships are as follows:

26.4.2.1 Positional Relationship Concerning the Eyes

The lower border of eyebrows is generally located on the supraorbital margin, the palpebral fissure falls approximately on the lower third of the orbital height, vertical level of the inner canthus corresponds approximately to 10 mm below the dacryon, and outer canthus corresponds approximately to 11 mm below the point at which the zygomaticofrontal suture crosses the orbital margin.

26.4.2.2 Positional Relationship Concerning the Nose

Height of the nasion-subnasale (external nose) corresponds to the height of nasion-subspinale (nasal aperture of the skull), the ala-to-ala measurement of the nose is greater than the width of the nasal aperture by 5 mm in caucasoids, 8 mm in negroids, and 6.5 mm in mongoloids.

26.4.2.3 Positional Relationship Concerning the Lips

Oral slit appears to coincide with the line formed by the bite, central point of the oral slit (stomion) falls 1–2 mm higher than the upper central incisal surface, and Oral angle (cheilion) generally lies on the junction of the canine and the first premolar.

Multiple Choice Questions

1. Cusp of Carabelli is a feature observed in the dentition of which population groups?

   (a) Caucasoids.
   (b) Mongoloids.
   (c) Negroids.
   (d) Australoids.
   Answer: (a)
2. Which of the following is the first deciduous tooth to erupt?
   (a) First molar.
   (b) Lower central incisor.
   (c) Upper central incisor.
   (d) Upper lateral incisor.
   Answer: (b)
3. Which of the following teeth is the best for use in Gustafson's method of dental age estimation.
   (a) First molar.
   (b) Central incisor.
   (c) Lateral incisor.
   (d) Second molar.
   Answer: (b)
4. According to Krogman, what is the accuracy with which the sex of an unknown individual be estimated using only their pelvic bones.
   (a) 80%
   (b) 90%
   (c) 95%
   (d) 98%
   Answer: (c)
5. Orbital margins of the female sex are ________.
   (a) Blunt.
   (b) Curved.
   (c) Rounded.
   (d) Sharp.
   Answer: (d).

References

Adovasio J, Dirkmaat D (1996) The role of archaeology in the recovery and interpretation of human remains from an outdoor forensic setting. In: Haglund W, Sorg M (eds) Forensic taphonomy. CRC, Boca Raton, FL

Akinbami BO (2014) Measurement of cephalic indices in older children and adolescents of a Nigerian population. Biomed Res Int 2014:527473. https://doi.org/10.1155/2014/527473

Albalawi A, Alam M, Vundavalli S et al (2019) Mandible: an indicator for sex determination—a three-dimensional cone-beam computed tomography study. Contemp Clin Dent 10:69. https://doi.org/10.4103/ccd.ccd_313_18

Bhavyaa R, Sujitha P, Muthu MS et al (2021) Prevalence of the cusp of Carabelli: a systematic review and meta-analysis. Ann Hum Biol 48:572–584. https://doi.org/10.1080/03014460.2022.2032339

Black SM, Ferguson E (eds) (2011) Forensic anthropology: 2000 to 2010. CRC, Boca Raton, FL

Blue MNM, Tinsley GM, Ryan ED, Smith-Ryan AE (2021) Validity of body-composition methods across racial and ethnic populations. Adv Nutr 12:1854–1862. https://doi.org/10.1093/advances/nmab016

Buretic-Tomljanovic A, Giacometti J, Ostojic S, Kapovic M (2007) Sex-specific differences of craniofacial traits in Croatia: the impact of environment in a small geographic area. Ann Hum Biol 34:296–314. https://doi.org/10.1080/03014460701211017

Cunningham C, Scheuer L, Black SM (2016) Developmental juvenile osteology, 2nd edn. Academic, Amsterdam

Fan F, Tu M, Li R et al (2020) Age estimation by multidetector computed tomography of cranial sutures in Chinese male adults. Am J Phys Anthropol 171:550–558. https://doi.org/10.1002/ajpa.23998

Fully G (1956) Une nouvelle methode de determination de la taille. Ann Méd Lég Criminol Police Sci Toxicol 35:266–273

Gupta S, Gupta V, Vij H et al (2015) Forensic facial reconstruction: the final frontier. J Clin Diagn Res JCDR 9:ZE26–ZE28. https://doi.org/10.7860/JCDR/2015/14621.6568

Gustafson G (1947) Microscopic examination of teeth as a means of identification in forensic medicine. J Am Dent Assoc 35:720–724. https://doi.org/10.14219/jada.archive.1947.0323

Haglund W, Connor M, Scott D (2001) The archaeology of contemporary mass graves. Hist Archaeol 35:57–69. https://doi.org/10.1007/BF03374527

Hunnargi SA, Menezes RG, Kanchan T et al (2008) Sexual dimorphism of the human sternum in a Maharashtrian population of India: a morphometric analysis. Leg Med Tokyo Jpn 10:6–10. https://doi.org/10.1016/j.legalmed.2007.05.011

İşcan MY, Steyn M (2013) The human skeleton in forensic medicine, 3rd edn. Charles C Thomas Publisher, LTD, Springfield, IL

Kanchan T, Chugh V, Chugh A et al (2021) Age estimation using third molar maturation based on Demirjian's criteria. Leg Med Tokyo Jpn 53:101959. https://doi.org/10.1016/j.legalmed.2021.101959

Kanchan T, Rastogi P (2009) Sex determination from hand dimensions of north and south Indians. J Forensic Sci 54:546–550. https://doi.org/10.1111/j.1556-4029.2009.01018.x

Krogman WM (1962) The human skeleton in forensic medicine. Am J Phys Anthropol 20:227–229

Li R (2021) Forensic biology: identification and DNA analysis of biological evidence. CRC, Boca Raton, FL

Ngeow WC, Aljunid ST (2009) Craniofacial anthropometric norms of Malaysian Indians. Indian J Dent Res 20:313–319. https://doi.org/10.4103/0970-9290.57372

Purkait R, Chandra H (2004) A study of sexual variation in Indian femur. Forensic Sci Int 146:25–33. https://doi.org/10.1016/j.forsciint.2004.04.002

Ruengdit S, Troy Case D, Mahakkanukrauh P (2020) Cranial suture closure as an age indicator: a review. Forensic Sci Int 307:110111. https://doi.org/10.1016/j.forsciint.2019.110111

Saul FP (1972) The human skeletal remains of altar de Sacrificios. An osteobiographic analysis. Pap Peabody Mus Archaeol Ethnol Harv Univ Camb Mass 63:3–123

Shedge R, Kanchan T (2019) Cranial sutures and age estimation—a few reflections. J Forensic Legal Med 61:144. https://doi.org/10.1016/j.jflm.2018.11.005

Shedge R, Kanchan T (2021) Forensic anthropology: an overview of the various avenues in human identification. J Indian Acad Forensic Med 43:300–301. https://doi.org/10.5958/0974-0848.2021.00077.4

Shedge R, Kanchan T, Garg PK et al (2020) Computed tomographic analysis of medial clavicular epiphyseal fusion for age estimation in Indian population. Legal Med 46:101735. https://doi.org/10.1016/j.legalmed.2020.101735

Shedge R, Kanchan T, Kumar Garg P et al (2021) Age estimation from sternebral fusion in an Indian population—a computed tomographic evaluation. Leg Med Tokyo Jpn 53:101951. https://doi.org/10.1016/j.legalmed.2021.101951

Suzuki M, Sakai T (1964) Shovel-shaped incisors among the living polynesians. Am J Phys Anthropol 22:65–71. https://doi.org/10.1002/ajpa.1330220117

Ubelaker D (2006) Introduction to forensic anthropology. In: Forensic anthropology and medicine: complementary sciences from recovery to cause of death. Springer, Cham, pp 3–12

Ubelaker DH (2018) Recent advances in forensic anthropology. Forensic Sci Res 3:275–277. https://doi.org/10.1080/20961790.2018.1466384

Ubelaker DH, Wu Y, Cordero QR (2019) Craniofacial photographic superimposition: new developments. Forensic Sci Int Synergy 1:271–274. https://doi.org/10.1016/j.fsisyn.2019.10.002

Warrier V, Kanchan T, Shedge R et al (2021) Computed tomographic age estimation from the pubic symphysis using the Suchey-brooks method: a systematic review and meta-analysis. Forensic Sci Int 325:110811. https://doi.org/10.1016/j.forsciint.2021.110811

Forensic Odontology

27

Rutwik Shedge, Anuj Bhardwaj, and Tanya Chauhan

Abstract

Teeth are used for cutting and crushing food. Animals also use their teeth for catching their prey and to defend themselves. Teeth are the hardest parts of animal body because of a layer of enamel on their surface. Odontology is the study of teeth and different factors that affect the formation of teeth. Forensic odontology is the application of various principles and knowledge of dentistry in the administration of law. Forensic odontology has major application in the identification of dead where other methods of identification are not possible, such as in cases of mass disasters, mutilated remains, and charred bodies. Other applications of odontology are age estimation of living and dead, bite mark analysis on the human body and foodstuff, and identification of victims of wild animal attacks. Identification through dental evidence is established by comparing dental remains with antemortem or post-mortem dental records. Therefore, dental evidence depends upon the availability of data. Our teeth keep changing with time, so if there is a large time difference between dental records and the dental evidence, it may cause a problem in identification. Therefore, it is less definite than evidence like fingerprints and DNA.

R. Shedge (✉)
School of Forensic Science, National Forensic Sciences University, Gandhinagar, Gujarat, India

A. Bhardwaj
CFSL Chandigarh, Chandigarh, India

T. Chauhan
Deparment of Forensic Science, National Forensic Sciences University, Delhi Campus, New Delhi, Delhi, India

A. Puri et al. (eds.), *Fundamentals of Forensic Biology*,
https://doi.org/10.1007/978-981-99-3161-3_27

Keywords

Odontology · Dentition · Enamel · Bite Mark · Age Estimation · Human identification

27.1 Definitions

Forensic Odontology: Application of science of odontology and dentistry for dissemination of justice.

Dentition: The number, condition, and arrangement of teeth in any individual.

Enamel: The hardest substance of the human body; thin outer covering of the crown of a tooth.

Bite mark: Impression created on a pliable surface by biting on it.

Disaster Victim Identification: Identification of victims of any natural disaster or a mass fatality.

27.2 Human Dentition

Dentition is number, arrangement, and shape of teeth in the mouth. It reflects the diet and way of life of an animal species. Human dentition is adapted to omnivorous diet. The human dentition is both heterodont and diphyodont (Wakamatsu et al. 2019). Heterodont dentition is a type in which the teeth of the animal are divided into discrete classes—incisors, canines, and molars. Diphyodont dentition is a type in which two successive sets of teeth are developed in an animal's life time. The first set of human teeth are called primary teeth or milk teeth or deciduous teeth and are 20 in number (Krogman 1962). They start erupting by 6 months and the eruption is completed by approximately 2.5 years of age. The second set of teeth are called permanent teeth or secondary teeth, and they replace primary teeth. The permanent teeth are 32 in number and are called so because no other teeth will appear in an individual's life time after them. First molars are the first permanent teeth to erupt at the age of 6 years, and the final teeth to erupt are the third molars, which may erupt between the ages of 17 and 25 years. However, third molars, also known as the wisdom teeth may not erupt at all in few individuals.

27.3 Type of Teeth

27.3.1 Anatomy of a Tooth

Each human tooth has two regions: a crown and roots. The crown is the portion of the tooth above the gingiva (gum) and is covered with a thin layer of enamel, the hardest substance of the human body (Gray and Lewis 2000). Enamel provides protection to the tooth, and makes it resistant to the vagaries of nature. The root is the

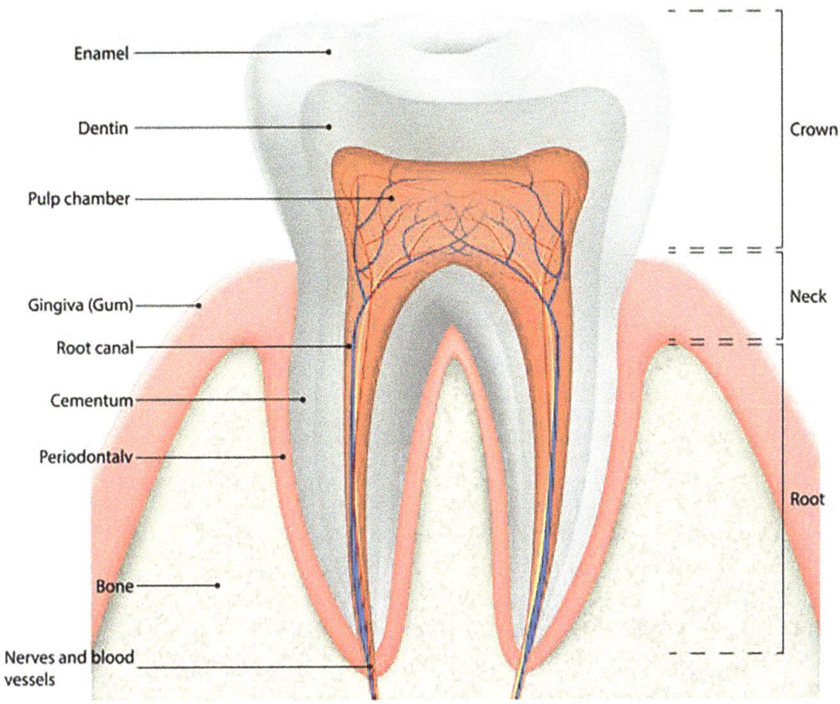

Fig. 27.1 Anatomy of Tooth

portion of the tooth underneath the gingiva that articulates with the alveolus through periodontal ligament. The root is covered with a mineralized tissue called the cementum. Underneath the crown's enamel and the root's cementum lies dentine. Dentine is a highly calcified tissue and provides rigidity and shape to the tooth. The innermost structure within the tooth is the dental pulp that receives blood vessels and nerves through an opening of the tooth. The dental pulp gradually tapers towards the apex of the tooth root and forms the root canal (Krogman 1962). The schematic representation of a human tooth is shown in Fig. 27.1.

27.3.2 Classification of Teeth

Teeth can be classified on the basis of function they perform while eating food. Based on this criterion, there are four types of teeth:

- Incisors: Chisel-shaped teeth present at the front of the jaw. They are used for biting and slicing up the food.

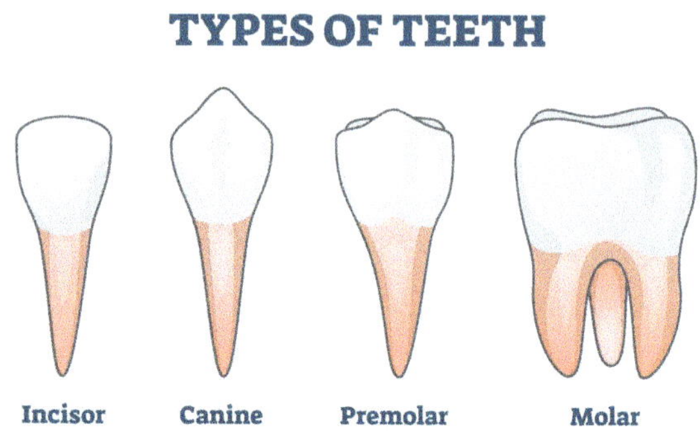

Fig. 27.2 Types of human teeth (The aforementioned figure is a licensable stock image. The illustrator should use this image to create their own graphical figure)

- Canines: These are the pointed teeth used for stabbing and holding the food. As humans are omnivorous, their canines are shorter and blunter in comparison to carnivorous animals.
- Premolars and Molars: They are present at the end of jaws. They have deep roots and ridges called cusps. These teeth are used to crush or grind the food. Premolars are slightly smaller and present in front of molars.

Figure 27.2 depicts each of the four human teeth.

27.3.3 Sides of a Tooth

There are 5 sides to each tooth (Krogman 1962). In Disaster Victim Identification (DVI), these sides of teeth and any of their characteristic features such as the presence of cavity or filling are of great importance. The different sides are:

- Buccal—towards cheek (used for posterior teeth such as premolars and molars).
- Labial—towards the lips (used for anterior teeth such as incisors and canines).
- Lingual—towards tongue (used for teeth of the mandible).
- Palatal—towards the palate (used for teeth of the maxilla).
- Mesial—towards mid line of body.
- Distal—away from mid line of body.
- Occlusal—chewing surface of tooth.

Figure 27.3 depicts the different sides of the human teeth.

27.3.4 Dental Formula

Arrangement of teeth is depicted using a dental formula. The dental formula shows type and number of teeth on each side of the jaw as well as both the jaws. For humans, the dental formula is 2 1 2 3 in case of permanent teeth and 2 1 0 2 in case of

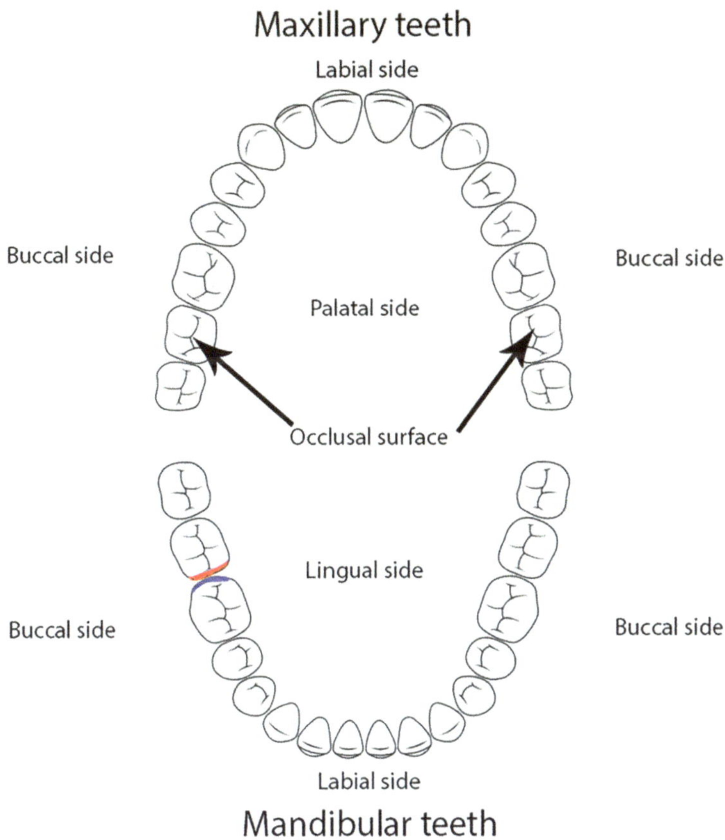

Fig. 27.3 The different sides of each tooth. The black arrows mark the occlusal surface which is the chewing surface for each tooth. Red marking highlights the mesial surface of the second molar, while the blue marking highlights the distal surface of the first molar

deciduous teeth. The first digit (2 for both the sets) states the number of incisors present, the second digit (1 for both the sets) indicates the number of canines present, the third digit (2 for permanent teeth and 0 for deciduous teeth) indicates the number of premolars present, while the last digit (3 for permanent teeth and 2 for deciduous teeth) indicates the number of molars present. There are no premolars or third molars in case of temporary teeth.

27.4 Dental Notation Systems

Dental notation systems, or tooth numbering systems, as their name suggests, are systems used to number each human tooth (Havale et al. 2015; Pemberton and Ashley 2017). There are multiple dental notation systems that exist, however, only a few are globally accepted.

27.4.1 Universal Dental Notation System

In the universal dental notation system, each of the permanent teeth are assigned a unique number that ranges from 1 to 32. The count starts from the right maxillary third molar to the left maxillary third molar, and then to the left mandibular third molar and finally the right mandibular third molar (Fig. 27.4). In case of the deciduous teeth, a unique letter is assigned to each tooth. The letter A is assigned to the right maxillary second molar, and alphabetical order is followed until the left maxillary second molar with the letter J. The left mandibular second molar is then assigned the letter K, and the alphabetical order is followed again until the right mandibular second molar which is assigned the letter T (Fig. 27.5).

27.4.2 FDI Dental Notation System

The Fédération Dentaire Internationale (FDI) or FDI World Dental Federation is the world's leading organization representing dentists and odontologists, and has its own dental notation system that is globally used. In this system, the permanent human dentition is divided into four quadrants, and each of the quadrants is assigned

Right Maxillary								Left Maxillary							
01	02	03	04	05	06	07	08	09	10	11	12	13	14	15	16
M3	M2	M1	PM2	PM1	C	LI	CI	CI	LI	C	PM1	PM2	M1	M2	M3
M3	M2	M1	PM2	PM1	C	LI	CI	CI	LI	C	PM1	PM2	M1	M2	M3
32	31	30	29	28	27	26	25	24	23	22	21	20	19	18	17
Right Mandibular								Left Mandibular							

Fig. 27.4 Universal dental notation system for permanent teeth

Right Maxillary					Left Maxillary				
01	02	03	04	05	06	07	08	09	10
M2	M1	C	LI	CI	CI	LI	C	M1	M2
M2	M1	C	LI	CI	CI	LI	C	M1	M2
20	19	18	17	16	15	14	13	12	11
Right Mandibular					Left Mandibular				

Fig. 27.5 Universal dental notation for deciduous teeth

Right Maxillary								Left Maxillary							
18	17	16	15	14	13	12	11	21	22	23	24	25	26	27	28
M3	M2	M1	PM2	PM1	C	LI	CI	CI	LI	C	PM1	PM2	M1	M2	M3
M3	M2	M1	PM2	PM1	C	LI	CI	CI	LI	C	PM1	PM2	M1	M2	M3
48	47	46	45	44	43	12	41	31	32	33	34	35	36	37	38
Right Mandibular								Left Mandibular							

Fig. 27.6 The FDI Dental Notation for permanent teeth

Right Maxillary					Left Maxillary				
55	54	53	52	51	61	62	63	64	65
M2	M1	C	LI	CI	CI	LI	C	M1	M2
M2	M1	C	LI	CI	CI	LI	C	M1	M2
85	84	83	82	81	71	72	73	74	75
Right Mandibular					Left Mandibular				

Fig. 27.7 The FDI Dental Notation for deciduous teeth

a unique number: Right maxillary quadrant is given the number 1, left maxillary quadrant 2, left mandibular quadrant 3, and right mandibular quadrant the number 4. Furthermore, each type of tooth is assigned its own number: central incisors are 1, lateral incisors are 2, canines are 3, first premolars are 4, second premolars are 5, first molars are 6, second molars are 7, and third molars are 8. Every tooth is thus uniquely identified by a combination of two numbers–the quadrant in which they lie, and their type. For example, the right maxillary third molar will be denoted as '18', while the left mandibular first premolar will be denoted as'35.' The notation for all the 32 permanent teeth is shown in Fig. 27.6.

In case of temporary teeth, similar style of notation is used, with all four quadrants having assigned the numbers 5, 6, 7, and 8 in the same order as the permanent dentition, and each type of tooth being assigned numbers as follows: central incisors are 1, lateral incisors are 2, canines are 3, first molars are 4, and second molars are 5. The notation for all 20 deciduous teeth is shown in Fig. 27.7.

27.5 Human Identification

One of the most vital utility of teeth in forensics is their capability for individualization of unknown human remains. Since teeth are covered by a layer of enamel, they are heat and chemical resistant to an extent, and thus are mostly preserved even in cases of charred remains and ancient remains (Bagdey et al. 2014; González-Colmenares et al. 2020) b. Teeth provide reliable identification when other means of identification may fail or not possible. Teeth play a vital role in identification of dead bodies of mass disasters like fire, explosion, flood, earthquake, and plane crash (Higgins and Austin 2013). Teeth also help in identification when dead bodies are charred, dismembered, mutilated, or in advanced stage of decomposition (Buchner 1985). Identification of an unknown individual can be done by comparing the dental remains with antemortem dental records. During comparison of dental remains with antemortem records, individual characteristics such as filings, dental caps, restorations, etc., of each tooth play an important role. The antemortem dental records may be present in form of a textual record, an orthopantomogram, or even in form of a dental cast (Fig. 27.8). A dental record may contain the following details:

- Total number of teeth.
- Arrangement of teeth.
- Position of teeth.
- Width of each tooth in the arch.
- Gap between teeth.

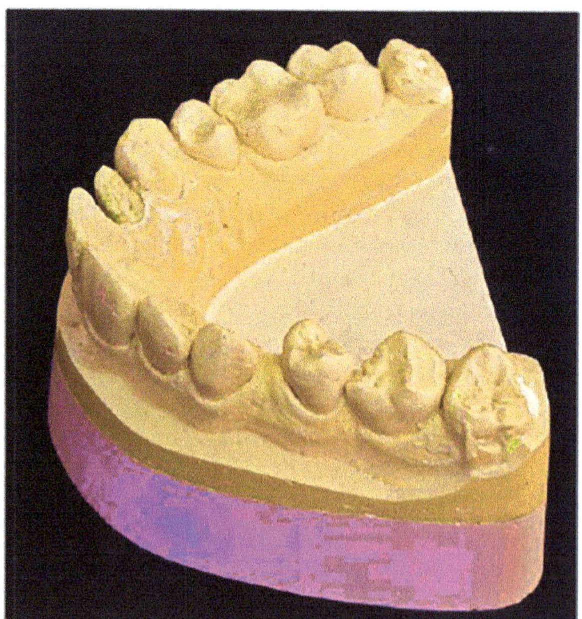

Fig. 27.8 A dental cast

- Number and type of filling done.
- Missing tooth position.
- Details of restoration procedures.

Even in cases where antemortem records are not available, an unknown individual may still be identified. Since the dental pulp within the tooth is extremely well preserved, DNA can be isolated from it even after a considerable time has passed. From the isolated DNA, multiple downstream analyses can be conducted to determine the ethnicity of the victim, their biological sex, their hair, eye, and skin colour (externally visible characteristics), as well as their STR profile for comparison with antemortem DNA samples, or profiles of the victim's parents.

27.5.1 Determination of Ethnicity

Using teeth, a forensic odontologist can determine the ethnicity of an unknown individual. The utility of teeth for determination of ethnicity of an unknown individual has been detailed in the previous chapter (Chap. 26).

27.5.2 Dental Age Estimation (DAE)

Teeth are one of the most reliable anatomical indicators of a person's age. They follow a predictable pattern of development and eruption that can be used to predict an individual's age.

27.5.2.1 DAE Using Eruption Sequence of Teeth

Since humans are diphyodonts, they have two sets of teeth—the first set being the temporary or the deciduous one, and the second set being the permanent one. The time of eruption of primary and secondary sets of teeth are detailed in Tables 27.1 and 27.2 respectively.

The third molar is an aberrant tooth, that may erupt any time between 17 and 25 years of age, or it may not erupt at all.

Table 27.1 Eruption of primary dentition in humans

Tooth	Eruption	Number of teeth
Central incisor (lower)	6–8 months	2
Central incisor (upper)	7–9 months	4
Lateral incisor (upper)	8–10 months	6
Lateral incisor (lower)	10–12 months	8
First molar	12–14 months	12
Canine	17–19 months	16
Second molar	20–30 months	20

Table 27.2 Eruption of secondary teeth in humans

Tooth	Eruption	Number of teeth
First molar	6–7 years	21–24
Central incisor	6–9 years	24
Lateral incisor	7–9 years	24
1st premolar	9–11 years	24
Second premolar	10–12 years	24
Canine	11–13 years	24
Second molar	12–14 years	25–28
Third molar	17–25 years	29–32

27.5.2.2 DAE Using Development Stages of the Tooth

Grossly, one can observe a tooth once it starts erupting from the gingiva (gum). However, eruption is one of the final stages in the development of teeth. Before the tooth comes out of the gum, it undergoes a complex development due to interactions between the oral cavity's ectoderm and the ectomesenchyme of the neural crest, which are responsible for production of enamel and other dental structures, respectively. Typically, there are 5 stages of odontogenesis—development of the tooth. The first stage is the bud stage in which the dental lamina develops into dental epithelial cells. These cells evolve into the tooth germ which consists of all the soft tissues necessary for the formation of tooth. The second stage is the cap stage in which the dental cells start forming the outer layers of the tooth and form a cap (enamel organ) that sits on the tooth bud. This cap will later form the cells required for formation of enamel. The dental papilla which is the remaining part of the tooth bud will then form dentine and the pulp. The bell stage is the third stage in which the enamel organ grows into a bell shape and its cells differentiate into four groups—inner enamel epithelium, outer enamel epithelium, stratum intermedium, and stellate reticulum. The shape of the crown is formed during this stage. Crown and root formation is the fourth stage of odontogenesis, in which the ameloblast cells and the odontoblast cells form the enamel and the dentine, respectively, while the dental papilla, the dental follicle, and Hertwig's epithelial root sheath form the roots of the tooth. The final stage is the eruption stage in which the tooth moves out into the oral cavity.

These different stages of tooth development can be observed radiologically using Orthopantomography, and since development for each tooth occurs at a predictable time, age estimation can be done using it. Multiple researchers over the last 50 years have come up with their own method of estimating an individual's age based on the tooth development (Nolla 1960; Haavikko 1970; Demirjian et al. 1973; Willems et al. 2001; Cameriere et al. 2004, 2008; Acharya 2011; Kanchan et al. 2021).

27.5.2.3 DAE Using Morphological Changes of the Tooth

While development and eruption of teeth are great indicators of an individual's age, they are completed by the age of 25 years. Beyond that, age can be estimated from the attritional changes of the teeth. Since the permanent teeth of humans are never replaced, they undergo certain morphological changes that can be used to estimate age (Kaur et al. 2015; Sultana et al. 2021). One such method that studies the morphological changes and provides models to estimate age is the Gustafson's method (Gustafson 1947). This method has been thoroughly described in the previous chapter (Chap. 26).

27.6 Bite Mark Analysis

Bite marks are any impressions created on a pliable surface through the action of biting. There are of great forensic relevance, and are often key evidences in cases of assaults, sexual assault, child abuse, abuse of the elderly, animal abuse, and victims of animal attack. They may even be found on a perpetrator of an assault, wherein the victim tried to defend themselves by biting the perpetrator (Ma et al. 2020). Bite marks can also be found on inanimate objects such as chewing gum, pieces of fruits, leather, ends of pens or pencils, etc. (Rivera-Mendoza et al. 2018).

27.6.1 Morphology of a Typical Bite Mark

A typical human bite mark consists of a circular or an oval mark, that is bruised, and a central region that is spared (Dorion 1982). The bite mark could have been made by only the anterior teeth (central incisors, lateral incisors, and canines) or it could have been made by more teeth. Depending upon the target substrate, both the jaws could be involved in leaving a mark, or, in cases wherein the victim's skin is partly covered by a fabric, only one jaw, or half portions of both the jaws will leave an imprint.

27.6.2 Injuries Associated with Bite Marks

A bite mark on the human body can have multiple injuries associated with it depending upon the bite force, nature of teeth, and the target area (Hollenback 2015). A bite mark may be characterised by abrasions which are injuries to the superficial layers of the skin. These occur when the occlusal or incisal surfaces of the teeth graze against the skin of the victim. Bruises may accompany bite marks as well. A bruise is caused due to extravasation of blood underneath the skin, without any break in the continuity of the skin. Bruised bite marks are accompanied with swelling and discolouration of blood underneath the skin. If the force of the teeth is too high, the skin may split underneath the force, and result in lacerations. Lacerations are mostly seen in areas which have underlying bony structures (cheeks, hands, etc.), and are marked by irregular edges, tissue tags, and bleeding. In certain cases, partial or complete avulsion may be seen as well. Whenever excessive bite force results in partial tearing out of a body part (mostly cartilaginous parts such as ears and nose), the bite mark is said to be a partial avulsion bite mark. In case a part of the cartilaginous tissue is completely detached from the body, such as ripping off of an ear or nose, etc., the bite mark is said to be an avulsion bite mark.

27.6.3 Collection of Evidence

27.6.3.1 Collection of Evidence from the Victim

Bite marks first need to be documented using examination of quality photographs. The camera is placed at a perpendicular perspective over the injury, and multiple photos are taken with and without an ABFO scale (Wright 1998; Pretty and Sweet 2001). ABFO stands for the American Board of Forensic Odontologists, and its ABFO No. 2 scale has become the standard for bitemark measurement and photography. While taking the photos, different light perspectives and sources may be used—reflected light, oblique light, IR light, UV light, etc. IR light is particularly helpful in understanding the extent of injuries in case of a bruised bite mark.

If there is a lot of blood over the bite mark, it has to be collected with a double swabbing method, air dried, and then kept in the swab's container. If the bitemark is devoid of blood, saliva may be directly collected using the double swabbing technique. One end of the swab is dipped in normal saline, and brushed over the bite mark. After the entire area is brushed, the other dry end of the sterile swab is brushed over the same area.

27.6.3.2 Collection of Evidence from the Suspect

In case there is suspicion on an individual for having bit another individual, evidences have to be collected from them. The first step in collection of evidences from the suspect is conduction of an extraoral and intraoral examination. In the extraoral examination, maximal opening of the mouth is measured, deviations in closing or opening of mouth are recorded, and presence of any scars or facial surgery are recorded. In the intraoral examination, number, nature, and condition of teeth are recorded. Salivary swabs are taken as well. The next step is taking dental impressions of the suspect. Impressions of both the casts are taken in using an alginate impression material and these are later casted using dental stone. The final stage in collection of evidence from suspect is the collection of sample bite. The suspect should be asked to bite on multiple sample materials similar to the actual case evidence, while trying to emulate the type of evidence bite mark (Naru 1997).

27.6.4 Examination of Bite Mark Evidence

The forensic odontologist should begin with examining the class characteristics of the bite mark. This begins with first identifying the evidence as a bite mark. In typical bite marks, this is comparatively easy as the impression of teeth may be found on the substrate. In cases of lacerated or avulsed bite marks, the presence of saliva can be an indicator that the injury is a bite mark. The bite mark is then classified according to its type and measurements are taken—of each arch if both are present, of each tooth if individual impressions of teeth are present, distance between the teeth impressions, etc. The bite mark is then compared with the suspect's sample bite, or the bite created using the casted teeth of the suspect. Comparison can either be done side by side, or photographs can be superimposed either manually or by using computer-generated overlays. Finally, the DNA from the saliva collected from the bite mark is compared with the DNA of the suspect to ascertain whether the suspect was indeed the perpetrator. In case there are any individualistic features such as impressions of supernumerary teeth, or talon's cusp, or rotated tooth, they are compared with the suspect's dentition.

Multiple Choice Questions

1. Which of the following teeth are responsible for slicing up food?
   (a) Incisors.
   (b) Canines.
   (c) Premolars.
   (d) Molars.
   Answer: (a)
2. The first permanent tooth to erupt is the ________.
   (a) Upper central incisor.
   (b) Lower lateral incisor.
   (c) First molar.
   (d) First premolar.
   Answer: (c)
3. According to the FDI Dental Notation, what number is assigned to the left mandibular third molar?
   (a) 18
   (b) 28
   (c) 38
   (d) 45
   Answer: (c)
4. Which scale is used to photograph bite marks?
   (a) Tape measure.
   (b) ABFO.
   (c) 6-inch ruler.
   (d) Vernier callipers.
   Answer: (b)
5. The dental impression of a suspect is taken using which of the following materials?
   (a) Plaster of Paris.
   (b) Plasticine.
   (c) Clay.
   (d) Plasticine.
   Answer (d)

References

Acharya AB (2011) Age estimation in Indians using Demirjian's 8-teeth method. J Forensic Sci 56: 124–127. https://doi.org/10.1111/j.1556-4029.2010.01566.x

Bagdey SP, Moharil RB, Dive AM et al (2014) Effect of various temperatures on restored and unrestored teeth: a forensic study. J Forensic Dent Sci 6:62–66. https://doi.org/10.4103/0975-1475.127777

Buchner A (1985) The identification of human remains. Int Dent J 35:307–311

Cameriere R, Ferrante L, Cingolani M (2004) Variations in pulp/tooth area ratio as an indicator of age: a preliminary study. J Forensic Sci 49:317–319

Cameriere R, Ferrante L, De Angelis D et al (2008) The comparison between measurement of open apices of third molars and Demirjian stages to test chronological age of over 18 year olds in living subjects. Int J Legal Med 122:493–497. https://doi.org/10.1007/s00414-008-0279-6

Demirjian A, Goldstein H, Tanner JM (1973) A new system of dental age assessment. Hum Biol 45: 211–227

Dorion RB (1982) Bite mark evidence. J Can Dent Assoc 48:795–798

González-Colmenares G, Calvo-Díaz L, Nastul-Enríquez M et al (2020) Effect of high temperatures on teeth fixed with an orthodontic bracket. An in vitro study. Forensic Sci Int 308:110182. https://doi.org/10.1016/j.forsciint.2020.110182

Gray H, Lewis WH (2000) Anatomy of the human body. Bartleby.com, New York

Gustafson G (1947) Microscopic examination of teeth as a means of identification in forensic medicine. J Am Dent Assoc 35:720–724. https://doi.org/10.14219/jada.archive.1947.0323

Haavikko K (1970) The formation and the alveolar and clinical eruption of the permanent teeth. An orthopantomographic study. Suom Hammaslaak Toim 66:103–170

Havale R, Sheetal BS, Patil R et al (2015) Dental notation for primary teeth: a review and suggestion of a novel system. Eur J Paediatr Dent 16:163–166

Higgins D, Austin JJ (2013) Teeth as a source of DNA for forensic identification of human remains: a review. Sci Justice 53:433–441. https://doi.org/10.1016/j.scijus.2013.06.001

Hollenback JK (2015) Bite-mark analysis. J Calif Dent Assoc 43:486

Kanchan T, Chugh V, Chugh A et al (2021) Age estimation using third molar maturation based on Demirjian's criteria. Leg Med Tokyo Jpn 53:101959. https://doi.org/10.1016/j.legalmed.2021.101959

Kaur P, Astekar M, Singh J et al (2015) Estimation of age based on tooth cementum annulations: a comparative study using light, polarized, and phase contrast microscopy. J Forensic Dent Sci 7: 215–221. https://doi.org/10.4103/0975-1475.172441

Krogman WM (1962) The human skeleton in forensic medicine. Am J Phys Anthropol 20:227–229

Ma XF, Jin M, Sun H, Mi CB (2020) Application status and Prospect of bite mark evidence in forensic odontology. Fa Yi Xue Za Zhi 36:369–373. https://doi.org/10.12116/j.issn.1004-5619.2020.03.014

Naru AS (1997) Methods for the analysis of human bite Marks. Forensic Sci Rev 9:123–139

Nolla C (1960) The development of the permanent teeth. J Dent Child 27:254–266

Pemberton MN, Ashley M (2017) The use and understanding of dental notation systems in UK and Irish dental hospitals. Br Dent J 223:429–434. https://doi.org/10.1038/sj.bdj.2017.731

Pretty IA, Sweet D (2001) Adherence of forensic odontologists to the ABFO bite mark guidelines for suspect evidence collection. J Forensic Sci 46:1152–1158

Rivera-Mendoza F, Martín-de-Las-Heras S, Navarro-Cáceres P, Fonseca GM (2018) Bite mark analysis in foodstuffs and inanimate objects and the underlying proofs for validity and judicial acceptance. J Forensic Sci 63:449–459. https://doi.org/10.1111/1556-4029.13586

Sultana A, Zainab H, Jahagirdar P et al (2021) Age estimation with cemental incremental lines in normal and periodontally diseased teeth using phase contrast microscope: an original research. Egypt J Forensic Sci 11:39. https://doi.org/10.1186/s41935-021-00254-2

Wakamatsu Y, Egawa S, Terashita Y et al (2019) Homeobox code model of heterodont tooth in mammals revised. Sci Rep 9:12865. https://doi.org/10.1038/s41598-019-49116-x

Willems G, Van Olmen A, Spiessens B, Carels C (2001) Dental age estimation in Belgian children: Demirjian's technique revisited. J Forensic Sci 46:893–895

Wright FD (1998) Photography in bite mark and patterned injury documentation—Part 1. J Forensic Sci 43:877–880

Forensic Botany

28

Neelam Arya

Abstract

The average person enjoys plants in a multiple ways in his day-to-day life. The plants are everywhere in nature and the study of scientific techniques used for plants to answer questions relevant to the legal system is termed forensic botany and plants may occur as forensic evidence at a crime scene which have the potential to provide links between crime scenes and individuals/suspect. In conventional method of investigation, generally the botanical evidences are overlooked as the agencies involved in criminal justice system are not much aware about the specific potential use and advantages of plant evidence but the subject knowledge may turn the routine observations into case-cracking evidences. However, plant research continues in academic and private research institutions continuously and new tools are being developed that can be eventually applied to forensic botanical evidence to aid in criminal investigation. Consulting with an expert in plant systematics is an excellent place for the legal investigator to start when trying to obtain line of investigation pertaining to botanical evidences. Some experts work with only a single group of plants and some work with many groups and have a good knowledge of the entire field of botany.

However, Many plant materials cannot be identified and differentiated to the species level by traditional morphological characteristics when botanical specimens are degraded and lack physical features. The use of new techniques based on DNA fingerprinting provide novel approaches to varietal identification which offer advantages over traditional morphological comparisons. DNA is unique to each organism, so knowing the order of the DNA allows scientists to identify its species.

N. Arya (✉)
Forensic Science Laboratory, Madhuban, Haryana, India

A. Puri et al. (eds.), *Fundamentals of Forensic Biology*,
https://doi.org/10.1007/978-981-99-3161-3_28

Botanical evidences includes pollens, spores, wood traces, etc. Wood can be found at crime scenes in many forms: as a murder weapon, as material used to hide a body, or as trace evidence from forced entry or vandalism. In the process, wood fragments get attached or linked to the culprit or victims. These fragments can often be identified and linked to standards from the crime scene. Due to ever increasing refinement in the analysis of botanical evidences, the forensic botany is emerging as very crucial approach in crime investigation.

Keywords

Species identification · Palynology · Botanical evidence · Plant DNA analysis · Forensic botany · Wood identification, pollens

28.1 Forensic Botany

Often the green background of the world is unseen and unnoticed, taken for granted, Quietly absorbing carbon dioxide and providing oxygen and food. Plants are everywhere in nature, essential for all human and animal existence, and may occur as forensic evidence at a crime scene which have the potential to provide links between crime scenes and individuals or other vital crime scene information, because the components and construction of a plant's body, and its ecological requirements, are particular to the species. Forensic botany is emerging as a novel approach in crime investigation and botanical samples can serve as key pieces of trace evidence.

Forensic botany is the study of plants and how they can relate to law and legal matters. Forensic botany is a subdiscipline of forensics that is still in its infancy. As law enforcement officers are not much informed about the science of plants, therefore, important plant evidence is frequently overlooked. Although plant research continues in academic and private research laboratories and new tools are being developed that can be eventually applied to forensic botanical evidence to aid in criminal investigation. It is now becoming a strong tool, as the law enforcement agencies, forensic scientists, and prosecutors are becoming more familiar with the applications of plant material to criminal and civil cases. The forensic aspects require an understanding of what is necessary for botanical evidence to be accepted as evidence in our judicial system. Forensics requires recognition of pertinent evidence at a crime scene or associated with victim/suspect, appropriate collection, validation of new forensic techniques, and admissibility criteria for courts. The methods and protocol range from simple techniques like microscopy to more technical molecular biology techniques like DNA sequencing.

The botanical evidence can place a person or object at a crime scene, verify or refute an alibi, help in determining time since death, the time of a crime, and the place where a crime occurred. In the absence of eyewitnesses who could testify the dynamics of any crime/event, the crime scene investigation is fundamental to clarify the hypothesis given by parties. During the scene analysis, if some traces of plant material are found, then analysis of these traces allow to reconstruct the incident. Even if this evidence, of course, is circumstantial, it can be useful in forensic cases, together with the other evidences, to reconstruct the dynamics of incidents.

28.1.1 Basic Plant Biology

A plant may be defined as any member of the kingdom Plantae. Plants can be large or small, flat or erect, flowering or nonflowering, green, or possessing other pigmentation. The plant biology deals with the anatomical and taxonomical classification of plants and utilizes standard nomenclature techniques to derive the scientific names of plants. Historically, the science of plant biology or botany has included all living organisms except animals, but it is clear that there is a major division of life between cells with a simple level of organization, the prokaryotes, and those with much more complex cells, the eukaryotes. Among eukaryotes, three main multicellular kingdoms are recognized: animals, plants, and fungi. The unifying characteristic features that define plants as different from other eukaryotes are-

- They are photosynthetic and obtain all their nutrients from inorganic sources, i.e. they are autotrophic.
- The photosynthetic pigment is chlorophyll, and in all plants except some algae, there are two forms, a and b, contained within chloroplast.
- The cells have a cell wall made predominantly of the cellulose.
- Vegetative structure and physiology is similar throughout the seed plants and there are many similarities with other vascular plants as well.

Consulting with an expert in plant systematics is an excellent place for the legal investigator to start when trying to obtain line of investigation pertaining to botanical evidences. Some experts work with only a single group of plants and some work with many groups and have a good knowledge of the entire field of botany.

28.1.2 Use of Botanical Evidence

Plant evidence can step into the spotlight as star witnesses, but only if the plant evidence interpreted by a trained botanist. Very few professionals involved with law enforcement have a background or training in botany. Typically, legal investigators should seek a forensic scientist/botanist with well-rounded training and experience who possesses knowledge of the various specialties within the botanical field in question. Not all forensic scientist/botanist have such training. As in many fields, most professional botanist have specialized in one particular area, and have only scant knowledge concerning the forensic possibilities within others (Fig. 28.1).

There are many kinds of botanists who can help with the interpretation of plant evidence by the various specialties like, in plant anatomy, plant morphology,

Fig. 28.1 Botanical evidence as private eye for investigators

ecology, physiology and genetics (DNA). Plants may be present as biological evidence in many ways:

- Seeds caught and carried in a pant cuff,
- Plant leaves and stems entangled and carried in a vehicle's undercarriage, wheel wells, etc.
- Grass stains on garments after a sexual assault,
- Clump of leaves within victim's palm,
- Presence of pollen grains on clothing,
- Traces of wooden log around the hit mark on any surface.

Thus, investigators would know the value of locating plant material on a suspect or victim, and then establishing the origin from which that plant came. On a microscopic level, this may be a reasonable outcome. A sample of leaves within victim's palm will catch the curious eye of an investigator. On the other hand, what about botanical evidence at the microscopic level, such as grains of pollens.? Sometimes the plants become associated with criminal activities; many homicides, accidental deaths, and suicide occurred through the inappropriate and deliberate use of plants to achieve harmful effects. But only if the investigator is aware of the potential existence of that evidence, he will do all efforts to search for it and the botanical evidence plays a key role in investigation. Finally, with limited or no training in botany, crime scene personnel will likely not know whether or not the plant material is indigenous to the area, relatively common or rare.

28.1.3 Pant Identification by DNA

Analysis of trace evidence by traditional forensic botany techniques have limitations of reproducibility and standardization from degraded samples. The absence of an

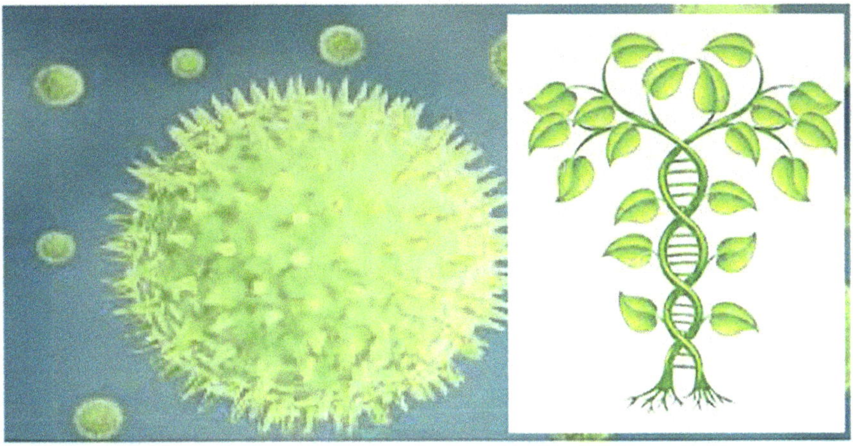

Fig. 28.2 Plant DNA profiling: strong tool for plant identification

accurate identification system remains the major obstacle to the present inability, to routinely and correctly identify trace botanical evidence. On the other hand, many plant materials cannot be identified and differentiated to the species level by traditional morphological characteristics when botanical specimens are degraded and lack physical features.

The use of new techniques based on DNA profiling/fingerprinting provide novel approaches to varietal identification which offer advantages over traditional morphological comparisons. By taking advantage of genetic code system, the more sophisticated plant DNA fingerprinting and other biomolecular techniques are being used to enhance the information gained from a plant evidence. DNA is unique to each organism, so knowing the order of the DNA allows scientists to identify its species. New DNA profiling/fingerprinting technology has decreased the time and cost of identifying the specie of a given organism. The effectiveness of source attribution of plant material will depend on, how unique and of a sample accurate the species is to geographic area and its genetic history. If a plant is very rare, source attribution may not be difficult; however, many plants require DNA testing to confirm that the evidentiary sample originated from a source plant, if generated by seed, or a source population if generated by clonal reproduction, which is a considerable potential approach in investigation of crime, accident, and suicide circumstances.

Plants have DNA, since they are living beings. The DNA in plant cells is found in the nucleus, the mitochondria, and the chloroplasts. Although chloroplasts and mitochondria contain some genetic material, the nucleus contains the majority of DNA in plant cells. DNA is found in the same shape in the cells of all plants as in animals (Fig. 28.2).

Some plant species are capable of self-fertilization, and some are nearly exclusively self-fertilizers. This means that a plant can be both mother and father to its offspring, a rare occurrence in animals. Plants are generally more capable of surviving as polyploids. Polyploid organisms have more than two sets of homologous chromosomes. For example, humans have two sets of homologous chromosomes, meaning that a typical human will have 2 copies each of 23 different chromosomes, for a total of 46. There are various methods for plant DNA fingerprinting like-

- Restriction Fragment Length Polymorphisms (RFLPs).
- Randomly Amplified Polymorphic DNAs (RAPDs).
- Amplified Fragment Length Polymorphism (AFLP).
- Simple Sequence Repeats (SSRs).

DNA profiling involves the extraction of DNA from plant cells for quantification and quality assessment. Different plant DNA extraction methods have been standardized and used in several kinds of applications. In general, plant DNA extraction and purification can be divided into six steps:

1. Tissue disruption/homogenization,
2. Cell lysis in DNA extraction buffer,
3. Separation of DNA from other cellular components,
4. DNA precipitation,
5. DNA washing,
6. DNA collection for downstream processing.

Although, it sounds biochemically simple, but the nature of the plant starting material can present unique technical challenges that can be difficult to overcome for even the experienced scientists. For DNA profiling of a plant variety, a set of molecular markers and a method to detect them are required. Two different sets of molecular markers detected with the same method will result in two different DNA profiles for a particular variety. In contrast, two different methods to detect the specific alleles of a given molecular marker set are expected to result in identical DNA profiles.

The number of markers should be balanced with the accuracy of the genotype required for the purpose, so the selection of marker and construction of a Species-Specific Database should be considered by following some general criteria-

- Repeatability and possible sources of molecular markers derived from different laboratories to score data.
- Molecular markers selected from newly generated sequence data.
- As far as possible, avoid the markers with **null** alleles.
- Good performing markers are preferred over complex marker profiles that are sensitive to interpretation as the clear answers allows the easier harmonization.
- Co-dominant markers are generally preferred over dominant markers as they have a higher discriminative power.
- The number of markers should be balanced with the accuracy of the genotype required for the purpose.

- Performance of molecular markers and genotyping methods are evaluated in a validation process and validated markers lead to harmonized results.
- Database construction should consider different levels of data processing and store the end results.
- Standardization of the detection method and technology is not required as long as the performance meets the quality criteria and the resulting DNA profiles are consistent.

DNA has revolutionized the field of criminology and hugely improved the functioning of the criminal justice system, as highly polymorphic DNA markers become increasingly available for a wide range of plant species and there will be increasing opportunities for applications to forensic investigations.

28.1.4 Forensic Palynology

Recently, several criminal and civil cases have depended on identification of plants from their pollen. Suspects in violent crimes have been directly linked to a crime scene by a specific pollen type which was unseen and unknown by them to be present on their clothing. **Palynology** is a scientific discipline concerned with the study of plant pollen, spores, and certain microscopic planktonic organisms. It is known to be a highly valuable, accurate, and effective means of criminal investigation which has been used by a number of experienced scientists to provide evidence in selected legislature. Spores and pollen grains have a number of morphological and ultra structural features. Pollen from different flowering plant species can be useful in criminal justice system as trace evidence. Palynology has been tested in court and has provided evidence for contact between object and place, location of clandestinely disposed human remains, and estimated times of body deposition, has differentiated murder and rape sites from deposition sites, and has provided provenance for objects and materials (Fig. 28.3).

Pollen and spores are produced in large numbers and dispersed over large areas by wind and water; they are recoverable in statistically significant assemblages. The structure of pollens and spores is extremely resistant to any external environments

Fig. 28.3 showing pollens and spores

more likely, heat and cold, washing, smudging, and degradation. Pollen grains may remain preserved for many years. If due attention is given to this important aspect, many cases can be solved easily. Pollen and spores provide important information which prove very crucial in crime investigation like-.

- Determine the geographical origin of a specimen.
- Create ties between the crime scene and individuals.
- test alibis.
- Determine possession or trade of prohibited or endangered species.
- Determine the travel history of an item.
- Determine if a crime scene is a primary or secondary scene and to identify unclaimed or unidentified bodies.
- Useful in linking the victim with the location where the crime occurred.

Pollen grains are produced in the anthers of flowers, differ in many ways, and can be characterized using microscopic techniques. They should be collected carefully because often these elements contain abundant pollen and spores. Samples can be recovered from a wide range of sources-

- Dirt collected from the clothing of victim/suspect.
- Pockets/cuffs.
- Skin/finger nails of living and deceased persons.
- Hair.
- Nasal passage.
- Shoes.
- Vehicle tires.
- Air filters in cars.
- On objects/stones/bricks.

There are three methods for obtaining useful pollen grain images. These three methods can be considered as semi-automated traditional methods used to detect the pollen grains. The identification of palynomorphs depended on traditional methods, such as scanning electron microscopy

- Transmitted-light microscopy (TLM),
- Wide field fluorescent method,
- Structured illumination (Apotome) method.

For example, the transfer of pollen to pristine shoes and to shoes that were previously worn at other localities can be analyzed palynologically. With the exception of one sample, the pollen found on the footwear had a characteristic signature, supporting the view of a general distinctiveness of pollen from individual sites and the concept of widespread palynological heterogeneity. Pollen can also help forensic scientists validate an alibi. Several years ago, a man was found dead on a hill. Eye witnesses helped police identify a suspect, but the suspect claimed that he had never been to the area of the crime and had never worn the sweater that witnesses reportedly saw him in; however, after swabbing the sweater, forensic scientist identified pollen of a hill area plant that occurs near the location where the victim was found dead. Although, this evidence is circumstantial, investigators were able to use palynology to discredit the suspect's alibi and show that it was very likely the suspect had visited that hill area where the victim was found dead. Pollen obtained from a suspect that matches that of a crime scene could basically suggest that the person has visited that area at some point recently, not necessarily suggesting that they have committed a crime.

As in any new application of reliable science or techniques, they may provide some advantages and disadvantages in their application in forensic investigation. It has become more accepted as a significant crime-solving tool. The advantages are outweighing the disadvantages because of the mechanism of how pollens are spread according to its small size and how it attaches to many objects, like surfaces, skin, and folds of clothes. On the other hand, this field has disadvantages because it lacks complete information, location, and techniques to collect the samples necessary to conduct investigations. The limited number of trained specialists in this field or even full-time available palynologist is also a problem, and there are no academic centers or forensic facilities that care to train a scientific staff in specified field.

28.1.5 Wood Identification

From time to time, wood plays an important role in solving criminal cases. Comparative wood anatomy and wood identification can play an important role in forensic science. Wood identification can provide significant supporting evidence during criminal investigations. However, it is still an underutilized field of investigation with its most common application limited to identifying specific as well as suspected wood traces. Wood samples vary in size and can range from trees and logs to pieces as small as individual wood fibers. Botanists can easily identify trees, especially when the leaves, flower, or fruits are present. Illegal logging and the illegal timber trade are major problems domestically and internationally, threatening not just individual species, but entire ecosystems. Using standard techniques, wood technologists, foresters, and wood anatomists can generally identify logs, timber, wood products, and pieces of wood bigger than a matchbox. As the size of the piece decreases, however, accurate determinations are more difficult. Very small fragments require special handling and identification techniques.

Wood can be found at crime scenes in many forms: as a murder weapon, as material used to hide a body, or as trace evidence from forced entry or vandalism.

In the process, wood fragments get attached or linked to the culprit or victims. These fragments can often be identified and linked to standards from the crime scene.

When gathering trace evidence, do not ignore wood fragments. They could contain clues that help solve the crime. In one form or another, in connection with a variety of crimes, wood fragments' use as evidence is limited primarily to these areas:

- Color, odor, and density: The naked eye examination is used mainly to visually determine the color, odor, weight, and unusual features or defects of the wood.
- Chemical test: such as the Froth and fluorescent tests, spot-test for aluminum, and sodium nitrite spot-test.
- light microscope: Standard micro techniques can be used to cut, stain, and mount the sections for analysis.
- Macroscopic examination includes physical matches between broken pieces of wood, carrying tool marks, etc.
- Wood anatomy: includes matching of growth patterns in different pieces or to the examination of a piece for clues and determine the extent of variation and to interpret whether the differences are the effects of environmental conditions or habitat, or whether they reflect genetic differences.
- Detector Dogs: The use of detector dogs for rapid-field identification of timber can be very useful for particular species of concern. However, a dog can only confirm if it detects the odor of a substance to which it has been trained to respond; it cannot determine the identity of other substances.

Wood may present some special challenges because of its fibrous nature. The identification of wood requires a knowledge of wood anatomy as well as skill in sectioning and mounting small particles for microscopical study. A piece of wood, no matter how small, retains its orientation with respect to the tree in which it originally grew. These directions are not difficult to locate on a tree or a large piece of wood, but become a challenge as the size of the specimen decreases. Since most pieces of wood, with diameters of the order of a toothpick, are elongated parallel to the fibrous elements of which they are composed, the easiest way to determine orientation is to cut a cross-section straight through the sliver using a sharp razor. The freshly cut surface is then held in the fingers (or if it is too small in a pair of forceps) and examined under a stereomicroscope. The directions of the rays will immediately reveal the true orientation of the fragment. The specimen is now oriented so that the required sections can be cut from it. If the piece of wood is large enough, it can either be mounted for sectioning in a microtome, or sections can be cut freehand with a microtome knife blade using a technique similar to that employed in peeling an apple. Particles of sawdust, which are too small to section, are sprinkled on a slide so that they do not lie on top of one another, boiled in glycerin-alcohol and observed directly under the microscope. Many of the sawdust particles will be cut in a generally radial or tangential section and, therefore, will often show enough characteristics to permit an identification to be made. While examination of forensic sample, it is essential that good known samples be obtained. A broken door and frame may produce particles from several different kinds of wood. If adequate known samples are not taken at the crime scene, the microscopist may erroneously eliminate questioned wood particles found on a sledge hammer suspected in a break-in, for instance, from being associated with the incident. This problem can become quite taxing in certain situations, such as when sawdust from a workshop or particle board are involved. The number of different woods which may be involved can mean that many samples may have to be prepared and examined over a relatively long period of time before a conclusion can be reached.

On occasion, identification may be carried down to the species level by an experienced microscopist well trained in wood anatomy. If the wood is not exotic, a microscopist trained in wood identification will recognize the common woods almost at sight, based on a few key characteristics. An exotic wood which is rare in commerce will normally make the best evidence. Particles of wood recovered from the scalp of a murder victim were identified as a species of Kokka, a rather rare tropical wood not commonly found on the market. An examination of a pool which belonged to the prime suspect showed that it was made from the same wood and had sustained damage to its thick end. This was the most important piece of physical evidence in the trial which resulted in the conviction of the defendant. **After identifying the wood trace,** it becomes necessary to interpret the results. The first consideration, aside from the integrity of the evidence itself, must be the accuracy of the identifications. These must be based on a sound knowledge of wood anatomy and comparison to reference specimens of standard woods. Whenever possible, identifications should be checked by a second analyst, also skilled in the techniques of wood identification. This is a general rule in forensic science and helps to ensure the quality of the results obtained. The next step must be to consider how common or rare the wood is. In this regard, it is important to know, or at least be able to look up, something about the uses of a particular wood. Even a relatively rare wood may be common in a certain application. Knowledge of the uses of different woods will help prevent mistakes of this kind. The final conclusion of a positive wood examination will be that the questioned and known woods are both the same type of wood.

Multiple Choice Questions

1. Forensic palynology refers to the examination of.
   (a) soil,
   (b) shoot and root,
   (c) flowers,
   (d) pollen and spores.
   Answer (d)
2. Scientist who first used the pollens in forensics.
   (a) Jeffry.
   (b) Paul Sear.
   (c) Bryant.
   (d) Horrocks.
   Answer (c)
3. Expertise of a forensic scientist must include the following.
   (a) Analysis of physical evidence.
   (b) Standard protocol of analysis.

(c) Expert witness.
(d) All of above.
Answer (d)

4. The age of tree can be estimated by.
   (a) Diameter of trunk.
   (b) Height of tree.
   (c) Number of annual rings.
   (d) biomass.
   Answer (c)
5. When exposed to air, which undergoes decay.
   (a) Dark wood.
   (b) Sap wood.
   (c) Light wood.
   (d) Hard wood.
   Answer (b)
6. The transfer of pollen from anther to stigma is.
   (a) Fertilization.
   (b) Pollination.
   (c) Diffusion.
   (d) All of above.
   Answer (b)
7. Can wood traces found stick on accidental vehicle be matched with wood found on scene of crime.
   (a) Yes.
   (b) Depends on environment condition.
   (c) Not at all.
   (d) In some cases.
   Answer (a)
8. Botanical evidence may be present on.
   (a) Victim's body and clothing.
   (b) Suspect vehicle and articles.
   (c) Stick on weapon of offence.
   (d) All of above.
   Answer (d)
9. Inappropriate and deliberate use of plants can cause homicides or accidental deaths.
   (a) Not possible.
   (b) Cannot comment.
   (c) Absolutely.
   (d) All of above.
   Answer (c)
10. Evan's blue dye stains the.
   (a) Live cells.
   (b) Dividing cells.

(c) Dead cells.
(d) None of above.
Answer (c)

Further Reading

Amankwaa AO, McCartney C (2021) The effectiveness of the current use of forensic DNA in criminal investigations in England and Wales. Wiley Interdiscip Rev 3(6):e1414. https://doi.org/10.1002/wfs2.1414

Aquila I, Ausania F, Serra A (2014) The role of forensic botany in crime scene investigation: case report and review of literature. J Forensic Sci 59:3

Bryant VM Jr, Jones GD (2006) Forensic palynology: current status of a rarely used technique in the Unites States of America. Forensic Sci Int 163:183

Craft KJ, Owners JD, Ashley MV (2007) Application of plant DNA markers in forensic botany: genetic comparison of Quercus evidence leaves to crime scene trees using microsatellites. Forensic Sci Int 165:65–70

Dunbar M, Murphy M (2009) DNA analysis of natural fibre rope. J Forensic Sci 54:1–6

Ishak S, Dormontt E, Young JM (2021) Microbes in forensic botany. Forensic Sci Med Pathol 17(2):297–307. https://doi.org/10.1007/s12024-021-00362-4

Thomas J. McClintock (2014) Forensic analysis of biological evidence. ISBN 9781466504561

Melton T, Holland M (2012) Forensic mitochondrial DNA analysis: current practice and future potential. Forensic Sci Rev 24:101–122

Mildenhall DC, Wiltshire PE, Bryant VM (2006) Forensic palynology: why do it and how it works. Forensic Sci Int 163(3):163–172

Milne L (2005) How Pollen Brought a Murderer to Justice. In: A grain of truth. Reel New Holland Publ, Sydney

Arya Neelam (2007) Forensic botany as a tool in crime investigation. Published in proceedings in XVIII all India forensic science Conference-2007, 270-73

Arya Neelam (2009) Green revolution: forensic botany as a Noval approach in crime investigation. Published in proceedings in XX all India forensic science Conference-2009, 15–20

Arya Neelam (2011) High potential of plant DNA analysis in forensics. Published in souvneir on biodiversity: challenges & opportunities

Rana AK (2018) Crime investigation through DNA methylation analysis: methods and applications in forensics. Egypt J Forensic Sci 8:7

Zhang ZL, Liang WB, Sun HB, Yang X, Ma LY (2021) Forensic application of plant evidence. Fa Yi Xue Za Zhi 37(1):87–90. https://doi.org/10.12116/j.issn.1004-5619.2019.490708

Forensic Entomology 29

Kamsalem Guite, Rutwik Shedge, Varsha Warrier, and Tanuj Kanchan

Abstract

Forensic entomology deals with the application of arthropods and insects in the investigation of legal matters. The areas of investigation include arthropods and insect infestation of raw and processed food items, invasion of human habitation and material structures, and infestation of dead and decomposing bodies. These forensic entomology branches are known as stored or food product entomology, urban entomology, and medico-legal entomology, respectively. Insects are known to exhibit a distinct pattern of succession in the decomposing body. The arrival of the first insect, the blowfly, sets off a biological clock that catalyzes the response of other insects to invade the corpse. The established successional pattern of insect infestation is useful in the determination of postmortem interval (PMI) in the medico-legal context. Moreover, medico-legal entomology is also used in the investigation of criminal acts such as rape, homicide, suicide, translocation of the dead body, drug abuse, abuse and neglect of the elderly, cruelty to animals, personal identification, insect species identification, and geographical origin of species. Factors such as temperature, presence of clothing and wrappings around the body, pH of the burial soil, and depth of burial can delay the arrival of insects and the rate of decomposition. The integration of scientific techniques in the field of forensic entomology has fundamentally expanded our understanding of these tiny creatures and enhanced their admissibility in the settlement of legal matters.

K. Guite (✉) · V. Warrier · T. Kanchan
Department of Forensic Medicine and Toxicology, All India Institute of Medical Sciences, Jodhpur, Jodhpur, Rajasthan, India

R. Shedge
School of Forensic Science, National Forensic Sciences University, Gandhinagar, Gujarat, India

A. Puri et al. (eds.), *Fundamentals of Forensic Biology*,
https://doi.org/10.1007/978-981-99-3161-3_29

Keywords

Forensic entomology · Stored · Food product entomology · Urban entomology · Medico-legal entomology · Successional pattern · Postmortem interval

29.1 Introduction

Entomology is the study of arthropods and insects. It is the branch of zoology that deals with the invertebrate phylum Arthropoda and the class Insecta. The word entomology is derived from the Greek word "*entomon*," meaning insect, and "*logos*," meaning discourse. Insects make up about 80% of all known animal species, more than a million, the largest group in the animal kingdom (Singh and Sachan 2007).

Entomology generally involves the study of the biology, habitation, mutation, and management of arthropods and insects in relation to the environment (Singh and Sachan 2007). Arthropods and insects are cosmopolitan in distribution, and they have high adaptability to various geographical areas. They play a vital role in the maintenance of ecological balance and the recycling of nature. Despite being numerous, less than 1% of the total known insect species are harmful to us (Omkar 2017). While some insects create problems for humans by destroying agricultural crops, carrying diseases, and sometimes causing fatal bites, others play a valuable role in the food, textiles, and cosmetic and medical industries (Vanin 2018).

In recent decades, arthropods and insects have been increasingly recognized by law to regulate legal matters. This is possible due to the extensive studies conducted by experts from various professions around the world. This introduced entomology into the field of crime solving, now known as forensic entomology.

Forensic entomology is a branch of forensic science and is associated with the utilization of arthropods and insects in a court of law (Lord and Rodriguez 1989). Forensic entomology can be divided into three categories: urban, store-product, and medico-legal entomology (Lord and Stevenson 1986; Catts and Goff 1992). Urban entomology deals with arthropods and insects that infest houses, buildings, and physical structures. Stored food/product entomology deals with arthropods and insects that infest crops, stored products, and processed food, and medico-legal entomology deals with arthropods and necrophagous insects in criminal cases such as rape, murder, estimation of PMI_{min} to name a few. Medico-legal entomology is the most discussed and researched area of forensic entomology (Magni et al. 2013). The utility of insects is diverse, and there is still much to be explored about them. This chapter focuses on the field of forensic entomology and its application to everyday life.

29.2 History of Forensic Entomology

The field of forensic entomology is ancient. The foundation of this field lies in the abundance of experimentation, research, and, above all, human curiosity to understand the purpose of the existence of these tiny creatures. The first documented case of forensic entomology came from China in 1235 (thirteenth century). Sung Tźu, a Chinese lawyer and death investigator, wrote in his medical law book, Hsi yuan chi lu "The washing away of wrongs," about the event of the crime. The case involved a stabbing homicide with a sickle. The criminal was convicted after a blowfly landed on the old sickle smeared with blood (Warrier and Shedge 2019).

In the late seventeenth century, the Italian physician and naturalist Francesco Redi confirmed that flies and maggots do not arise from meat, but from eggs laid by adult flies that were attracted to decaying flesh. In the mid-eighteenth century, the Swedish naturalist, Carolus Linnaeus, introduced the binomial nomenclature in 1735 (Gennard 2012). In the nineteenth century, Louis-François Bergerette, a French physician, made the first estimate of the postmortem interval on a corpse from his investigation of a mummified child found trapped in a chimney in 1850 (Benecke 2007).

In the following decades, investigations into death using insects continued. In 1878, a French pathologist Paul Camille Hippolyte Brouardel worked with Monsieur Perier and Jean Pierre Megnin. Together, they investigated the death of a newborn child infested with arthropods. Thereafter, Megnin published his books *Fauna des Tombeaux* (1887) and *La Faune des Cadavres* (1894) in which he described eight successive waves of insects colonizing dead bodies on land and two successional waves colonizing dead bodies in soil (Benecke 2007; Rivers and Dahlem 2014).

In 1881, 10 years before Megnin published his book, German physician Hermann Reinhard conducted an extensive study of insects in buried remains. His research led to the detection of Phorid flies, also known as Coffin flies, inhabiting buried or trapped remains (Gennard 2012). Around this time, in the United States, Murray Galt Motter examined the decomposition patterns of unearthed human remains and identified the role of different insect species, the effect of the depth of burial, and the soil conditions at the burial sites (Benecke 2007). In the twentieth century, Jean-Henri Casimir Fabre (1823–1915) published a collection of information on carrion beetles and necrophagous fly species. He named these collections, the accumulation of his research, *Souvenirs Entomologiques* (Souvenirs of Insect Life). Then, in 1916, J.M. Aldrich published "Sarcophaga and Allies." Twenty years later, D.G. Hall published "The Blowflies of North America" (Rivers and Dahlem 2014). After more than three decades, Adel S. Kamal's publication in 1958 characterized 13 species of necropsy insects, adding to the existing knowledge of insects of scientific importance (Kamal 1958). In 1965, Jerry Payne conducted the first experiment to study insect succession on the decomposing body using pig carcasses. Through subsequent efforts and initiatives by various other scientists, forensic entomology was recognized as a branch of forensic science and consequently accepted by the courts. In 1996, the American Board of Forensic Entomology (ABFE) was established. Later, the European Association for Forensic Entomology (EAFE) was founded in 2002 and the North American Forensic Entomology Association (NAFEA) organized its first meeting in Las Vegas in 2002 (Anderson 2014).

Recent advances in science and technology and their subsequent incorporation into the field of forensic entomology have increased our understanding of arthropods and insects. As a result, the importance of forensic entomology in criminal investigations will continue to grow.

29.2.1 Stored Food/Product Entomology

Stored food/product entomology deals with the investigation of insect pests found in raw and processed food items. The food items include cereals, pulses, seeds, nuts, dry fruits, and other kinds of durable items (Hagstrum and Subramanyam 2009). Typically, forensic cases of stored product entomology involve pest infestation at the time of harvesting, storing, processing, selling, purchasing, and consuming of the food items (Defilippo et al. 2017).

29.2.2 Urban Entomology

Urban entomology is the branch of forensic entomology that deals with arthropods and insects in and around human habitation (Rivers and Dahlem 2014). Here, the term "human habitation" not only comprises houses and infrastructures, but also the environment in which people live (Catts and Goff 1992). Urban entomology includes not only structural elements but also the problems of insect invasion of patients in healthcare facilities, pets, and farm animals (Keil 2015). Moreover, urban pests affect not only our well-being, but also man-made structures, landscapes, food security, and livestock (Gordon 2020). Furthermore, these pests make irritating noises and destroy our clothes, carpets, books, food items, furniture, etc. (Singh and Sachan 2007). They also cause painful bites and stings. Some feed on blood, while others transmit parasites and pathogens, causing entomophobia (Robinson 2005). Exposure to insect pests can induce allergic reactions (Gondhalekar 2019). Moreover, their bites/stings can cause discomfort and can sometimes be fatal.

29.3 Medico-Legal Entomology

When a dead body is recovered, the question of what, where, how, when, who, and why needs to be addressed by forensic experts. The methods of estimating the postmortem interval (PMI) using the normal changes in the body have limitations in that they can estimate the time since death within the first 72 h (Hall et al. 2012; El-Kady 1999). In such situations, arthropods and insects become the best alternative to answer the above questions. The first recorded case of medico-legal entomology happened in China in the thirteenth century. Since then, the application of insect knowledge in the medico-legal field has attracted considerable interest from researchers of various professional backgrounds around the world. Insect evidence is often involved in crimes such as rape, suicide, murder, and poaching. Also, there are deaths due to drug abuse and consumption of poisons and pesticides. Moreover, insect infestations on living individuals or animals are also relatively common. In all criminal activity where the evidence entails insect material, the nature of investigation falls under the domain of medico-legal entomology. Forensic medical entomology or forensic medico-legal entomology, popularly known as medicocriminal entomology (Hall 1990; Hall and Huntington 2009), is the branch of forensic entomology that deals with the utilization of arthropods and insects to estimate the minimum postmortem interval (PMI_{min}), determine the cause of death, and relate the evidence to the primary and secondary crime scenes (Catts and Goff 1992; Tüzün et al. 2015). Medico-legal entomology has the highest application among the three branches of forensic entomology (Magni et al. 2013).

The application of entomology to criminal investigation is myriad. Through their activity, insects can reveal abuse, neglect, and cruelty toward children, the elderly, and livestock. Besides, insects can also disclose the presence of wounds on a body based on their area of localization. Insects are also helpful in identifying if the murder is executed in a particular area and the corpse is relocated to a different geographical area (Charabidze et al. 2017). They can also assist to ascertain that a particular vehicle is being used to transport the dead body. Additionally, the presence of insect species from a particular geographic area on a cadaver in an area where that species is absent can provide evidence of translocation of the corpse. Furthermore, non-native insects and pollens recovered on the dead body can be analyzed for comparison with available data of the nearby geographical region. This can give a clue to the location of the primary crime scene (Volckaert 2020; Pinto et al. 2021). These miniature insects are highly adaptive and can withstand immolation (Malainey and Anderson 2020). As a result, they can provide evidence to indicate the cause of death in a fire or explosion case.

Blowflies (Calliphoridae) are usually the first insect to arrive on a dead body (Malejko et al. 2020). They are the pioneers of the successive pattern of insect colonization because they initiate the system that triggers the response of other insects and arthropods to the decomposing body. Megnin was the first scientist to observe the pattern of insect colonization on the dead body (Smith 1986). He reported that insects exhibit eight successive waves on land and two successive waves in soil. The necrophagous insects underwent four stages, i.e., egg, larval, pupal, and adult stages to attain maturation. The knowledge of the waves of insect successional patterns and the stages of insect development is paramount as they form the core application of insects in medico-legal entomology, most importantly, in the estimation of the PMI_{min} of a recovered body.

The criminal, often after the murder, tried to hide the victim's body by burial in a shallow trench. Since the decomposed body is rarely encountered, necrophagous insects developed a strong sense of smell and locomotion for their survival (Lord and Rodriguez 1989). If insects discover the buried body and gain access to it, depending upon the condition of the burial soil (whether the soil is loose or compact), the presence of physical material such as a coffin, clothing and wrappings around the

Fig. 29.1 Adult blue blowfly (*Calliphora vicina*). (Pest and Diseases Image Library, Bugwood.org)

Fig. 29.2 Adult green bottle fly (*Lucilia* spp.). (David Cappaert, Bugwood.org)

body, pH of the soil, and depth of burial, they can effectively colonize and decompose the body (Jordan and Tomberlin 2017; Mabika et al. 2014; Wang et al. 2017).

The most important insects in medico-legal entomology belong to the order Diptera, which includes blowflies (family Calliphoridae) and flesh flies (family Sarcophagidae). The species differ in abundance from region to region, habitat to habitat, and from season to season (Lord and Rodriguez 1989). There is a greater variety of arthropod species in the tropics than in urban areas of human habitation (Azwandi et al. 2013). Also, the insect species present may differ geographically. Other forensically significant arthropods include beetles (order Coleoptera), moths (order Lepidoptera), and mites (class Acari) (Figs. 29.1, 29.2, 29.3, 29.4, 29.5, 29.6, and 29.7) (Malejko et al. 2020).

Fig. 29.3 Adult flesh fly (Sarcophagidae)

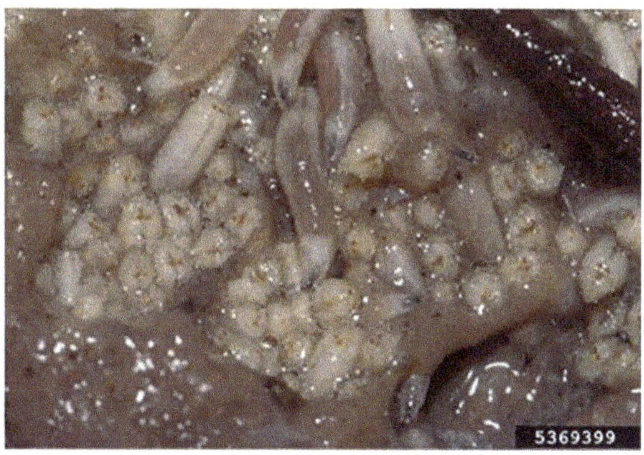

Fig. 29.4 Blowfly larvae. (Susan Ellis, Bugwood.org)

29.3.1 Important Insects in Medico-Legal Entomology

Arthropods and insects are ubiquitous and have a vast geographical habitat range. Insects are the first organisms to show up on an exposed dead body. The first insect to arrive, the adult blowfly, lays about 300 eggs at a time on the host (Sarwar 2020). Soon after hatching, the larvae feed on the host and assist in the release of volatile gases that attract other necrophagous insects. The feeding stage of the larvae is so intense that they devour the corpse within a few days or weeks. Sarwar (2020)

Fig. 29.5 Adult Dermestid beetle. (Daniel R. Suiter, University of Georgia, Bugwood.org)

Fig. 29.6 American carrion beetle (*Necrophilia americana*). (Gary Alpert, Harvard University, Bugwood.org)

mentioned that under favorable conditions, larvae can devour 60% of the human body in less than a week. To describe the numerous eggs produced by a single adult fly and the rigorous feeding habit of the larvae, Linnaeus once mentioned that three adult flies would decimate a horse as fast as a lion would (Benecke 2007). While the blowflies are oviparous, the flesh flies are larviparous. Though there are more than a million insect species in the world, only a handful of them is relevant to medico-legal entomology. Some of these forensically significant insects are listed in Table 29.1.

29.3.2 Classification of Forensic Arthropods and Insects

The insects that feed on dead bodies can be divided into four ecological groups: necrophagous species, predators, and parasite species that feed on necrophagous

Fig. 29.7 Adult skin beetle (*Trox unistriatus*). (Pennsylvania Department of Conservation and Natural Resources Forestry, Bugwood.org)

insects, omnivorous species that feed on the corpse and other insects, and adventive species (Smith 1986).

1. Necrophagous species: These species include Diptera such as Calliphorids and Sarcophagids, and Coleoptera such as Silphids and Dermestids. They are the most forensically significant group because of their primary association with the corpse (Hall et al. 2012).
2. Omnivores species: These species feed on the dead body and other infesting insects. The presence of enormous omnivorous species may cause delayed decomposition due to the dietary habit of other necrophagous insects (Early and Goff 1986). This group includes species of ants, wasp, and beetles.
3. Predators and parasites species: This group is considered second only to necrophagous species in terms of forensic importance (Smith 1986). Arthropods and insects in this group include species of Coleoptera (Silphids, Staphylinids, Histerids), Diptera (Calliphorids), Hymenoptera that feed on immature flies, and Mites (Macrochelids, Parasitids, Parholaspids, and Uropodids) that predates on other mites, nematodes, and insects (Catts and Goff 1992).
4. Adventive species: This group includes arthropod species that benefit from the corpse for their habitation. Arthropods in this group include spiders (which may

Table 29.1 Insects of forensic importance

Flies (Order Diptera)	
Blow flies (family Calliphoridae)	*Calliphora vicina* (Robineau-Desvoidy), *Calliphora vomitoria* (Linnaeus), *Chrysomya albiceps* (Wiedemann), *Chrysomya megacephala* (Fabricius), *Chrysomya rufifacies* (Macquart), *Cynomya mortuorum* (Linnaeus), *Lucilia illustris* (Meigen), *Phormia regina* (Meigen), *Lucilia sericata* (Meigen)
Flesh flies (family Sarcophagidae)	*Sarcophaga bullata* (Parker), *Sarcophaga haemorrhoidalis* (Fallen), *Sarcophaga argryostoma*, *Sarcophaga crassipalpis*, *Sarcophaga peregrina*, *Sarcophaga albiceps*, *Sarcophaga africa*, *Sarcophaga caerulescens*
Muscid flies (family Muscidae)	*Hydrotaea aenescens* (Wiedemann), *Hydrotaea leucostoma* (Wiedemann), *Musca domestica* (Linnaeus), *Muscina stabulans* (Robineau-Desvoidy), *Musca autumnalis* (De Geer), *Musca sorbens*
Skipper flies (family Piophilidae)	*Piophila casei* (Linnaeus), *Prochyliza* sp. (Walker)
Dung flies (family Scathophagidae)	*Scatophaga stercoraria* (Linnaeus)
Black scavenger flies (family Sepsidae)	*Sepsis* sp.
Small dung flies, minute scavenger flies (family Sphaeroceridae)	*Poecilosomella angulata* (Thomson)
Soldier flies (family Stratiomyidae)	*Hermetia illucens* (Linnaeus)
Humpbacked flies or scuttle flies (family Phoridae)	*Megaselia scalaris* (Loew), *Conicera tibialis* (Schmitz)
Moth flies, sand flies, and owl midges (family Psychodidae)	*Psychoda alternata* (Say)
Beetles (Order Coleoptera)	
Carrion beetles (family Silphidae)	*Heterosilpha ramosa* (Say), *Necrodes surinamensis* (Fabricius), *Necrophilia americana* (Linnaeus), *Nicrophorus orbicollis* (Say), *Oiceoptoma noveboracense* (Forster), *Thanatophilus sinuata* (Fabricius), *Thanatophilus rugosus* (Fabricius), *Nicrophorus vespilloides* (Herbst)
Skin beetles, leather beetles, hide beetles, carpet beetles, larder beetles (family Dermestidae)	*Dermestes ater* (DeGeer), *Dermestes maculatus* (DeGeer), *Dermestes lardarius*, *Dermestes frischii* (Kugelann)
Rove beetles (family Staphylinidae)	*Creophilus maxillosus* (Gravenhorst), *Platydracus fossator* (Gravenhorst), *Platydracus maculosus*
Clown beetles (family Histeridae)	*Hister noma* (Erichson), *Saprinus pensylvanicus*, *Arcinops pumilo* (Ericson), *Saprinus lugens* (Erichson)
Checkered beetles (family Cleridae)	*Necrobia rufipes* (DeGeer), *Korynete scaeruleus*

(continued)

Table 29.1 (continued)

Hide beetles (family Trogidae)	*Trox suberosus* (Fabricius), *Trox unistratus* (Beauvaris)
Scarab/dung beetles (family Scarabaeidae)	*Deltochilum gibbosum gibbosum* (Fabricius), *Phanaeus vindex* (MacLeay), *Deltochilum brasiliense* (Castelnau), *Eurysternus parallelus* (Castelnau), *Coprophanaeus (Megaphanaeus) ensifer* (Germar)

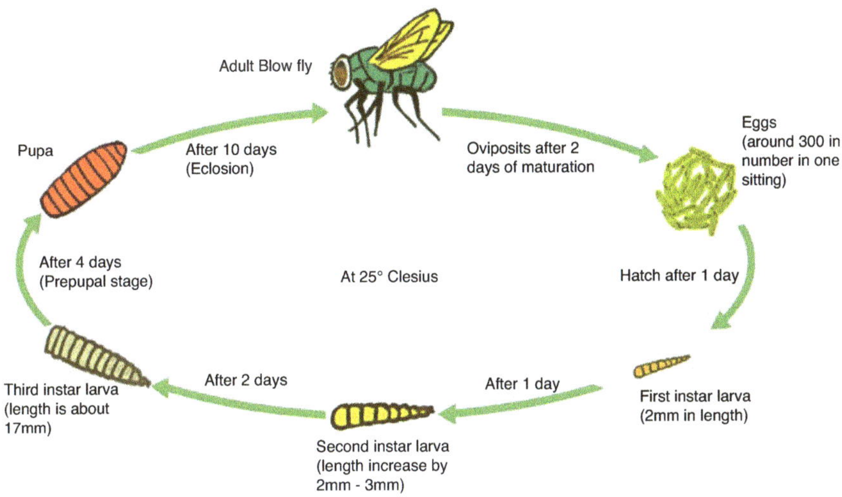

Fig. 29.8 Life cycle of blowfly

become incidental predators), centipedes, springtales, pill bugs, and mite species Acaridae, Lardoglyphidae, and Winterschmidtiidae (Smith 1986; Goff 1989).

29.3.3 Life Stages of Blowfly

The life cycle of a blowfly is a speedy process and usually spans between 2 and 4 weeks to attain maturation depending on the species (Cruz 2006; Dimitrov et al. 2020). The rate of development is temperature-dependent (Chen et al. 2019), i.e., the higher the temperature, the faster the growth and vice versa (Cruz 2006; Anderson 2014; Sarwar 2020). Blowfly exhibits four stages of development (Hall et al. 2012): the egg stage, the larva (maggot) stage, the pupa stage, and the adult stage (Dellinger and Day 2015). A brief description of the four developmental stages is given below (Fig. 29.8):

29.3.3.1 The Egg Stage

The eggs are small and slightly elongated. Their size ranges from 2 to 3 mm, and they look whitish yellow. They occur in clusters, especially in body openings, and are visible to the naked eye. The number of eggs is relatively less in winter, and they are concealed deep within the nostrils or underneath the eyelids of the host. The eggs hatch within 1–3 days. However, the time of hatching depends upon the species and the existent environmental conditions (Lord and Rodriguez 1989).

29.3.3.2 The Larva (Maggot Stage)

Larvae are the newborns from the eggs. They have a tapering anterior end that contains a pair of mouth hooks used for feeding and movement. The larvae have a pair of flattened structures called spiracles for breathing. The larval stage represents the active feeding and growing stage. The growth and development of the larvae involve three instars: first, second, and third instars, which are separated by a period of molting of the skin (Hall et al. 2012). The first instar larva has a pair of D-shaped posterior spiracles, while the second and third instar larvae have a pair of fan-shaped anterior spiracles that are dorsally situated in the thorax. At each stage of molting, the larva sheds the outer layer and internal cephalopharyngeal skeleton (mouthparts), and the lining of the tracheal system. Therefore, each larva in each larval stage has a new set of spiracular slits and mouthparts (Anderson and Cervenka 2002). The larvae mature within several days to several weeks depending upon the ambient temperature, the larval body temperature, and the heat emitted by the larval masses.

29.3.3.3 The Pupal Stage

At the end of the third instar, the larvae cease feeding and migrate to the nearby areas. They slow down their activity and contract their body into a shorter, thicker form that will eventually become the puparium (Cruz 2006). They bury themselves inside the soil or hide under surrounding objects. They secrete a hard outer skin or casting around their body known as pupae and undergo a transformation called metamorphosis. The pupal cases are football-shaped, reddish to dark brown, and can survive for more than a hundred years (Lord and Rodriguez 1989). Generally, the pupal cases are found buried in the soil beneath the corpse and around its remains.

29.3.3.4 The Adult Stage

After complete metamorphosis or transformation inside the pupae, the adult fly emerges from the pupae to continue the life cycle (Figs. 29.9, 29.10, and 29.11).

29.3.4 Fauna of Decomposing Body

The decomposition of a dead body is a long process. In general, decomposition lasts between a few weeks to several months and, in some cases, even years depending upon the environmental conditions. In their experimental study using pig carcasses, Tullis and Goff (1987) noticed five waves of successional arthropods.

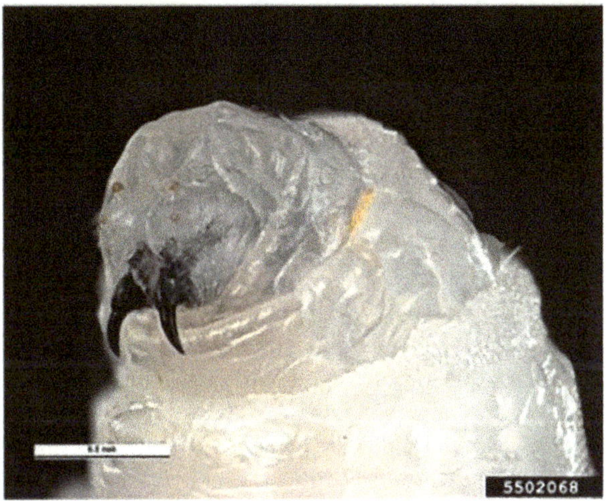

Fig. 29.9 Larva of flesh flies (*Sarcophaga* spp.). (Pest and Diseases Image Library, Bugwood.org)

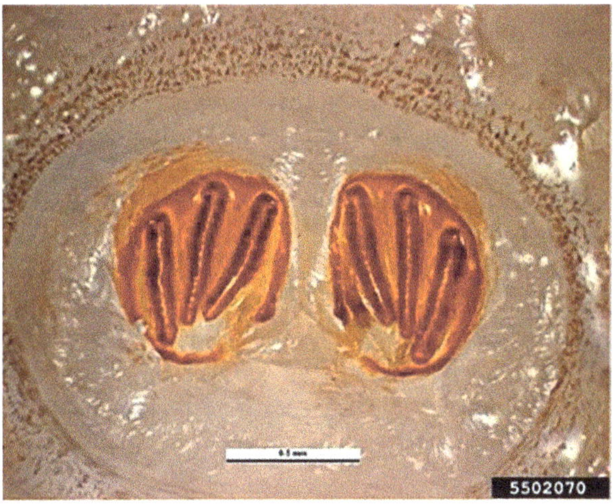

Fig. 29.10 Larva of flesh flies (*Sarcophaga* spp.). (Pest and Diseases Image Library, Bugwood.org)

Fig. 29.11 Pupa of Catalpa sphinx (*Ceratomia catalpae*). (Herbert A. "Joe" Pase III, Texas A&M Forest Service, Bugwood.org)

29.3.4.1 Fresh Stage (1–2 Days)

This stage begins immediately after death and ends with the onset of bloating. Although autolysis has begun, there is minimal morphological change. Arthropods observed in this stage include Calliphorids such as *Chrysomya megacephala* (F.) and *Chrysomya rufifacies*, *Drosophila (Sophophora)* sp., *Diploneura peregrina*, and *Creophilus maxillosus* (L.).

29.3.4.2 Bloated Stage (2–7 Days)

Putrefaction starts at this stage, and bloating occurs due to the action of anaerobic bacteria. The activity of colonizing insect maggots and the process of putrefaction increase the internal temperature of the carcass. This stage ends with the deflation of the carcass due to penetration of the abdominal wall by the feeding larvae. This stage harbors the largest number of adult Diptera. The predominant species were *C. megacephala* and *C. rufifacies*. Other insect species include *Parasarcophaga misera*, *Nabis lusciosus* (White), *C. maxillosus,* and *Thyreocephalus albertisi* (Fauvel).

29.3.4.3 Decay Stage (5–13 Days)

Most adult calliphorids leave the carcass at this stage. There is an increase in the internal temperature of the decomposing carcass to higher than the ambient temperature. The rise in internal temperature conforms to the development of putrefactive odor. This increase in temperature is due to the heat released by the accumulation of numerous maggots feeding on the carcass. The rise in temperature is followed by a rapid drop in the internal temperature to match the ambient temperature. At the end of this stage, the emitted odor diminishes. There is also a significant reduction in the weight of the carcass because of the feeding larvae and their subsequent relocation post-feeding to nearby places for pupation. This stage was dominated by the larvae of insects, which include the *Chrysomya* species, *C. rufifacies*, *C. megacephala*, *Ophyra* spp., *Ophyra aenescens* (Wiedemann), *Ophyra chalcogaster*, and *Atherigona orientalis* (Schiner).

29.3.4.4 Post-Decay Stage (10–23 Days)

This stage marks the departure of the matured larvae from the carcass. The carcass remain contains bones, hair, cartilage, tissues, and by-products of decay (BOD). The BOD acts as a source of food for visiting arthropods. The weight of the carcass continues to decline but not at a substantial level. Common arthropods found in this stage include *Leptocera punctipennis*, *Leptocera bifrons*, *Puliciphora lucifera* (Dahl), *P. australis*, and *Psychoda pseudoalternata* (Williams) and Dermaptera.

29.3.4.5 Remains Stage (18–90 Days and Above)

At this stage, the BOD is exhausted, and the remains consist of bones and a few cartilages. The transition from the post-decay stage to the remains stage is gradual. There is a sharp decline in the Dipteran population. Some of the entomofauna encountered in this stage include Psycholids, Dermaptera, Isopoda, and Ants.

29.3.5 Factors Affecting Insect Colonization in a Decomposing Body

The colonization of the decomposing body by arthropods and insects involves inter-species and intra-species interaction and their interaction with the physical environment. Broadly considered, species interaction in nature is influenced by two factors: biotic and abiotic (Jordan and Tomberlin 2017). The biotic factors include competition, predation, commensalism, pathogens, and resource quality. The abiotic factors include nutrition, temperature, humidity, Ph, salinity, and resource quality (Jordan and Tomberlin 2017). While biotic interaction is crucial for maintaining the balance between species, abiotic interaction essentially determines their chance of existence and continuation of the life cycle. The biogeoclimatic zone, also known as the geographic zone, is an important element that influences insect distribution. The biogeoclimatic zone includes both the biotic and abiotic factors such as vegetation, habitat, soil type, and meteorological condition such as temperature, humidity, and rainfall (Mabika et al. 2014). Of all the factors affecting the life cycle of insects, temperature appears to be the most crucial.

Insects are poikilothermic, meaning their body temperature changes with the ambient temperature (Hofer et al. 2017). Singh and Sachan (2007) noted that most insects survive anywhere between 0 and 50 °C, but assumably no single species can thrive across this range. The optimum temperature range for most species is between 22 and 38 °C. However, some species can survive temperatures above these ranges. For instance, some dipteran larvae can survive at 55 °C or more, but certain beetles can evolve at about 0 °C (Singh and Sachan 2007). Different insect species have different temperature ranges in which they can survive, outside of which they die. Insects are also known to tolerate high or low temperatures during certain stages of their life cycle. For example, many insects can survive in much colder temperatures in winter than in summer. Tropical species are generally less cold-hardy than temperate species. Terrestrial insects can usually tolerate a wider range of temperatures than aquatic insects. This is because the range of temperature change for terrestrial habitats is typically larger than that for aquatic habitats (Singh and Sachan 2007). Temperature also affects adult insect activity patterns, such as the timing of egg laying (Fauvergue et al. 2008) and larval development (Ames and Turner 2003). In addition to temperature and rain (Archer 2004), competition for food can either accelerate or retard insect colonization of decomposing bodies (Azwandi et al. 2013).

Seasons also contribute to the rate of body decomposition and the arrival of necrophagous insects. In general, the hot season attracts more insects to a corpse than the cold season. Low temperatures during the cold season can impede the hatching of eggs (Iancu et al. 2015). Structures such as coffins, clothing, body wrappings, and closed indoor scenes act as insulation from the outside environment (Hall et al. 2012; Lowe et al. 2013). The presence of these elements limits insects' access to the corpse and delays its decomposition. Furthermore, the clothed-covered part of the body dries slower than the unclothed part, resulting in fewer maggots over an extended period of time (Voss et al. 2011). Interestingly, the pH of the burial soil can either hasten or delay the decomposition rate. Higher pH (alkaline) soil tends to retard decomposition and vice versa (Lowe et al. 2013). The depth at which a body is buried is another factor that needs to be considered when determining the time of insect colonization. Buried corpses take a longer time to decompose. The diversity of the necrophagous insect species is found to vary with soil depth (Rysavy and Goff 2014). In general, arthropod diversity increases (Smith 1986; Simon 2019) with depth and with the long-term burial of cadavers (Simon 2019). The size of the decomposing body also determines the diversity of insect fauna (Silva et al. 2014) and the possibility of raising offspring by necrophagous insects as small carcasses provide little to no chance of generating offspring (Perotti and Braig 2009; Wang et al. 2017). In a study of organophosphate poisoning by Gunatilake and Lee Goff (1989), the investigators observed that the presence of Malathion in the tissues delayed insect colonization on the corpse for several days.

The insect colonization of decomposing bodies is enigmatic in the aquatic environment. There is a disparate pattern of insect succession in water and on land. Forensically significant aquatic insect includes Mayflies (Ephemeroptera), Caddisflies (Trichoptera), and True flies (Diptera) that is chiefly represented by Midges (Chironomidae) and Black flies (Simuliidae). These insect species are not obligatory carrion feeders; rather, they utilize the submerged flesh for food and nests (Amendt et al. 2009). Although seasonal changes appear to influence the pattern of aquatic insect succession, there seem to exist some confounding elements to it (Anderson and Hobischak 2004). Extensive research is imperative to further our understanding of aquatic insects for medico-legal use.

29.3.6 Estimation of Postmortem Interval (PMI)

The necrophagous species, blowflies (Diptera), are the first insect to arrive at a scene of death. They are considered the most useful evidence for the estimation of PMI_{min} (Hall et al. 2012). In warm weather, blowflies (Calliphoridae) can arrive within a few seconds or minutes (El-Kady 1999). Shortly after death, the body starts to break down through the process of autolysis. During this process, various gases such as hydrogen sulfide, carbon dioxide, ammonia, methane, sulfur dioxide, and hydrogen are released from the body (Smith 1986). The sulfur-based compounds attract insects to the body, while the ammonium-based compounds induce oviposition (Elden Elfeky et al. 2017). Flies have efficient sensory organs that help them detect these gases even within far-off distances (El-Kady 1999).

Upon arrival, blowflies would either begin laying eggs or, at first, feeding on the fluid emitted from the corpse. They then began laying eggs in body cavities such as the mouth, nose, ears, eyes, and around the exposed anus and genitals. This happens because blowflies cannot penetrate the skin due to their soft tongue. The body cavities provide a favorable environment for larval growth and development (Lord and Rodriguez 1989). The larvae grow in a semi-fluid environment and go through stages of transformation until maturity. Blowflies' activity triggered a biological clock that facilitated the continuous arrival of other arthropod and insect species. The analysis of the larval development and the foreseeable course of colonization of dead bodies by insects form the principal of PMI estimation (Catts and Goff 1992; Anderson and Cervenka 2002).

Conventionally, the postmortem interval is estimated from the changes on the external surface of the body. These include livor mortis, rigor mortis, lividity, and changes in the eye. However, when the body is in an advanced state of decomposition and the identity of the external appearances cannot be accessed, insects offer the best interpreting resource for determining the PMI. In general, there are two methods of estimating PMI using insects. The first is the pattern of successive waves of insect colonization and the second is the age and development of maggots (Anderson n.d.; Anderson and Cervenka 2002). The succession pattern data provide information on the range of time, while the development data provide information on the period of time that the body was colonized by insects (Wang et al. 2019). Because the duration of insect colonization does not provide accurate PMI (Amendt et al. 2007), the development data are modified to represent a segment of the total PMI. This refined concept is known as minimum postmortem interval (PMI_{min}) and is explained as the time between the colonization of the corpse by insects and the discovery of the oldest immature insect, i.e., when it was first laid as an egg or as larvae in some species (Villet et al. 2009).

If a body died between a month and a year or more, data on the successional waves of insect colonization are used. However, if the body died 1 month before discovery, maggots' age and their development are used as the time frame (Anderson n.d.). The latter is done by measuring the length of the oldest larvae and comparing it with reference data (Volckaert 2020). If blowfly eggs are encountered on a dead body at night or early in the morning, it is presumed that the body died the previous day or earlier (El-Kady 1999).

When entomological evidence is involved in a crime, it is important to gather data on insect activity, ambient temperature, barometric pressure, humidity, rain, depth of burial, and the presence of scavengers (Nihal Açikgöz 2016). This is crucial because higher temperatures (Elden Elfeky et al. 2017) and rainfall accelerate the decomposition process (Hofer et al. 2020).

The collection of data on the past ambient temperature from the nearest weather station is needed to reconstruct the rate of decomposition and period of insect succession (Michalski and Nadolski 2018). Precise temperature data at a crime scene are also critical for making accurate PMI_{min} and estimating the pre-appearance interval (PAI) of insect succession patterns (Hofer et al. 2017). The most reliable variable for the determination of PMI_{min} is the accumulated degree day (ADD) and the accumulated degree hours (ADH) based on the measurement of temperature (Nihal Açikgöz 2016). Alternatively, in the absence of these data, examination of the cuticular bands in blowfly (Tyndale-Biscoe and Kitching 1974), measurement of pteridine levels in adult screw-worm fly (Bernhardt et al. 2017), and assessment of the changes in the chemical composition of the empty puparia (Bajerlein et al. 2018; Wydra and Matuszewski 2021) can be used for estimating the PMI_{min}.

29.3.7 Entomology in Child Abuse, Neglect of the Elderly, and Cruelty to Animal

Medico-legal entomology is not only confined to the investigation of the dead. It also finds its utility in the living. It can deliver information on the location of the wound, the time the wound is inflicted, and if the nature of the wound characterizes abuse or neglect (Brundage and Byrd 2016). This knowledge is especially useful in the investigation of neglected and abused children and the elderly. The presence of head lice and maggots, often in huge numbers on a person, is a sign of neglect and lack of hygiene (Volckaert 2020). The infestation of injured living tissues by insect larvae is called myiasis. Zumpt (1965) defined myiasis as "the infestation of live human and vertebrate animals with dipterous larvae, which, at least/or a certain period, feed on the host's dead or living tissue, liquid body substances, or ingested food." Myiasis occurs due to the presence of untreated wounds or injuries. The maggots feed on the dead and dying tissues of the individual and can cause extensive damage if not attended to on time. By measuring the maggot size or length, it is possible to estimate the minimum time lapse since the individual child (e.g., dirty and soiled diapers) or elderly (e.g., infested bed sores) was last attended to (Anderson n.d.).

Insects also play an essential role in the investigation of cruelty to animals. Fur, in the absence of a wound, is the first area to be colonized by insects in neglected animals (Brundage and Byrd 2016). Traditionally, the presence of insect maggots (myiasis) in a living animal in a particular area is an indication of injury. The time of colonization is estimated either from the stages of larval development or the succession of insect species and is an indication of the period the animal has been neglected. Interestingly, the insects that cause myiasis are the same as those that colonize the decomposing bodies. Blowfly species namely the New World screwworm fly (*Cochliomyia hominivorax*) and the Old World screwworm fly (*Chrysomya bezziana*) are obligate parasites that cause myiasis. Other myiasis causing Old Worm insects include Window Gnat (*Anisopus fenestralis* Scopoli), *Phryne fenestralis* Lindner, *Psychoda Latreille*, *Trichoptera* Meigen, *Megaselia Rondani*, Common House fly (*Musca domestica* Linnaeus), *Lucilia sericata*, *Lucilia porphyrina* (Walker), *Sarcophaga haemorrhoidalis* (Fallen), *Sarcophaga albiceps* (Meigen), and *Gasterophilus intestinalis* (De Geer) (Zumpt 1965).

29.3.8 Geographical Habitat of Some Insects of Forensic Significance

Insects have a vast geographical distribution. Although cosmopolitan in distribution, they have their own ecological niche where they are found in abundance. However, the distribution of species is not uniform and local density differs from region to region (Charabidze et al. 2017). Their activity and richness are also determined by the season (Benbow et al. 2013; YuBo et al. 2018; Zou et al. 2022), vegetation (MacLeod and Donnelly 1957; Hwang and Turner 2009), and abundance of light

(Holdaway 1933; Barenbaum 1985) or shade (Nuorteva 1964; Barenbaum 1985). Vasconcelos and Araujo (2012) reported that the necrophagous insect species including *Hystriocnema plinthopyga* (Wiedemann, 1830), *Oxysacordexia avuncular, O. fluminensis* (Lopes, 1946), *O. intona* (Curran and Walley, 1934), *O. modesta* (Lopes, 1946), *Oxyvinia excisa* (Lopes, 1950), *Chloroprocta idioidea* (Robineau-Desvoidy, 1830), *Hemilucilia segmentaria* (Fabricius, 1805), *H. semidiaphana* (Randoni, 1850), and *Lucilia sericata* (Meigen, 1826) were exclusive to the forest of northeastern Brazil, whereas the urban areas were strongly associated with the Dipteran species such as *O. simplicoides* Lopes (1933), *O. thornax* (Walker, 1849), *Sarcodexia lambens* (Wiedmann, 1830), *Biopyrellia bipuncta* (Wiedemann, 1830), *Brontaea delecta* (Wulp, 1896), *Hydrotaea nicholsoni* (Curran, 1939), *Ophyra aenescens* (Wiedemann, 1830), and *L. cuprina*. Contrarily, the investigation of the death of 26 men by Pujol-Luz et al. (2006) reported the presence of only one fly species, *P. fulvinota* in the Brazilian Amazon rainforest. In the United States, Jeong et al. (2022) conducted a survey of Tennessee's abundant insect species and found *Phormia regina* (Meigen) and *Lucilia coeruleiviridis* (Macquart) to be the major insect species in the region.

Fremdt and Amendt (2014) highlighted the potential use of *Sarcophaga subvicina* Baranov and *Sarcophaga variegata* (Scopoli) as an indicator of summer urban habitation and *S. albiceps* as an indicator of rural habitation in the Frankfurt region of Germany. In addition, the authors also noted a strong association between *Sarcophaga caerulescens* and rural habitats and between *Sarcophaga similis* and urban habitats. Moore et al. (2022) stated that from the empty pupae of insects, the cuticular hydrocarbons (CHCs), being species-specific, can be used to identify and distinguish the geographical origin of *C. vicinia* from Germany, Spain, Norway, and England. They also emphasized the possibility of distinguishing *L. sericata* within Germany, which is isolated by a geographical range of 70 km, through the chemical analysis of cuticular hydrocarbons. Furthermore, Byrne et al. (1995) admitted that cuticular hydrocarbons can be used to distinguish species of *Phormia regina* from three geographic populations from Tucannon River and Lyle Grove, Washington, and Rensselaer, Indiana. An analysis of actual cases in insect records from 1993 to 2007 by Hodecek and Jakubec (2022) reported that *Calliphora vicina* was the dominant species throughout the Swiss season. In another study conducted in Switzerland by Cherix et al. (2012), the researchers found that *S. caerulescens, S. similis,* and *S. africa* were the dominant species in indoor death scenes, while *S. argyrostoma* dominated outdoor death scenes. In the middle east, in Central Iran, *Calliphora vicina*, *Lucilia sericata*, *Musca domestica*, *Wohlfahrtia nuba*, and *Chrysomya albiceps* were observed to be the dominant insect species (Mozaffari et al. 2020).

Several studies of these necrophagous insects have also been carried out in Asia, particularly in China. Shen et al. (2021) noted that *Protophormia terraenovae* (Robineau-Desvoidy, 1830) is the dominant year-round necrophagous insect species in the Lhasa (Qinghai-Tibetan Plateau) region of China. In another study conducted by Zou et al. (2022) in Shenyang, China, the researchers observed that in summer and autumn, the species of Phoridae and Platystomatidae were dominant at a burial depth of 30 and 60 cm. Moreover, the authors also reported that *Platystoma mandschuricum* (Enderlein, 1937) and *Aleochara puberula* (Klug, 1833) were the first insects to colonize the carcass in the spring season and the insect species to colonize the carcass for the longest time, extending their infestation till the winter months. A study conducted by Song et al. (2022) on Hainan Island, China, found that the dominant insect species was *Chrysomya megacephala* (Fabricius, 1794), followed in descending order by *Chrysomya rufifacies* (Macquart, 1843), *Hemipyrellia ligurriens* (Wiedemann, 1830), *Boettcherisca peregrine* (Robineau-Desvoidy, 1830), *Parasarcophaga dux* (Thomson, 1868), *Parasarcophaga misera* (Walker, 1849), *Synthesiomyia nudiseta* (Wulp, 1883), and *Ophyra chalcogaster* (Wiedemann, 1924). Due to the ongoing global warming and climate change, these insect species are likely to diversify their habitats and geographical distributions in the coming years. Because insects are resistant and highly adaptive, their unique ability to acclimatize to change can have implications for their service in forensic investigations, particularly in the identification of the relocation of bodies and the accompanying elements.

29.3.9 Insects as Indicators for Identifying Translocation of Dead Body, Determining the Presence of Wounds, and Tracing the Origin of Drugs

In recent decades, there has been a substantial increase in crime and insects provide valuable assistance in the investigation of such cases. Generally, crime scenes have blood stains, and in some instances, the victim's body has excretory materials due to the relaxation of the sphincter muscles during brutal death (Anderson 2014). The odor emanating from the blood and excreta attracted insects to the area. If the body is then transported elsewhere for disposal, the odor will further attract insect communities from the area to infest the corpse. This is the fundamental logic underlying the use of insects in determining whether a corpse was transported or disturbed after disposal, but the bulwark of forensic entomology is based on a practical approach. Therefore, this section is devoted to case studies involving insect evidence in the investigation of criminal activities set earlier in this chapter.

Nuorteva et al. (1967) and Smith (1986) presented a case where insects help link the possible area of death before the body has been moved to a different place. The body of a deceased woman was found on the floor near a closed window in her flat in Helsinki on the sixth of September 1964. The death occurred due to heart failure, 1 month prior to the discovery. Larvae collected from the skin surface during the autopsy on eight September 1964 were reared to adulthood. Upon maturity, the insect species were identified as *Calliphora vicina* (Robineau-Desvoidy) and *Lucilia sericata* (Meigen). However, the presence of the *Lucilia sericata* species suggested that the cadaver was transported to cities in southern Finland or an islet in the small archipelago of Finland, as the range of *Lucilia sericata* is restricted to these regions of Finland only. Through the analysis of insect evidence, it was concluded that the woman did not die in the house, but somewhere in other areas of Finland where the insect *Lucilia sericata* is abundant.

Another case was reported by Smith (1986) related to insects in the translocation and disturbance of a discarded body. The headless body of a woman was discovered among gorse and bracken in Devon, England, in September 1983. The significant absence of Calliphora larvae and puparia suggested that the body has been preserved in a warm, dry place, probably indoors, given the abundance of larvae and pupae of Ophyra species. When the head was recovered, it has a huge number of Calliphora species, but only one Ophyra species. Further interrogation of the suspect revealed that the woman was shot dead and kept in a sauna for 5 months. The body was then disposed of in the woods (where it was found) with the head removed at the scene and stored in a plastic bag in the boot of a car.

Gail and Gaudet (1999) and Cruz (2006) described a case where insects helped determine the occurrence of body translocation. The decomposing body of a woman was discovered in a dumping site in Maryland, the USA, in July 1984. The body was exposed to open daylight. Upon examination, the investigators observed that there were no species of *Lucilia sericata*, a green bottle fly that prefers lighted areas. However, the body was colonized by *Phormia regina*, a species of black blowfly that prefers shaded areas. From this insect evidence, the investigators concluded that the body was moved from a shaded area to a well-lighted area where it was finally discarded.

An interesting case presented by Anderson (2014) involved the remains of a man found in a shallow grave in the Lower Mainland area of British Columbia. The well-skeletonized body indicated that the body had been colonized by blowflies for an extended period of time. However, there was no evidence of puparial remains in the upper half of the body. A search for insect evidence on the second day of investigation uncovered multiple pupal cases associated with the lower half of the body. This suggested that the upper half of the body had decomposed at the same site as the lower half but had later been moved to a secondary location after skeletonization, probably because part of it had been exposed. The evidence indicated that the body was disturbed by reburying the upper half of the body several weeks after the original burial.

Another incident involving the translocation of a body as a cover-up for a crime was reported by Erzinclioglu (2003). Dr. Anthony Samson Perera was a specialist lecturer in Oral Biology at the University of Leeds, School of Dentistry, UK. He murdered and dismembered his 13-year-old adopted daughter Nilanthi from Sri Lanka. He buried part of her remains at his residence in Wakefield, West Yorkshire, England. Dr. Perera also preserved some of Nilanthi's bones in enamelware, stainless steel, and a coffee jar in the laboratory where he worked. A criminal investigation was initiated following a complaint made by Perera's neighbors and his co-worker, Frank Ayton. Forensic investigation revealed the presence of human vertebrae and rotting flesh in three indoor plant pots at his home. Human bones were also recovered under the living room floor. A search of the garden further revealed a recently disturbed burial spot behind the garage containing human remains. In addition, a microscopic examination of the mites in the decaying flesh unearthed from the garden and from beneath the living room floor revealed that parts of the body from the garden were later moved under the living room floor. Furthermore, the presence of larvae (endemic to human settlements) on bodies recovered from the garden, beneath the living room floor, and laboratory, indicated that the body parts were derived from the same individual. Moreover, the presence of larvae of spring-breeding fly species corresponds to the time around the disappearance of Nilanthi. The insect evidence helped convict Dr. Anthony Samson Perera, who was sentenced to life imprisonment on charges of manslaughter.

Not only can insects tell a "real story," but they also provide valuable information about drug trafficking. Illegal drugs are often manufactured in one country and sold in another, so knowing where the drug was made can be important. The case presented by Cruz (2006) demonstrated the utility of insects in identifying drug trafficking. Marijuana grown in the higher elevations of Colombia was trafficked to America. At the time of their entry into the United States, the drugs were seized by the Border Patrol agents. Further analysis of the marijuana bricks revealed that a unique spider found only in the mountainous highlands of Colombia was present in a pound of marijuana brick. This allows the investigator to track the source of the drug.

Another case from New Zealand reported by du Plessis and der Walt (2004) presented an example of how forensic entomology can be applied to an illicit drug investigation. Sixty insect specimens were found in two separate cannabis seizures. Of these, only one insect species was known to exist in New Zealand, while the other eight insect species were typical of Asia. By investigating the nature and extent of the overlap, investigators were able to determine that the cargo originated from the "Tenasserim region between the Andaman Sea in the west and Thailand in the east." It can even be assumed that cannabis was harvested near streams and lakes with nearby fig trees and termite nests. As a result of this evidence, one of the suspects admitted to trafficking the drugs.

Insects can also provide clues about the presence of wounds on corpses. One instance has been reported by Anderson (2014). A few days after going out for a walk, the body of a young woman was found. Her body had many maggots on her chest and palms. Regrettably, no forensic entomologist was summoned to investigate the incident and the victim was buried and listed as an uncertain death. When Dr. Bill Rodriguez, a Forensic Anthropologist was shown the photographs, he noticed that maggot activity was predominantly on the chest and palms rather than the face, suggesting the presence of scars in those areas. Moreover, the presence of maggots on the palms, an area where insects rarely colonize due to the hard skin, indicated the presence of defensive wounds. Based on this evidence, a court-ordered exhumation was carried out and the remains were reconsidered. Re-examination of the corpse revealed numerous stab wounds to the chest and severe lacerations on the hands that nearly severed one thumb. The case was subsequently revived as murder.

In another case described by Anderson (2014), insects help connect the perpetrator of the crime to the time it was committed. In a Chicago suburb case, a woman was raped by a man wearing a ski mask. The rape occurred in early to mid-summer. A suspect was identified, and a search of his house revealed the presence of a ski mask. The suspect admitted to possessing the mask, but he swore he had not worn it since last winter, months before the rape. However, investigators noticed several plant parts on the mask, including Cockleburs. When a forensic entomologist opened the Cockleburs, it was found to contain a live caterpillar. From the analysis of the life cycle of the caterpillar, the mask was found to be outdoors by the beginning of the summer of the same year. When presented with the evidence, the suspect confessed to committing the crime.

29.3.10 Entomotoxicology

Entomotoxicology is the qualitative and quantitative determination of toxic substances consumed through accidental, suicidal, or homicidal acts (Malejko et al. 2020). It also deals with the investigation of explosive compounds from larvae feeding on corpses (Cruz 2006), and the effects of toxins on the growth and development of insects (Gagliano-Candela and Aventaggiato 2001). Entomotoxicology involves finding the cause of death through the analysis of insects and larvae for the presence of drugs, poisons, pesticides, insecticides, and metals such as Mercury, Cadmium, and Thallium. (Gosselin et al. 2011). Insects are found in abundance and their puparial cases remain for a long time for analysis even when other sources such as tissues are no longer available (Gagliano-Candela and Aventaggiato 2001; Bourel et al. 2001). Besides, drugs tend to remain more stable in insects than in decaying human tissues (Kintz et al. 1990). The presence of drugs such as cocaine (Goff et al. 1989), heroin (Goff et al. 1991), methamphetamine (Goff and Lord 2009), amitriptyline (Goff et al. 1993), and 3,4 methylenedioxymethamphetamine (MDMA) (Goff et al. 1997) and morphine (Hédoui et al. 1999) in the food source is known to increase the larval size and either increase or decrease pupation time and larval mortality. The first report of the accumulation of metals such as copper, zinc, iron, and calcium in the tissues of house flies was made by Sohal and Lamb in the 1970s (Sohal and Lamb 1977, 1979). Later, in 1980, *J.C. Beyer* and colleagues (Beyer et al. 1980) detected the presence of drugs in fly larvae, and this subsequently became the foundation of entomotoxicology.

Blowfly, flesh fly, and beetle are the insects that are mainly used for toxicological analysis. The larvae, puparial cases, adult insects, exuviae, beetle fecal material (frass), fly predators, and scavengers are collected for toxicological analysis (Gagliano-Candela and Aventaggiato 2001). The method of extraction and preparation of insect samples is similar to the method used for human tissue (Kintz et al. 1990). Though there are various methods of extraction techniques, solid-phase extraction is known to give the best-purified organic toxicant from aqueous entomological extract (Gagliano-Candela and Aventaggiato 2001). The analysis and identification of the toxicological substance are done with the help of instruments such as high-performance liquid chromatography (HPLC), gas chromatography–mass spectrometry (GCMS), and liquid chromatography–mass spectrometry (LCMS), radioimmunoassay (RIA), and enzyme-linked immunosorbent assay (ELISA) (Moody 2006; Chophi et al. 2019).

29.3.11 Collection and Preservation of Insect Specimens

Proper collection and correct preservation of entomological artifacts are critical for insect species identification and PMI_{min} estimation. The investigators should be familiar with forensically relevant arthropods and insects. Moreover, it is necessary that precautions be taken at all steps of sample collection and handling. Essential items for collecting and storing insect specimens include forceps, teaspoons, vials or vented storage tubes, ethanol, a small sieve, a handheld insect capture net, protective clothing (disposable suits, gloves, shoe covers, face masks), tapes, notebook, pens, pencils, and digital camera (Hall et al. 2012). Lord and Rodriguez (1989) specified the process for the collection and preservation of insect specimens. They are discussed as follows.

29.3.11.1 Procedure

1. A sufficient number of representative insect species should be collected, consisting of larvae, pupae, and adults. It is important to catch the adults first before they flee the scene. A hand net should be used when capturing flying adults.

   Captured adults may be retained for analysis. When transferring adults directly for analysis, a stock solution of 70% ethanol or isopropyl alcohol and water in a 1:1 ratio should be used. This is necessary because isopropyl alcohol concentrations above 70% make the specimen brittle. If the specimen is stored in formalin, it must be transferred to 70% isopropyl alcohol as soon as possible.
2. Always wear gloves. Maggots on and inside the carcass can be collected with a spoon, forceps, or fingers. If the specimen is small (less than 5 mm), it can be collected with a small artist's paintbrush dipped in a preservation solution. Immature beetles do not need to be reared for identification and can be added directly to the preserving liquid.
3. Maggots under the carcass can be collected by scooping the top layer of soil with a spoon and storing it in a refrigerated plastic bag. Essentially, the purpose of using a refrigerated plastic bag is to slow larval growth and development.
4. In situations of advanced decomposition, it is important to collect soil at a suitable depth below where the corpse is placed. It is also important to collect a large amount of soil samples around the body. The soil should be stored in a plastic bag and refrigerated for further analysis.
5. In the case of bones, the cavities in the bone should be carefully examined. The skull can be examined with forceps and a light source after being placed on a white sheet or large piece of white paper. You can use water to wash the skull with a collection strainer or gauze underneath to capture maggots and insects.
6. Collected representative maggots should be divided into two groups—one group for immediate preservation and the other group grown to adulthood for species identification.
7. Place the to-be-preserved group in boiling water (approximately 76 or 170 °F) for 2–3 min. Then transfer to storage solution (70% ethanol).

8. The to-be-grown for identification group should be kept in a carton box, quarter to half filled with vermiculite or moist soil. This group should not be kept in sealed plastic bags or vials for more than 12 h.
9. Use the fastest mode of transportation to reach the rearing facility. Ordinary postal services are not recommended for this purpose but can be used in the absence of rapid transit.
10. At the rearing facilities, maggots can be successfully grown using animal liver and muscles taken from the carcass. Forceps must be used to transfer the maggots to the feed. Maggots are best kept in an environment that mimics the environment in which they were collected.
11. Beakers or small glass bowls with a diameter of 8–10 cm are suitable for rearing maggots. Each growing container should contain about 15–25 maggots.
12. Containers for preserved and grown specimens should be labeled in the order with the date of collection, time of collection, location of remains (along with a complete set of photographs of the crime scene), area of the body infested, name, address, and collector's phone number.
13. Maggots' activity is to be closely monitored and records kept of its size and instar stage. The feed can be provided as needed. Mature maggots move downward into the food material and pupate.
14. Adult flies emerge from the puparial case. Depending on the ambient temperature, this process may take weeks or months. It is important to place standard insect-rearing cases around the growing container to prevent adult insects from escaping.
15. Give the newly emerged flies a small amount of Gatorade soaked in cotton wool for 24 h. This will harden the outer skin and will allow for accurate identification of the insect species.
16. Later, these new adults can be kept in 70% ethanol or pinned to an insect box for storage.

29.4 Limitations of Forensic Entomology

Although the applications of entomology to forensics are massive, the discipline, like any other branch of science, is not without flaws. Volckaert (2020) and Catts and Goff (1992) pointed out the drawbacks of forensic entomology. Some of the highlights were stated as follows:

1. The whole process of specimen collection is an exhausting process.
2. Insect species may die before proper analysis. Also, the state of the corpse might make it impossible to gather live insects.
3. Lack of proper facilities for safe collection and proper transport can hinder the efficient collection of specimens.
4. Sometimes, there can be a potential absence of live insects if the remains are relatively ancient.

5. Maggots rearing is a time-consuming process and requires a certain temperature and humidity for their proper growth and development.
6. The estimated PMI_{min} is a mere representation of arthropod activity and may not necessarily correspond to the total postmortem interval.
7. The commonly used successional wave pattern of colonization to estimate PMI_{min} is not viable in cold climates and in situations where the body is frozen.
8. Improper handling or careless collection of specimens can damage or kill specimens.
9. The method of victim disposal has a large impact on insect colonization or lack thereof. For example, frozen bodies must be exposed to an open environment for some time to attract insects.
10. The colonization of a dead body by insects is considerably determined by the depth of burial and the physical environment of the corpse.
11. The estimation of accurate PMI_{min} does not depend only on the successful identification of the insect species. Many factors such as temperature, humidity, soil pH, presence of clothing, and wrappings play an equally important role in the determination of the PMI_{min}.
12. Since the distribution and activity of insects are geographically and seasonally determined, there can be lots of complications in calculating the PMI_{min}.

29.5 Acarology

Mites and ticks are not insects. They belong to the order Acari or class Acarina and are distinguished by the presence of chelicerae in the mouthpart and pedipalp (Singh and Sachan 2007). They are found in fresh and salt water, in our homes, and even in our bodies (Perotti et al. 2009). They are particularly useful when factors such as temperature and physical barriers retard the decomposition of the body by delaying insect invasion of the corpse. The fact that mites and ticks are present in humans and animals and their obligatory dependence on them made their utility in forensic investigation advantageous. Forensic acarology is part of forensic entomology that deals with the legal investigation of mites found on the dead in violent criminal cases (Rasmy 2007). Like insects, mites and ticks can proliferate and colonize the body. In 1894, Megnin witnessed the presence of mites on the dead body in the first wave of succession and their climax invasion in the sixth wave (Smith 1986; Perotti et al. 2009).

Some mites are phoretic, meaning that they depend on other species for their migration (Perotti and Braig 2009). They are believed to outnumber other arthropod and insect species that scavenge a dead body. Mites can arrive at the carcasses by walking, air currents, material transfer, and specific carrier host (Perotti et al. 2009). Each phoretic mite species utilizes a specific insect host for migration to arrive at its feeding destination, i.e., the corpse (Tüzün et al. 2015). For instance, blowflies are the specific host for the phoretic mites Prostigmata, Astigmata, and Mesostigmata. Mites can reproduce much faster than their carrier insects (Perotti et al. 2009). They are usually known to be transmitted in their transitional stage of development. Once they arrive at their destination, they leave the carrier insect and attain maturation shortly after. These mites feed on the eggs and larvae of the infesting insects on the corpse (Perotti and Braig 2009; Tüzün et al. 2015). During the early stages of decomposition, Macrochelid (mites that form a phoretic association with beetles and muscids) are sighted in abundance. Because of their specific affinity to arthropod and insect species for dispersion, they can provide information as to which insect has arrived even when the insects are not available. This can essentially contribute to finding the environmental condition of the crime scene and estimating the PMI_{min}.

Mites are also used as an indicator of human habitat, workplace conditions, and association with animals. Some mites from the order Metastigmata are non-phoretic (Tüzün et al. 2015). Also known as Ticks, they infest living vertebrates by sucking blood. The presence of these non-phoretic mites on a corpse can be an indication that the living individual was infected with this mite when alive. DNA analysis of these feeding mites can be used to determine the identity of the decedent. Mites that commonly infest humans include the family Demodicidae, which are present in the hair follicles, and the family Sarcoptidae, which are present in the skin and our clothes. They can survive for about 2 weeks after the death of their host (Tüzün et al. 2015). This can be useful when estimating the PMI_{min} and can provide other valuable information even if insect activity is impeded by physical barriers around the corpse.

29.6 Recent Advancements in Forensic Entomology

29.6.1 Wildlife Forensic Entomology

Wildlife exploitation in terms of poaching has become a global concern because of the extensive damage incurred by it (Obour et al. 2016). To combat the illegal poaching and trading of wildlife across international borders, the Convention on International Trade in Endangered Species of Wild Fauna and Flora (CITES) is established in 1973. Wildlife forensic entomology is the legal use of insects to investigate and solve illegal wildlife poaching. Wild animals are poached for their valuable body parts. For example, elephants are killed for their tusks, rhinoceros for their horns, pangolins for their scales, tigers for their skins, claws, and teeth (UNODC, World Wildlife Crime Report 2020), and bears for their gall bladders (Hall and Huntington 2009). These animal products are traded within countries and across international borders illegally. A proper understanding of the arthropods and insect species distribution in a geographical location can help relate investigators to the area where the poaching occurred. For instance, the species of blue bottle fly *Calliphora alaskensis* is found in the cold northern regions of Canada, Alaska, Colorado, and Wyoming, while the species of green bottle fly *Lucilia Mexicana* is found in the warm areas in the southern regions of Texas through Brazil (Brundage and Byrd 2016). The presence of these fly species on animals outside their geographic range indicates that the animal is not endemic to the region of recovery.

29.6.2 DART-HRMS Technology

Direct analysis in real-time high-resolution mass spectrometry (DART-HRMS) technique involves thermal desorption, penning ionization, and transferring of charge from metastable helium into the mass spectrometer by helium-heated gas (Doué et al. 2015). The technique is non-destructive and does not require sample preparation. Furthermore, the technique enables the instant differentiation of various insect species including Calliphoridae, particularly *Calliphora vicina*, *Lucilia sericata*, *L. coeruleiviridis*, and *Phormia regina* species as well as the Phoridae and Sarcophagidae families from the amino acid profile of the insect eggs (Giffen et al. 2017).

29.6.3 Molecular Identification of Insect Species

Molecular methods are used to determine insect species when morphological determination is not possible due to damage to the specimen. Molecular identification of insect species can be done from the larvae and does not require the rearing of the larvae to adulthood (Chen et al. 2004). The technique involves the sequencing of the mitochondrial DNA, particularly the mitochondrial cytochrome C oxidase subunit I (COI), cytochrome oxidase II (COII), and t-RNA leucine of the larvae (Wells and Stevens 2008; Cainé et al. 2009; Kavitha et al. 2012, 2013; Park et al. 2018b). Besides, the molecular technique also utilizes single-nucleotide polymorphisms (SNPs) (Jang et al. 2019) and nuclear DNA markers such as the internal transcribed spacer 2 (ITS2) (Park et al. 2018a) for species identification.

29.6.4 Human DNA Identification from Larval Gut Contents

Larvae feeding on decomposing human tissue store their feed in the gut. By analyzing their gut contents with molecular techniques, the mitochondrial DNA (mtDNA) of humans can be extracted from the hypervariable segment 2 (HV2) of mtDNA of the larvae (Wells et al. 2001). This can help in the identification of the decedent when the morphological features are badly decomposed and physical identification is not possible.

29.6.5 Virtual Forensic Entomology

The method involves the incorporation of micro-CT to visualize and comprehend the internal anatomical changes inside the puparial case of the developing pharate adult (Richards et al. 2012). The technique requires staining the pupae with iodine prior to observation under low contrast resolution. This novel approach has the advantage in that it is relatively fast, non-destructive, and can process multiple specimens at a time. The method can also be used for estimating PMI_{min} when the insect evidence involves pupae.

Multiple Choice Questions

1. Forensic entomology is the study of?
   (a) Arthropods
   (b) Pollen grains
   (c) Diatoms
   (d) Decomposition
   Answer: A
2. Which of the following insect families is the first to reach a cadaver?
   (a) Beetles
   (b) Mites
   (c) Aphids
   (d) Flies
   Answer: D
3. Where are clumps of eggs of flies most commonly observed in a dead body?
   (a) In the hair
   (b) Near natural orifices
   (c) Underarms
   (d) Sole of the feet
   Answer: B
4. Presence of large number of larvae over the chest of a cadaver indicates ________
   (a) Signs of lack of hygiene
   (b) Presence of open wounds
   (c) Heart disease
   (d) Asphyxial death
   Answer: B
5. Study of mites and ticks is called?
   (a) Entomology
   (b) Palynology
   (c) Limnology
   (d) Acarology
   Answer: D

References

Amendt J, Campobasso CP, Gaudry E et al (2007) Best practice in forensic entomology- standards and guidelines. Int J Legal Med 121:90–10. https://doi.org/10.1007/s00414-006-0086-x

Amendt J, Zehner R, Johnson DG, Wells J (2009) Future trends in forensic entomology. In: Amendt J et al (eds) Current concepts in forensic entomology. Springer, New York, NY, pp 353–368

Ames C, Turner B (2003) Low temperature episodes in development of blowflies: implications for postmortem interval estimation. Med Vet Entomol 17:178–186. https://doi.org/10.1046/j.1365-2915.2003.00421.x

Anderson GS (2014) Forensic entomology. In: James SH, Nordby JJ, Bell S (eds) Forensic science: an introduction to scientific and investigative techniques, 4th edn. CRC Press/Taylor & Francis Group, Boca Raton, FL
Anderson GS (n.d.) Forensic entomology: the use of insects in death investigations. https://www.sfu.ca/~ganderso/forensicentomology.htm#_ftnref1. Accessed 28 Sept 2022
Anderson GS, Cervenka VJ (2002) Insects associated with the body: their use and analyses. In: Haglund WD, Sorg MH (eds) Advances in forensic taphonomy: method, theory and archaeological perspectives. CRC Press, Boca Raton, FL
Anderson GS, Hobischak NR (2004) Decomposition of carrion in the marine environment in British Columbia, Canada. Int J Legal Med 118:206–209. https://doi.org/10.1007/s00414-004-0447-2
Archer MS (2004) Rainfall and temperature effects on the decomposition rate of exposed neonatal remains. Sci Justice 44(1):35–41. https://doi.org/10.1016/s1355-0306(04)71683-4
Azwandi A, Keterina N, Owen H (2013) Adult carrion arthropod community in a tropical rainforest of Malaysia: analysis on three common forensic entomology animal models. Trop Biomed 30: 481
Bajerlein D, Taberski D, Matuszewski S (2018) Estimation of postmortem interval (PMI) based on empty puparia of *Phormia regina* (Meigen) (Diptera: Calliphoridae) and third larval stage of *Necrodes littoralis* (L.) (Coleoptera: Silphidae) – advantages of using different PMI indicators. J Forensic Legal Med 55:95–98. https://doi.org/10.1016/j.jflm.2018.02.008
Barenbaum MR (1985) Ninety-nine more maggots, mites, and munchers. University of Illinois Press, Urbana, IL; Chicago, IL, pp 10–46
Benbow ME, Lewis AJ, Tomberlin JK, Pechal JL (2013) Seasonal necrophagous insect community assembly during vertebrate carrion decomposition. J Med Entomol 50:440–450. https://doi.org/10.1603/ME12194
Benecke M (2007) A brief survey of the history of forensic entomology/Ein kurzer Streifzug durch die Geschichte der forensischen Entomologie. Acta Biologica Benrodis 14:15
Bernhardt V, Hannig L, Kinast R et al (2017) Quantitative pteridine fluorescence analysis: a possible age-grading technique for the adult stages of the blow fly *Calliphora vicina* (Diptera: Calliphoridae). J Insect Physiol 98:356–359. https://doi.org/10.1016/j.jinsphys.2017.03.002
Beyer JC, Enos WF, Stajić M (1980) Drug identification through analysis of maggots. J Forensic Sci 25(2):411
Bourel B, Tournel G, Hedouin V et al (2001) Morphine extraction in necrophagous insects remains for determining ante-mortem opiate intoxication. Forensic Sci Int 120:127–131. https://doi.org/10.1016/s0379-0738(01)00428-5
Brundage A, Byrd JH (2016) Forensic entomology in animal cruelty cases. Vet Pathol 53(5): 898–909. https://doi.org/10.1177/0300985816651683
Byrne AL, Camann MA, Cyr TL et al (1995) Forensic implications of biochemical differences among geographic populations of the black blow fly, *Phormia Regina* (Meigen). J Forensic Sci 40:372. https://doi.org/10.1520/jfs13789j
Cainé LM, Corte Real F, Saloña-Bordas MI et al (2009) DNA typing of Diptera collected from human corpses in Portugal. Forensic Sci Int 184:e21. https://doi.org/10.1016/j.forsciint.2008.10.016
Catts EP, Goff ML (1992) Forensic entomology in criminal investigations further annual reviews. Annu Rev Entomol 37:253–272
Charabidze D, Gosselin M, Hedouin V (2017) Use of necrophagous insects as evidence of cadaver relocation: myth or reality? PeerJ 5:e3506. https://doi.org/10.7717/peerj.3506
Chen WY, Hung TH, Shiao SF (2004) Molecular identification of forensically important blow fly species (Diptera: Calliphoridae) in Taiwan. J Med Entomol 41:47–57
Chen W, Yang L, Ren L et al (2019) Impact of constant versus fluctuating temperatures on the development and life history parameters of *Aldrichina grahami* (Diptera: Calliphoridae). Insects 10:184. https://doi.org/10.3390/insects10070184

Cherix D, Wyss C, Pape T (2012) Occurrences of flesh flies (Diptera: Sarcophagidae) on human cadavers in Switzerland, and their importance as forensic indicators. Forensic Sci Int 220:158–163. https://doi.org/10.1016/j.forsciint.2012.02.016

Chophi R, Sharma S, Sharma S, Singh R (2019) Forensic entomotoxicology: current concepts, trends and challenges. J Forensic Legal Med 67:28–36

Cruz AM (2006) Crime scene intelligence: an experiment in forensic entomology. The National Defense Intelligence College Press, Washington, DC

Defilippo F, Calzolari M, Bonilauri P (2017) Stored-product forensic entomology: study of development of *Plodia Interpunctella* (Lepidoptera: Pyralidae) and use of data in determining time intervals of food infestation. Poster presented at the 14th Meeting European Association for Forensic Entomology, Treviso, Italy, 7–10 June 2017

Dellinger TA, Day E (2015) Blow flies Diptera: Calliphoridae. Virginia Cooperative Extension Virginia Tech. Virginia State University. https://www.pubs.ext.vt.edu/content/dam/pubs_ext_vt_edu/ENTO/ENTO-134/ENTO-134-PDF.pdf. Accessed 3 Oct 2022

Dimitrov RE, Amendt J, Rothweiler F, Zehner R (2020) Age determination of the adult blow fly *Lucilia sericata* (Diptera: Calliphoridae) through quantitative pteridine fluorescence analysis. Foren Sci Med Pathol 16:641–648. https://doi.org/10.1007/s12024-020-00295-4

Doué M, Dervilly-Pinel G, Pouponneau K et al (2015) Direct analysis in real time - high resolution mass spectrometry (DART-HRMS): a high throughput strategy for identification and quantification of anabolic steroid esters. Drug Test Anal 7:603–608. https://doi.org/10.1002/dta.1727

Early M, Goff ML (1986) Arthropod succession patterns in exposed carrion on the island of O'ahu, Hawaiian Islands, USA. J Med Entomol 23(5):520–531

Elden Elfeky AK, Moharm IM, El Aswad BEDW (2017) Role of house fly in determination of post-mortem interval: an experimental study in albino rats. Ain-Shams J Forensic Med Clin Toxicol 28:133–143

El-Kady EM (1999) Problems facing application of forensic entomology. Pak J Biol Sci 2:280

Erzinclioglu Z (2003) Maggots, murder, and men: memories and reflections of a forensic entomologist. St. Martin's Griffin, New York, NY

Fauvergue X, Lo Genco A, Lo Pinto M (2008) Virgins in the wild: mating status affects the behavior of a parasitoid foraging in the field. Oecologia 156:913–920. https://doi.org/10.1007/s00442-008-1037-7

Fremdt H, Amendt J (2014) Species composition of forensically important blow flies (Diptera: Calliphoridae) and Flesh Flies (Diptera: Sarcophagidae) through space and time. Forensic Sci Int 236:1–9. https://doi.org/10.1016/j.forsciint.2013.12.010

Gagliano-Candela R, Aventaggiato L (2001) The detection of toxic substances in entomological specimens. Int J Legal Med 114:197–203

Gail A, Gaudet J (1999) A practical exercise in forensic entomology. RCMP Gazette 53:10–14

Gennard D (2012) Forensic entomology: an introduction, 2nd edn. Wiley-Blackwell, London

Giffen JE, Rosati JY, Longo CM, Musah RA (2017) Species identification of necrophagous insect eggs based on amino acid profile differences revealed by direct analysis in real time-high resolution mass spectrometry. Anal Chem 89:7719–7726. https://doi.org/10.1021/acs.analchem.7b01708

Goff ML (1989) Gamasid mites as potential indicators of postmortem interval. In: Channabasavanna GP, Viraktamath CA (eds) Progress in acarology, vol 1. Brill, Leiden, pp 443–450

Goff ML, Lord WD (2009) Insects as toxicological indicators and the impact of drugs and toxins on insect development. In: Byrd JH, Castner JL (eds) Forensic entomology: the utility of arthropods in legal investigations, 2nd edn. CRC Press, Boca Raton, FL

Goff ML, Omori AI, Goodbrod JR (1989) Effect of cocaine in tissues on the development rate of *Boettcherisca Peregrina* (Diptera: Sarcophagidae). J Med Entomol 26:91–93. https://doi.org/10.1093/jmedent/26.2.91

Goff ML, Brown WA, Hewadikaram KA, Omori AI (1991) Effect of heroin in decomposing tissues on the development rate of *Boettcherisca Peregrina* (Diptera, Sarcophagidae) and implications of this effect on estimation of postmortem intervals using arthropod development patterns. J Forensic Sci 36:537–542

Goff ML, Brown WA, Omori AI, LaPointe DA (1993) Preliminary observations of the effects of amitriptyline in decomposing tissues on the development of *Parasarcophaga ruficornis* (Diptera: Sarcophagidae) and implications of this effect to estimation of postmortem interval. J Forensic Sci 38:316–322

Goff ML, Miller ML, Paulson JD et al (1997) Effects of 3,4-methylenedioxymethamphetamine in decomposing tissues on the development of Parasarcophaga ruficornis (Diptera: Sarcophagidae) and detection of the drug in postmortem blood, liver tissue, larvae, and puparia. J Forensic Sci 42:276–280

Gondhalekar AD (2019) 2018 Highlights of urban entomology. J Med Entomol 56:1188–1193

Gordon JR (2020) Urban entomology highlights from 2019 help create integrated pest management plans. J Med Entomol 57(5):1342–1348. https://doi.org/10.1093/jme/tjaa121

Gosselin M, Wille SMR, Fernandez MMR et al (2011) Entomotoxicology, experimental set-up and interpretation for forensic toxicologists. Forensic Sci Int 208:1–9

Gunatilake K, Lee Goff L (1989) Detection of organophosphate poisoning in a putrefying body by analyzing arthropod larvae. J Forensic Sci 34:714. https://doi.org/10.1520/jfs12698j

Hagstrum DW, Subramanyam B (2009) A review of stored-product entomology information sources. Am Entomol 55(3):174–183

Hall RD (1990) Medicocriminal entomology. In: Catts EP, Haskell NH (eds) Entomology and death: a procedural guide. Joyce's Print Shop, Inc, Clemson, SC

Hall RD, Huntington TE (2009) Introduction: perceptions and status of forensic entomology. In: Byrd JH, Castner JL (eds) Forensic entomology: the utility of arthropods in legal investigations, 2nd edn. CRC Press, Boca Raton, FL

Hall M, Whitaker A, Richards C (2012) 8 Forensic entomology. In: Grant NM, Roberts J (eds) Forensic ecology handbook: from crime scene to court, 1st edn. Wiley, pp 111–140

Hédoui V, Bourel B, Martin-Bouyer L et al (1999) Morphine perfused rabbits: a tool for experiments in forensic entomotoxicology. J Forensic Sci 44(2):347–350

Hodecek J, Jakubec P (2022) Spatio-temporal distribution and habitat preference of necrophagous Calliphoridae based on 160 real cases from Switzerland. Int J Legal Med 136:923–934. https://doi.org/10.1007/s00414-021-02769-8

Hofer IMJ, Hart AJ, Martín-Vega D, Hall MJR (2017) Optimising crime scene temperature collection for forensic entomology casework. Forensic Sci Int 270:129–138. https://doi.org/10.1016/j.forsciint.2016.11.019

Hofer IMJ, Hart AJ, Martín-Vega D, Hall MJR (2020) Estimating crime scene temperatures from nearby meteorological station data. Forensic Sci Int 306:110028. https://doi.org/10.1016/j.forsciint.2019.110028

Holdaway FG (1933) Differential behaviour of *Lucilia sericata* Meig. and *Lucilia Caesar* L. in natural environments. J Anim Ecol 2:263. https://doi.org/10.2307/962

Hwang CC, Turner BD (2009) Small-scaled geographical variation in life-history traits of the blowfly *Calliphora vicina* between rural and urban populations. Entomologia Experimentalis et Applicata 132:218–224. https://doi.org/10.1111/j.1570-7458.2009.00891.x

Iancu L, Carter DO, Junkins EN, Purcarea C (2015) Using bacterial and necrophagous insect dynamics for post-mortem interval estimation during cold season: novel case study in Romania. Forensic Sci Int 254:106–117. https://doi.org/10.1016/j.forsciint.2015.07.024

Jang H, Shin SE, Ko KS, Park SH (2019) SNP typing using multiplex real-time PCR assay for species identification of forensically important blowflies and flesh flies collected in South Korea (Diptera: Calliphoridae and Sarcophagidae). Biomed Res Int 2019:6762517. https://doi.org/10.1155/2019/6762517

Jeong Y, Weidner LM, Pergande S et al (2022) Biodiversity of forensically relevant blowflies (Diptera: Calliphoridae) at the anthropology research facility in Knoxville, Tennessee, USA. Insects 13:109. https://doi.org/10.3390/insects13020109

Jordan HR, Tomberlin JK (2017) Abiotic and biotic factors regulating inter-kingdom engagement between insects and microbe activity on vertebrate remains. Insects 8(54):1–19

Kamal AS (1958) Comparative study of thirteen species of sarcosaprophagous Calliphoridae and Sarcophagidae (Diptera) I. Bionomics Ann Entomol Soc Am 51:261–271. https://doi.org/10.1093/aesa/51.3.261

Kavitha R, Nazni WA, Chye T et al (2012) Molecular identification of blow flies recovered from human cadavers during crime scene investigations in Malaysia. Malay J Pathol 34:127

Kavitha R, Tan TC, Lee HL et al (2013) DNA typing of Calliphorids collected from human corpses in Malaysia. Trop Biomed 30:119

Keil CB (2015) Research needs for forensic entomology in Ecuador. REMCB 36:71–78

Kintz P, Tracqui A, Mangin P (1990) Toxicology and fly larvae on a putrefied cadaver. J Foren Sci Soc 30:243–246

Lord WD, Rodriguez WC (1989) Forensic entomology: the use of insects in the investigation of homicide and untimely death. Prosecutor 22(3):41–48

Lord WD, Stevenson JR (1986) Directory of forensic entomologists, 2nd edn. Defense Pest Information Analysis Center, Walter Reed Army Medical Center, Washington, DC

Lowe AC, Beresford DV, Carter DO et al (2013) The effect of soil texture on the degradation of textiles associated with buried bodies. Forensic Sci Int 231:331–339. https://doi.org/10.1016/j.forsciint.2013.05.037

Mabika N, Masendu R, Mawera G (2014) An initial study of insect succession on decomposing rabbit carrions in Harare, Zimbabwe. Asian Pac J Trop Biomed 4(7):561–565. https://doi.org/10.12980/APJTB.4.2014C1031

Macleod J, Donnelly J (1957) Some ecological relationships of natural populations of calliphorine blowflies. J Anim Ecol 26:135. https://doi.org/10.2307/1786

Magni P, Guercini S, Leighton A, Dadour I (2013) Forensic entomologists: an evaluation of their status. J Insect Sci 13:78

Malainey SL, Anderson GS (2020) Effect of arson fires on survivability of entomological evidence on carcasses inside vehicle trunks. Forensic Sci Int 306:110033. https://doi.org/10.1016/j.forsciint.2019.110033

Malejko J, Deoniziak K, Tomczuk M et al (2020) Puparial cases as toxicological indicators: bioaccumulation of cadmium and thallium in the forensically important blowfly *Lucilia sericata*. Front Chem 8:586067. https://doi.org/10.3389/fchem.2020.586067

Michalski M, Nadolski J (2018) Thermal conditions in selected urban and semi-natural habitats, important for the forensic entomology. Forensic Sci Int 287:153–162. https://doi.org/10.1016/j.forsciint.2018.03.042

Moody DE (2006) Immunoassays in forensic toxicology. In: Encyclopedia of analytical chemistry. Wiley, New York, NY, pp 1–34. https://doi.org/10.1002/9780470027318.a1109

Moore H, Lutz L, Bernhardt V et al (2022) Cuticular hydrocarbons for the identification and geographic assignment of empty puparia of forensically important flies. Int J Legal Med 136:1791–1800. https://doi.org/10.1007/s00414-022-02786-1

Mozaffari E, Saghafipour A, Arzamani K et al (2020) Geographical distribution, biodiversity, and species richness of medically important necrophagous flies in Central Iran. J Med Entomol 57:377–381. https://doi.org/10.1093/jme/tjz203

Nihal Açikgöz H (2016) Malpractices concerning collection of forensic entomological evidence and importance of standardization in Turkey. Foren Res Criminol Int J 3(1):237–241. https://doi.org/10.15406/frcij.2016.03.00080

Nuorteva P (1964) Differences in the ecology of *L. caesar* (L.) and *L. illustris* (Meig.) (Diptera, Calliphoridae) in Finland. Wiad Parazytol 10:583–587

Nuorteva P, Isokoski M, Laiho K (1967) Studies on the possibilities of using blowflies (Dipt.) as medico-legal indicators in Finland. Annales Entomologici Fennici 33:217–225

Obour R, Asare R, Ankomah P, Larson T (2016) Poaching and its potential to impact wildlife tourism: an assessment of poaching trends in the Mole National Park in Ghana. Am J Transplant 3:169–192. https://doi.org/10.30958/ajt.3-3-1

Omkar (2017) Industrial entomology. Springer, New York, NY
Park JH, Kim YH, Ham CS et al (2018a) Molecular identification of forensically important calliphoridae and sarcophagidae species using ITS2 nucleotide sequences. Forensic Sci Int 284:1–4. https://doi.org/10.1016/j.forsciint.2017.12.017
Park JH, Shin SE, Ko KS, Park SH (2018b) Identification of forensically important Calliphoridae and Sarcophagidae species collected in Korea using SNaPshot multiplex system targeting the cytochrome C oxidase subunit I gene. Biomed Res Int 2018:2953892. https://doi.org/10.1155/2018/2953892
Perotti MA, Braig HR (2009) Phoretic mites associated with animal and human decomposition. Exp Appl Acarol 49:85–124. https://doi.org/10.1007/s10493-009-9280-0
Perotti MA, Goff ML, Baker AS et al (2009) Forensic acarology: an introduction. Exp Appl Acarol 49:3–13. https://doi.org/10.1007/s10493-009-9285-8
Pinto J, Magni P, O'Brien RC, Dadour I (2021) Forensically relevant blow flies (Diptera: Calliphoridae) of Central Connecticut, USA. Forensic Sci Int 327:110940. https://doi.org/10.1016/j.forsciint.2021.110940
du Plessis M, der Walt LM-V (2004) Forensic entomology: relevant to legal dispute resolution? JJS 29(3):100–121
Pujol-Luz JR, Marques H, Ururahy-Rodrigues A et al (2006) A forensic entomology case from the Amazon Rain Forest of Brazil. J Forensic Sci 51:1151–1153. https://doi.org/10.1111/j.1556-4029.2006.00217.x
Rasmy A (2007) Forensic acarology: a new area for forensic investigation. Acarines: J Egypt Soc Acarol 1:5–6. https://doi.org/10.21608/ajesa.2007.4983
Richards CS, Simonsen TJ, Abel RL et al (2012) Virtual forensic entomology: improving estimates of minimum post-mortem interval with 3D micro-computed tomography. Forensic Sci Int 220:251–264. https://doi.org/10.1016/j.forsciint.2012.03.012
Rivers D, Dahlem G (2014) The science of forensic entomology. Wiley-Blackwell, London
Robinson W (2005) Urban insects and arachnids a handbook of Urban Entomology. Cambridge University Press, Cambridge
Rysavy NM, Goff ML (2014) Preliminary observations of arthropods associated with buried carrion on Oahu. J Forensic Sci 60(2):462–467. https://doi.org/10.1111/1556-4029.12643
Sarwar M (2020) Typical flies: natural history, lifestyle and diversity of Diptera. IntechOpen, London. https://doi.org/10.5772/intechopen.91391
Shen X, Jiang Y, Shang Y et al (2021) A preliminary study of necrophagous flies (Insecta: Diptera) in the Lhasa Region (Qinghai–Tibet plateau), China. J Med Entomol 59:467–479. https://doi.org/10.1093/jme/tjab187
Silva ÂZ, Hoffmeister CH, Anjos VA et al (2014) Necrophagous Diptera associated with wild animal carcasses in southern Brazil. Revista Brasileira de Entomologia 58:337–342. https://doi.org/10.1590/s0085-56262014000400001
Simon TM (2019) Forensic entomology and wildlife crime- the role of Blowfly larvae. Int Res J Biol Sci 8(9):1–7
Singh R, Sachan GC (2007) Elements of entomology, 1st edn. Rastogi Publications, India
Smith KGV (1986) A manual of forensic entomology. British Museum (Natural History), London
Sohal RS, Lamb RE (1977) Intracellular deposition of metals in the midgut of the adult housefly, Musca Domestica. J Insect Physiol 23:1349–1354. https://doi.org/10.1016/0022-1910(77)90156-1
Sohal RS, Lamb RE (1979) Storage-excretion of metallic cations in the adult housefly, *Musca Domestica*. J Insect Physiol 25:119–124. https://doi.org/10.1016/0022-1910(79)90089-1
Song T, Zhang P, Tang K et al (2022) A preliminary study on the distribution of necrophagous flies on Hainan Island, China. J Forensic Sci 67:741–748. https://doi.org/10.1111/1556-4029.14937
Tullis K, Goff ML (1987) Arthropod succession in exposed carrion in a tropical rainforest on O'ahu Island, Hawai'i. J Med Entomol 24:332–339
Tüzün A, Murat HG, Bilgili G (2015) Forensic acarology. Türk Bilimsel Derlemeler Dergisi 8(1):54–56

Tyndale-Biscoe M, Kitching RL (1974) Cuticular bands as age criteria in the sheep blowfly Lucilia cuprina (Wied.) (Diptera, Calliphoridae). Bull Entomol Res 64:161–174

UNODC (2020) World wildlife crime report 2020: Trafficking in protected species. UN, New York, NY. https://doi.org/10.18356/9789210052252

Vanin S (2018) Forensic entomology: an overview. Crime Secur Soc 1:1. https://doi.org/10.5920/css.2018.05

Vasconcelos SD, Araujo MC (2012) Necrophagous species of Diptera and Coleoptera in northeastern Brazil: state of the art and challenges for the forensic entomologist. Revista Brasileira de Entomologia 56:7–14. https://doi.org/10.1590/s0085-56262012005000014

Villet MH, Richards CS, Midgley JM (2009) Contemporary precision, bias and accuracy of minimum post-mortem intervals estimated using development of carrion-feeding insects. In: Amendt J et al (eds) Current concepts in forensic entomology. Springer, New York, NY, pp 109–137. https://doi.org/10.1007/978-1-4020-9684-6_7

Volckaert H (2020) Current applications and limitations of forensic entomology. Themis: Res J Just Stud Foren Sci 8:4. https://doi.org/10.31979/themis.2020.0804

Voss SC, Cook DF, Dadour IR (2011) Decomposition and insect succession of clothed and unclothed carcasses in Western Australia. Forensic Sci Int 211:67–75. https://doi.org/10.1016/j.forsciint.2011.04.018

Wang Y, Ma MY, Jiang XY et al (2017) Insect succession on remains of human and animals in Shenzhen, China. Forensic Sci Int 271:75–86. https://doi.org/10.1016/j.forsciint.2016.12.032

Wang M, Chu J, Wang Y et al (2019) Forensic entomology application in China: four case reports. J Forensic Legal Med 63:40–47. https://doi.org/10.1016/j.jflm.2019.03.001

Warrier V, Shedge R (2019) LETTER TO THE EDITOR Forensic entomology: applications in investigations. J Indian Acad Forensic Med 41(3):219

Wells JD, Stevens JR (2008) Application of DNA-based methods in forensic entomology. Annu Rev Entomol 53:103–120

Wells JD, Introna F, Di Vella G et al (2001) Human and insect mitochondrial DNA analysis from maggots. J Forensic Sci 46(3):685–687. https://doi.org/10.1520/jfs15022j

Wydra J, Matuszewski S (2021) The optimal post-eclosion interval while estimating the post-mortem interval based on an empty puparium. Foren Sci Med Pathol 17(2):192–198. https://doi.org/10.1007/s12024-020-00328-y

YuBo Z, Yue Z, ZhengXiang Z et al (2018) Seasonal diversity of necrophagous flies in Anshun, Guizhou. Acta Parasitologica et Medica Entomologica Sinica 25(2):101–106

Zou T-L, Feng D-X, Huang G-Y et al (2022) Species composition and succession of necrophagous insects on small buried baits in China. J Med Entomol 59:1182–1190. https://doi.org/10.1093/jme/tjac045

Zumpt F (1965) Myiasis in man and animals in the old world: a textbook for physicians, veterinarians and zoologists. Butterworth and Co (Publishers) Ltd, London

Wildlife Forensics 30

Arjun Rao Isukapatla, Prachi Yadav, and Moumita Sinha

Abstract

Wildlife forensics is a multidisciplinary field that combines the techniques and methods of various sciences, including biology, chemistry, and genetics to address the illegal trade and exploitation of wildlife species. The ultimate aim of wildlife forensics is to provide scientific evidence to support investigations and enforcement of wildlife crimes and to contribute to the conversation and protection of wildlife populations. This field plays a crucial role in addressing trafficking, which has a significant impact on biodiversity, ecosystems, and human health.

Keywords

Wildlife · Wildlife Protection Act · Geographic · mtDNA · DNA barcoding

30.1 Wildlife

Wildlife broadly refers to the diverse array of non-domesticated creatures, animals, plants, and other organisms that play an important role in nature. It also includes a wide variety of species, such as mammals, birds, reptiles, fish, and insects. Wildlife

A. R. Isukapatla (✉)
Department of Life Sciences, Christ University, Bengaluru, Karnataka, India
e-mail: arjun.rao@christuniversity.in

P. Yadav
Department of Forensic Science, Guru Ghasidas Vishwavidyalaya, Bilaspur, Chhattisgarh, India

M. Sinha
Department of Forensic Science, Kalinga University, Raipur, Chhattisgarh, India
e-mail: moumita.sinha@kalingauniversity.ac.in

A. Puri et al. (eds.), *Fundamentals of Forensic Biology*,
https://doi.org/10.1007/978-981-99-3161-3_30

builds and maintains the balance of ecosystem and yet many species are also important for the livelihoods of human and cultural values. In contrast to this, numerous human activities such as habitat destruction, pollution, and overhunting play a major threat to wildlife populations around the globe. Many efforts are necessary to conserve and protect wildlife and their habitats for biodiversity preservation and for conserving the planet health. The idea of wildlife can challenge different thoughts and emotions in people. For some, wildlife symbolizes nature beauty and diversity and the importance to preserve it for forthcoming generations. Others may think that wildlife can be a source of food or economic gains or a potential threat to human safety or property. Many have deep emotions to specific species, ecosystem, nature, and a sense of reverence, wonder, and curiosity when observing wildlife and natural habitats, especially when they are aware of the threats that wildlife is facing in recent times. Contrary, some people may feel that conserving wildlife is not the priority and that human needs should take precedence and not see the importance of protecting wildlife and preserving their habitat. Overall, views on wildlife are complex and may shape by cultural, economic, and personal reasons.

30.1.1 Wildlife Crime

Wildlife crime refers to illegal activities related to wild animals and plants, such as poaching, illegal trade, smuggling, and habitat destruction (Cooper and Cooper 2008; Lawton and Cooper 2009). Wildlife crime can have a significant impact on the survival of threatened and endangered species and can also disrupt the balance of ecosystems. After the illegal smuggling of narcotic drugs and arms trafficking, the third stands for wildlife products and related materials in any form (Sahajpal et al. 2009; Sreepada et al. 2002). Wildlife crime is often organized and perpetrated by criminal networks, which can make it difficult to combat. However, there are many efforts being made to combat wildlife crime, including the use of forensic science to investigate crimes, the development of laws and regulations to protect wildlife, and the use of education and awareness campaigns to reduce demand for wildlife products. International cooperation is also critical in addressing wildlife crime, as many of the illegal activities cross borders and require collaboration among different countries and organizations.

30.1.2 Few Wildlife Crimes Entire the Globe Are Mentioned Below

Poaching Poaching is one of the most significant wildlife crimes, and it is driven by the demand for wildlife products such as ivory, rhino horn, and animal parts used in traditional medicine. Illegal trade in wildlife is also a major problem, as it fuels the black market for wildlife products. This can include the illegal trade of live animals, such as exotic pets, and also trade in products derived from wild animals, such as fur, leather, and other body parts.

Smuggling Smuggling is another form of wildlife crime, as it involves the illegal transport of wildlife and wildlife products across national or international borders. Habitat destruction is also a major problem, as it can make it difficult for wildlife to survive and reproduce and can also disrupt the balance of ecosystems.

Illegal Trade Illegal trade of wildlife, also known as wildlife trafficking, is the buying, selling, and transportation of wild animals and plants, as well as their products and derivatives, in violation of national and international laws and regulations.

Habitat Destruction Habitat destruction refers to the destruction or degradation of natural habitats such as forests, wetlands, and grasslands, which can make it difficult for wildlife to survive and reproduce. Habitat destruction is a major threat to wildlife populations around the world and is driven by a variety of human activities such as urbanization, agriculture, mining, logging, and dam construction. Deforestation is one of the most significant causes of habitat destruction.

Endangered Species Endangered species are wild animals and plants that are at risk of extinction. The term "endangered" is used to describe a species that is in danger of dying out completely. The International Union for Conservation of Nature (IUCN) maintains a Red List of Threatened Species, which categorizes species into different levels of risk, with "Endangered" being the second highest level of risk.

Trophy Hunting Trophy hunting refers to the practice of hunting wild animals for the purpose of obtaining a part of the animal (such as a head, antlers, or tusks) as a trophy or souvenir. This practice is controversial, as some argue that it is a necessary tool for conservation and population management, while others argue that it is cruel and inhumane and can contribute to the decline of certain species.

Wildlife Crime Syndicate A wildlife crime syndicate refers to a group of individuals or organizations that are involved in illegal activities related to wildlife. These activities can include illegal hunting, poaching, trafficking of wildlife and wildlife products, and other forms of wildlife exploitation. Wildlife crime syndicates often operate on a large scale and can be highly organized, using methods such as bribery, corruption, and violence to evade law enforcement.

30.1.3 Wildlife Forensics

Wildlife forensics is an associated field of crime investigation that uses the scientific procedures and principles to examine, classify, and compare the recovered pieces of evidence from the crime scene. Numerous ways of poaching, killing, hunting of wild animals and illegal cutting, and trading of plants for medicinal purposes make ways to deteriorate the wild species round the globe. This can include the analysis of physical evidence such as DNA, fingerprints, and trace evidence, as well as the examination of wildlife products such as meat, ivory, and fur. Wildlife forensics can be used to identify suspects, link suspects to crimes, and establish the origin and movement of wildlife products. The field of wildlife forensics is interdisciplinary, drawing on expertise from a variety of fields such as biology, genetics, chemistry, and criminology. The goal of wildlife forensics is to help protect wildlife by providing scientific evidence to support law enforcement efforts to combat wildlife crime. In modern times, the emerging field of wildlife forensics has constantly progressed with new tools and technologies for addressing crimes against wildlife. The advanced use of technologies such as fingerprinting identification, hybrid DNA analysis, toxicology, ballistics, and cyber forensic has upgraded the law enforcement ability and conservator groups to examine, investigate, and prosecute wildlife crimes.

- Fingerprints can play a role in wildlife investigation by providing a means of identifying suspects who have left their fingerprints at a crime scene, or on items related to the crime such as weapons or illegally obtained wildlife products. Fingerprints can also be used to identify poachers by matching fingerprints found on illegal hunting equipment, poaching traps, or other objects found at the crime scene.
- DNA is commonly used in wildlife forensics as a tool for identifying individual animals and determining their relationships with one another. This can be used for a variety of purposes, such as tracking the movement of threatened and endangered species, investigating wildlife crimes, and determining the origin of products made from wildlife parts, such as ivory or meat. DNA analysis can also be used to study the genetic diversity and population structure of wildlife populations, which can enlighten conservation efforts.
- For forensic toxicology in wildlife examination, the samples can be examined from wildlife for the presence of toxic chemicals, such as pesticides, heavy metals, and pollutants. This can help to determine the cause of death in cases of wildlife poisoning, as well as to identify the source of the poison. Toxicology can also be used to monitor the health of wildlife populations and detect any changes in the levels of pollutants in the environment. This information can then be used to implement conservation measures to protect wildlife from exposure to toxic chemicals.
- Forensic ballistics in wildlife investigation can be used to match bullets or bullet fragments found at a crime scene to a specific firearm and to determine the trajectory of the bullet, which can help investigators determine the location of the shooter. This information can be used to identify suspects and build a case against them. Forensic ballistic techniques can also be used in cases of poaching, where wildlife species are hunted illegally, and bullets or fragments of bullets are found at the crime scene, and the bullet or fragment can be matched to the firearm used by the poacher, which can help to identify and prosecute the poacher and also help to recover the illegally obtained wildlife species.
- Cyber forensics can play a role in wildlife investigation by helping to uncover and analyze digital evidence related to wildlife crimes, such as illegal hunting or

trafficking of endangered species. This can include analyzing data from surveillance cameras, social media, and other online sources to identify suspects and gather information about their activities. Additionally, cyber forensics can be used to track the movement of illegal wildlife products through digital supply chains and to identify and shut down online marketplaces that facilitate such crimes.

In addition, the use of forensic techniques can also help in understanding the extent of the illegal trade of wildlife products and the impact of these activities on wild populations. Furthermore, the use of remote sensing and GIS analysis can also be useful in detecting illegal activities like illegal logging and habitat destruction and also in monitoring wildlife populations.

30.2 Scope of Wildlife Forensics

In a worldwide scenario, wildlife crime is fascinating huge monetary gains for those who participate in this illegal smuggling but unknowingly at the cost of many species extinction (Johnson et al. 2014). From the wildlife crime monetary benefit latest estimates to USD 30 billion per year, this not includes the illegal trade in timber, which alone costs more than USD 5 billion dollars and illegal fishing from USD 20–25 billion dollar approximately (Linacre and Ciavaglia 2017). The cost for the wildlife species black market may increase for those species, which are listed in the protection category under the Convention on the International Trade in Endangered Species of Wild Flora and Fauna (CITES) (Schneider 2008). Wildlife forensics is the application of forensic science techniques to investigate crimes against wildlife. This can include the illegal trade in endangered species, poaching, and the illegal use of wildlife products. The scope of wildlife forensics includes the collection and analysis of physical evidence, such as DNA, blood, and other biological samples, as well as the examination of wildlife products and packaging materials to determine their origin. Wildlife forensic scientists also use techniques such as stable isotope analysis and trace element analysis to track the movement of animals and their products. Additionally, wildlife forensics also includes the use of forensic techniques to combat wildlife crime, such as the identification of poached animals and the development of new technologies to aid in the detection of illegal wildlife trade.

Forensic science is crucial in wildlife investigation as it allows for the identification and prosecution of individuals who engage in illegal activities related to wildlife. It can provide valuable evidence to support criminal cases and can also aid in identifying and dismantling criminal networks involved in wildlife trafficking. Forensic techniques used in wildlife investigation include DNA analysis, which can identify the species of an animal and link a suspect to a specific animal or animal product; trace evidence analysis, which can identify the origin of an animal or animal product; and veterinary forensic pathology, which can determine the cause of death of an animal. Forensic science also helps in wildlife conservation, as it can be used to identify illegal trade routes and understand the full extent of the problem, providing evidence to support conservation efforts and help to protect endangered species. Furthermore, it also allows investigators to understand the motivations of poachers and traffickers, which can inform more effective countermeasures.

30.2.1 Importance of Wildlife Forensics

Wildlife forensics plays a critical role in protecting and conserving wildlife populations and their habitats. The illegal trade in wildlife products is a major threat to many species, and wildlife forensic techniques can be used to identify, track, and prosecute those responsible for these crimes. One of the key uses of wildlife forensics is in the identification of poached animals and illegally traded wildlife products. By analyzing physical evidence, such as DNA, blood, and other biological samples, forensic scientists can match a product or animal to a specific location or population and determine if it was obtained illegally. This information can be used to prosecute those responsible for wildlife crimes and to help protect endangered species. Another important use of wildlife forensics is in the development of new technologies and methods to detect and prevent illegal wildlife trade. For example, wildlife forensic scientists are currently working on developing DNA testing methods that can be used to quickly and accurately identify different species of animals and their products, making it easier to detect illegal trade. Moreover, wildlife forensics is a unique field that helps to bridge the gap between conservation, law enforcement, and forensic science by providing scientific evidence for court cases. Furthermore, it helps in the protection and conservation of biodiversity and sustainable use of natural resources.

30.2.2 Ethics in Wildlife Forensics

Ethics play an important role in wildlife forensics, as it involves the collection, preservation, and analysis of evidence from wild animals, which can raise a number of ethical considerations. Some of the key ethical issues that may arise in wildlife forensics include the following.

Animal Welfare Wildlife forensics may involve the collection of samples from live animals or from animals that have been killed, which raises concerns about the welfare of these animals. It is important to ensure that any collection of samples is done in a way that minimizes harm to the animal.

Respect for Wildlife Wildlife forensics may also involve the examination of evidence from animals that have been killed, which raises concerns about respect for wildlife. It is important to ensure that the collection and examination of evidence are done in a way that is respectful of the animals and their remains.

Privacy Wildlife forensics may also involve the collection and analysis of DNA samples from wild animals, which raises concerns about the privacy of these animals. It is important to ensure that any collection and analysis of DNA samples are done in a way that respects the privacy of the animals.

Conservation Wildlife forensics may also involve the investigation of crimes against wildlife, which raises concerns about the conservation of wildlife. It is important to ensure that the investigation of crimes against wildlife is done in a way that promotes conservation and the protection of wildlife.

Transparency and Impartiality Wildlife forensics investigation should be transparent and impartial, avoiding any bias or corruption that may impede the process and integrity of the investigation.

30.3 Wildlife Protection Act 1972

The Wildlife Protection Act (WPA) of 1972 is an Indian law that provides for the protection of wild animals, birds, and plants. The act is intended to protect India's biodiversity and natural heritage, and it is enforced by the country's forest department. The act has been amended several times over the years to include more animals and plants under protection and to increase penalties for wildlife crimes. The act also provides for the creation of special courts for the trial of offenses under the act, which helps to speed up the process of prosecution of wildlife crimes. The Wildlife Protection Act of 1972 is an important law that helps to protect India's wildlife and biodiversity, and it serves as a model for other countries looking to develop similar legislation. Under the act, hunting and trapping of wild animals are prohibited, and trade in wild animals and their products is regulated. The Act also provides for the creation of protected areas, such as national parks, wildlife sanctuaries, and closed areas, where hunting and other activities that could harm wildlife are prohibited. The Wildlife Protection Act of 1972 also includes provisions for the regulation of trade in wildlife and wildlife products and the seizure and forfeiture of illegally obtained wildlife and wildlife products. Penalties for violations of the act include fines and imprisonment.

Some of the key measures that India has taken to protect its wildlife under Wildlife Protection Act 1972 include the following.

Protected Areas India has a large network of protected areas, including national parks, wildlife sanctuaries, and biosphere reserves, which provide critical habitat for many of the country's species of wild animals, birds, and plants.

Wildlife Crime Control Bureau A dedicated agency is established to combat wildlife crime, and it helps in the investigation of wildlife crimes and the seizure and forfeiture of illegally obtained wildlife and wildlife products.

Ban on Wildlife Trade India has banned the trade of certain wildlife products, such as tiger bone and rhinoceros horn, in an effort to protect these species from extinction.

Community-Based Conservation India has also implemented community-based conservation programs, which involve local communities in the management and conservation of wildlife and their habitats. These programs have helped to reduce human–wildlife conflict and to build support for conservation among local communities.

The Wildlife Protection Act (WPA) of 1972 is divided into six schedule lists that provide varying degrees of protection. Poaching, smuggling, and illegal trade in animals listed in Schedules 1–4 are all prohibited. Animals included in the schedule are completely protected from hunting, and trade and commerce involving such animals is strictly regulated. The WPA 1972 governs wildlife conservation and protection in India. India is the first country in the world to include measures for environmental preservation and conservation in its constitution.

The WPA 1972 is categorized into six different schedules (Table 30.1) that protect and preserve wildlife animals, plants, and the ecosystem. The animals listed in the schedule are defined as protected from any illegal activities like hunting, poaching, trading, and commerce of these animals. India stood first to take safety measures for preserving and conserving the environmental habitats.

30.3.1 Conservation Projects

Conservation projects are efforts to protect and preserve natural resources and habitats, such as forests, wetlands, and wildlife. These projects can take many forms, including habitat restoration, reintroduction of endangered species, and protection of keystone species. Additionally, conservation projects may also focus on reducing human impacts on the environment, such as by promoting sustainable land use practices and reducing pollution. Conservation projects are documented to utilize the evolutionary theory, which helps in current and future wildlife challenges, and also for wildlife survival since the key challenges are global warming, pollution, wildlife threat, and hunting. The projects can be undertaken by government agencies, non-profit organizations, or private companies and may be funded by a variety of sources, including government grants, private donations, and corporate sponsorships. Few of the conservation projects are planned under WPA 1972.

30.3.1.1 Project Snow Leopard

Project Snow Leopard is a conservation program launched by the government of India in 2009 to protect and conserve the endangered snow leopard and its ecosystem in the Himalayan region of the country. The project aims to improve the conservation status of the snow leopard through a combination of measures such as habitat protection, prey base enhancement, community participation, and research. The

Table 30.1 Different schedules under Wildlife Protection Act

S. no.	Schedule	Definition	Animals
1.	Schedule I	Deals with endangered species and requires strict protection. Species defined in this schedule are prohibited from hunting in India, except if they become threat to human life or in case of disabled or diseased beyond recovery	Andaman wild pig, blackbuck, brow-antlered deer, Himalayan brown bear, cheetah, Chinese pangolin, clouded leopard, four-horned antelope, bison or gaur, golden langur, Indian lion, Kashmir stag, leopard or panther, musk deer, rhinoceros
2.	Schedule II	Deals with the animals also needed high protection with trading prohibited either as animal, or part of animals	Assamese macaque, Himalayan crestless porcupine, Himalayan newt or salamander, pig-tailed macaque, stump-tailed macaque, wild dog or dhole, spiny-tailed lizard, or sanda
3.	Schedule III	Deals with non-endangered species include protected species where hunting is prohibited but the punishments are less severe than the first two schedules	Barking deer or muntjac, bharal, chital, hyaena, nilgai, sambar, wild pig, sponges
4.	Schedule IV	As mentioned in Schedule III	A five-striped palm squirrel, Himalayan mouse hare, hedgehog
5.	Schedule V	Deals with animals that are termed "vermin" and can be hunted	Common crow, fruit bats, mice, rats
6.	Schedule VI	Deals with the cultivation of a specific plant and restricted from Illegal possession, sale, and transportation	Beddomes' cycad, blue vanda, kuth, ladies slipper orchids, pitcher plant, red vanda

project is being implemented in 12 states in India, including Himachal Pradesh, Jammu and Kashmir, and Uttarakhand, which are known to be snow leopard habitats. Schedule I of the WPA 1972 and the International Union for Conservation of Nature (IUCN) declared the species under "Vulnerable category." The species listed in the Convention on International Trade in Endangered Species (CITES) and the Convention on Migratory Species (CMS), which defines the highest conservation of the species both under national and International borders. The protection program is established under the Global protocol of Global Snow Leopard and Ecosystem Protection Program.

30.3.1.2 Project Tiger

Project Tiger is a wildlife conservation program launched by the government of India in 1973 to protect the Bengal tiger and its habitats. The project aims to ensure a viable population of tigers in the wild, while also preserving areas of biological importance as a natural heritage for the benefit of all living beings. The project is being implemented in various tiger reserves across the country, which are managed by the National Tiger Conservation Authority (NTCA). These reserves provide a protected area for tigers to breed and live, while also promoting conservation and protection of the species' natural habitats. Project Tiger also includes a comprehensive monitoring system and anti-poaching efforts to ensure the safety of tigers and other wildlife in the reserves. The project is considered one of the most successful conservation programs in the world. Measures are taken for the conservation of tigers under the WPA 1972.

Core Zones

The designated areas are exclusively used for the conservation of tiger but should not invade the scheduled tribes' rights or other forest inhabitants. Additionally, the zone and areas should be free from forest actions and disturbances. The timber collection, grazing of cattle, and other human activities are not allowed within this zone.

Buffer Zones

This zone acts as an additional habitat for the tigers and also offers co-existence of human activities. This zone is parallel to the core tiger habitat area.

30.3.1.3 Project Elephant

Project Elephant is a conservation program launched by the Government of India in 1992 to protect Indian elephants and their habitats. The main objectives of the program include protection of elephants, their habitats, and corridors, as well as mitigation of human–elephant conflicts. The program also focuses on research and monitoring of elephant populations and the development of elephant-friendly infrastructure. It is important to mention that human activities such as agriculture, urbanization, and infrastructure development have led to the fragmentation and degradation of elephant habitats and corridors, causing human–elephant conflicts and threatening elephant populations. Elephant corridors are narrow strips of land that connect fragmented habitats, allowing elephants to move safely between them. In India, elephant corridors are used by wild elephants to migrate between habitats in search of food, water, and mates. These corridors are important for the survival and genetic diversity of elephant populations. The Indian government, along with various conservation organizations, has identified and is working to protect several elephant corridors across the country (Table 30.2). Some of the important elephant corridors in India include the following.

30.3.1.4 Project Hangul

Project Hangul is a conservation program aimed at protecting the Hangul deer (Cervus elaphus hanglu), also known as the Kashmir stag, which is a subspecies of the European red deer. The program is run by the government of the India and state of Jammu and Kashmir, in collaboration with local communities and conservation organizations. The Hangul deer is considered one of the most critically endangered deer species in the world, with only around 150 remaining in the wild. Project Hangul is an important conservation effort to protect the remaining population of the species and to ensure its survival for future generations. The main objectives of the program include the following.

Table 30.2 Elephant corridors in India

S. no.	Elephant corridor	State	Connects to
1	The Rajaji-Corbett Elephant Corridor	Uttarakhand	Rajaji National Park and the Corbett National Park
2	The Kutch-Pachmarhi Elephant Corridor	Gujarat	Pachmarhi Forest in Madhya Pradesh
3	The North Bank Elephant Corridor	Assam	Dibru Saikhowa National Park
4	The Nilgiri Elephant Corridor	Tamil Nadu	Sathyamangalam Wildlife Sanctuary
5	The Anechowkur Elephant Corridor	Karnataka	Talakaveri Wildlife Sanctuary

Protection of the Hangul Deer The program aims to protect the deer from poaching and habitat loss, which are the main threats to the species.

Habitat Restoration The program works to restore the natural habitat of the Hangul deer, which includes wetland, grassland, and forest habitats.

Monitoring The program conducts regular monitoring of the population of the Hangul deer and its habitat to assess the effectiveness of conservation efforts.

Research The program conducts research on the biology and ecology of the Hangul deer to understand the species better and to make informed conservation decisions.

Community Engagement The program involves local communities in conservation efforts, through awareness campaigns and alternative livelihood programs.

Law Enforcement The program works with law enforcement agencies to combat poaching and to protect the deer from illegal hunting.

Ecotourism The program promotes ecotourism as a means of generating income for local communities and raising awareness about the importance of conservation.

30.3.1.5 Crocodile Conservation Project

A crocodile conservation project is a program aimed at preserving and protecting crocodile populations and their habitats. This can include activities such as monitoring crocodile populations, habitat restoration, research, education and awareness campaigns, and law enforcement to prevent illegal hunting and trade. The ultimate goal of a crocodile conservation project is to ensure the long-term survival of crocodile species and the ecosystems they inhabit.

Directions for Crocodile Conservation Project Conduct a comprehensive assessment of the current crocodile population and distribution: This will provide a baseline for monitoring population trends and identifying areas where conservation efforts need to be focused.

Develop a conservation plan: This should include specific goals and objectives, as well as strategies for achieving them.

Implement Habitat Restoration and Protection Measures This could include activities such as removing invasive species, restoring wetlands, and protecting nesting sites.

Conduct Research to Better Understand Crocodile Biology and Ecology This will provide valuable information for making informed conservation decisions.

Educate and Raise Awareness About Crocodile Conservation This can include public education campaigns, working with local communities, and training law enforcement to better protect crocodiles and their habitats.

Enforce Laws and Regulations to Prevent Illegal Hunting and Trade This is critical to protecting crocodile populations from unsustainable hunting and poaching.

Monitor and Evaluate the Success of Conservation Efforts Regular monitoring will help identify areas where the conservation plan needs to be adjusted and track progress toward achieving conservation goals.

30.3.1.6 Project Rhino

Project Rhino was launched in 2005 to protect and conserve the rhinos. The Indian Rhino Vision 2020 project aimed to achieve a wild population of at least 3000 greater one-horned rhinos spread across seven protected areas in the Indian state of Assam by 2020. Kaziranga National Park, Pobitora National Park, Orang National Park, Manas National Park, Laokhowa Wildlife Sanctuary, Burachapori Wildlife Sanctuary, and Dibru Saikhowa Wildlife Sanctuary are the seven protected areas.

30.3.1.7 Indian Rhino Vision 2020

The Department of Environment and Forests, State of Assam collaboration with The Bodo Autonomous Council as an active partner, executed The Indian Rhino Vision 2020. World Wildlife Fund WWF-India, WWF AREAS (Asian rhino and elephant action strategy), and The IRF (International Rhino Foundation), along with number of NGOs, are actively participating in the program. The prime focus and vision are to increase the one-horned rhino population from around 2000–3000 and to protect them from poaching and illegal hunting, and their distribution across seven protected areas to safeguard them for long-term sustainability is also one of the major visions of the program.

30.4 Pieces of Evidence and Identification

Different markers used worldwide for species identification in wildlife forensics are morphological studies, microscopic examinations, anatomical landmarks, taxonomic identification, phylogeographic analysis, serological methods, and genetic markers. This individual identification is based on hair characterization, long bones, tooth morphology, pug marks, and related traits (Bell and Machin 2011). In wildlife forensics, the pieces of evidence used for species identifications are processed meats (Fortajada et al. 2021), shark fins (Chapman et al. 2003), egg shells (Coghlan et al. 2012), animal hairs (Ahmed et al. 2018), bones (Gouda et al. 2020), ivory (Kitpipit et al. 2016), rhinoceros horns (Hsieh et al. 2003), turtle shell (Yadav et al. 2021), feathers (Dove and Koch 2011), and fish scales (Kumar et al. 2007).

30.4.1 Geographical Origin in Wildlife Forensics

Identifying the geographic origin of animals in cases of import and export of animal or animal body parts is crucial. Identifying the origin of wildlife samples can be proceeded by isotope comparison using methods like inductively coupled plasma mass spectrometry (ICP-MS) and isotope ratio mass spectrometry (IRMS) (Tobe 2009). In routine wildlife forensics and other scientific disciplines, using the isotope analysis to identify the trace pieces of evidence, to track the movements of animals, and to regulate the geographic origin of individual animals is possible (Bowen et al. 2005). Radioactive isotope analysis is a technique used in wildlife forensics to determine the origin or movement of wildlife and their products, such as meat, ivory, and fur. The method involves measuring the levels of naturally occurring radioactive isotopes, such as carbon-14 and strontium-90, in the sample. These isotopes have different levels in different geographic regions due to variations in geology and climate, which makes it possible to identify the region of origin. Additionally, the ratios of different isotopes in the sample can provide information on the age of the animal, diet, and migration patterns. This information can be used in investigations related to wildlife trafficking and poaching illegal trade in endangered species. Detecting the stable isotopes in animal soft tissue can possibly result to identify the precise location of an animal shortly after death. The results can be varied based on the types of tissue and the type of animal from which it originates. The type of tissue material uses for such analysis is imperative as the different types of tissues can incorporate the different results at different time intervals (Ogden et al. 2004). Stable isotope analysis is applicable to wide range of applications in forensic science and can be used to validate byproducts like meat, milk, and cheese (Renou et al. 2004; Boner and Förstel 2004) and for different migration patterns of animals and birds (Hobson and Wassenaar 1996; Chamberlain et al. 1996).

Remote sensing and GIS (geographic information systems) analysis can be used in wildlife forensics to help identify and track the movement of animal populations, as well as to monitor and manage habitats. Remote sensing can be used to gather data on the environment and the animals that inhabit it, such as through the use of satellite imagery and aerial photography. GIS can then be used to analyze these data and create maps and models that can be used to understand the distribution and movements of animal populations, as well as to identify potential threats to their habitats. This information can be used to help protect and conserve animal populations and to aid in the investigation of wildlife crimes.

Remote sensing can be used in wildlife forensics to gather data on the environment and the animals that inhabit it. These data can be analyzed using GIS (geographic information systems) to create maps and models that can be used to understand the distribution and movements of animal populations, as well as to identify potential threats to their habitats. One example of the use of remote sensing in wildlife forensics is the use of satellite imagery to monitor and track the movement of endangered species. High-resolution satellite imagery can be used to identify and map the locations of specific animals or groups of animals, such as elephants or rhinos. This information can be used to understand their migration patterns, population densities, and potential threats to their habitats, such as illegal poaching or habitat destruction. Aerial photography can also be used to gather data on the environment and wildlife. This can be used to identify and map wildlife habitats, such as wetlands or forests, and to monitor changes in these habitats over time. This information can be used to understand how human activities, such as land development or resource extraction, may be impacting wildlife populations. Another use of remote sensing in wildlife forensics is the use of thermal imaging cameras to detect and track the movement of animals at night. This can be particularly useful for monitoring nocturnal animals, such as bats, or for identifying and tracking illegal poaching activities.

30.4.1.1 Advantages of Geographic Information Systems in Wildlife Forensics

There are several advantages of using geographic information systems (GIS) in wildlife forensics, including the following.

Data Integration GIS allows for the integration and analysis of multiple types of data, such as satellite imagery, aerial photography, and GPS tracking data, which can be used to create more detailed and accurate maps and models of animal populations and habitats.

Spatial Analysis GIS provides a wide range of spatial analysis tools, such as spatial querying, overlay, and modeling, which can be used to understand the distribution and movements of animal populations and the potential threats to their habitats.

Visualization GIS allows for the creation of maps and 3D models that can be used to visualize and analyze data, making it easy to understand and communicate information on wildlife populations and habitats.

Historical Data Management GIS allows for the management and analysis of historical data over time, which can be used to understand changes in animal populations and habitats over time.

Collaboration and Sharing GIS makes it easy to share and collaborate on data and analysis with other researchers, conservation organizations, and law enforcement agencies.

Cost-Effective GIS is relatively low-cost and easy to use, making it a valuable tool for conservation efforts and the investigation of wildlife crimes.

Predictive Modeling GIS can be used to create predictive models that can be used to identify potential threats to animal populations and habitats and to plan conservation strategies.

30.4.2 Morphological and Microscopic Examination in Wildlife Forensics

Morphological examination is a traditional method of species identification in wildlife forensics that involves the visual examination of physical characteristics, such as anatomy, size, shape, and coloration. This method is used to identify species based on the unique morphological characteristics of different taxa, such as feathers, scales, fur, and bones. Morphological examination is often used as a first step in species identification, especially in cases where DNA analysis is not feasible or practical, or in conjunction with DNA analysis to provide additional information for species identification. It is also used to determine the age, sex, and health status of individual animals and to provide information on the cause of death in cases of wildlife mortality. Despite advances in DNA-based methods, morphological examination remains an important tool in wildlife forensics due to its versatility and ease of use in a wide range of sample types and conditions.

Hair morphology examination is a subcategory of morphological examination in wildlife forensics that focuses on the physical characteristics of hair samples, such as shape, color, and structure. Hair samples can provide valuable information for species identification, as well as for determining the age, sex, and health status of individuals. In addition, hair samples can provide information on the individual's geographic location and diet, which can be used to infer the movements and behavior of wildlife. A morphological examination can be performed by visual examination of the hair under a microscope or by using specialized techniques, such as light microscopy, scanning electron microscopy, and Fourier transform infrared spectroscopy. These techniques can be used to examine the cuticular scale pattern, shaft shape, and pigment distribution of hair samples, which can provide species-specific information. Hair morphology examination is often used in combination with other methods, such as DNA analysis, to provide a more complete picture of the species and individual.

The other type of morphology examination can be of Horns, which is a subcategory of morphological examination in wildlife forensics that focuses on the physical characteristics of horn samples, such as shape, size, color, and structure. Horns are important structures in many species of wildlife, serving various functions such as defense, display, and competition for mates. As a result, they can provide valuable information for species identification and population management. Horn examinations can be performed by visual examination or under a microscope or by using specialized microscopic techniques, such as scanning electron microscopy, and X-ray diffraction. These techniques can be used to examine the internal structure, size, and shape of horns, which can provide species-specific information. In wildlife forensics, horns examination is particularly useful in cases of illegal trade and poaching of protected or endangered species, where the horns are highly valued for their use in traditional medicines, ornamental items, and other commercial purposes. Horn morphology examination can also provide important information for population management and conservation, by helping to determine the age and sex of individuals and by providing information on the geographic origin and migration patterns of wildlife.

Teeth examination is a subcategory of morphological examination in wildlife forensics that focuses on the physical characteristics of teeth samples, their size, shape, and structure. Teeth are important structures in many species of wildlife, serving various functions such as feeding, eating, and for prey. As a result, they can provide valuable information for species identification and population management. These techniques can be used to examine the internal structure, size, and shape of teeth, which can provide species-specific information. Teeth morphology examination is often used in combination with other methods, such as DNA analysis, to provide a more complete picture of the species and individual. In wildlife forensics, teeth morphology examination is particularly useful in cases of illegal trade and poaching of protected or endangered species, where the teeth are highly valued for their use in traditional medicines, ornamental items, and other commercial purposes. Teeth morphology examination can also provide important information for population management and conservation, by helping to determine the age and sex of individuals and by providing information on the geographic origin and migration patterns of wildlife.

Hair morphology refers to the physical structure and characteristics of an animal's hair. In wildlife forensics, hair morphology is often used to identify the species of an animal, as well as the individual, based on differences in hair structure such as medulla pattern, scale pattern, shaft diameter, and pigment distribution. By examining the hair under a microscope, forensic scientists can compare it to reference hair samples to determine the species and sometimes even the individual source of the hair sample. Hair is commonly used in wildlife forensics as a means of species identification (Sahajpal and Goyal 2009). Hair morphology can be used to solve crimes involving wildlife, such as poaching or illegal hunting. Wildlife forensics uses microscopy to examine physical evidence from wildlife crime scenes, such as hairs, bones, scales, and feathers. Microscopes help identify the species of origin and provide evidence linking a suspect to a crime scene. Techniques used in wildlife forensics microscopy include light microscopy, scanning electron microscopy (SEM), and transmission electron microscopy (TEM).

The cuticle is the outermost layer of hair shaft, and its structure is important in wildlife forensics for species identification. The cuticle structure of an animal hair can vary depending on the species, and it can provide valuable information for species identification. Some of the features of the cuticle structure that can be used for species identification include the shape and pattern of the scales, the thickness of the cuticle, and the presence or absence of pigment in the cuticle. The cortex is the main structural part of an animal hair shaft, and it provides important information for species identification in wildlife forensics. The cortex structure can vary depending on the species, and it can contain important information such as pigment, medulla patterns, and cortical fibrils. The presence or absence of pigment, as well as its distribution, can be used to identify the species of an animal. The medulla pattern, which is the innermost part of the cortex, can also be used for species identification, as different species may have unique medulla patterns. Cortical fibrils, which are the structural proteins within the cortex, can also be used for species identification, as their arrangement and size can vary between species. The compositional pattern of the pigmentation can vary widely in different animals. It is denser toward the medulla. In animal, pigments are often found in solid masses called ovoid bodies, especially in dogs and cattle. Some of the cuticular and medullary patterns of wild and domestic animals are shown in (Fig. 30.1). Microscopy of hair involves the examination of hair shafts and cuticles under a microscope to determine the species of origin and to compare the hair sample with known reference samples. Light microscopy is used to examine the overall structure of the hair, including the shape and size of the shaft and the pattern of the cuticle. Scanning electron microscopy (SEM) is used to examine the surface characteristics of the hair in greater detail, such as the scale pattern (Chernova 2014). Transmission electron microscopy (TEM) can provide high-resolution images of the hair's internal structure, including the cortex and medulla. These techniques can help forensic scientists determine the species of origin of hair found at a crime scene and can provide evidence linking a suspect to the scene.

30.4.3 DNA-Based Identification in Wildlife Forensics

Wildlife forensics mainly involves DNA identification for illegal trades, hunting animals, and the different types of wildlife crime identification. Microscopic identification or morphological determination cannot be justified for wildlife material identification where the materials' presence may be in other forms. In such cases, molecular-based approaches or DNA-based identification tools can be useful for examinations (Linacre and Tobe 2011). The universal method opted for DNA-based analysis is mitochondrial DNA (mtDNA) sequencing for specimen identification of either animals or animal parts with a wide range of markers. Such highly precise way of identification methods are truly based upon the type of identification materials such as specific region of animal and plant material, taxonomic level identification

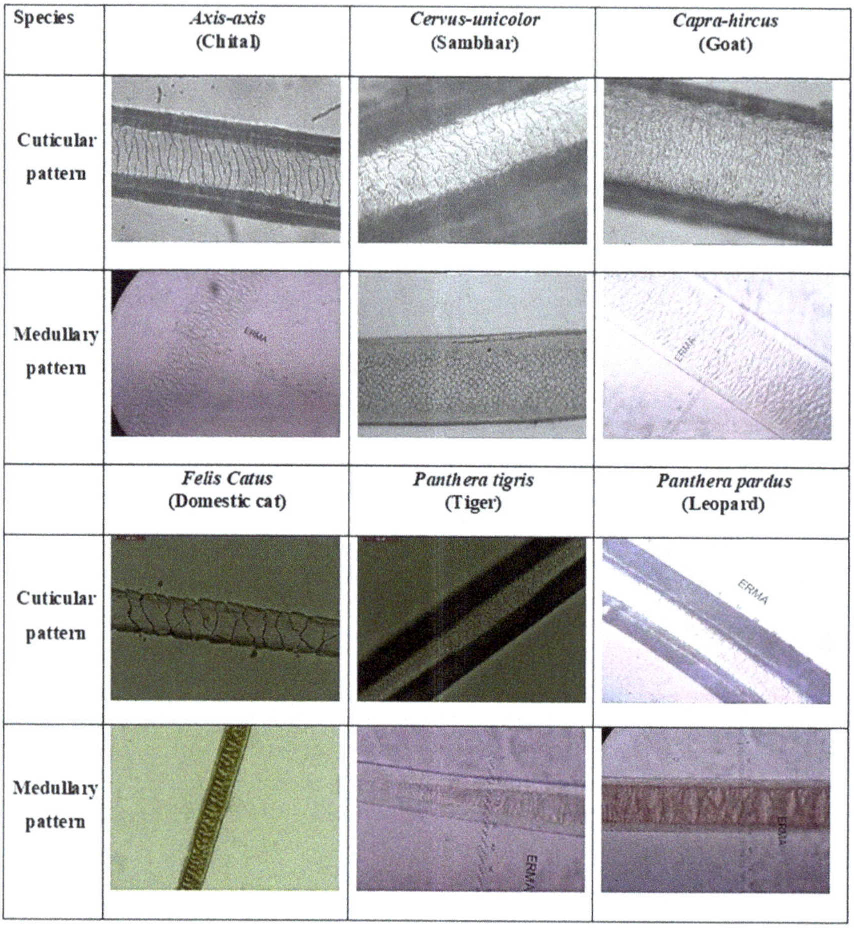

Fig. 30.1 Cuticular and medullary patterns of wild and domestic animals

for inter- and intra-species level, applications of new DNA markers, phylogenetic tree-based approaches, and applying the next generation sequencing (NGS) models. The high standards and requirements for DNA-based identification is an approach to truly identify the species of animal and plant with a specific set of either autosomal or nuclear DNA markers. Usage of molecular biology techniques in wildlife material investigation is found more reliable than the older techniques like anatomical landmark-based identification, morphological characterization, microscopic features, and serological markers. In wildlife forensics, the DNA-based methods were utilized and found to be a more reliable technique for species identification, which includes random amplified polymorphic DNA (RAPD), restriction fragment length polymorphism (RFLP), amplified fragment length polymorphism (AFLP), PCR amplification, usage of species-specific DNA markers, and DNA sequencing.

These DNA markers not only limit species identification but can be found useful to relationship identity between different and common species.

30.4.3.1 Mitochondrial DNA

Mitochondrial DNA (mtDNA) is often used in wildlife forensics due to its high copy number, making it easier to detect and analyze than nuclear DNA. mtDNA is also useful in species identification as it is maternally inherited and evolves more quickly than nuclear DNA, allowing for distinct differences between species to be identified. In wildlife forensics, mtDNA analysis is often used to identify the species of an unknown sample, to match an individual animal to a population, or to trace the origin of a product made from animal parts (such as ivory or fur). However, mtDNA analysis restricts to wildlife forensics, such as the possibility of homoplasy (similar mtDNA sequences in different species) and the need for reference databases for comparison. Ideally, mtDNA typing in wildlife forensics is commonly applied due to the regular usage in molecular taxonomy and phylogenetic analysis for species identification, the availability of universal primers, and due to the highly degraded samples identification in wildlife encounter (Johnson et al. 2014). Comparatively, the nuclear marker has a low amplification and success against the mtDNA, which is not much precisely used as in wildlife material identification (Tobe and Linacre 2008). Other genes such as cytochrome b (cyt b) (Tobe and Linacre 2010), the cytochrome oxidase 1 (CO1) (Ferri et al. 2009), two ribosomal RNA (Guha and Kashyap 2006), and the control (D-loop) region (Pun et al. 2009) are the highly influenced mitochondrial loci preferred in taxonomic identification and phylogenetic analysis in wildlife forensics. Some highly specific mitochondrial 12S (Balitzki-Korte et al. 2005; Melton and Holland 2007) and 16S ribosomal RNA loci (rRNA) (Mitani et al. 2009) markers are also found to be highly utilized for wildlife species identification. DNA barcoding is widely used and considered one of the most useful tools in wildlife forensics for the identification of different species including cooked and processed materials (Ferri et al. 2009). Mitochondrial region-specific cytochrome C oxidase (CO1) is referred to as the barcoding region sequence for different species of animal identifications (Fig. 30.2) (Dawnay et al. 2007).

30.4.3.2 Single-Nucleotide Polymorphisms (SNP)

Single-nucleotide polymorphisms (SNPs) are genetic variations in which a single-nucleotide (A, T, C, or G) in the DNA sequence differs between individuals. In wildlife forensics, SNPs can be used as genetic markers to establish the identity of an individual animal and determine its relatedness to other individuals within a population (Ogden 2011). SNPs have several advantages over other types of genetic markers, such as microsatellites or minisatellites, in wildlife forensics. They are abundant in the genome, allowing for the simultaneous analysis of many markers in a single reaction. They are also highly informative, providing high discrimination power for individual identification and population structure analysis (Martinsohn and Ogden 2009). SNPs can be used in various wildlife forensics applications, such as tracking the movement of animals, investigating wildlife crimes, and determining the relatedness of individuals within a population. They are also useful for

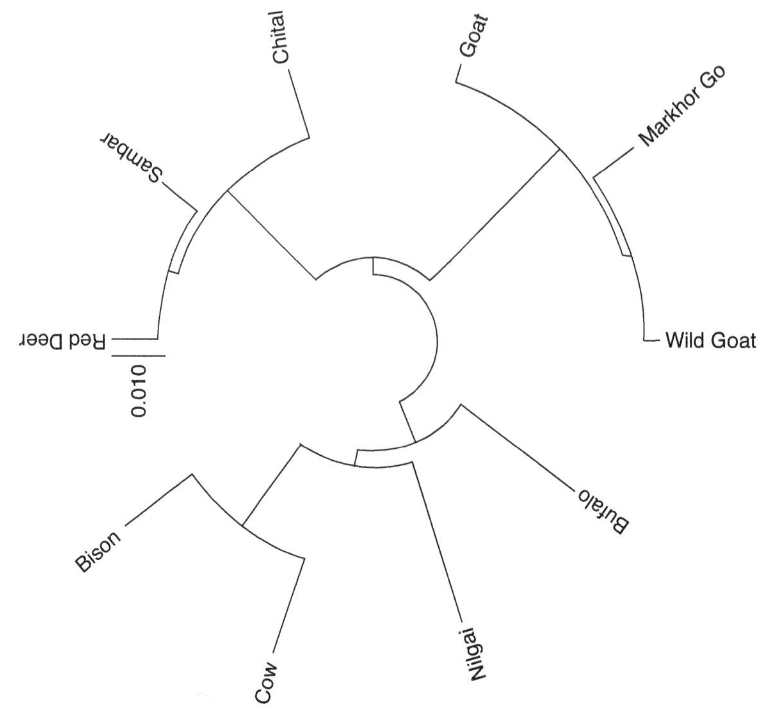

Fig. 30.2 Maximum-likelihood tree based on a short sequence of 16S rRNA. Cluster 1: cow and bison, Cluster 2: nilgai and buffalo, Cluster 3: wild goat, markhor goat, and goat, Cluster 4: red deer, sambhar, and chital

population genetics studies, as they can provide information on the genetic diversity, structure, and demographic history of a population (Sobrino et al. 2005). The use of SNPs in wildlife forensics has been applied to a wide range of species, including mammals, birds, fish, and reptiles, and has become an important tool in the fight against wildlife trafficking and illegal trade, as well as in conservation biology and management.

30.4.3.3 Cytochrome c Oxidase and Cytochrome b

Cytochrome c oxidase subunit 1 (CO1) is a gene that is commonly used in wildlife forensics. It encodes for a subunit of the cytochrome c oxidase complex, which is involved in cellular respiration. The CO1 gene is often used as a DNA barcode in species identification and has become a standard marker in forensic investigations of wildlife species (Stock et al. 2009), especially in cases of illegal trade and poaching. The high genetic variability of the CO1 gene across species, combined with the ease of DNA extraction and amplification, makes it a valuable tool for species identification and differentiation. Cytochrome b (cyt b) is another gene that is commonly used in wildlife forensics for species identification and differentiation. Cyt b is a mitochondrial gene that encodes for a protein involved in cellular respiration. It is often used in combination with other genetic markers, such as CO1 and the ribosomal RNA genes, to confirm species identification. The cyt b gene is highly conserved within a species but varies enough between species to be used as a genetic marker. The high degree of variability of the cyt b gene makes it a valuable tool in forensic investigations of wildlife species, especially in cases of illegal trade and poaching.

30.4.3.4 12S and 16S Ribosomal RNA

The 12S ribosomal RNA (12S rRNA) gene is another marker that is commonly used in wildlife forensics. Like the CO1 gene, it is useful for species identification and differentiation. The 12S rRNA gene is a component of the ribosome and is highly conserved within a species (Janczewski et al. 1995), but varies enough between species to be used as a genetic marker. It is especially useful in forensic investigations involving the identification of animal products and derivatives, such as meat, leather, and traditional medicines. The 12S rRNA gene is also used in phylogenetic studies to infer evolutionary relationships among species. The 16S ribosomal RNA (16S rRNA) gene is also used in wildlife forensics, particularly in species identification and differentiation. It is a component of the ribosome and, like the 12S rRNA gene, is highly conserved within a species but varies enough between species to be used as a genetic marker. The 16S rRNA gene is widely used in microbial ecology and has been applied to various wildlife forensic investigations (Jogayya et al. 2013), including the identification of bacteria in food and other animal-derived products. In addition to species identification, the 16S rRNA gene is also used in phylogenetic studies to infer evolutionary relationships among species (Kitano et al. 2007) and to infer the taxonomic classification of microorganisms.

30.4.3.5 D-Loop Region

The D-loop region is a non-coding region of the mitochondrial DNA (mtDNA) that is commonly used in wildlife forensics for species identification and differentiation. The D-loop region is highly variable between species and is commonly used in conjunction with other mtDNA markers (Saccone et al. 1987), such as cytochrome b, to increase the power of species identification. The D-loop region is often used in forensic investigations of wildlife species, especially in cases of illegal trade and poaching, due to the high degree of variability (Brown et al. 1986) and the ease of DNA extraction and amplification from a variety of sample types. Additionally, the D-loop region is useful in phylogenetic studies to infer evolutionary relationships among species and in population genetic studies to infer patterns of genetic diversity and structure within and among species.

30.4.3.6 Microsatellites or Short Tandem Repeats (STRs)

Microsatellites are DNA sequences that repeat a small number of times within a genome and vary greatly between individuals. In wildlife forensics, microsatellites are used as genetic markers to establish the identity of an individual animal and determine its relatedness to other individuals within a population. Microsatellites are useful in wildlife forensics because they are highly variable, allowing for the creation of unique DNA fingerprints for individual animals. This can be useful in tracking the movements of animals, investigating wildlife crimes, and determining the relatedness of individuals within a population. In wildlife forensics, STRs are used to identify individuals within a species. This is accomplished by analyzing the number of repeats at specific loci (known as “markers”) across the genome. Because STRs are highly variable between individuals, even within the same species, they can be used to distinguish between individuals (Poetsch et al. 2001) and track the movement of wildlife or wildlife products. STR analysis is widely used in forensic investigations, including cases of illegal trade, poaching, and poaching of protected or endangered species (Hoff-Olsen et al. 2001). STR analysis can also be used to track the movement of wildlife products, such as meat, leather, and traditional medicines, to help enforce wildlife protection laws. Microsatellites are also useful for population genetics studies, as they can provide information on the genetic diversity, structure, and demographic history of a population. The use of microsatellites in wildlife forensics has been applied to a wide range of species, including mammals, birds, fish, and reptiles. It has become an important tool in the fight against wildlife trafficking and illegal trade, as well as in conservation biology and management. Microsatellites have been widely adopted as the preferred genetic markers for individual identification in many species and are widely used in wildlife forensics worldwide.

30.4.3.7 Minisatellites or Variable Number of Tandem Repeats (VNTRs)

Minisatellites also known as variable number of tandem repeats (VNTRs) are DNA sequences that repeat many times within a genome and vary greatly between individuals. They are useful in wildlife forensics as they provide a unique DNA fingerprint for each animal, which can be used to identify specific individuals, track their movements, and establish their relatedness to other animals. Animal samples in criminal cases rarely include substantially undamaged nuclear DNA, which is essential for VNTR analysis (Ogden et al. 2009). Minisatellites are particularly useful in species determination where traditional DNA markers, such as microsatellites, are not informative enough for individual identification (Kashyap et al. 2004). Minisatellites have been applied in a variety of species, including fish, birds, mammals, and reptiles, and have become an important tool in the fight against wildlife trafficking and illegal trade. The use of minisatellites in wildlife forensics can provide valuable information on population structure and diversity, which is important for conservation biology and management.

30.4.3.8 Messenger RNA

Messenger RNA (mRNA) is a type of RNA molecule that plays a key role in the expression of genes. In wildlife forensics, mRNA analysis can be used to determine the presence of specific genes in a sample and to gain insight into the biological processes occurring within an animal. One common application of mRNA analysis in wildlife forensics is the identification of species. By analyzing the mRNA expression patterns of specific genes, researchers can determine the species of an animal based on its genetic signature. This approach can be particularly useful when traditional DNA-based methods are not feasible, such as when working with degraded or low-quality DNA samples. In addition to species identification, mRNA analysis can also be used to study the effects of environmental factors on wildlife populations, such as exposure to pollutants or changes in habitat quality. By analyzing mRNA expression patterns, researchers can determine the impact of these factors on the biology of wildlife species, providing valuable information for conservation and management efforts.

30.4.3.9 DNA Barcoding

DNA barcoding is a method used in wildlife forensics to identify species based on a short, standardized segment of DNA. This technique has become increasingly popular in recent years for its accuracy, speed, and cost-effectiveness in species identification. By comparing the DNA sequences obtained from a sample with a reference database of known species, it is possible to determine the identity of an animal, even from small or degraded samples. DNA barcoding has been applied in various wildlife forensic applications, such as solving wildlife crimes, monitoring illegal trade, and supporting conservation efforts. The most commonly used DNA barcoding segment in wildlife forensics is the mitochondrial cytochrome c oxidase subunit I (COI) gene (Vences et al. 2005). The COI gene provides a fast, reliable, and cost-effective way to identify species, as it is highly conserved within species but divergent between species. This makes it an ideal barcode for species identification and has been adopted by the Barcode of Life Data System (BOLD) as the reference barcode for animals. Additionally, other DNA barcoding segments, such as the ribosomal internal transcribed spacer (ITS) region, have also been used in wildlife forensics, but COI remains the most widely used and accepted barcode in the field.

30.5 Conclusion

Wildlife forensics is a field that uses scientific methods to investigate crimes against wildlife, including poaching, trading, and hunting of endangered species. It involves collecting, analyzing, and interpreting physical pieces of evidence from wildlife crime scenes to help enforce wildlife protection laws and regulation. The ultimate goal of wildlife forensics is to protect and conserve wildlife populations and to bring those responsible for illegal activities to justice.

References

Ahmed YA, Ali S, Ghallab A (2018) Hair histology as a tool for forensic identification of some domestic animal species. EXCLI J 17:663

Balitzki-Korte B, Anslinger K, Bartsch C, Rolf B (2005) Species identification by means of pyrosequencing the mitochondrial 12S rRNA gene. Int J Legal Med 119(5):291–294

Bell B, Machin S (2011) The impact of migration on crime and victimisation. UK Government, London

Boner M, Förstel H (2004) Stable isotope variation as a tool to trace the authenticity of beef. Anal Bioanal Chem 378(2):301–310

Bowen GJ, Wassenaar LI, Hobson KA (2005) Global application of stable hydrogen and oxygen isotopes to wildlife forensics. Oecologia 143(3):337–348

Brown G, Gadaleta G, Pepe G, Saccone C, Sbisà E (1986) Structural conservation and variation in the D-loop-containing region of vertebrate mitochondrial DNA. J Mol Biol 192:503–511

Chamberlain CP, Blum JD, Holmes RT, Feng X, Sherry TW, Graves GR (1996) The use of isotope tracers for identifying populations of migratory birds. Oecologia 109(1):132–141

Chapman DD, Abercrombie DL, Douady CJ, Pikitch EK, Stanhopen MJ et al (2003) A streamlined, bi-organelle, multiplex PCR approach to species identification: application to global conservation and trade monitoring of the great white shark, Carcharodon carcharias. Conserv Genet 4: 415–425

Chernova OF (2014) Scanning electron microscopy of the hair medulla of orangutan, chimpanzee, and man. Dokl Biol Sci 456(1):199

Coghlan ML, White NE, Parkinson L, Haile J, Spencer PB, Bunce M (2012) Egg forensics: an appraisal of DNA sequencing to assist in species identification of illegally smuggled eggs. Foren Sci Int Genet 6(2):268–273

Cooper JE, Cooper ME (2008) Introduction to veterinary and comparative forensic medicine. Blackwell, Oxford

Dawnay N, Ogden R, McEwing R, Carvalho GR, Thorpe RS (2007) Validation of the barcoding gene COI for use in forensic genetic species identification. Forensic Sci Int 173:1–6

Dove CJ, Koch SL (2011) Microscopy of feathers: a practical guide for forensic feather identification. Microscope 59(2):51

Ferri G, Alu M, Corradini B, Licata M, Beduschi G (2009) Species identification through DNA "barcodes". Genet Test Mol Biomark 13(3):421–426

Fortajada EB, Fontanilla IKC, De Ungria MCA (2021) Incorporating animal forensics in routine meat inspection in the Philippines. Foren Sci Int Anim Environ 1:100020

Gouda S, Kerry RG, Das A, Chauhan NS (2020) Wildlife forensics: a boon for species identification and conservation implications. Forensic Sci Int 317:110530

Guha S, Kashyap VK (2006) Molecular identification of lizard by RAPD & FINS of mitochondrial 16s rRNA gene. Legal Med 8(1):5–10

Hobson KA, Wassenaar LI (1996) Linking breeding and wintering grounds of neotropical migrant songbirds using stable hydrogen isotopic analysis of feathers. Oecologia 109(1):142–148

Hoff-Olsen P, Jacobsen S, Mevag B et al (2001) Microsatellite stability in human post-mortem tissues. Forensic Sci Int 119:273–278

Hsieh HM, Huang LH, Tsai LC, Kuo YC, Meng HH et al (2003) Species identification of rhinoceros horns using the cytochrome b gene. Forensic Sci Int 136:1–11

Janczewski DN, Modi WS, Stephens JC, O'Brain SJ (1995) Molecular evaluation of mitochondrial 12S RNA and cytochrome b sequences in the Pantherine lineage of Felidae. Mol Biol Evol 12: 690–707

Jogayya KN, Meganathan PR, Dubey B, Haque I (2013) Mitochondrial 16S ribosomal RNA gene for forensic identification of crocodile species. J Forensic Legal Med 20:334–338

Johnson RN, Wilson-Wilde L, Linacre A (2014) Current and future directions of DNA in wildlife forensic science. Foren Sci Int Genet 10:1–11

Kashyap VK, Sitalaximi T, Chattopadhyay P et al (2004) DNA profiling techniques in forensic analysis. Int J Hum Genet 4:11–30

Kitano T, Umetsu K, Tian W, Osawa M (2007) Two universal primer sets for species identification among vertebrates. Int J Legal Med 121:423–427

Kitpipit T, Thanakiatkrai P, Penchart K, Ouithavon K, Satasook C, Linacre A (2016) Ivory species identification using electrophoresis-based techniques. Electrophoresis 37(23–24):3068–3075

Kumar R, Singh PJ, Nagpure NS, Kushwaha B, Srivastava SK et al (2007) A non-invasive technique for rapid extraction of DNA from fish scales. Indian J Exp Biol 45:992–997

Lawton MPC, Cooper JE (2009) Wildlife crime scene visits. Appl Herpetol 6:29–45

Linacre A, Ciavaglia SA (2017) Wildlife forensic science. In: Handbook of forensic genetics: biodiversity and heredity in civil and criminal investigation. World Scientific Publishing, Singapore, pp 449–471

Linacre A, Tobe SS (2011) An overview to the investigative approach to species testing in wildlife forensic science. Investig Genet 2(1):1–9

Martinsohn JT, Ogden R (2009) Fishpoptrace: developing Snp-based population genetic assignment methods to investigate illegal fishing. Foren Sci Int Genet 2:294–296

Melton T, Holland C (2007) Routine forensic use of the mitochondrial 12S ribosomal RNA gene for species identification. J Forensic Sci 52(6):1305–1307

Mitani T, Akane A, Tokiyasu T, Yoshimura S, Okii Y, Yoshida M (2009) Identification of animal species using the partial sequences in the mitochondrial 16S rRNA gene. Leg Med 11(Suppl 1): S449–S450

Ogden R (2011) Unlocking the potential of genomic technologies for wildlife forensics. Mol Ecol Resour 11(Suppl. 1):109–116

Ogden LJE, Hobson KA, Lank DB (2004) Blood isotopic (δ13C and δ15N) turnover and diet-tissue fractionation factors in captive Dunlin (Calidris alpina pacifica). Auk 121(1):170–177

Ogden R, Dawnay N, McEwing R (2009) Wildlife DNA forensics: bridging the gap between conservation genetics and law enforcement. Endanger Species Res 9:179–195

Poetsch M, Seefeldt S, Maschke M et al (2001) Analysis of microsatellite polymorphism in red deer, roe deer, and fallow deer – possible employment in forensic applications. Forensic Sci Int 116:1–8

Pun KM, Albrecht C, Castella V, Fumagalli L (2009) Species identification in mammals from mixed biological samples based on mitochondrial DNA control region length polymorphism. Electrophoresis 30(6):1008–1014

Renou J-P, Deponge C, Gachon P, Bonnefoy J-C, Coulon J-B, Garel J-P, Vérité R, Ritz P (2004) Characterization of animal products according to geographic origin and feeding diet using nuclear magnetic resonance and isotope ratio mass spectrometry: cow milk. Food Chem 85(1):63–66

Saccone C, Attimonelli MA, Sbisà E (1987) Structural elements highly preserved during the evolution of the D-Loop containing region in vertebrate mitochondrial DNA. J Mol Evol 26: 205–211

Sahajpal V, Goyal SP (2009) Microscopic examinations in wildlife investigations. In: Forensic science in wildlife investigations. CRC Press, Boca Raton, FL, pp 33–74

Sahajpal V, Goyal SP, Raza R, Jayapal R (2009) Identification of mongoose (genus: Herpestes) species from hair through band pattern studies using discriminate functional analysis (DFA) and microscopic examination. Sci Justice 49:205–209

Schneider JL (2008) Reducing the illicit trade in endangered wildlife: the market reduction approach. J Contemp Crim Justice 24:274

Sobrino B, Brion M, Carracedo A (2005) SNPs in forensic genetics: a review on SNP typing methodologies. Forensic Sci Int 154:181–194

Sreepada RA, Desai UM, Naik S (2002) The plight of Indian sea horses: need for conservation and management. Indian Acad Sci 82:377–378

Stock F, Edwards CJ, Bollongino R, Finlay EK, Burger J et al (2009) Cytochrome b sequences of ancient cattle and wild ox support phylogenetic complexity in the ancient and modern bovine populations. Anim Genet 40:694–700

Tobe SSS (2009) Determining the geographic origin of animal samples. In: Forensic science in wildlife investigations. CRC Press, Boca Raton, FL, pp 141–170

Tobe SS, Linacre AM (2008) A technique for the quantification of human and non-human mammalian mitochondrial DNA copy number in forensic and other mixtures. Foren Sci Int Genet 2(4):249–256

Tobe SS, Linacre A (2010) DNA typing in wildlife crime: recent developments in species identification. Foren Sci Med Pathol 6:195–206
Vences M, Thomas M, van der Meijden A, Chiari Y, Vieites DR (2005) Comparative performance of the 16S rRNA gene in DNA barcoding of amphibians. Front Zool 2:1–12
Yadav P, Kumar A, Sahoo S, Yadav N, Hussain SA, Gupta SK (2021) Identification of Gangetic turtles based on species-specific variations on mitochondrial cyt b and nuclear Cmos genes. Foren Sci Int Anim Environ 1:100035

Evidence and Identification

31

Rahul Ravindra Darunde, Hansi Bansal, and Avinash Puri

Abstract

A relatively recent area of criminal investigation is wildlife forensics. Its objectives are to correlate the evidence from crime scenes to a suspect and a victim, in this case, an animal, by using scientific methods to inspect, classify, and compare the evidence from crime scenes. It is the application of different techniques of forensic science to investigate crimes related to wild animals, including poaching, illegal trade in wildlife products, and other wildlife-related crimes, to identify the species, origin, and history of the evidence, and to link it to the crime scene or suspects. Poaching, often known as killing wild creatures legally exempt from hunting, is one of the most serious offenses that wildlife forensic specialists look into. The use of wildlife forensics in criminal investigations has proven to be a quick, accurate, and dependable method with comprehensive coverage and simple accessibility. The resolution of taxonomic conflicts, the identification of spatiotemporal genetic divergence, the study of evolutionary history, origins, and even endemism have all benefited from it.

Keywords

Endangered · Habitats · Climate · Genetic divergence · Wildlife trade

R. R. Darunde
Department of Forensic Science, Medi-Caps University, Indore, India

H. Bansal
Department of Forensic Science, Government Institute of Forensic Science, Nagpur, India

A. Puri (✉)
Department of DNA and Forensic Biology, Regional Forensic Science Laboratory, Indore, Madhya Pradesh, India

A. Puri et al. (eds.), *Fundamentals of Forensic Biology*,
https://doi.org/10.1007/978-981-99-3161-3_31

31.1 Wildlife Forensics

Wildlife forensic science has emerged as a crucial tool for implementing laws about the illegal trade of endangered and protected animals. The illegal wildlife trade is responsible for the exploitation of several animal and plant species and the destruction of habitats (Ogden and Linacre 2015). The introduction of exotic species, climate change, soil degradation, and ecological imbalance are all made more accessible by the illegal wildlife trade. The commerce in wildlife is not a recent development; it has existed for many centuries. Ivory was employed in ancient European cultures for tools and jewelry. Due to their quest for ivory, the Romans decimated the elephant herds in northern Africa and Asia Minor (Keskin et al. 2022). The illegal taking, trading, exploiting, possessing, or killing of animals or plants in violation of local, national, or international regulations is known as wildlife crime. The recent increase in wildlife crime, especially wildlife trafficking, poses a severe threat to the environment, local and national economy, public health, security, and even the criminal justice system (Kurland et al. 2017). Although forensic analysis has long been used in wildlife law enforcement, interest in the field has grown significantly during the past 10 years (Linacre et al. 2011).

Identification of the species of origin of carcasses is necessary in forensic cases (Dalton and Kotze 2011). DNA profiling, toxicology, animal pathology, ballistics, and fiber analysis are a few analytical techniques used to investigate crimes against wildlife under the umbrella term "wildlife forensics" (Huffman and Wallace 2012). Recognizing the species represented in the evidence is a crucial first step in any inquiry into a wildlife crime. While certain species are strictly protected and can face severe fines for being killed or traded commercially, others have no legal protection. As a result, establishing a taxonomic identity is necessary to evaluate whether a wildlife crime has been committed (Trail 2021). Morphological or DNA investigation often uses wildlife evidence to identify the species present. Using physical characteristics found in the data, such as avian feather patterns, mammalian dentition, or reptilian scale counts, trained and competency-tested professionals identify species (Lyman 2019). After illicit drug sales and the sale of firearms, the wildlife trade is the third-largest illegal enterprise in the world. While animal carcasses are easily identified at the species level, wildlife artifacts are stripped of identifying traits at their crime scenes, making identifying both legally and criminally trafficked wildlife species and objects challenging for wildlife forensic experts (Sahajpal et al. 2009).

31.2 Physical Evidence

Any object, material, or physical entity connecting the victim, suspect, and the crime scene is considered physical evidence. The hardest part of processing a crime scene is identifying the evidence, gathering it, and keeping it safe so that it can be examined further in a laboratory. A forensic laboratory veterinary pathologist will often collect the carcass and identify the cause of death. Once an item of evidence is identified for collection, a placard with a unique number is set up next to it. Prior to collection, evidence is then photographed and mapped (Linacre 2009). Some of the commonly found physical evidence in wildlife crime scenes is as follows.

31.2.1 Hair

An essential part of species identification at a crime scene is looking at the hair of the animals. The root and shaft of the hair can be distinguished. The guard hairs found on most mammal species flatten (shield) at the distal end (away from the skin). Some species have guard or fur hairs that appear zigzagged. When creating a reference checklist and looking through unidentified samples, these general forms or profiles should be carefully noted because they aid in focusing the research on the most likely family, genus, or species (Linacre 2009). The guard hair, one of the four primary types of hair identified in mammals, is crucial for determining an animal's species (Knecht 2012). For the characterization of different traits, it is strongly advised to examine the hair shaft, including its medulla (internal center), cortex (transitional layer), and cuticle (outer covering), using proper magnifying tools such as compound microscopes or scanning electron microscopes (Bell 2011; Ahmed et al. 2018).

31.2.2 Claw

A claw is a pointed, curved nail or talon that can be seen on the feet of several animals, including cats, birds, and lobsters. Hunting, self-defense, and gripping or holding onto objects are the usual uses of claws. Keratin is used to make claws in mammals. Some animals, like cats, can extend or retract their retractable claws at will. In many cultures and legends, the use of claws as a symbol of strength or fury is also every day. A clear indication of poaching for the illicit sale of the taken items is the discovery in a conservation area of the carcass of a protected animal whose claws had been removed (Huffman and Wallace 2012). A stereo microscope can be used to investigate the surface morphology of a claw at low magnification (Linacre et al. 2011).

31.2.3 Ivory

Elephant ivory and other skeletal artifacts are increasingly traded illegally because they are valued as ornamental objects. African elephants (Loxodonta cyclotis and L. africana) and Asian elephants (Elephas maximus) are currently in danger, and the ivory trade is the primary cause (Lee et al. 2013). Dentine is a material that includes ivory, resembling bones (Espinoza and Mann 1992). Ivory has a distinctive microstructural characteristic known as the Schreger patterns, which have measured and identifiable angles. All three varieties of elephants, including mammoths, exhibit a distinct variance in the mean angles of the Schreger pattern. It might serve as a criterion for separating various types of ivory. Identifying and separating ivory from horn, hoof, and shells using FTIR and Raman's spectroscopy based on their band patterns at different wavelengths is depicted through studies (Singh et al. 2006). Ivories and other artifacts can be used to date using FTIR and Raman's spectroscopy (Espinoza and Mann 1993).

31.2.4 Bones

Hunters and poachers keep bones, horns, or hooves as trophies and decorative things. As they are curled for use as ornaments, statues, and altered artifacts, bones are also referred to as "poor man's ivory." Joints, general size and shape, dentition, ligament insertion, and other characteristics of bones that identify species' characteristics are typically used to identify bones. Since most species have distinctive dental structures, the likelihood of correct identification is significantly increased by the presence of dental structures (Sims et al. 2011). Zooarchaeology commonly establishes the identity of animals through their remains or fossil materials and bone pieces. Other characteristics used to identify mammal bones include vascular tissues, tiny nutrition canals, and gritty particles inside the bone matrix (Espinoza and Mann 1992). Bones can be used to identify morphological species. However, doing so can be difficult because the bones' diagnostic features are subject to aging, and the origin of the species is unclear (Sims et al. 2011). Recently, the bones of lions and tigers were separated using morphological identities and a quick real-time PCR approach (Dalton et al. 2020).

31.2.5 Mitochondrial DNA

Molecular markers are collections of DNA sequences with known chromosomal locations whose inheritance can be tracked for genetic diversity or DNA typing (Kumar et al. 2018; Coghlan et al. 2012; Green et al. 2019; Nishant et al. 2017; Mosa et al. 2019). Mitochondrial markers and nuclear markers are the two primary markers utilized in wildlife forensic science (Mitra et al. 2018; Jan and Fumagalli 2016). Mitochondrial DNA (mtDNA) is a collection of genetic materials that are genetically conserved and located in the mitochondria of the cell. Taxonomic, phylogenetic, forensic, and conservation organizations have joined forces to develop an advanced, reliable, and reasonably priced DNA sequencing method to eliminate these restrictions and tie up loose ends (Green et al. 2019). It has several uses in phylogenetic investigations (Kumar et al. 2020; Toren et al. 2016). In the process, incomplete data for 29 non-chordate species were found, along with several other things (Boore 1999). The National Centre for Biotechnology Informatics (NCBI) most recently made 9034 genome sequences available out of the 10,249 animal mitochondrial genomes (Zhou et al. 2020). For efficient profiling of wild species, mitochondrial DNA markers such as cytochrome b (Cyt b), cytochrome c oxidase subunit I, II, and III (COI), ATP synthase 6 (ATP6), NADH dehydrogenase 3 (ND3), NADH dehydrogenase 4 (ND4), NADH-ubiquinone oxidoreductase chain 4 L (ND4 L) gene, 16S, 12S rRNA, tRNAs, and the control region (D-loop) are used (Mitra et al. 2018; Detwiler 2019; Abbas et al. 2020). COI is the most often used mtDNA marker overall due to its benefits, including its small size, wide variety, ready-to-use universal primers, and durability (De Mandal et al. 2014; Linacre et al. 2011).

31.2.6 Nuclear DNA

Short tandem repeat (STR) profiles of nuclear DNA based on an animal species are primarily used in using nuclear DNA markers in individual identification. Especially for domesticated mammals, using the short tandem repeat (STR) locus is a good tool for determining samples' levels (Linacre et al. 2011). Amplified fragment length polymorphism (AFLP), random amplified polymorphic DNA (RAPD), and short sequence repeats (SSRs) or microsatellites are a few of the regularly employed nuclear markers (Mitra et al. 2018). Markers like RAPD and AFLP stand out from the competition because they can identify and distinguish between various animal species without requiring prior molecular knowledge (Arif et al. 2011). The most often utilized nuclear DNA markers are microsatellites. With bi-allele or multi-allele representing an individual or a population, respectively, microsatellites are co-dominant markers. These genetic markers are polymorphic and highly adaptable for molecular fingerprinting since they are simple to amplify using PCR (Arif et al. 2011; Arif and Khan 2009).

31.2.7 Firearms and Ammunition

During the postmortem examination, it is essential to remove the pellets or bullets from the carcass in cases where firearms were used to determine the type of firearms used. To identify the firearm used, pellets or bullets should be delivered to the Forensic Laboratory for the Ballistic expert's analysis. If someone is suspected of shooting an animal, it is advised to wash their hands and check their fingers for any gunshot residue. Along with the ballistics examination report and fingerprints, this is crucial evidence to prove that the accused shot (Negi 2013; Warlow 2004).

31.2.8 Digital Devices

The accused/suspect's mobile phone devices should be checked for stored phone numbers, and a list of those numbers should be made. The laptops and computers should also be checked for information like a map and other facts about pertinent topics like information on sanctuaries or national parks, and email files of the agreements. A laptop should only be used in public with witnesses present. The suspect/accused should be questioned regarding passwords used to access email accounts, computers, etc. The devices should be promptly turned off after inspection, and their battery should be removed. These devices are required to be seized for expert examination (Negi 2013).

31.2.9 Hoof Marks

The hoof is the protective outer covering of the foot of certain animals, such as horses, cattle, deer, pigs, and many other mammals. Hooves are made of keratin, the same substance that makes up human nails and hair. In wildlife forensic investigations, hooves can identify the animal species from which a sample came and provide information about the animal's age, sex, and habitat. In addition, the analysis of hooves can be used to identify the cause of death, such as starvation, disease, or trauma (Edwards et al. 1998).

31.3 Analysis of Wildlife Evidence

Analyzing wildlife forensic evidence is crucial in investigating and prosecuting wildlife crimes. Some of the fundamental techniques used in the analysis of wildlife forensic evidence include the following.

31.3.1 Isotope Analysis

Isotope analysis is used to study the ecology and biology of wild animals, including their diet, migration patterns, and habitat use. The technique involves analyzing the isotopes, or different forms, of elements found in an animal's tissue, such as its bones, hair, or feathers. Isotope analysis is non-destructive and is applied to many species and samples, making it a valuable tool for wildlife conservation and management. It is used to study the population dynamics, habitat use, diet, and migration patterns of wild animals, which can help conservationists better understand and protect wildlife (Bowen et al. 2005).

31.3.2 DNA Analysis

DNA analysis typically involves extracting DNA from a sample, amplifying a specific region of the DNA using polymerase chain reaction (PCR), and then analyzing the amplified DNA using techniques such as gel electrophoresis, capillary electrophoresis, or DNA sequencing. Mitochondrial DNA is abundant in cells, easy to extract, and has a high degree of variation among different species, making it an ideal target for PCR amplification and DNA sequencing. STRs are highly variable among individuals and populations and can be used to track the movement of illegal products. DNA analysis can be used to identify the species of origin of an illegal product, such as ivory, meat, or fur. It can also be used to determine the animal's or plant's geographical origin, which can help investigators identify the source of the illegal trade. The validity of the DNA analysis results depends on the DNA sample's quality and quantity, the choice of DNA markers, and the accuracy of the PCR amplification and DNA sequencing methods (Lorenzini 2005; Kanokwongnuwut et al. 2020).

31.3.3 Entomological Analysis

Entomological analysis in wildlife forensics involves the study of insects, particularly their life cycles and stages, to determine the time of death, or the postmortem interval, in human or animal death investigations. It can be used in wildlife crime investigations, such as illegal hunting or poaching, to help establish the time of death of an animal and also to provide information about the location where the animal was killed (Byrd and Tomberlin 2019).

31.3.4 Microscopy

Microscopic analysis in wildlife forensics involves the examination of small samples or particles using a microscope to identify the species or origin of the sample. This can include analyzing hair, feathers, scales, bones, and other biological materials. Different microscopes are commonly used in wildlife forensic investigations, including light, fluorescence, transmission electron, scanning electron, and Fourier transform infrared microscopes. Light microscopy examines hair, feathers, scales, and other small samples. Fluorescence microscopy uses fluorescent dyes to highlight certain features of a sample, such as specific types of cells or proteins. It is used to examine hair, feathers, and other small samples. Transmission electron microscopy (TEM) examines small samples, such as bone, teeth, or scales. Scanning electron microscopy (SEM) is used to examine small samples, such as hair, feathers, and scales, and provide detailed information about the surface features of the sample. Fourier transform infrared spectroscopy (FTIR) is used to identify the species of animal from which a sample came and to detect the presence of certain chemicals or substances that may be used in wildlife crime (Edwards et al. 1998; Gouda et al. 2020).

31.3.5 Chromatography

Chromatography is an instrumental technique used to separate and analyze different components of a mixture. In wildlife forensics, various types of chromatography can be used to analyze samples from wildlife crime scenes, such as blood, tissue, feathers, or other organic materials. Gas chromatography (GC) and liquid chromatography (LC) are two commonly used types of chromatography in wildlife forensics. GC is used to separate and analyze volatile compounds, such as pesticides or pollutants, while LC is used to separate and analyze non-volatile compounds, such as drugs or toxins. Other types of chromatography that are used in wildlife forensics include high-performance liquid chromatography (HPLC), ion chromatography (IC), and size exclusion chromatography (SEC). Gas chromatography–mass spectrometry (GC-MS) and liquid chromatography–mass spectrometry (LC-MS) are powerful tools for identifying and quantifying illegal drugs, pesticides, and other chemicals in wildlife samples (Bell 2011; Gouda et al. 2020).

31.3.6 Computer Forensics

Computer forensics in wildlife refers to using digital forensic techniques to investigate crimes and other illegal activities related to wildlife. It includes the analysis of computers, smartphones, and other digital devices that may contain evidence of illegal hunting, poaching, or wildlife trafficking. Forensic examiners use various techniques to extract data from digital devices, such as disk imaging, file carving, and keyword searching. They also use specialized software to analyze the data, such as GPS mapping tools to track the movement of poachers or social media analysis tools to identify networks of wildlife traffickers. It can also be used in wildlife conservation, for example, to prevent illegal logging and trade of wildlife and study the wildlife population for conservation (Negi 2013).

31.3.7 Ballistic Analysis

Ballistic analysis can be used in wildlife crime investigations, such as illegal hunting or poaching, to help establish a link between a suspect and a crime. Ballistic analysis in wildlife forensics involves the examination of firearms and ammunition to identify the type of weapon used in a crime, as well as to determine the origin of the weapon and to link it to specific crimes. Analysis of bullets involves comparing the characteristics of bullets recovered from a crime scene to those fired from a specific weapon to determine if the weapon was used in the crime. Bullet characteristics that can be analyzed include the rifling marks, land and groove marks, and caliber. Analysis of cartridge cases involves comparing the characteristics of firing pin impressions, extractor marks, and ejector marks (Warlow 2004).

31.3.8 Toxicology

Toxicological analysis in wildlife forensics involves the examination of biological samples, such as blood, tissue, or organs, to detect the presence of toxic substances or pollutants. It detects many toxic substances, including pesticides, heavy metals, and other pollutants. These substances can harm wildlife and are helpful as evidence in criminal investigations. Toxicology analysis can also detect the presence of drugs, such as tranquilizers and opioids, which are used illegally to capture wildlife for trade or other purposes (Everatt et al. 2019).

31.3.9 Geographical Information System (GIS)

GIS is a powerful tool that allows scientists to combine, analyze, and visualize spatial data, such as maps, satellite imagery, and other data layers. GIS is used in wildlife forensics to help identify the origin of an animal or its products, such as meat or ivory. It is used to map the location of crime scenes, illegal hunting sites, and routes of illegal trade and to identify patterns and trends in wildlife crime. One way GIS can be used in wildlife forensics is to map the distribution of different animal species and their isotopic signatures, which can help to identify the geographic origin of illegally traded wildlife products, such as ivory or shark fins, and track down the locations where the animals were poached (Bowen et al. 2005; Linacre and Ciavaglia 2017).

31.3.10 Radiocarbon Dating

Radiocarbon dating (carbon-14 dating) determines organic materials' age. It measures the amount of carbon-14, a radioactive isotope of carbon, present in a sample. It is used to determine the age of an animal or plant, which can help to identify illegal hunting methods or to track the illegal trade of certain species like ivory. In wildlife forensics, radiocarbon dating is used to determine the age of animal remains or other organic materials found at a crime scene. This information helps solve wildlife-related crimes, such as poaching or illegal hunting (Uno et al. 2013; Cerling et al. 2016).

31.3.11 3D Scanning

3D scanning in wildlife forensics involves using 3D scanning technologies, such as laser scanning or structured light scanning, to create detailed 3D models of biological samples, such as bones, teeth, or other body parts. These models can be used to identify the species of animal from which the sample came, as well as to identify specific individuals within a population. It can also create 3D models of other body parts, such as footprints, tracks, or scats, to identify animal species and study their behavior and ecology (Košinová et al. 2022).

31.3.12 Dental Analysis

In wildlife forensics, dental analysis is used to identify poached animals, study the population dynamics and health of wildlife, and track the movement of animals.

Dental characteristics such as tooth shape, size, number, and wear patterns are used to identify the species of an animal. Dental development and wear patterns can determine the age of an animal. It can provide vital information for the conservation and management of wildlife populations and criminal investigations related to wildlife (Viciano et al. 2022).

31.3.13 Metabolomics

Metabolomics is a field of study that involves analyzing the complete set of small molecules, known as metabolites, present in an organism. In wildlife forensics, metabolomics can be used to study the metabolic profile of animals and provide information on their diet, health, and exposure to pollutants or other environmental factors. It is used to study the exposure of animals to pollutants or other environmental factors. Metabolites can be used as biomarkers to detect exposure to pollutants, such as pesticides or heavy metals, which can help detect illegal activities, such as illegal hunting or wildlife trafficking (Luo et al. 2020).

It is imperative to note that not all of these techniques are suitable for all types of evidence or all cases. The selection and use of these techniques should be based on the case's specific circumstances and the forensic analyst's expertise.

31.4 Significance of Wildlife Forensics

Investigation and prosecution of wildlife crimes depend heavily on wildlife forensics. Some of the principal advantages of wildlife forensics are as follows:

- Determining whether or not a product is made from illegal wildlife: Wildlife forensics techniques, such as DNA analysis, can be used to identify the type of an animal or plant and can assist investigators in figuring out whether or not a product is unlawful.
- Tracing the origin of illegal products: Wildlife forensics methods, such as isotope analysis, can be used to track the provenance of illegal wildlife products and assist investigators in locating the illegal trade's origin.
- Establishing a link and proving criminal intent: Wildlife forensics methods like fingerprint and blood spatter analysis can produce tangible proof of criminal intent.
- Promoting international cooperation: To fight wildlife crime worldwide, wildlife forensics techniques can be utilized to exchange information and coordinate activities between various nations and organizations.
- Providing scientific proof in court: Wildlife forensics can produce proof that can be utilized in court to help prosecute wildlife offenders.
- Supporting conservation efforts: By identifying the species that are becoming extinct and determining the cause, wildlife forensic techniques can assist conservationists in taking preventive action.

- Closing the legal loopholes: Using wildlife forensic techniques, the legal foundation to combat wildlife crime can be strengthened by finding the weaknesses in current laws and regulations.
- Preventing wildlife crime: Using wildlife forensics can help prevent wildlife crime by making it more likely that offenders will be found out and brought to justice.
- Preventing the extinction of endangered species: Wildlife forensics methods can be used to locate and monitor the illegal trade in threatened and endangered species.
- Providing insights into crime syndicates: Wildlife forensics can help detectives break up criminal networks involved in wildlife crime by revealing their organizational structure and methods of operation.
- Facilitating international trade: Wildlife forensics can help ensure that international trade in wildlife products is legal and sustainable and can help prevent the illegal trade in endangered species.
- Advancement of scientific knowledge: By supplying new data on species, their distribution and population trends, and the effects of human activity on wildlife, wildlife forensics can contribute to the advancement of scientific knowledge.
- Preventing unlawful hunting: Wildlife forensics can be utilized to locate and apprehend those who engage in illegal hunting.
- Determining the effects of climate change: Wildlife forensics can determine the effects of climate change on different wildlife species and their habitats, which can then be used to guide management and conservation efforts.
- Raising public awareness: Wildlife forensics can be used to inform people about the need to protect wildlife and the repercussions of wildlife crime.

References

Abbas G, Nadeem A, Javed M, Ali MM, Aqeel M, Babar ME et al (2020) Mitochondrial cytochrome-b, cytochrome-c and D-loop region based phylogenetic and diversity analysis in blackbuck (Antilope cervicapra). Kafkas Üniversitesi Veteriner Fakültesi Dergisi 26(1):25

Ahmed YA, Ali S, Ghallab A (2018) Hair histology as a tool for forensic identification of some domestic animal species. EXCLI J 17:663

Arif IA, Khan HA (2009) Molecular markers for biodiversity analysis of wildlife animals: a brief review. Anim Biodivers Conserv 32(1):9–17

Arif IA, Khan HA, Bahkali AH, Al Homaidan AA, Al Farhan AH, Al Sadoon M, Shobrak M (2011) DNA marker technology for wildlife conservation. Saudi J Biol Sci 18(3):219–225

Bell LS (2011) Forensic science in support of wildlife conservation efforts—morphological and chemical approaches (global trends)—. Foren Sci Rev 23(1):31–36

Boore JL (1999) Animal mitochondrial genomes. Nucleic Acids Res 27(8):1767–1780

Bowen GJ, Wassenaar LI, Hobson KA (2005) Global application of stable hydrogen and oxygen isotopes to wildlife forensics. Oecologia 143(3):337–348

Byrd JH, Tomberlin JK (eds) (2019) Forensic entomology: the utility of arthropods in legal investigations. CRC Press, Boca Raton, FL

Cerling TE, Barnette JE, Chesson LA, Douglas-Hamilton I, Gobush KS, Uno KT et al (2016) Radiocarbon dating of seized ivory confirms rapid decline in African elephant populations and provides insight into illegal trade. Proc Natl Acad Sci 113(47):13330–13335

Coghlan ML, White NE, Parkinson L, Haile J, Spencer PB, Bunce M (2012) Egg forensics: an appraisal of DNA sequencing to assist in species identification of illegally smuggled eggs. Foren Sci Int Genet 6(2):268–273

Dalton DL, Kotze A (2011) DNA barcoding as a tool for species identification in three forensic wildlife cases in South Africa. Forensic Sci Int 207(1–3):e51–e54

Dalton DL, Kotzé A, McEwing R, De Bruyn M, Mnisi C, Mwale M (2020) A tale of the traded cat: development of a rapid real-time PCR diagnostic test to distinguish between lion and tiger bone. Conserv Genet Resour 12(1):29–31

De Mandal S, Sanga Z, Nachimuthu SK (2014) Metagenomic analysis of bacterial community composition among the cave sediments of Indo-Burman biodiversity hotspot region. PeerJ 2: e631v1

Detwiler KM (2019) Mitochondrial DNA analyses of Cercopithecus monkeys reveal a localized hybrid origin for C. mitis doggetti in Gombe National Park, Tanzania. Int J Primatol 40(1): 28–52

Edwards HGM, Hunt DE, Sibley MG (1998) FT-Raman spectroscopic study of keratotic materials: horn, hoof and tortoiseshell. Spectrochim Acta A Mol Biomol Spectrosc 54(5):745–757

Espinoza EO, Mann MJ (1992) Identification guide for ivory and ivory substitutes. WWF, Gland

Espinoza EON, Mann MJ (1993) The history and significance of the Schreger pattern in proboscidean ivory characterization. J Am Inst Conserv 32(3):241–248

Everatt KT, Kokes R, Lopez Pereira C (2019) Evidence of a further emerging threat to lion conservation; targeted poaching for body parts. Biodivers Conserv 28(14):4099–4114

Gouda S, Kerry RG, Das A, Chauhan NS (2020) Wildlife forensics: a boon for species identification and conservation implications. Forensic Sci Int 317:110530

Green ME, Appleyard SA, White W, Tracey S, Devloo-Delva F, Ovenden JR (2019) Novel multimarker comparisons address the genetic population structure of silvertip sharks (Carcharhinus albimarginatus). Mar Freshw Res 70(7):1007–1019

Huffman JE, Wallace JR (2012) Wildlife forensics: methods and applications. John Wiley & Sons, New York, NY

Jan C, Fumagalli L (2016) Polymorphic DNA microsatellite markers for forensic individual identification and parentage analyses of seven threatened species of parrots (family Psittacidae). PeerJ 4:e2416

Kanokwongnuwut P, Kirkbride KP, Linacre A (2020) Detecting latent DNA in wildlife forensic science investigations. Sci Justice 60(4):358–362

Keskin BB, Griffin EC, Prell JO, Dilkina B, Ferber A, MacDonald J et al (2022) Quantitative investigation of wildlife trafficking supply chains: a review. Omega 115:102780

Knecht L (2012) The use of hair morphology in the identification of mammals. In: Wildlife forensics. Wiley, New York, NY, pp 129–143

Košinová K, Turek J, Cukor J, Linda R, Häckel M, Hart V (2022) The application of 3D imaging as an appropriate method of wildlife craniometry: evaluation of accuracy and measurement efficiency. Animals 12(23):3256

Kumar V, Sharma N, Singal K, Sharma A (2018) A wildlife forensic study for the species identification of Indian blackbuck through forensically informative nucleotide sequencing (FINS). Biosci Biotechnol Res Asia 15(1):175

Kumar V, Tyagi K, Chakraborty R, Prasad P, Kundu S, Tyagi I, Chandra K (2020) The Complete Mitochondrial Genome of endemic giant tarantula, Lyrognathus crotalus (Araneae: Theraphosidae) and comparative analysis. Sci Rep 10(1):1–11

Kurland J, Pires SF, McFann SC, Moreto WD (2017) Wildlife crime: a conceptual integration, literature review, and methodological critique. Crime Sci 6(1):1–15

Lee EJ, Lee YH, Moon SH, Kim NY, Kim SH, Yang MS et al (2013) The identification of elephant ivory evidences of illegal trade with mitochondrial cytochrome b gene and hypervariable D-loop region. J Forensic Legal Med 20(3):174–178

Linacre A (ed) (2009) Forensic science in wildlife investigations. CRC Press, Boca Raton, FL

Linacre A, Ciavaglia SA (2017) Wildlife forensic science. In Handbook of forensic genetics: biodiversity and heredity in civil and criminal investigation. World Scientific Publishing, Singapore, pp 449–471

Linacre A, Gusmão L, Hecht W, Hellmann AP, Mayr WR, Parson W et al (2011) ISFG: recommendations regarding the use of non-human (animal) DNA in forensic genetic investigations. Foren Sci Int Genet 5(5):501–505

Lorenzini R (2005) DNA forensics and the poaching of wildlife in Italy: a case study. Forensic Sci Int 153(2–3):218–221

Luo X, Meng J, Chen X, Cheng L, Yan S, Gao L et al (2020) Metabolomics-based study reveals the effect of lead (Pb) in the culture environment on Whitmania pigra. Sci Rep 10(1):1–10

Lyman RL (2019) Assumptions and protocol of the taxonomic identification of faunal remains in zooarchaeology: a North American perspective. J Archaeol Method Theory 26(4):1376–1438

Mitra I, Roy S, Haque I (2018) Application of molecular markers in wildlife DNA forensic investigations. J Foren Sci Med 4(3):156

Mosa KA, Gairola S, Jamdade R, El-Keblawy A, Al Shaer KI, Al Harthi EK et al (2019) The promise of molecular and genomic techniques for biodiversity research and DNA barcoding of the Arabian Peninsula flora. Front Plant Sci 9:1929

Negi SB (2013) Wildlife crime investigation a hand book for wildlife crime investigation officers. Wildlife Crime Control Bureau, Ministry of Environment and Forests Government of India, New Delhi

Nishant K, Vrijesh KY, Ajay KR (2017) Wildlife forensic: current techniques and their limitations. J Foren Sci Criminol 5(4):402

Ogden R, Linacre A (2015) Wildlife forensic science: a review of genetic geographic origin assignment. Foren Sci Int Genet 18:152–159

Sahajpal V, Goyal SP, Raza R, Jayapal R (2009) Identification of mongoose (genus: Herpestes) species from Hair through band pattern studies using discriminate functional analysis (DFA) and microscopic examination. Sci Justice 49(3):205–209

Sims ME, Baker BW, Hoesch RM (2011) Tusk or bone? An example of ivory substitute in the wildlife trade. Ethnobiol Lett 2:40–44

Singh RR, Goyal SP, Khanna PP, Mukherjee PK, Sukumar R (2006) Using morphometric and analytical techniques to characterize elephant ivory. Forensic Sci Int 162(1–3):144–151

Toren D, Barzilay T, Tacutu R, Lehmann G, Muradian KK, Fraifeld VE (2016) MitoAge: a database for comparative analysis of mitochondrial DNA, with a special focus on animal longevity. Nucleic Acids Res 44(D1):D1262–D1265

Trail PW (2021) Morphological analysis: a powerful tool in wildlife forensic biology. Foren Sci Int Anim Environ 1:100025

Uno KT, Quade J, Fisher DC, Wittemyer G, Douglas-Hamilton I, Andanje S et al (2013) Bomb-curve radiocarbon measurement of recent biologic tissues and applications to wildlife forensics and stable isotope (paleo) ecology. Proc Natl Acad Sci 110(29):11736–11741

Viciano J, López-Lázaro S, Tanga C (2022) Post-mortem dental profile as a powerful tool in animal forensic investigations—a review. Animals 12(16):2038

Warlow T (2004) Firearms, the law, and forensic ballistics. CRC Press, Boca Raton, FL

Zhou W, Qiu LZ, Liu H, Deng HF, Yue LX, Gao Y (2020) Corrigendum to "Notch1-mediated histone demethylation of HCN4 contributes to aconitine-induced ventricular myocardial dysrhythmia" [Toxicol. Lett. 327 (2020) 19-31]. Toxicol Lett 331:257–258

Forensics in Bioterrorism

32

Monisha Samuel and Rutwik Shedge

Abstract

Terrorism can be defined as any method employed to threaten and terrorize large groups of individuals, governments, armies, or society as a whole. Thus, one may accept, in the milieu of historical analysis of bioterrorism, that it involves the usage of different biological agents by any person or groups, including political or military personnel and official states, motivated by various reasons (be they political, religious, or other ideological objectives), in order to attain such consequences (Barras and Greub, Clin Microbiol Infect 20:497–502, 2014). Bioterrorism covers a very broad spectrum of issues, from catastrophic terrorism with mass casualties, to microevents via biological agents, which leads to various results like civil conflict, disturbance, illness, infirmities, and death (Barras and Greub, Clin Microbiol Infect 20:497–502, 2014).

Keywords

Bioterrorism · Biological warfare · Biological weapons · Bio war · Public health emergency · Forensics · Microorganisms · Bacteria · Toxins · Plague · Viruses

Definitions

Bioterrorism: The use of infectious agents or other harmful biological or biochemical substances as weapons of terrorism.

Biological Weapons: They are either microorganisms like virus, bacteria, or fungi, or toxic substances produced by living organisms that are produced and released deliberately to cause disease and death in humans, animals, or plants.

Plague: An epidemic disease causing a high rate of mortality.

M. Samuel (✉) · R. Shedge
Department of Forensic Science, National Forensic Sciences University, Tripura Campus, Agartala, Tripura, India

A. Puri et al. (eds.), *Fundamentals of Forensic Biology*,
https://doi.org/10.1007/978-981-99-3161-3_32

Quarantine: A restraint upon the activities or communication of persons or the transport of goods designed to prevent the spread of disease or pests.

Epidemics: A widespread occurrence of an infectious disease in a community at a particular time.

32.1 Brief History

The historical study of biological warfare (BW) and bioterrorism is enormously tough, and any conclusions in this respect have to be obtained with extreme care, because of various concomitant reasons: first, the lack of consistent scientific data regarding alleged biological attacks, especially before the dawn of modern microbiology. The second is the polemical situations surrounding any alleged bioterrorist attack, within which the accessible documents become prone to multiple political influences and manipulations, which makes it very difficult to interpret accurately. The third possible reason is also the historical distance of primeval stories about biological attacks, leading to misinterpretation when read with contemporary lenses (Christopher et al. 1997; Joshua 1999). Thus, given such factors, it is easy to understand why it may become taxing for historians to distinguish natural epidemics from alleged biological attacks.

32.1.1 Use of Biological Weapons During Ancient Times, Middle Ages, and Colonial Period

Infectious diseases and other biological weapons have been identified for their potential effect on people from as early as the fourteenth century BC (Table 32.1). Beginning from pre-historic times, the Melanesian tribesman (actual Vanuatu) employed arrowheads contaminated with tetanus (Carus 2017). The delivery of diseased rams potentially infected with tularemia by Hittites might have formed the first documented example of biological warfare. This had been a strategy to weaken their enemies (Trevisanato 2007). In the sixth century BC, as recorded by the Greek historian Herodotus, the Scythian archers are used to infect their arrows by plunging them in a concoction of rotting cadavers and human blood. According to the interpretation of recent times, this mixture might possibly be a source of *Clostridium perfringens* and *Clostridium tetani* (Grmek 1979). In the third century BC, the army commander Hannibal of Cartagena destroyed his adversary's fleet, which belonged to King Eumenes II of Pergamon with pots full of venomous snakes. Such classic examples have been reported by historians where the use of arrows and other objects infected with different infectious or toxic products obtained from animal parts or plants has been prevalent in order to destroy the enemy (Barras and Greub 2014). Another event documented is the poisoning of wells by Emperor Barbarossa with human bodies in Tortona, Italy. Also, another event was when Jean, Duke of Normandy, cast dead horses over the wall into the castle of Thun l'Eveque, captured by the Englishman (Geissler and Moon 1999).

Table 32.1 Examples of biological warfare during the ancient times and Middle Ages

Year	Events
Fourteenth century BC	The Hittites send rams infected with tularemia to their adversaries
Sixth Century BC	The Scythian archers were immersing their arrows in a mixture of blood and decaying corpses to infect them
1155	Barbarossa polluted water wells with human remains, Tortona (Italy)
1346	In the beleaguered city of Caffa, Mongols toss the remains of plague victims over the walls (Crimea)
1422	The Lithuanian army throw manure, garbage, and diseased victims during the siege of the town of Carolstein (Bohemia)
1495	Spanish sell wine to their French adversaries mixed with leprosy patients' blood, Naples (Italy)
1500	Pizarro provided variola-infected garments to the native populations in South America
1650	The Polish army attacks its adversaries by firing them with rabid canine saliva.
1710	Russian soldiers launch plague cadavers over Swedish soldiers at Reval (Estonia)
1797	To spread malaria among the enemy, the Napoleonic troops flooded the fields around Mantua, Italy
1863	During the American Civil War, Confederates sold garments made from the remains of patients with smallpox and yellow fever to Union forces

In the Middle Ages, a more famous example has been the siege of Caffa (now known as Feodosia in Ukraine/Crimea), a Genovese base on the Black Sea coast, by the Mongols. Tartar (Mongol) army catapulted dead bodies of plague fatalities over the city walls of Caffa to attack the Genoese army with an epidemic of bubonic plague (Geissler and Moon 1999). There has been speculation that quite plausibly this was how the plague was transmitted by the Mongols by throwing diseased cadavers with catapults into the Mediterranean ports and throughout Europe. Though given the highly complex epidemiology of plague, this interpretation of the Black Death, which might have killed more than 25 million people, starting from a very specific and localized origin, remains highly controversial. Similarly, it is provocative as to whether the effect of throwing infected bodies could have been the only cause of the outburst of an epidemic. However, this occurrence of the use of corpses in order to infect a population remains a milestone in the history of bioterrorism (Geissler and Moon 1999; Wheelis 2002). Other similar incidents have been the siege of the Bohemian city of Carolstein by Lithuanian troops in 1422. The Lithuanian army hurled corpses of those who died in battle along with manure and garbage into the Bohemian town of Karlstein (Robertson and Robertson 1995). In 1495, it has been documented that the Spanish peddled wine contaminated with the blood of leprosy patients to their French adversaries in Naples (Italy). In 1500, Pizarro offered variola-infected clothing to the South American native communities (Noah et al. 2002). In 1650, it was witnessed that the Polish fired saliva from rabid dogs in the direction of their enemies. In 1710, the Russian army catapulted dead remains over the Swedish troops in Reval (Estonia) (Robertson and Robertson 1995). In 1797, the Napoleonic armies further flooded the plains around Mantua (Italy) to augment the spread of malaria (Cohen et al. 2003).

During the ensuing centuries, smallpox represented one of the most commonly used biological weapon of occidental war and colonial history. Introduced in the American mainland by the Europeans, it was explicitly employed several times as a way to infect Native Americans. During the American civil war, the Confederate's troop sold yellow fever and smallpox-infected clothing to Union troops (Cohen et al. 2003) They also contaminated water supplies for the Union forces with animal cadaver (Wheelis 2002).

32.1.2 Use of Biological Weapons During the Modern Era

The end of the nineteenth century and development of the germ theory of disease and the foundation of microbiology by Louis Pasteur and Robert Koch led to the true modern era of biological warfare. With the onset of World War I, many nations like Germany and France (to a lesser scale) developed secret BW programs, such as the infection of animal feed with Bacillus anthracis or Burkholderia mallei in order to infect the enemy. German troops sold horses and mules diseased with glanders and anthrax to their Allies (Noah et al. 2002). Sheep infected with glanders and anthrax were peddled to Russian, Britain, and Indian armies by the Germans as well. Further German troops also endeavored to spread cholera in Italy and plague in St. Petersburg (Poupard and Miller 1992). Thus, between late 1914 and early 1915, Germany pioneered the dissemination of biological weapons and was initiating a true BW campaign, using different biological pathogens against several countries. Whatever the success of these programs might have been, the threat of BW along with the terror of chemical warfare being used on the battlefield, became, for the first time in history, a foremost political concern at the international level. As a result, the Geneva Protocol for the Prohibition of the Use in War of Asphyxiating, Poisonous or Other Gases, and of Bacteriological Methods of Warfare was approved in 1925. The Geneva Protocol very clearly prohibited the use of biological weapons, but did not prevent their research and production. Thus, states that had ratified the Geneva Protocol continued researching and developing intense programs on biological weapons (Christopher et al. 1997).

During the interwar period, under the command of Major General Shiro Ishii, the first biological weapon research facility at the Tokyo Army's medical school, obscured as part of the Kwantung Army's Water Purification Department, began its research program on biological agents. Although the center lacked technical sophistication, this program had the manpower. More than 3000 military personnel along with hundreds of medical doctors and technicians were recruited for research and development. More than 150 research centers, including the infamous unit 731 considered responsible for human experimentation during the Second Sino-Japanese War, were spread throughout the Japanese empire, from Manchuria to Indonesia. This research program, until today, is considered one of the most appalling programs, which performed atrocious experiments on humans. It is estimated that at least 3000 prisoners of war (Chinese, Koreans, Mongolians, Soviets, Americans, British, and Australians) were supposedly used as test subjects in unit 731. These experiments included inoculation with many pathogens, weapons tests, and germ warfare attacks (Barras and Greub 2014). As a result of the dissemination of pathogens during the Japanese BW program, it is estimated that thousands of civilians perished as a direct consequence of the attacks. Moreover, the incapability of controlling the dissemination caused the death of several Japanese soldiers during attacks (Barras and Greub 2014).

The American BW program was a covert operation implemented in 1942 after suspicion of the use of BW agents by Germany and Japan (Barras and Greub 2014). Non-pathogenic bacteria like *Bacillus globigii* and *Serratia marcescens* were used for open-air field testing along with the spreading of bacterial aerosols in various public places. At its peak, the program involved more than 3000 people and multiple research and development facilities. Research with more than 30 agents including bacteria (e.g., *B. anthracis*, *Brucella suis*, *Coxiella burnetii*, and *Francisella tularensis*), toxins (e.g., botulin and staphylococcal enterotoxin B), and viruses (e.g., Venezuelan equine encephalitis and yellow fever virus) were conducted (Joshua 1999). The program was officially dismantled in 1969, and the facilities were converted for defense purposes only. Further to this, the Biological Weapon and Toxin Convention (BWTC) was signed in the year 1972. However, amidst this the country had successfully developed ten different BW agents and designed a sophisticated delivery system (Khardori and Kanchanapoom 2005).

In the 1980s, one dramatic example of this can be provided by the Rajneesh cult, a sect that in 1984 purposefully poisoned salad bars with Salmonella typhimurium in a number of restaurants in Dalles, Oregon. In addition to carrying out its infamous sarin gas attack in the Tokyo subway in March 1995, another religious cult known as Aum Shinrikyo was also developing, at the same time, a program on crude biological weapons containing Clostridium botulinum and B. anthracis, but with no proof of effectiveness (Noah et al. 2002). Another famous attack was the "anthrax letters case," which caused 751 cases, 45 of which required hospitalization. This appears to be one of the very few verified incidents of biological terrorism following World War II. Over the course of the autumn, a number of letters were written to journalists or elected politicians. Five of the 22 anthrax-infected individuals died as a direct result of anthrax or its complications. The specific strain employed was linked to a laboratory run by the US Army at Fort Detrick, but the attackers are still at large (Jaton and Greub 2014). This case demonstrates that BW is still a huge concern to the public and it must be handled seriously and addressed at the individual and political levels as well. Additionally, it highlights the significance of being adequately prepared for the occurrence of such an event.

32.2 Investigative Response to Bioterrorism

A wide-ranging approach to coping with bioterrorism must integrate efforts to avert the proliferation of biological weapons along with designing methods for detecting underground biological weapons programs; strategies for preventing their use in case biological weapons do multiply; and mechanisms for defending civilian and military populations where deterrence is unsuccessful. The emphasis in this multifaceted approach should be on the investigation and defense strategies, simply because the growth of biological weapons is tough to control and covert biological weapons programs are difficult to track. The deterrence thus will likely be less effective. Consequently, the way to move forward is by improving intelligence and information management, the forensic community should be focused on improving investigative responses along with defenses against biological weapons. The means to achieve it encompass the rapid detection of biological agents together with preclinical, clinical, and agricultural monitoring and diagnosis.

To begin with, increased vigilance in the intelligence community could reduce the unplanned spread of knowledge that may aid in bioweapon planning by terrorist organization. Voluntary international and national exertions to promote the exchange of biotechnology-related information could easily advance security and safety in the handling, storage, and transportation of sensitive biological substances and equipment. Understanding the area of biotechnology, where various potential bioagents are developed, and also becoming familiar with certain cultural techniques, processes, and procedures are extremely essential in identifying potential threat. In such cases, expert scientists and technologists can offer immense help to the community.

The next thing is to design a systemic approach and target the completion of various activities in a phased manner prior and through an incident for a successful response to a bioterrorist attack. The various phases are as follows (Government of India, Ministry of Home Affairs 2006).

32.2.1 Preparedness Phase

This phase encompasses actions to be taken by various agencies and response teams to ensure an essential state of preparedness. These include assessment of the laboratory facilities and their constant upgradation. The hospitals should also be regularly evaluated. Healthcare professionals, rapid response team (RRT), and quick response medical team (QRMT) who would be the first responders should be provided with adequate training so as to sustain and manage a case of an imminent attack. Ensuring the availability of adequate stocks of medicines and vaccines is very essential. Along with all this, proper coordination with security organization in order to organize mock drills for health professionals, government departments, animal husbandry, security, law enforcing, and other agencies so as to evaluate their preparedness levels so they can act promptly in case of an attack is of utmost importance. To be able to prepare contact details so that communications are unimpeded during an attack is also another vital task in this initial phase of response. The public also should be kept aware of imminent attacks in a sensitive manner so as to not create unnecessary panic but also encourage reporting in a voluntary manner.

32.2.2 Early Warning Phase

Various activities including case characterization, notification, compilation, and analysis of epidemiological data are the main part of the surveillance system's early warning phase. Further, in order to assess the severity and scope of any potential attack and to put effective measures in place, public health epidemiologists require to identify the attack as quickly as possible and conduct an investigation.

32.2.3 Notification Phase

Any unusual sickness or typical symptoms seen in unusually high numbers must be reported to the appropriate authorities. Rapid epidemiological investigations, swift laboratory support for diagnosis confirmation, isolation, quarantine, keeping healthcare facilities prepared for potential casualty management, and establishing public health infrastructure for control are all initiatives in this phase.

32.2.4 Response Phase

Rapid epidemiological examination, prompt laboratory support, mass casualty management, and the implementation of preventive, curative, and targeted control measures are all part of this phase's operations. These are meant to stop the disease from spreading further. Considering and understanding the time, place, and distribution of those affected, the routes of transmission, the impact on important infrastructure and healthcare facilities, the agencies and organizations involved in the response, and the public health responders, local, state, and federal emergency operation centers for disease management, are among other factors, which are extremely critical before launching a response. Establishing contact with and planning alongside health department staff members who are in charge of emergency response to keep efficient tracking of every contact and activity that follows is another important step in this phase.

Further to this, establishing specific, quantifiable, and realistic early health response targets based on the evaluation of the problem and devising a plan of action should be prioritized. Assigning duties and keeping a record of all actions also help in this stage. This is finally followed by implementing the plan of action. The RRTs/QRMTs have to further inspect the outbreak or upsurge in the disease prevalence. There is a collection of samples, and a state or national testing laboratory has to be assigned. Hospitals have to be informed so they can receive the patients and provide them with adequate care. Immediate relief centers or temporary hospitals are to be set up if required. Measures for quarantine and disease control have to be implemented. Treatment guidelines should be communicated to all parties concerned as soon as the disease has been recognized. The identified laboratory is required to create a standard operating procedure (SOP) for laboratory testing, which is then distributed for adoption to all other hospital laboratories and district hospitals. Specific laboratory reagents are to be supplied to all the designated testing centers and hospitals. To avoid any panic, the "Dos and Don'ts" list is disseminated through print and electronic media. Hospitals take the necessary precautions for isolation, quarantine, waste disposal, and personal safety. An impact assessment team is essential for evaluating the impact of the outbreak on humans, animals, and plants.

32.2.5 Recovery Phase

The setbacks caused by the bioterrorist attack are mended, and the preparation plans for the future incorporating the lessons learned through the attack are evaluated. The harm done to the public health facilities is repaired, as are the necessities used during the response phase. There are announcements made to the public on the return to normal. The RRTs gather and examine data to find the shortcomings in the application of the response measures. The future contingency plan is then updated with the required revisions.

32.3 Tracing of Prospective Bioterrorism Agent Through SNP, VNTR, and Whole-Genome Sequence Analysis

Finding the origin of a disease, especially in cases of potential bioterrorist attacks, can save many lives. It enables actions to be taken to stop the disease from spreading further and also track the perpetrator. Even though the process of transmission for many human infections is well understood, it is frequently challenging to pinpoint the precise origin of a disease outbreak using laboratory techniques (Schürch and Siezen 2010). Today, microbial forensics has evolved into a wide range of laboratory analytical techniques and methodologies, going far beyond those used in medical diagnoses or epidemiological investigations, such as molecular sequencing, biochemical analyses, microbial cultures, electron microscopy, mass spectrometry, and crystallography (Echeonwu et al. 2018). As previously discussed, human diseases can be brought on by many different biological agents. For example, it is usually hard to distinguish between all bacterial strains involved in a single outbreak using the current techniques of generating DNA fingerprints of various bacterial agents of an infectious disease. Therefore, this makes it impossible to track the disease's spread, but the recent use of next-generation sequencers and whole-genome sequencing tools has enabled genetic information of all isolates to be analyzed and thus is a modern answer to this challenge.

Thus, these modern molecular methods that have been used in what is known as molecular epidemiology to identify the origins of microbial disease epidemics have significant source tracing of potential threats. Some of these genetic markers are single-nucleotide polymorphisms (SNPs), insertions and deletions, repetitive sequences, mobile elements, housekeeping genes, pathogenicity islands, structural genes, resistance and virulence genes, whole-genome sequences, sexual and asexual reproduction, conjugation, transduction, horizontal gene transfer, lysogeny, gene conversion, gene duplication, recombination, rearrangements, and mutational hotspots, and they are very commonly used today to trace the origin of various microbial strains (Budowle et al. 2005). Major nucleic acid-based technologies used for this purpose include multilocus sequence typing (MLST). Further that has been enhanced in the form of ribosomal multilocus sequence typing, multilocus variable number tandem-repeat (VNTR) analysis (MLVA), clustered regularly interspaced short palindromic repeats (CRISPRs), whole-genome sequencing, and microarray. Apart from these, next-generation sequencing in recent times has momentously and positively impacted the skills of forensic genetics. Some of them are enlisted below.

32.3.1 Using Single-Nucleotide Polymorphisms (SNP) for Source Tracing

SNP stands for single-nucleotide polymorphism, a type of single-nucleotide variation that occurs at certain locations in the genome. These variances are found to be in different extent in various populations. SNPs are present in every genome, including that of humans, plants, and microorganism. Due to their large benefits in parentage testing, SNPs have attracted the attention of forensic researchers. SNPs are excellent for examining and identifying damaged samples with small amplicons since they have very low mutation rates. SNPs offer useful details on the geographic origin and unique identification of unknown people, plants, and microbes (Sobrino et al. 2005).

Yersinia pestis, the cause of the deadly plague, has been elaborately studied. One such study showed that due to the evolutionary similarity between *Y. pestis* and *Y. pseudotuberculosis*, it is difficult to differentiate them by techniques like MLST, whereas the same scientists were able to find 76 conservative SNPs from 3250 orthologous coding sequences in their work. This was based on a comparative study of three different *Y. pestis* genomes that represented various pathogen biovars (Achtman et al. 1999). Further whole-genome sequencing analysis has shown an even greater number of SNPs, which would discriminate even better when utilized for genetic typing (Morelli et al. 2010).

32.3.2 Multilocus Variable Number Tandem Repeat Analysis in Source Tracing

Tandem repeat genomic polymorphism in mammals has been very beneficial in genetic mapping and is still an integral part of forensic DNA fingerprinting. These repetitive sequences are regularly classified based on the varying ranges of base pair numbers. They are further divided into satellites (megabases of DNA in association with chromatin), minisatellites (repeat units of about 6–100 bp across hundreds of base pairs), or microsatellites (with repeat units of between 1 and 5 bp across a few tens base pairs).

It has been demonstrated that MLVA can both categorize strains that are distantly related and tell them apart as well. Researchers have used multiple VNTR loci to detect significant variations across these loci from various *Y. pestis* strains. A large number of *Y. pestis* isolates were successfully divided into 61 unique genotypes by Pourcel et al. They have been able to suitably distribute the three biovars among its three major branches as well. They have also proposed that a subset of seven VNTR markers would be sufficient for rapid comparison of a new strain with their collection and can be achieved very easily with a Web-based facility (Klevytska et al. 2001; Pourcel et al. 2004).

Another such study reported the use of MLVA technique to investigate and effectively trace the origin of a 1993 aerosolization of *B. anthracis* spores by the Aum Shinrikyo cult group. The publication highlighted the use of eight VNTR markers and a missing pXO2 gene plasmid as markers that were able to identify the attack strain as the Sterne vaccine strain of *B. anthracis*. This was locally available then in Japan for veterinary purpose (Keim et al. 2001).

32.3.3 Whole-Genome Sequence Analysis for Source Tracing

Today, it is easily feasible to find several genetic changes using whole-genome sequencing analysis, including SNPs, gene loss/gain, gene rearrangement, VNTRs, and CRISPRs. Sequencing results in in-depth data regarding the variety of bacterial genomes and thus enables an improved source-tracing analysis. Whole-genome sequence typing (WGST) analyzes variations across the entire genome using the fast-developing next-generation sequencing technology. Source-tracing databases could be developed that comprise whole-genome sequences of hundreds of bacterial strains using this technology. This will help in shortening the time span required for source tracing and contribute to the assessment of potential strains that can be manipulated for bioterrorism purpose (Yang and Keim 2012).

32.4 Conclusion

In conclusion, terrorists are more dedicated and have access to readily available resources, information, and technology that can be used to create bioweapons that can cause significant harm to civilizations that are unprepared and vulnerable. Bioterrorism, biocrime, and biowarfare are likely to recur since similar incidents have happened frequently occurred in the past. Today, various microorganisms are more easily accessible, and technology to develop them as weapons are more plausible (Jansen et al. 2014).

The development and/or validation of every step of the forensic investigation process, from sample collection through result interpretation, is necessary for an effective program that could thwart such attacks. Beyond the conventional forensic laboratory and its practitioners, additional existing and developing skills would also be required. More laboratories must be implemented in researching new procedures and constantly comparing them to those that have been accepted by the forensic DNA science community (Broussard 2001). This will guarantee the required quality when analyzing evidence and the results will be more readily accepted in the court of law. The biggest challenge today is to be able to characterize bioweapons using our knowledge and forensic skills. Microbial Forensics could thus play an enhanced role in the coming years and can serve as an effective tool in the resolution of a biocrime or an act of bioterrorism.

Multiple Choice Questions

1. Infectious diseases have been identified for their potential effect on people from as early as _____ century
   (a) Tenth century
   (b) Fourth century
   (c) Fourteenth century
   (d) Eighth century
   Answer: C
2. In 2001, 22 cases were reported in the USA where spores of a microorganism were sent through mail. It caused thousands of people to be affected. Identify the microorganism.
   (a) Bacillus anthracis
   (b) Clostridium botulinum
   (c) Streptomyces griseus
   (d) Staphylococcus aureus
   Answer: A
3. Yersinia pestis is the causative agent for which disease
   (a) Botulism
   (b) Bubonic plague
   (c) Smallpox
   (d) Hay fever
   Answer: B
4. Which microorganism was used as a bioterrorism weapon in the Oregan attack in the year 1984?
   (a) Salmonella typhimurium
   (b) Clostridium botulinum
   (c) Bacillus anthracis
   (d) Yersinia pestis
   Answer: A
5. SNPs stands for
   (a) Single nitrogen polymerase
   (b) Single-nucleotide polymorphism
   (c) Single-nucleoside polymorphism

   (d) Start nucleotide polymorphism
   Answer: B
6. Minisatellites are defined as
   (a) Repeat units of between 100 and 500 bp across a few tens base pairs
   (b) Repeat units of about 6–100 bp across hundreds of base pairs
   (c) Repeat units of between 1 and 5 bp across a few tens base pairs
   (d) Repeat units of 250–2000 bp across hundreds of base pairs
   Answer: B
7. Dead bodies of plague fatalities were thrown over the city walls of Caffa to attack the Genoese army with an epidemic of _____ .
   (a) Anthrax
   (b) Bubonic plague
   (c) Typhoid
   (d) Pneumonia
   Answer: B
8. Which phase restores the public health facilities and replenishes the necessities used during the response phase?
   (a) Recovery phase
   (b) Preparedness phase
   (c) Early warning phase
   (d) Response phase
   Answer: A

References

Achtman M, Zurth K, Morelli G et al (1999) Yersinia pestis, the cause of plague, is a recently emerged clone of Yersinia pseudotuberculosis. Proc Natl Acad Sci U S A 96:14043–14048. https://doi.org/10.1073/PNAS.96.24.14043

Barras V, Greub G (2014) History of biological warfare and bioterrorism. Clin Microbiol Infect 20: 497–502. https://doi.org/10.1111/1469-0691.12706

Broussard LA (2001) Biological agents: weapons of warfare and bioterrorism. Mol Diagn 6:323–333. https://doi.org/10.2165/00066982-200106040-00013

Budowle B, Johnson MD, Fraser CM et al (2005) Genetic analysis and attribution of microbial forensics evidence. Crit Rev Microbiol 31:233–254. https://doi.org/10.1080/10408410500304082

Carus WS (2017) A short history of biological warfare: from pre-history to the 21st century. Government Printing Office, Washington, DC

Christopher GW, Cieslak TJ, Pavlin JA, Eitzen EM (1997) Biological warfare. A historical perspective. JAMA 278:412–417

Cohen A, Robenshtok E, Rotman E, Sagi R (2003) The history of biological warfare. Human experimentation, modern nightmares and lone madmen in the twentieth century. EMBO Rep 4: S47–S52. https://doi.org/10.1038/SJ.EMBOR.EMBOR849

Echeonwu B, Nwankiti O, Chollom S, Olawuyi K (2018) Bioterrorism threat: a review of microbial forensics source-tracing of some bioterrorism agents. J Foren Sci Med 4:161. https://doi.org/10.4103/JFSM.JFSM_5_18

Geissler E, Moon JEC (1999) Biological and toxin weapons: research, development and use from the middle ages to 1945. SIPRI Chemical and Biological Warfare Studies, vol 18. SIPRI, Solna. 296 pp

Government of India, Ministry of Home Affairs (2006) Standard Operating Procedures (SOP) for responding to a terrorist attack using biological agents. Government of India, New Delhi

Grmek M (1979) Ruses de guerre biologiques dans l'Antiquité. Rev Etud Grec 92:141–163. https://doi.org/10.3406/REG.1979.4222

Jansen HJ, Breeveld FJ, Stijnis C, Grobusch MP (2014) Biological warfare, bioterrorism, and biocrime. Clin Microbiol Infect 20:488–496. https://doi.org/10.1111/1469-0691.12699

Jaton K, Greub G (2014) Clinical microbiologists facing an anthrax alert. Clin Microbiol Infect 20: 503–506. https://doi.org/10.1111/1469-0691.12682

Joshua L (1999) Biological weapons: limiting the threat. MIT Press, Cambridge, MA

Keim P, Smith KL, Keys C et al (2001) Molecular investigation of the Aum Shinrikyo anthrax release in Kameido, Japan. J Clin Microbiol 39:4566–4567. https://doi.org/10.1128/JCM.39.12.4566-4567.2001

Khardori N, Kanchanapoom T (2005) Overview of biological terrorism: potential agents and preparedness. Clin Microbiol Newsl 27:1–8. https://doi.org/10.1016/J.CLINMICNEWS.2005.01.001

Klevytska AM, Price LB, Schupp JM et al (2001) Identification and characterization of variable-number tandem repeats in the Yersinia pestis genome. J Clin Microbiol 39:3179–3185. https://doi.org/10.1128/JCM.39.9.3179-3185.2001

Morelli G, Song Y, Mazzoni CJ et al (2010) Yersinia pestis genome sequencing identifies patterns of global phylogenetic diversity. Nat Genet 42:1140–1143. https://doi.org/10.1038/NG.705

Noah DL, Huebner KD, Darling RG, Waeckerle JF (2002) The history and threat of biological warfare and terrorism. Emerg Med Clin N Am 20:255–271

Poupard JA, Miller LA (1992) History of biological warfare: catapults to capsomeres. Ann N Y Acad Sci 666:9–20. https://doi.org/10.1111/J.1749-6632.1992.TB38020.X

Pourcel C, André-Mazeaud F, Neubauer H et al (2004) Tandem repeats analysis for the high resolution phylogenetic analysis of Yersinia pestis. BMC Microbiol 4:22. https://doi.org/10.1186/1471-2180-4-22

Robertson AG, Robertson LJ (1995) From Asps to allegations: biological warfare in history. Mil Med 160:369–373. https://doi.org/10.1093/MILMED/160.8.369

Schürch AC, Siezen RJ (2010) Genomic tracing of epidemics and disease outbreaks. Microb Biotechnol 3:628–633

Sobrino B, Brión M, Carracedo A (2005) SNPs in forensic genetics: a review on SNP typing methodologies. Forensic Sci Int 154:181–194. https://doi.org/10.1016/J.FORSCIINT.2004.10.020

Trevisanato SI (2007) The 'Hittite plague', an epidemic of tularemia and the first record of biological warfare. Med Hypotheses 69:1371–1374. https://doi.org/10.1016/J.MEHY.2007.03.012

Wheelis M (2002) Biological warfare at the 1346 Siege of Caffa. Emerg Infect Dis 8:971–975. https://doi.org/10.3201/EID0809.010536

Yang R, Keim P (2012) Microbial forensics: a powerful tool for pursuing bioterrorism perpetrators and the need for an international database. J Bioterror Biodef 2:1–8. https://doi.org/10.4172/2157-2526.S3-007

Hair Analysis

33

Avinash Puri and Neelam Arya

Abstract

Hair is extremely important physical evidence being the most common biological material, which plays a crucial role in a criminal investigation. Hair evidence is generally associated with crimes involving physical contacts such as rape, murder, road accidents, and trafficking and may be found at the crime scene, on the body and clothes of victim/suspect, or attached to a weapon of offense, tool, or vehicle. Even thefts of animals cause the shedding of animal hairs on the person or vehicle involved in the offense. There are many differences between human and animal hairs. The comparison and identification of human and animal hair can provide investigators with valuable information for potential leads. Morphological and microscopic comparison is considered a reliable tool for the specific identification of the characteristics found in hair.

With the advancement of technology and awareness, hair analysis is a widely accepted tool to determine color, race, sex, age, metal poisoning, illegal wildlife trade, sexual assault, and disputed maternity and paternity matters. Hair is actually unique to the individual is very probable since, if it were not, it would be the exception to the general rule of biological individuality. The molecular analysis and microscopic comparison of morphological characteristics of hair have been accepted both scientifically and legally for decades.

A. Puri (✉)
Department of DNA and Forensic Biology, Regional Forensic Science Laboratory, Indore, Madhya Pradesh, India

N. Arya
Forensic Science Laboratory, madhuban, Haryana, India

A. Puri et al. (eds.), *Fundamentals of Forensic Biology*,
https://doi.org/10.1007/978-981-99-3161-3_33

Keywords

Hair identification · Scalp hair analysis · Drug detection from hair · Hair DNA profiling · Hair as bioindicator

33.1 Hair Analysis

Hair is found on most of the limbs of human beings and animals. In all natured crimes such as murder, sexual assaults, traffic accidents, and hanging, hair is likely to detach from the scalp, and other areas of the body and get transferred from one person to another or on objects at the scene of crime. Hair is extremely important physical evidence being the most common biological material, which plays a crucial role in a criminal investigation. It must be collected in every case in which it occurs. In murder cases, hairs may readily get attached/stick to the weapon of offense. The examination of such hairs aid in establishing the involvement of weapon in the crime committed. In cases like kidnapping, human trafficking, and robbery, hairs may provide crucial evidence of contact of the accused/suspect with the scene of crime, the victim, the vehicle, the weapon or article of clothing, etc. Hair examination may be used for associative and investigative purposes and to provide information for crime scene reconstruction. The comparison and identification of human and animal hair can provide investigators with valuable information for potential leads. Morphological and microscopic comparison is considered a reliable tool for the specific identification of the characteristics found in human and animal hair.

Because of the small size of this evidence, hair may be difficult to find. Care and patience are required to conduct a thorough search for this evidence. However, the hairs are very resistant to decomposition and putrefaction thats why the hair remains intact as an evidence for a longer period of time. Hair is usually used to study the characterization of the known sample versus the questioned hairs recovered from the scene of crime to verify if they are from a common source for the establishment of a relation between crime and the criminal. It is not possible to determine the person that a strand of hair come from. Because hair studies are slow and often difficult, there has been a tendency to minimize the importance of hair as evidence. Even though much worthwhile information can be determined through hair examination. Significant information such as species, location of growth on the body, hair treatment, and hair disease is some of the types of information that can be determined. Care must be taken at all times to avoid altering the hair itself or the debris carried on it during such a preliminary observation. Perhaps it is even more important to avoid contamination of the sample as several police investigators are not aware of the technicality of it. Toxic elements such as arsenic and thallium are preferably deposited in hair and other keratinous appendages like nails. Hair is often analyzed for the detection of arsenic poisoning because the metabolic process takes arsenic direct to the hair. Although the hair also absorb arsenic and hold tightly., so that it cannot be assumed that hair analysis is always capable of indicating poisoning from arsenic. It has been long known that other elements like lead, silica, iron, and

Fig. 33.1 Human vs animal hair

phosphorous exist in small quantities in some hair. The distribution of such elements in hairs of different persons has been found to be highly individual (Fig. 33.1).

33.1.1 Human Hair Characteristics

There are four types of human hair, they are primordial, lanugo, vellus, and terminal hair. Primordial and lanugo hairs develop at three and 6 months in the baby during of gestation period before the birth. Vellus hairs are fine and lack medullae inside the cortex and are present all over the body. Terminal hairs are prominent in appearance and hard in structure and are present on the scalp/head, eyebrows, eyelashes, face, armpits, and around genital organs.

Hairs are complex fibrous outgrowths projecting from the skin that possess different microscopic characteristics such as shape, color, and root appearance. Human hair consists of a bundle of fibrils each of which is composed of protein known as keratin. It contains certain excretory products. Environmental or external contamination may cause a covering over the hair with dust, dirt, or oily substances. Morphologically, a single hair strand on a macro-scale has a root, a shaft, and a tip. The proximal-most portion of the hair is root. The main portion of the hair is known as the shaft. The distal-most portion of the hair is known as the tip. The tip may be natural or cut. If the tip bears clean cut, it may have been served with a sharp instrument/article. If the cut is stepped or irregular or split in appearance, it is likely to have been served by a blunt instrument/article. The shaft, i.e., length of the hair may point to the body from which it came. A cross section of a hair may indicate the part of body from which it originated. If it is triangular in shape with three concave sides, it may be beard or mustache hair. Hair from the torso is usually oval- or kidney-shaped. Hairs from the head are generally round, but curly hairs are sometimes oval. An examination of the root can tell the microscopic characteristic of hairs such as

- Color.
- Hair form/texture.
- Length.
- Diameter.
- Pigmentation.
- Structure.
- Variability in growth pattern.

Hair varies from region to region on the body of the same person. Forensic scientists distinguish six types of hair on the human body and each hair type has its own shape and characteristics as follows:

1. Head hair.
2. Eyebrows and eyelashes.
3. Beard and mustache hair.
4. Underarm hair.
5. Auxiliary or body hair.
6. Pubic hair.

The experience in examining hair and the study of its characteristics will supply far more information than can be obtained by the study of any classification pattern/scheme. The identification has been focused majorly on morphological differences rather than on physical properties. Using characteristics such as color, curliness, and thickness may help identify different human hairs. Two of these properties, i.e., refractive index and density indicate great extension of their use in the future. The refractive index of a pure compound is a constant and is an extremely valuable factor in the identification of the compound. In case of hair, which is not a pure compound and is not even homogenous, it is apparent that the refractive index is an empirical value and its determination must be performed in a prescribed and reproducible manner in order to achieve significant result. Studies of the density of hair have been far less extensive than studies of refractive index. Because the determination and use of density as a comparative factor permit a hair to be picked from among sample taken from perhaps as many as eight or ten persons, the relation of the density of the hair to sex, age, and origin is at present almost completely unknown.

Internally, there are three main structural layers in a hair, namely the cuticle, the cortex, and the medulla. Examination of each of these parts gives useful information. The outer layer, or **cuticle**, consists of overlapping scales, with the free ends of the scales directed toward the tip of the shaft. Just beneath the cuticle is the **cortex**, made up of compact, elongated cells and often containing pigment granules made of

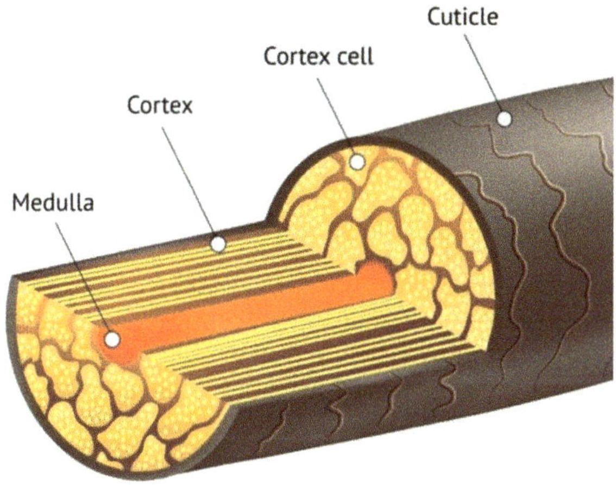

Fig. 33.2 Structure of hair

melanin present in the hair follicles that are originally produced by melanocytes, which appear like nerve cells. Inside or the cortex of a hair shaft is also different in both inter- and intra-species as the medulla and pigmentation vary according to the localities. When differentiating human and animal hair, it is necessary to look at these features in detail. The central core of a hair shaft is the **medulla**, composed largely of air spaces. Medulla is one of the more valuable morphological identifying characteristics. In human and certain anthropoids the medulla is narrow, occupying only about one-third of the shaft width. Medulla is classified as continuous, intermittent, or fragmental. Human hair normally shows a fragmental medulla, but many individual hairs show no medullary material at all, and sometimes a single hair may have an almost continuous medulla. There is evidence to indicate that only hairs above a certain diameter will contain medullation and the larger the hair the more nearly continuous the medullation will be. The ratio of medulla diameter to that of the shaft is known as the medullary index. The following information can be obtained from the examination of hair (Fig. 33.2):

- Is the sample a hair or any other material?
- Are the hairs of human or animal origin?
- From what part of the body have they originated?
- Does the hair come from a man or woman?
- How old is the person from whom the hairs have originated?
- Does the hair belong to a certain individual?
- Are the hairs dyed?
- Does it fall naturally or not?

In humans, body hair is mostly reduced, and it does not play as large a role in temperature regulation as it does in other animals. When humans are born, they have about five million hair follicles, only 2 percent of which are on the head. This is the largest number of hair follicles a human will ever have. As a human ages, the density of hair decreases.

33.1.2 Animal Hair Characteristics

Human and animal hair are structures that have a similar appearance, but structurally they differ. Animal hair is of three types: vibrissae, bristle, and wool. All those three types are very important for their lifestyles as they involve different functions. Identifying whether the hair is human or animal is the first step in forensic hair analysis. Animal hair is sometimes very important as physical evidence, and its distinction from human hair and its identification as to species are very important matters. Experienced forensic experts have little difficulty in making the distinction at any time, when police are required to make sure that any hair whether human or animal in origin. The obtained information may be employed to establish facts pertinent to the investigation. Human and animal hairs show similarities in having an outer cuticle, cortex, and medulla. The outer surface of the hair is covered by scales. Though there are similar morphological features, the scale pattern provides distinguishing characteristics between animal and human hairs. The scales of an animal's hair show many distinctions such as coronal and spinous patterns, whereas in the case of humans the scale of "imbricate" pattern. Human hairs usually have a thin or an absent medulla region. Animal hairs usually have thick medullae. Human and animal hairs can be differentiated by color, tip, cuticle, root, cortex, and other special characteristics. Various morphological characters like the shaft diameter and its variations along the length, spatial variation, cross-sectional shape, and pigment distribution are often found suitable for the identification of human and animal hair. The key difference between human and animal hair is that human hair does not stop growing; hence, it is longer while animal hair stops growing when it reaches a certain length; therefore, it is shorter (Table 33.1).

The next step is comparing the found hair to known suspects or animals present at the crime scene. This analysis can be subjective based on the person's experience with identifying the shape of the medulla. As per the necessity of the case, many times simple identification of animal or human origin is sufficient in providing potential leads.

33.2 Comparison Process and Analysis

Morphological examination of hair samples is the first step in forensic hair comparisons. The main medico-legal concerns with hair examination include identification of the species of origin, ascertainment of the hair's provenance from the body, and, finally, comparison of the control hair sample from the victim to the hair sample from the crime scene. Though it is not possible to definitely identify a sample of hair originating from a particular person's head, unequivocal determination of human origin can be established based on microscopic examination of the hair's

Table 33.1 Differences between human hair and animal hair

Characteristics	Human hair	Animal hair
Cuticular scales	– Flattened – Small with irregular margins	– Projecting large and in various patterns
Medulla	– Fragmental discontinuous or may be absent – Diameter usually less than 1/3 of the shaft diameter – Thinner	– Discontinuous or even continuous – Diameter is usually more than half of the shaft diameter – Thicker
Shaft	– Diameter between 50 and 150 microns	– Diameter less than 25 or more than 3000 microns.
Pigment	– Rarely absent – Evenly distributed	– May be absent – Present near medulla
Length	Much longer	Shorter
Growth	Does not stop growing	Stop growing when reaches a certain length
Color	Consistent from root to tip	May change several times from root to tip

cuticle, cortex, medulla, and pigment granules. Hair proceeds through three stages as it develops. The first stage is called the anagen stage and lasts approximately 1000 days. Eighty to ninety percent of all human hair is in the anagen stage. This is the period of active growth when the cells around the follicle are rapidly dividing and depositing materials within the hair. The second stage, i.e., catagen stage, follows as the hair grows and changes. The catagen stage accounts for about 2 percent of all hair growth and development. The third and final stage is the telogen stage. During this stage, the hair follicle is dormant or resting and hairs are easily lost. About 10 to 18 percent of all hairs are in the telogen stage. There is no pattern as to which hairs on the head are in a particular stage at any time. Forensic scientists usually study the hair's scale pattern and appearance of the medulla to identify hair of unknown origin and perform three major types of hair analysis:

1. Testing the hair shaft for drugs in a person's system.
2. Analyzing DNA collected from the root of the hair
3. Viewing Hair under a Microscope to Determine Its Origin

It is only in the last few years that serious attention has been given to the forensic use of information obtained from hair analysis in the investigation.

In crimes involving struggle, a victim may have hairs in his fist. In sexual assault cases, pubic hairs may be found on the genitals of the accused and victim. Therefore, the search should be concentrated on places where one can expect to find hairs left by criminals. Samples of questioned hair will most commonly be found in vacuum sweepings or will be picked from a suspect's clothing. In order to examine hair evidence collected in an investigation, it is a must to take exemplars from the individuals involved. A court order may be required to obtain hair specimens from a defendant. Hair may be collected by the medico-legal doctor or investigator.

The standard hairs are collected from the head, pubes, and other relevant body areas of the individual suspected of being the source of the questioned hairs, which are necessary for comparison with questioned hairs.

Initially, color is ascertained by visual and microscopic examination and the length and other broad characters are noted. The microscopic examination provides a good deal of information about the shape and size of the cuticular scales, appearance, and diameter of the medulla in relation to the shaft diameter. Attempts aimed at identifying the intrinsic biochemical markers like the blood group substance, the proteins, or their byproduct. The electrophoretic pattern of polypeptides of hair from different primates has been found species-specific and applied to the study of proteins and enzymes in hair root sheath. Conventional chemical methods for the study of hair are destructive of the evidence. The applicability of such physic-chemical method appears to be promising for use in the future. With the advancement of forensic DNA typing, microscopic hair comparisons and DNA analysis can be complementary and provide information on the source of hair. Hair analysis is done by evaluating hair structure and DNA from cells attached to the root of the hair. It can be used to find out if people are related. Forensic hair analysis can be done to help identify a person who may have been present at a crime scene.

Human hair contains two types of DNA. Nuclear DNA is found in the tissue at the root of a strand of hair, while mitochondrial DNA is found in the shaft of hair itself. Initially, only nuclear DNA could be extracted and analyzed, but today, both types of DNA can be used to help identify individuals. If a hair is deemed not suitable for nuclear DNA analysis, it can be sent for mitochondrial DNA testing. A root does not have to be present in order for a hair to be suitable for mitochondrial DNA testing.

Extensive work is yet necessary for universal acceptance of these approaches to the examination of even small pieces of hair. All these techniques are suggestive of hair identification in the criminal justice system.

Multiple Choice Questions

1. Which preservative is used for preserving hair?
   (a) Rectified.
   (b) Norman saline.
   (c) Buffer.
   (d) No preservative is required.
   Answer (d)
2. Innermost layer of human hair is.
   (a) Cortex.
   (b) Scale.
   (c) Medulla.
   (d) Cuticle.
   Answer (c)
3. Which of the gland secrets oil in hair?
   (a) Sebaceous gland.
   (b) Parotid gland.

   (c) Pituitary gland.
   (d) All of the above.
   Answer (a)
4. The type of hair present on an unborn baby.
   (a) Coarser hair.
   (b) Vellus hair.
   (c) Terminal hair.
   (d) Lanugo hair.
   Answer (d)
5. Hair samples collected from the scene of crime from different spots must be packaged.
   (a) Collectively.
   (b) Separately.
   (c) Hardly matters.
   (d) Whatever is easy.
   Answer (b)
6. DNA profiling of hair is possible.
   (a) From hair tip.
   (b) From cut hair.
   (c) From rooted hair.
   (d) All of the above.
   Answer (c)
7. Study of hair is known as.
   (a) Histology.
   (b) Palynology.
   (c) Trichology.
   (d) Pathology.
   Answer (c)
8. Difference between human hair and animal hair can be done on the basis of.
   (a) Size.
   (b) Medulla.
   (c) Diameter of shaft.
   (d) All of the above.
   Answer (d)

References

Butler JM (2010) Fundamentals of forensic DNA typing. Academic, Cambridge, MA, p 101

Chojnacka K, Gorecka H (2006) The effect of age, sex, smoking habit and hair color on the composition of hair. Environ Toxicol Pharmacol 22:52–57

Cortellini V, Carobbio A, Brescia G (2019) A comparative study of human and animal hairs: microscopic hair comparison and cytochrome c oxidase. J Forensic Sci Med 5:20–23

Dahiya MS and Yadav SK (2013) Elemental composition of hair and its role in forensic investigation. Scientific Report https://doi.org/10.4172/sciencereports722

Elizabeth AG and David RF, 2005.**A** Simplified method for mitochondrial DNA extraction from head hair shafts. J Forensic Sci 50:5

Fatma A, Madkour, Abdelsabour-Khalaf M (2021) Scanning electron microscopy of the nasal skin in different animal species as a method for forensic identification. Microsc Res Tech 85(5):1643

Harrison S, Sinclair R (2003) Haircoloring, permanent styling and hair structure. J Cosmet Dermatol 2:180–185

Kempson IM, Skinner WM, Kirkbride KP (2007) The occurrence and incorporation of copper and zinc in hair and their potential as bioindicators : a review. J Toxicol Environ Health B10:611–622

Patle S, Bagchi D, Singh KP (2022) Comparative study of morphology and keratin levels in hair from deer and goat. J Threat Taxa 14(12):22346

Pilli E (2012) Human and animal hair in forensic evidence: problems, troubleshooting and workarounds. J Forensic Sci:1–58

Rogers GE (2004) Hair follicle identification and regulation. Int J Dev Biol 48:163–170

Scientific Working Group on Materials Analysis (SWGMAT) (2005) Forensic human hair examination guidelines

Tafaro JT (2000) Use of microscopic postmortem changes in anagen hair roots to associate questioned hairs with known hairs and reconstruct events in two murder cases. J Forensic Sci 45:495–499

Taylor AS (2007) The principles and practice of medical jurisprudence. Churchill Livingstone, London

Forensic DNA Database

34

Rutwik Shedge, Aditi Iyengar, Monisha Samuel, and Tanya Chauhan

Abstract

In the previous chapters we learned about how DNA is extracted using different methods and how its PCR amplification and STR typing is done. We receive results of these methods in the form of an electropherogram—which gives us the details about STR polymorphisms present in the DNA of a particular sample. If we have different samples, for example, the blood of a suspect and the blood and tissues recovered from underneath the fingernails of the victim, we can simply extract and profile the DNA from both the samples and compare them to find out whether they belong to the same individual. However, in certain cases, we do get the perpetrator DNA, but without anything to compare it with. In the same case as above, if the police do not have a suspect pool, any DNA profile generated from the tissue recovered from victim's fingernails will have to be stored until a suspect is identified. Thus, the first DNA databases were created, which were essentially repositories of DNA of convicted criminals, and were used as tools to assist law enforcement individuals in solving crimes. Whenever any unknown biological

R. Shedge (✉)
School of Forensic Science, National Forensic Sciences University, Gandhinagar, Gujarat, India

A. Iyengar
Impact Science, Cactus Communications, Mumbai, India

M. Samuel
School of Forensic Science, National Forensic Sciences University, Gandhinagar, Gujarat, India

Department of Forensic Science, National Forensic Sciences University, Delhi Campus, New Delhi, Delhi, India

T. Chauhan
Department of Forensic Science, National Forensic Sciences University, Tripura Campus, Agartala, Tripura, India

A. Puri et al. (eds.), *Fundamentals of Forensic Biology*,
https://doi.org/10.1007/978-981-99-3161-3_34

sample was recovered by the law enforcement agencies, its DNA was profiled and compared with the profiles present in the DNA database.

Keywords

DNA database · DNA profile · CODIS · Extended CODIS · NDNAD

34.1 Definitions

DNA database: A repository of DNA profiles collected from unknown perpetrators, crime scenes, missing individuals, relatives of missing individuals, convicts, and accused individuals.

CODIS: Combined DNA Index System or the national DNA database of the United States of America.

NDNAD: the National DNA Database of United Kingdom.

34.2 History of Forensic DNA Databases

34.2.1 NDNAD

In 1994, the Criminal Justice and Public Order Act was passed by the British parliament, and this laid the foundation stone for the first ever forensic DNA database: the National DNA Database (NDNAD) (Wallace 2006). The Criminal Justice and Public Order Act allowed the police to take DNA samples from anyone charged with a "recordable" offence without their consent, and to search the database for any matching profiles (Wallace 2006). These recordable offences included almost all the crimes, from homicide to even being drunk, begging, and participating in an illegal protest. The NDNAD was declared to be functional in 1995 but was restricted to only the most heinous crimes (homicides, sexual assaults, violent crimes, etc.) because of the lack of funds and resources. By 2020, the NDNAD had 6.6 million profiles stored within it.

34.2.2 DPD

In 1996, the New Zealand Police in collaboration with the Institute of Environmental Science and Research (ESR) created a central DNA database known as the DNA profile databank (DPD) that had a collection of DNA profiles from convicted offenders and volunteers (Harbison et al. 2001). The DPD never included any profiles from suspects and had profiles only of individuals convicted with offences that had more than 7 years of imprisonment. The profiles within the DPD are never removed unless the conviction is quashed.

34.2.3 Austrian DNA Intelligence Database

In 1997, the Austrian DNA Intelligence Database was established, and it consists of two units: the executive branch located in the ministry of Interior in Vienna and the laboratory present in the Institute of Legal Medicine, University of Innsbruck. Buccal swabs of any individual suspected of commission of recordable crimes are taken, and DNA profile is generated from it (Parson et al. n.d.). The profile is removed only after acquittal.

34.2.4 The Dutch DNA Database

Netherlands made its own DNA database operational in 1997 and is administered by the Secretary-General of the Dutch Forensic Institute (NFI). Complete, partial, and mixed profiles are stored in the database (Amelung et al. 2020a). DNA of only those individuals, whose profile is needed for investigation of a particular crime, whose penalty may be imprisonment of more than 4 years, is stored in the database. Also, the profile of the convicted individual is removed after 20–30 years of conviction (Li 2021).

34.2.5 The German DNA Database

Germany established its DNA database in 1998 under the leadership of its Minister of Interior (Amelung et al. 2020b). By 2018, there were about 900,000 profiles stored within the database. DNA profiles of all individuals accused of commission of crimes which have a penalty of more than 1 year imprisonment are entered in the system. The profiles are removed only if the accused is found not guilty, or after 5–10 years of conviction in case of convicts exhibiting good behaviour in the prisons (Li 2021).

34.2.6 DNA Databases in the Western Balkans

In the early 1990s, the identification of human skeletal remains from mass graves was undertaken in the countries of western Balkan region – Bosnia and Herzegovina, Croatia, Macedonia, and Kosovo (Marjanović et al. 2011). The scientists leading these missions of humanitarian forensics faced a lot of issues while trying to develop the biological profile of victims and started working towards the construction of DNA databases. In 1998, Slovenia became the first western Balkan country to establish a forensic DNA database. In Slovenia, the DNA profile of any individual suspected of having committed a recordable crime is added to the database (Li 2021). Croatia followed Slovenia's lead and established its own DNA database under the aegis of Mister of Interior in 2001. The countries of Bosnia and Herzegovina, Croatia, Montenegro, and Macedonia have drafted action plans and are currently in the process of establishing their DNA databases (Marjanović et al. 2011).

34.3 DNA Database Entries and Expansion

The criteria for the entry of a DNA profile in the DNA database varies from country to country. As seen in the history section of the current chapter, a few countries such as New Zealand and Netherlands add the DNA profile of only convicts who have been jailed for crimes that have a punishment of imprisonment of 7 and 4 years, respectively. On the other hand, in countries such as Austria and Switzerland, the DNA profiles of even the suspects of a crime are added to the national database. In a country like the United States of America, the criteria for entry of a DNA profile to the database depends upon not only the federal laws, but also the laws of the state in which the crime has occurred. 31 of 50 states of the USA have DNA arrestee laws, 29 out of these 31 states collect DNA for some state felonies, while 8 out of 31 collect DNA from arrestees for certain misdemeanours too. Out of these 31states, 8 states apply their laws to juveniles as well (Samuels et al. n.d.).

DNA databases are constructed with the goal of assisting the law enforcement agencies in investigation of crimes. They are used to compare DNA profiles from crime scenes with DNA profiles of known offenders, or profiles generated from other crime scenes. Traditionally, the DNA profile in question includes a typical autosomal STR panel along with amelogenin. However, a new strategy that is being employed currently to increase the scope of comparison is to expand the DNA profiles of the DNA database. This expansion entails DNA typing beyond the core 13 STR + amelogenin sex locus, by using Y-STRs (Short Tandem Repeats on the Y chromosome), mitochondrial DNA typing (mtDNA), SNP profiling (Single Nucleotide Polymorphisms) on autosomes, and whole genome sequencing (WGS). Y-STRs have their applications in paternity testing, identification of missing male individuals, identification of male DNA in a mixture of male and female DNA (cases of sexual assaults), identification of number of male perpetrators in a sexual assault case, etc. Mitochondrial DNA typing has its utility in maternity testing and identification using shed and cut hair shafts, etc. SNPs are useful in identification when the forensic evidence consists of degraded or low copy number DNA. Whole genome sequencing may prove to be the best solution for forensic case work as it can provide the most data (autosomal STRs, Y-STRs, SNPs, etc.) from a single sample.

Expansion of DNA databases also entails expansion of the DNA profile pool. For instance, earlier samples were collected from only the convicted felons in violent crimes for DNA profiling and entry into the database. However, as times progress, newer crimes are being included in the list for which DNA profile from the convict is necessary. Also, in some countries, the DNA is collected from suspects of a crime, and the profile is maintained within the national DNA database unless the individual is acquitted or removed from the suspect pool. In some countries, it is also suggested that the database should include, along with the perpetrator, accused, and suspect profiles, the general population profiles. This will ensure quick identification of victims of abduction, kidnapping, missing individuals, and even major accidents and mass disasters (Wickenheiser 2022).

34.4 CODIS

34.4.1 CODIS Infrastructure

CODIS, or the Combined DNA Index System, is the national DNA database of the Unites States of America and was established by the Federal Bureau of Investigation (FBI) through the Congressional DNA Identification act in 1998 (Panneerchelvam and Norazmi 2003). CODIS identified 13 STR loci (Table 34.1) that were implemented in 1998 for DNA profiling purposes. CODIS has three hierarchical levels: LDIS, SDIS, and NDIS (Crouse et al. 2019; Li 2021).

LDIS stands for the Local DNA Index System and is maintained in all the local laboratories managed by the police departments and Sheriff offices. Any DNA profile generated at these law enforcement departments at a local level are uploaded in the LDIS, and through LDIS, they are transmitted to the SDIS and the NDIS. SDIS stands for the State DNA Index System and is present in each of the 50 states of USA. The SDIS is maintained by a state forensic science laboratory. SDIS allows any one local law enforcement agency to compare DNA profiles generated in that local area across all the other DNA profiles generated by the different local agencies in any state. SDIS also allows for the transmission of DNA profiles from LDIS to NDIS, and any reports or messages from the NDIS to the LDIS. NDIS, or the National DNA Index System is the highest level of CODIS, and has the DNA profiles from all the LDIS and SDIS, along with any DNA profiles that the federal labs generate. NDIS is administered and maintained by the FBI. FBI can also allow

Table 34.1 STR markers used in CODIS

Sl. No.	Locus	STRs
1	AMEL	Amelogenin
2	CSF1O	Human c-fms protooncogene of CSF-1 receptor
3	D3S1358	
4	D5S818	
5	D7S820	
6	D8S1179	
7	D13S317	
8	D16S539	
9	D18S51	
10	D21S11	
11	FGA	Fibrinogen alpha chain
12	TH01	Tyrosine hydroxylase 1
13	TPOX	Human thyroid peroxidase
14	VWA	Von Willebrand factor

foreign governments to compare their DNA profiles in the CODIS upon proper request by the foreign government to FBI or the INTERPOL (International Criminal Police Organization).

34.4.2 CODIS Indices

CODIS organizes all the DNA profiles entered in it into categories known as indices. DNA profiles of individuals convicted of crimes are added to the Convicted Offender Index, while the DNA profiles of individuals that have been arrested for commission of a crime, are added to the Arrestee index. The criteria for inclusion of a DNA profile in the arrestee index varies from state to state. The DNA profiles generated from evidences found at the scene of crime in forensic cases are stored in the Forensic Index. The FBI has also established the NMPDD or the National Missing Person DNA Database, for easier identification of missing persons and unidentified skeletal remains. The NMPDD consists of three subcategories: the Missing Person Index for storing DNA profiles of missing persons, the Unidentified Human Remains index for storing DNA profiles generated from unknown dead bodies and remains, and the Biological Relatives of the Missing Person Index which stores the DNA profiles of relatives of missing persons that have voluntarily donated their samples.

34.5 Forensic Investigation Using Database Searches

The goal of developing and maintaining a DNA database is to provide the law enforcement agencies with a tool for further forensic investigation. The whole point is to keep a repository of DNA profiles generated from convicts, arrestees, and different scenes of crime, so that comparison can be made between a DNA profile generated from a fresh case and the profiles stored in the repository. Broadly, there are three scenarios in which forensic investigation is done using databases:

34.5.1 Case-to-Offender Searches

The hypothesis in case-to-offender searches is that the DNA profile generated from a sample left by the perpetrator can be matched to a suspect or convict of some other crime. For example, consider a case where in a female victim has been sexually assaulted and killed in an overgrown farmland. The investigators find upon closer examination, some tissue, and blood underneath the victim's fingernails. The forensic scientists working this case manage to generate a DNA profile from the blood and tissue found underneath the victim's fingernails. The DNA profile generated is entered in the national DNA database of the country, and the investigators find a match with an individual that was convicted 10 years ago for physical assault of another female. This suspect had been released 3 years ago and has since then been loose. Thus, forensic DNA database search helped the investigators in finding the perpetrator of the case they are investigating.

34.5.2 Case-to-Case Searches

The hypothesis in case-to-case searches is that a serial offender may leave DNA at each of their different crime scenes, and matching the profiles of different samples in different scenes will help investigators in concluding that the multiple crimes have been committed by the same individual. For example, consider a case series of sexual assaults and homicides. At the first scene of crime, investigators find semen sample near the inner thighs of the female victim. They send the collected biological fluid to the forensic science laboratory, where a DNA profile is generated and stored in the DNA database. A few months later, a similar scene of crime is identified where another female victim has been sexually assaulted and murdered. Semen is again recovered from the person of the victim and sent for DNA profiling. The DNA profile is then stored in the national DNA database. When the DNA profiles of semen recovered from scene 1 and scene 2 are compared through the DNA database, it is observed that both the semen samples belonged to the same individual. Thus, two cases are thus linked using case-to-case searches.

34.5.3 Familial Searches

Familial searches are employed in cases of missing persons, when the antemortem DNA profile of the missing individual is absent. In this case, if an unknown dead body is recovered, a DNA profile is generated from the body and stored in the DNA database. Eventually, DNA profiles of suspected missing person's relatives are generated and compared with the missing person's DNA profile stored in the database. This can allow for repatriation of the victim.

Multiple Choice Questions

1. The first ever national DNA database was established in which of the following countries?
   (a) Austria.
   (b) Moldova.
   (c) United Kingdom.
   (d) United States of America.
   Answer: (c)
2. CODIS stands for.
   (a) Combined DNA identification system.
   (b) Combined DNA index system.
   (c) Combined DNA identification service.
   (d) Combined DNA international service.

Answer: (b)

3. The highest level in the CODIS infrastructure is __________.
   (a) LDIS.
   (b) SDIS.
   (c) NDIS.
   (d) NDNAD.
   Answer: (c)
4. The STR panel used for DNA profiling in CODIS consists of ____________.
   (a) 13 autosomal STR + Amel locus.
   (b) 16 autosomal STR + Amel locus.
   (c) 19 autosomal STR + Amel locus.
   (d) 22 autosomal STR + Amel locus.
   Answer: (a)
5. Which of the following is not a CODIS index?
   (a) Arrestee index.
   (b) Convicted offender index.
   (c) Suspect index.
   (d) Forensic index.
   Answer: (c)

References

Amelung N, Granja R, Machado H (2020a) The Netherlands. Pp 73–87

Amelung N, Granja R, Machado H (2020b) Germany. Pp 55–71

Crouse CA, Bauer L, Sessa T et al (2019) Combined DNA index system (CODIS)-based analysis of untested sexual assault evidence in Palm Beach County Florida. Forensic Sci Int Synergy 1: 253–270. https://doi.org/10.1016/j.fsisyn.2019.09.005

Harbison SA, Hamilton JF, Walsh SJ (2001) The New Zealand DNA databank: its development and significance as a crime solving tool. Sci Justice 41:33–37. https://doi.org/10.1016/S1355-0306(01)71846-1

Li R (2021) Forensic biology: identification and DNA analysis of biological evidence. CRC, Boca Raton, FL

Marjanović D, Konjhodžić R, Butorac SS et al (2011) Forensic DNA databases in Western Balkan region: retrospectives, perspectives, and initiatives. Croat Med J 52:235–244. https://doi.org/10.3325/cmj.2011.52.235

Panneerchelvam S, Norazmi MN (2003) Forensic DNA profiling and database. Malays J Med Sci MJMS 10:20–26

Parson W, Steinlechner M, Berger B, et al (n.d.) Laboratory Strategy of the Austrian DNA Intelligence Database. 2

Samuels J, Dwyer A, Halberstadt R, Lachman P Collecting DNA from Juveniles. 67

Wallace H (2006) The UK national DNA database: balancing crime detection, human rights and privacy. EMBO Rep 7:S26–S30. https://doi.org/10.1038/sj.embor.7400727

Wickenheiser RA (2022) Expanding DNA database effectiveness. Forensic Sci Int Synergy 4: 100226. https://doi.org/10.1016/j.fsisyn.2022.100226

Diatom Testing

35

Preetika M. Chatterjee

Abstract

Whenever a dead body is discovered in any water body, the prime question to be dealt by a forensic pathologist is to determine whether the death is due to drowning or it is the result of disposal after death. It is challenging to determine the cause of death in such cases and to differentiate a death due to drowning or other reasons. Drowning signs like pale skin colour, occurrence of froth in mouth and nostrils are clearly apparent in recent deaths due to drowning, however, with the passage of time these signs tend to disappear due to prolonged immersion of the dead body in the water body and advancing decomposition process. In case of drowning, water aspirated by the person into his lungs ultimately enters the main blood stream via alveoli in addition to the other micro-organisms and small impurities present in the water. Forceful inspiration and expiration causes the development of microscopic tears in the alveolar wall and the micro-organisms and micro-particles travel through these microscopic tears into the blood stream and ultimately to multiple organ sites and get deposited in the capillaries of these organs. These particles include a group of algae, commonly known as diatoms which tend to remain in the biological tissues and organs for longer periods of time and play a significant role in the determination of death due to drowning.

Keywords

Diatoms · Drowning · Death due to drowning · Postmortem drowning

P. M. Chatterjee (✉)
Department of Forensic Science, MATS University, Raipur, Chhattisgarh, India
e-mail: drpreetikac@matsuniversity.ac.in

A. Puri et al. (eds.), *Fundamentals of Forensic Biology*,
https://doi.org/10.1007/978-981-99-3161-3_35

35.1 Use of Diatoms in Drowning

Whenever a dead body is discovered in any water body, the prime question to be dealt by a forensic pathologist is to determine whether the death is due to drowning or it is the result of disposal after death. It is challenging to determine the cause of death in such cases and to differentiate a death due to drowning or other reasons. Drowning signs like pale skin colour, occurrence of froth in mouth and nostrils are clearly apparent in recent deaths due to drowning, however, with the passage of time these signs tend to disappear due to prolonged immersion of the dead body in the water body and advancing decomposition process. In case of drowning, water aspirated by the person into his lungs ultimately enters the main blood stream via alveoli in addition to the other micro-organisms and small impurities present in the water. Forceful inspiration and expiration causes the development of microscopic tears in the alveolar wall and the micro-organisms and micro-particles travel through these microscopic tears into the blood stream and ultimately to multiple organ sites and get deposited in the capillaries of these organs. These particles include a group of algae, commonly known as diatoms which tend to remain in the biological tissues and organs for longer periods of time and play a significant role in the determination of death due to drowning.

Diatoms specifically pertaining to crime scene evidences and victims are studied under a sub-field of fresh water ecology called Forensic Limnology. It includes various methods of collection, analysis, identification and comparison to standard database in order to estimate the diatom colonies ratio available in the sample with those of crime scene locations. In forensic cases related to secondary crime scene, the criminals sometime dump the dead body into the water body upon commission of crime either to dispose off the body or to simulate cause of death to be due to drowning. Macroscopic findings like plume of froth on mouth and nostrils, heavy lungs and emphysema, Paltaufs spots, eodema aquosum, frothy trachea and pleural effusions are major indicators of death due to drowning. However, these symptoms diminish with the passage of time and effect of putrefaction. Diatoms are resistant to decomposition and in cases of deaths due to drowning in skeletonized bodies or bodies in advance decomposition stages, are the only prediction factors left with respect to ant mortem or postmortem drowning. It is done by comparing diatom density and nature or type of diatom species present in the organ sample to that of the control water sample obtained from the water body site of recovery of the drowned body. Since its discovery, diatom test has been the most important test used in forensic case scenario for the detection of deaths due to drowning. Diatoms were detected in the lung fluid for the first time by Hoffmann et al. followed by its successful detection in the blood stream and parenchymal organs by Incze, who also reported the entrance of diatom into bloodstream through lungs. Tamasaka et al. discovered diatoms in organ tissues and bone marrow directly representing ante mortem aspiration of water during drowning (Fig. 35.1).

Diatoms bear acid-resistant silica shells called frustules which enable its separation from the organ tissues using the extraction method of acid digestion. These frustules bear morphological characters which are well preserved and apparent even

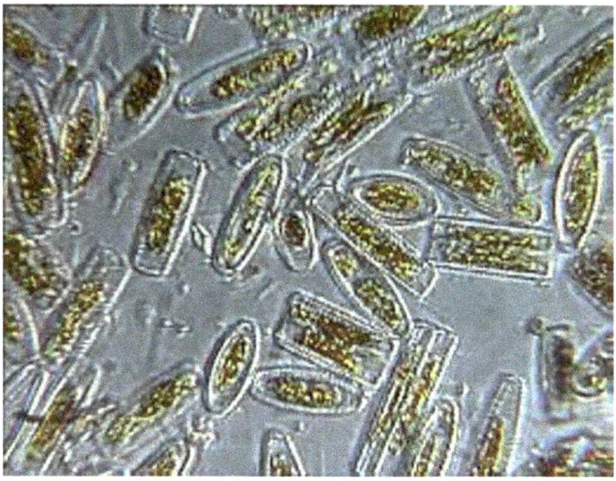

Fig. 35.1 Diatom colony of *Diatoma vulgaris* spp. (Source: Harding et al., 2007)

after rigorous chemical digestion. Diatom species which are unique and specific to certain habitats are best suited to be used as evidence in linking a crime scene to the victim. However, it is not the case in every crime scenario as most of the different diatom species are available in varied aquatic environments. None the less, the diatom species vary in abundance from water body to water body as it is highly sensitive towards environmental changes such as temperature, nutrients, pH, etc. Thus the site of drowning can be determined owing to comparison of diatom species occurrence ratio or abundance patterns obtained from the tissues with those derived from the suspected water body site samples. Diatom test can be further used for the identification of individuals, clothing or belonging recovered from the sites of drowning. The major factors pertaining to diatom test are concentration of diatoms in lungs in addition to the development of river monitoring programme in the area of investigation. Medicolegal investigation of drowning deaths are specifically benefitted by the continuous fresh water site monitoring and a comprehensive species level inventory development of diatom flora at such sites.

Qualitative and quantitative distribution of diatoms in human body is variable and thus is an important distinguishing factor for ante mortem and postmortem drowning. The easily available organs like lungs and stomach shoe higher diatom density owing to the process of passive percolation through nostrils or mouth. The density of diatoms undergoes decrease in concentration by factor of 10–100 during passage from the drowning medium into lungs or stomach. However, this density dip reaches the factor of 100–1000 upon transition from lungs to blood, kidneys, liver or bone marrow. Deeper the organ tissue lesser will be the diatom deposition density. However, contrary situation may arise in case of prolonged intake of contaminated water from natural water bodies. It can also be utilized to identify the geographic origin in case the unknown dead bodies. Hence, the validity of diatom test based on shape and number of valves derived from the tissue sample varies from 20 to 40 diatoms per 5 gm of bone marrow, 5 complete diatoms from 100 μL of tissue sediment of brain, kidney, liver, 37 diatom valves in heart tissue, 20 diatom valves in

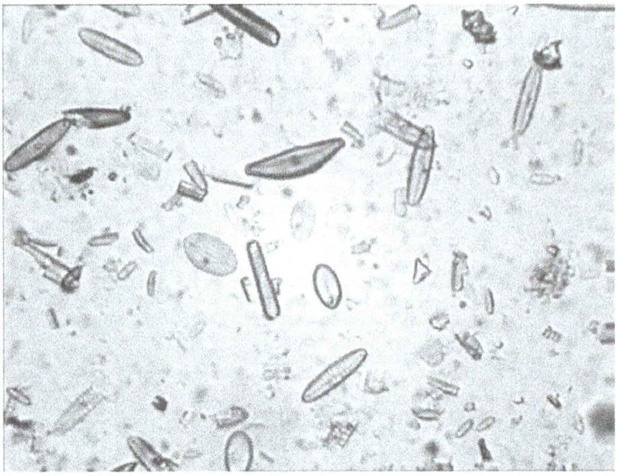

Fig. 35.2 Diatoms under light microscope. (Source: Harding et al., 2007)

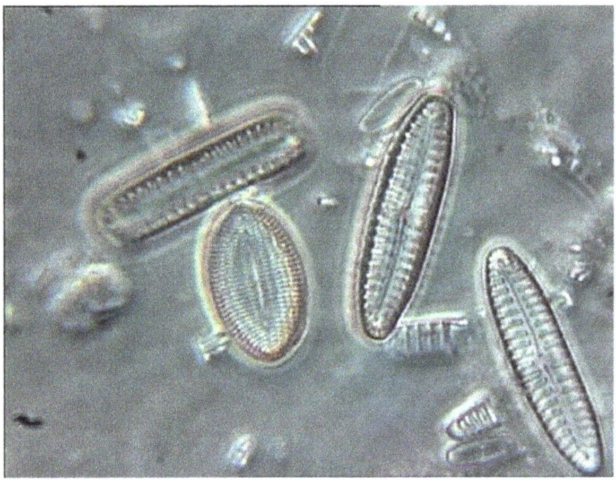

Fig. 35.3 Diatoms under Phase contrast microscope. (Source: Harding et al., 2007)

liver tissue and lung tissue to positively include or exclude results due to contamination. Species composition of diatoms varies in different sites of drowning based on the factors such as salinity, temperature, pH, inter-species competition. Thus, the species compositions pertaining to forensic samples are helpful in the prediction of geographic locations, habitat and time of year. Diatoms found particularly in certain restricted locations or site-specific endemic diatom species can be used as site identification markers in suspected drowning cases. In case of putrefied drowning cases, time of year can be predicted to a considerable extent. Hence, comparison of diatom species derived from suspected drowning site and organ tissue of drowned victim, based on nature, number and distribution of diatoms is indicative of death due to drowning or otherwise (Figs. 35.2, 35.3 and 35.4)

Sometimes instant death in water body due to weak heart conditions, pulmonary or circulatory system of the victim may decrease the time required for drowning, hence, decrease the amount of water inspired. Only small sized diatoms or valve

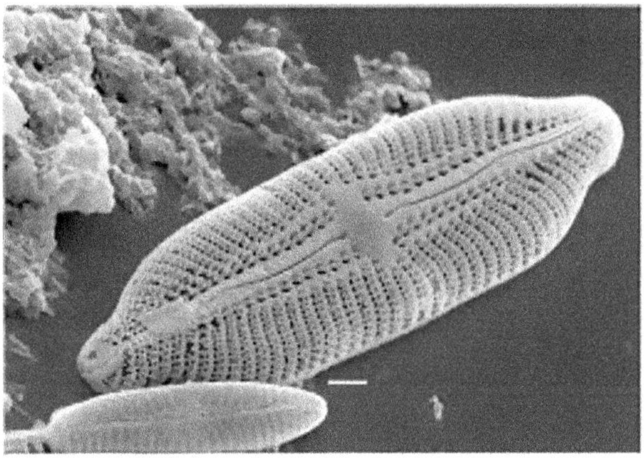

Fig. 35.4 Diatom under Scanning electron microscope. (Source: USGS.gov)

fragments can enter the organ tissues of drowning victims within the size range of about <110 μm. Contamination from the reagents, glass containers may add to the ambiguity of the test. Thus, diatom free water should be used while handling diatom samples to prevent it from exogenous diatom contamination along with careful handling of clothes and skin during autopsy to prevent contamination of internal organs from them. All these factors contribute towards negative results or cause ambiguity of diatom test, if not taken care of properly, lowering down its evidentiary value in the court of law. Thus, all these factors should be kept in mind and dealt with while handling diatom samples from both the body of the victim as well as the suspected site of drowning.

35.2 Structure of Diatoms

Diatoms mostly exist singly, although some may combine to form colonies. They are usually yellowish or brownish in colour and lack any flagella. Diatoms give a slimy or mucilaginous feel. They appear in seston or suspended form. Diatoms secrete a silicaceous outer covering or cell walls called frustules or test, marked with minute pores or depressions allowing the organism to interact with its environment. Locomotion and attachment to substrata or other cells to form colonies is facilitated by secretion of mucilage from the cell wall. The frustules are made up of two thecas which are generally in the form of two valves one slightly smaller than the other in order to facilitate proper fitting together of the two thecas one inside the other. When present in abundance in the water body, diatoms impart a golden brown color and thus are also known as Golden brown algae (Fig. 35.5).

They are unique in a way that silica which being virtually inert and indestructible, settles down in the bed of the water body after the death of the organism and ultimately forms the diatomaceous earth. In other words, the diatomaceous earth is formed by continuous deposition of silicaceous skeleton of diatoms. These silicaceous parts provide the basis of information regarding classification and

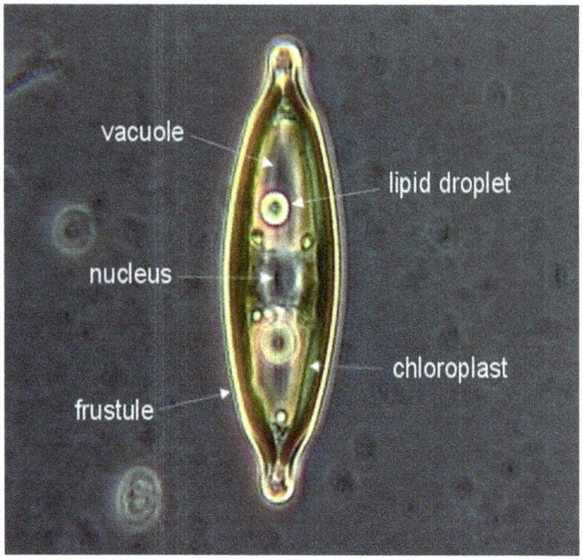

Fig. 35.5 Structure of a typical Diatom. (Source: Harding et al., 2007)

identification of these diverse organisms. The size range of diatoms, in general, varies from 2 to 200 μm in diameter or length, however, they can be as large as up to 2 mm long in certain species. Diatoms contain many photo pigments like Chlorophyll a and c which are stacked type having 3–5 thylakoid membranes along with carotenoid namely fucoxanthin present in plastids. They reproduce mostly by simple cell division although, certain species also shoe sexual mode of reproduction. The cell wall or the frustules are made of silica (SiO_2) which is often intricate and bears highly patterned morphological characters including a variety of pores, ribs, spines, marginal ridges and elevations which are characteristic to different genera and species and thus can be used to predict the same in a suspected diatom sample. Diatoms are highly polymorphic and may attach together by the means of a gelatin like material forming colonies which appear like ribbons. Motile diatoms are diamond or boat shaped.

35.3 Classification of Diatoms

Diatoms belong to kingdom Protista class Bacillariophyceae which are unicellular, autotrophic and eukaryotic organisms although some heterotropic or symbiotic species are also present in specific habitat. They commonly belong to genera such as Chaetoceros, Pseudonitzschia and Skeletonema. They inhibit almost every water body like springs, rivers, ponds, lakes, ditches in addition to freshwater, brackish water and marine waters. They can also be detected in terrestrial habitats like wet rocks, mosses and soils or even caves in addition to microhabitat like solid substrata, damp sediments and stems of rooted vegetation. They have highly complex taxonomy including more than 2,00,000 known species. However, many more are yet to

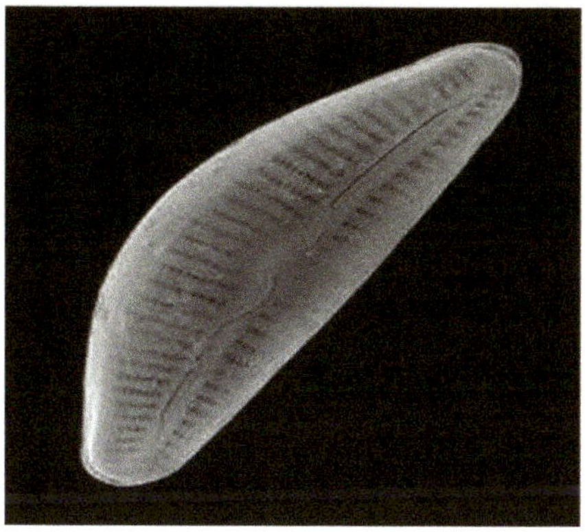

Fig. 35.6 Pennate Diatom

Fig. 35.7 Centrale Diatom. (Source: Science Photo Library)

be identified and described. Based on the shape of frustules, the diatoms can be further classified into Centrales and Pennales where the centrales are radially symmetrical while the pennales are bilaterally symmetrical. The centric species have circular shape in adaptation to watercolumn habitat of phytoplankton while pennate species prefer benthic habitats with rarely suspended in the water column (Figs. 35.6, and 35.7).

35.4 Sample Collection

See Fig. 35.8.

Tools required for Diatom Sample collection.

Simple tools like plastic tray/clean 2 L dishes with lids, wide mouth sample bottles of 100 mL and zip-lock plastic bags are required for storing the collected diatom sample. For collection toothbrush or other similar brush, knife or spoon, waterproof marking pen or labels, envelopes for collection of soil samples, plankton

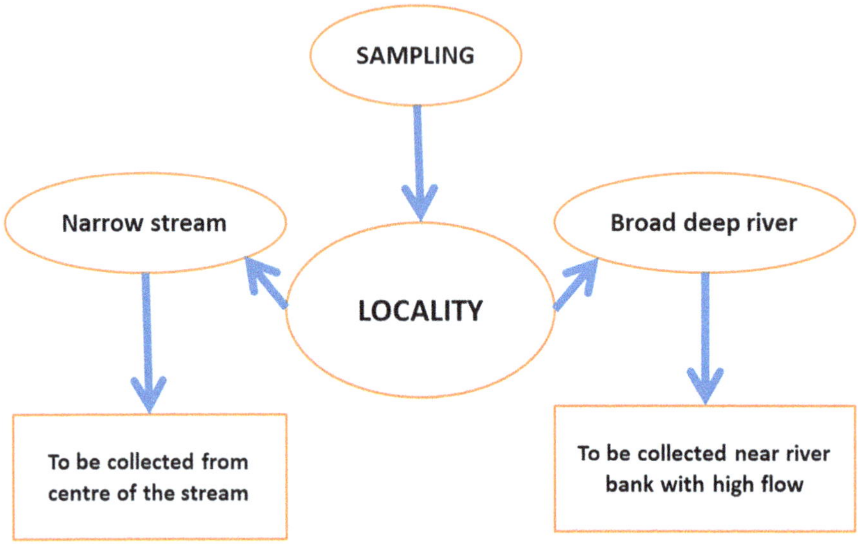

Fig. 35.8 Sample collection from drowning site

net, depth gauge, pipettes, latex tube fitted 100 mL syringe, Perspex tubes of 25 mm diameter in addition to field notebook, record forms and camera for documentation.

Acid digestion method of diatom extraction was first applied in the year 1904 and had been widely used since then irrespective of various controversies like its ambiguous nature causing deviations in cause of death, its entry into the body from other ways apart from lungs and so on. Still, the diatom test is recognized as one of the prescribed methods in drowning death cases. For this the forensic pathologist distinguishes the different diatom species on the basis of morphological examination and pattern analysis microscopically. Tissue samples need to be completely destroyed for extraction and detection of the diatoms from organ tissue samples. Diatom frustules are resistant against the mucus of the respiratory system and embolize from the circulatory system into internal organs. Bone being a closed system requires the blood circulation to transport the diatoms into the marrow, where other organs such as the lungs may become invaded passively. Hence, commonly used sample for acid digestion in diatom test is bone marrow as it supports the ante mortem drowning hypothesis and is least affected by contamination due to postmortem submersion. However, other organ tissues like liver, lungs, sternum bone, kidney and brain can also be utilized. The process starts with the removal of bone marrow with a spatula into a flask followed by addition of approx. 50 mL nitric acid. Then this suspension is boiled in a fuming hood on hot plate for approximately 48 h. Upon cooling, the mixture is centrifuged and washed with double distilled water. The sediment so obtained is then placed on a microscopic slide and examined under the microscope specifically the phase contrast microscope for the presence of diatoms. Different organs have been used as samples for detection of diatoms using acid digestion method like bone marrow, lungs and sternum bone, kidneys,

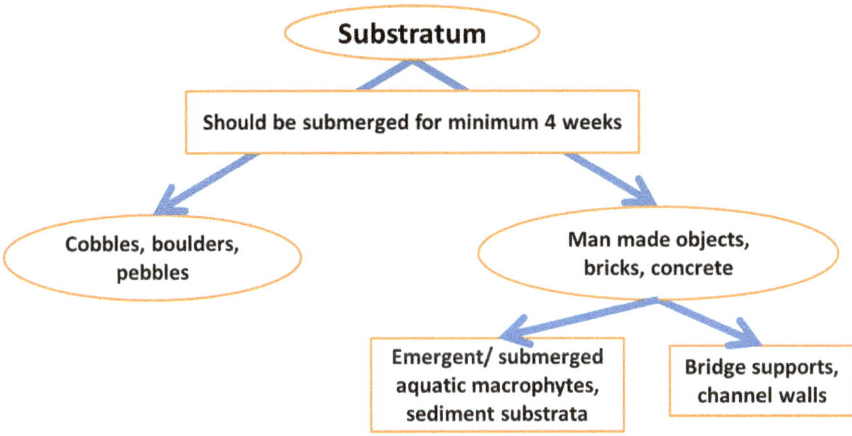

Fig. 35.9 Types of Substratum for sample collection from drowning site

brain, liver. Other than nitric acid, H_2SO_4, HNO_3 and H_2O_2 can also be used as solvent for digestion of soft tissues such as kidneys, brain, liver and lungs. H_2O_2 has been reported to be best suited for the digestion of clothing samples aiding the forensic investigations. Soluene-350 has been tested to be best suited for samples pertaining to sea water drowning. Water sample derived from the drowning site is also analysed for the presence of diatoms which are then compared with the sample tissue diatom species. For this 1–2 L of water sample is obtained from the site of suspected drowning and added with a few drops of formalin. This solution is left undisturbed overnight, decanted and the concentrate available at the bottom is subjected to microscopic examination and comparison (Fig. 35.9).

Acid digestion method, however, may cause destruction of diatom structure. In certain species such as *Chaetoceraceae, Corethronaceae, Bacteriastraceae, Rhizospleniaceae* and *Leptocylindraceae*, the frustules are thinly silicified and are easily digested by the use of acids, thus giving a notion of diatom absent during analysis. However, this can be overcome or minimized by using Lefort aqua regia, i.e., 3:1 ratio of nitric acid and hydrochloric acid. Conventional methods of acid digestion suffer from the limitations of being time consuming, dangerous and strenuous. Enzymatic digestion method to solublize soft tissue using Qiagen proteinase K enzyme in addition to Qiagen buffer ATL and 5 N HCl is rather faster and simpler method of diatom extraction with respect to suspected drowning cases. Sensitivity of molecular biology techniques in the contemporary scenario is so high that 16S RNA subunits of ribosomal RNA present in the planktonic DNA present in human tissues can be detected in case of drowned victims. Samples pertaining to highly putrefied samples can be easily and rapidly diagnosed with the use of electric impedance spectroscopy. Another technique in this list can be fluorimetry which can locate or isolate diatom species present in tissue or bone samples using luminescence and fluorescent tags which can be highly useful in comparison and differentiation of diatom species found at the suspected drowning sites from those present naturally. In addition, NCS, Protosol, ultrasonic irradiation and membrane filtration are some other methods which can be used in place of acid digestion in both fresh water as well as sea water diatom species. Detection of diatoms in the bone marrow samples is a strong indicative of that the person was alive and breathing at the time of drowning.

35.5 Microscopic Analysis and Identification of Diatom

For morphological examination of diatoms light microscope or phase contrast microscopes are generally used. However, for detailed image of diatoms both SEM and TEM are best suited. The electron microscope provides taxonomical details with species distinctions which are sometimes so minute to be visualized by light microscope. Nowadays, dark phase or electron microscopy is the major methodology utilized in the analysis of diatoms. TEM (transmission electron microscope) is capable of noticing the fine and delicate details present in the diatom species even in less silicified species such as *Chaetoceraceae, Corethronaceae, Bacteriastraceae, Rhizospleniaceae* and *Leptocylindraceae*. SEM (scanning electron microscope) provides the best visualization of the complete diatom frustules and the gross morphology of both internal as well as external features of the diatom. Unequivocal occurrence of diatoms on the microscopic slide show positive results in diatom test. This depends on the season and month of drowning in addition to the type of water body. No concordance of diatom species present in organ tissue and putative drowning site is an indicative of effect due to natural currents of human activity in displacement of the body from the actual site of drowning or death. SEM not only does not alter the morphology of the diatoms, it also is simple, relatively rapid and enables the diagnosis of drowned bodies under specific conditions with permanent photographic records.

Further Reading

Aghayev E, Thali MJ, Sonnenschein M, Hurlimann J, Kilchoer T et al (2005) Fatal streamer accident; blunt force injuries and drowning in post-mortem MSCT and MRI. Forensic Sci Int 152(1):65–71

Ago K, Hayashi T, Ago M, Ogata M (2011) The number of diatoms from the lungs and other organs in drowning deaths in bathwater. Legal Med 13(4):186–190

Ajay R, Sakshi M (2018) Significance of diatoms in diagnosis of drowning deaths: a review. Peer Re J Foren Gen Sci 1(5):1. https://doi.org/10.32474/PRJFGS.2018.01.000121

Anand TP, Unmesh AK (2016) Diatom test: a reliable tool to assess death by drowning? Int J Res Med Sci 4(5):1479–1484

Anon (2009) pseudocryptic species in the esturine benthic diatom navicula phyllepta (bacillariophyceae). J Phycol 45:1278–1289

Auer A, Mottonen M (1988) Diatoms and drowning. Z Rechtsmed 101(2):87–98

Bortolotti F, Del Balzo G, Calza R, Valerio F, Tagliaro F (2011) Testing the specificity of diatom test: search for false positives. Med Sci Law 21:7–10

Buri S, Saini OP, Garg R, Purohit RN, Soni K (2015) Detection of diatoms in autopsy of normal population vis a vis drowning. IJSAR 2(6):22–28
Bychkov D, Linder N, Turkki R et al (2018) Deep learning based tissue analysis predicts outcome in colorectal cancer. Sci Rep 8:3395
Cameron NG (2004) The use of diatom analysis in forensic geosciences. In: Pye K, Croft DJ (eds) Forensic geosciences: principles, techniques and applications, vol 232. Geological Society, London, pp 277–280
Carballeira R, Vieira DN, Febrero-Bande M et al (2018) A valid method to determine the site of drowning. Int J Legal Med 132:487–497
Chen XG, Zhang J, Huang Y et al (2013) Diatom taxa identification based on single-cell isolation and rDNA sequencing. Forensic Sci Int Genet Suppl 4:e308–e309
Coudray N, Ocampo PS, Sakellaropoulos T et al (2018) Classification and mutation prediction from nonsmall cell lung cancer histopathology images using deep learning. Nat Med 24:1559–1567
Cruz-Roa A, Gilmore H, Basavanhally A et al (2017) Accurate and reproducible invasive breast cancer detection in whole-slide images: a deep learning approach for quantifying tumor extent. Sci Rep 7:46450
DiGiancamillo A, Domeneghini C, Gibelli D et al (2011) Diatom extraction with HCl from animal tissues: a technical note. Legal Med 13:268–271
Ertosun MG, Rubin DL (2015) Automated grading of gliomas using deep learning in digital pathology images: a modular approach with ensemble of convolutional neural networks. AMIA Annu Symp Proc 2015:1899–1908
Farrugia A, Ludes B (2011) In: Viera DN (ed) Diagnostic of drowning in forensic medicine, forensic medicine-from old problems to new challenges. INTECH, Rijeka, Croatia, pp 53–60
Fucci N (2012) A new procedure for diatom extraction in the diagnosis of drowning. Clin Exp Pharmacol 2(1):1–3
Fukushima K, Miyake S (1982) Neocognitron: a new algorithm for pattern recognition tolerant of deformations and shifts in position. Pattern Recogn 15:455–469
Yadav G, Mishra MK, Gupta AK, Shailesh (2013) Identification of site specific diatom at Yamuna River of Allahabad. IOSR J Pharm 8(2):87–89
Giri BS, Tripathi CB, Cowdhary YB (1993) Characterization of drowning by diatom test. Indian J Med Res 98:40–43
Hasle GR, Syvertsen EE (1997) Marine diatoms. In: Tomas Carmelo R (ed) Identifying marine phytoplankton. Academic, San Diego, pp 5–385
Hofmann E (1878) Textbook of legal medicine.:629
Horton BP (2007) Diatom and forensic science. Paleontological study papers, vol 13. The Paleontological Society, Baltimore, pp 13–22
He F, Huang D, Liu L, Shu X, Yin H et al (2008) A novel PCR-DGGE-based method for identifying plankton 16S rRNA for the diagnosis of drowning. Forensic Sci Int 176(2–3):152–156
Hurlimann J, Feer P, Elber F, Niederberger K, Dirnhofer R et al (2000) Diatom detection in the diagnosis of death by drowning. Int J Legal Med 114(1–2):6–14
Incze G (1942) Fremdkorper in Blutkreislauf Ertrunkener. Zentralbl Allg Pathol Anat 79:176
Jian J, Wan L, Shao Y et al (2019) Postmortem chest computed tomography for the diagnosis of drowning: a feasibility study. Forensic Sci Res 6:152. https://doi.org/10.1080/20961790.2018.1557386
Kakizaki E, Sonoda A, Shinkawa N et al (2019) A new enzymatic method for extracting diatoms from organs of suspected drowning cases using papain: optimal digestion and first practical application. Forensic Sci Int 297:204–216
Kakizaki E, Yukawa N (2015) Simple protocoal for extracting diatoms from lung tissues of suspected drowning cases with in 3h: first practical application. Forensic Sci Int 251:179–185
Kane M, Fakunaga T, Maeda H, Nishi K (2000) Phylogenetic analysis of picoplankton in Lake Biwa and application to legal medicine. Electophoresis 21(2):351–354
Knight B (1996) Immersion deaths. In: Forensic pathology, 2nd edn. Arnold Publication, London, p 390

Krstic S, Duma A, Janevska B, Levkov Z, Nikolova K et al (2002) Diatom in forensic expertise of drowning. A Macedonian experience. Forensic Sci Int 127(3):198–203
Kumar M, Deshkar J, Naik SK, Yadav PK (2012) Diatom test-past, present and future: a brief review. IJRRMS 2(3):28–32
Lecun Y, Bengio Y, Hinton G (2015) Deep learning. Nature 521:436–444
Levin EA, Morgan RM, Scott KR et al (2017) The transfer of diatoms from freshwater to footwear materials: an experimental study assessing transfer, persistence, and extraction methods for forensic reconstruction. Sci Justice 57:349–360
Li Z, Liu X, Yu Y et al (2019) Barcoding for diatoms in the Yangtze River from the morphological observation and 18S rDNA polymorphic analysis. Forensic Sci Int 297:81–89
Ludes B, Coste M, North N, Doray S, Tracqui A et al (1999) Diatom analysis in victim's tissues as an indicator of the site drowning. Int J Legal Med 112(3):163–166
Ludes B, Coste M, North N et al (1999) Diatom analysis in victim's tissues as an indicator of the site of drowning. Int J Legal Med 112:163–166
Ludes B, Coste M, Tracqui A, Mangin P (1996) Continuous river monitoring of the diatoms in the diagnosis of drowning. J Forensic Sci 41(3):425–428
Lunetta P, Modell JH (2005) Macroscopical, microscopical, and laboratory findings in drowning victims, vol 3. Humana Press, New York, NY, pp 3–77
Lunetta P, Penttil€a A, H€allfors G. (1998) Scanning and transmission electron microscopical evidence of the capacity of diatoms to penetrate the alveolocapillary barrier in drowning. Int J Legal Med 111:229–237
Lunetta P, Penttila A, Hallfors G (1998) Scanning and transmission electron microscopical evidence of the capacity of diatoms to penetrate in the alveo-capillary barrier in drowning. Int J Legal Med 111(5):229–237
Magrey AH, Raj M (2014) Role of diatoms in forensic diagnosis of drowning cases from Jammu & Kashmir, India. Biosci Biotech Res Comm 7(1):72–77
Mann DG, Droop SJM (1996) Biodiversity, biogeography and conservation of diatoms. Hydrobiologia 336(1–3):19–32
Matsumoto H, Fukui Y (1993) A simple method for diatom detection in drowning. Forensic Sci Int 60(1–2):91–95
Ming M, Meng X, Wang E (2007) Evaluation of four digestive methods for extracting diatoms. Forensic Sci Int 170:29–34
Modi JP (1996) Textbook of medical jurisprudence & toxicology, 21st edn. NM Tripathi Private Ltd, Mumbai, pp 196–220
Mohanty NK, Rao B (1964) The diagnosis of death from drowning with particular reference to diatom method. J Indian Acad Forensic Sci 3:21–27
Network ACN, Related SOR, Rank T et al (2006) A fast learning algorithm for deep belief nets. Neural Comput 18:1527–1554
Pachar JV, Cameron JM (1993) The diagnosis of drowning by qualitative and quantitative diatom analysis. Med Sci Law 33(4):291–299
Parmar P, Rathod GB, Rathod S, Parikh A (2014) Nature helps to solve the crimes-diatoms study in case of drowning death. Int Arch Integr Med 1(3):58–65
Pathak A, Mangal HM (2009) Decomposition: cast a shadow over the drowning deaths. J Indian Acad Forensic Med 31(2):112–117
Peabody AJ (1980) Diatoms and drowning a review. Med Sci Law 20(4):254–261
Pedraza A, Bueno G, Deniz O et al (2017) Automated diatom classification (part B): a deep learning approach. Appl Sci 7:460
Piette MHA, De Letter EA (2006a) Drowning: still a difficult autopsy diagnosis. Forensic Sci Int 163(1–2):1–9
Pollanen MS (1998a) Diatoms and homicide. Forensic Sci Int 91(1):29–34
Pollanen MS (1998b) Forensic diatomology and drowning. Elsevier, Amsterdam
Porawski R (1966) Investigation of occurrence of diatoms in organs in death from various causes. J Forensic Med 13(4):134–137

Racz E, Konczol F, Toth D et al (2016) PCR-based identification of drowning: four case reports. Int J Legal Med 130:1303–1307
Ranga Rao GSRKG, Jakkam S, Prasad GKV (2014) A comprehensive study of drowning in and around Kakinada, two years retrospective study. Sch J Appl Med Sci 2(4D):1397–1401
Rohn EJ, Frade PD (2006) The role of diatoms in medico-legal investigations II: a case for the development and testing of new modalities applicable to the diatom test for drowning. Forensic Exam 15(4):1
Sankhla MS, Kumari M, Nanadan M, Kumar R (2016) Forensic identification of drowning death by the use of diatom analysis. Int J Dev Res 6(9):9491–9493
Sasidharan A, Resmi S (2014) Forensic diatomology: a review. Peer Rev E J Health Sci 1(3): JS002B
Saukko P, Knight B (2015) Knight's forensic pathology, 4th edn. CRC, Boca Raton, FL
Scott KR, Morgan RM, Jones VJ, Cameron NG (2014) The transferability of diatoms to clothing and the methods appropriate for their collection and analysis in forensic geosciences. Forensic Sci Int 241:127–137
Scott KR, Morgan RM, Jones VJ et al (2014) Reply to A. Dragutinovic, A reply to: the transferability of diatoms to clothing and the methods appropriate for their collection and analysis in forensic geoscience. Forensic Sci Int 241:127–137
Shen X, Liu Y, Xiao C et al (2019) Analysis of false-positive results of diatom test in the diagnosis of drowning—would not be an impediment. Int J Legal Med 133:1819–1824
Shiwei M, Feng F, Dong X, Seese R, Wang Z (2010) A contributory diagnosis of drowning in putrefactive corpses using the electric impedance spectroscopy. Rom J Leg Med 18(4):283–288
Sidari L, Nunno ND, Costantinides F, Melato M (1999) Diatom test with Soluene 350 to diagnose drowning in seawater. Forensic Sci Int 103(1):61–65
Singh M, Kulshrestha P, Satpathy DK (2004) Synchronous use of maggots and diatoms in decomposed bodies. J Indian Acad Forensic Med 26:121–124
Singh R, Singh R, Kumar S, Thakar MK (2006) Forensic analysis of diatoms-a review. Anil Aggarwal Internet J Forensic Med Toxicol 7(2):1
Sitthiwong N, Ruangyuttikarn W, Vongvivach S, Peerapornpisal Y (2014) Detection and identification of diatoms in tissue samples of drowning victims. Chiang Mai J Sci 41(5):1020–1031
Siver PA, Lord WD, Mccarthy DJA (1994) Forensic limnology: the use of freshwater algal community ecology to link suspects to an aquatic crime scene in southern New England. J Forensic Sci 39(3):847–853
Sumiko A, Miwako S, Hidemasa N et al (2003) A novel PCR method for identifying plankton in cases of death by drowning. Med Sci Law 43:23–30
Suto M, Kato N, Abe S et al (2009) PCR detection of bacterial genes provides evidence of death by drowning. Legal Med 11:S354–S356
Tamasaka L (1949) Diatom content of bone marrow in corpses in water. Orv Hetil 16:509–511
Timperman J (1972) The diagnosis of drowning-a review. J Forensic Sci 1(4):397–409
Uchiyama T, Kakizaki E, Kozawa S et al (2012) A new molecular approach to help conclude drowning as a cause of death: simultaneous detection of eight bacterioplankton species using real-time PCR assays with TaqMan probes. Forensic Sci Int 222:11–26
Vanelslander B, Creach V, Vanormelingen P, Ernst A, Chepurnov VA, Sahan E, Muyzer G, Stal LJ, Vyverman W, Sabbe K (2009) Ecological differentiation between sympatric pseudocryptic species in the estuarine benthic diatom Navicula Phyllepta (Bacillariophyceae) 1. J Phycol 45 (6):1278–1289
Vanormelingen P, Verleyen E, Vyvermn W (2008) The diversity and distribution of diatoms: from cosmopolitanism to narrow endemism. Biodivers Conserv 17(2):393–405
Verma K (2013) Role of diatoms in the world of forensic science. J Forensic Res 4:181–184
Vinayak V, Goyal MK, Mishra V, Rai N (2010) Diatoms as great forensic tool in investigating of deaths due to drowning: a case study. J Forensic Med Toxicol 27(1):51–54
Wang H, Liu Y, Zhao J, Sunlin H, Wang Y et al (2014) A simple digestion method with a Lefort aqua Regia solution for diatom extraction. J Forensic Sci 60(1):227–230

Williams DM, Kociolek JP (2017) Historical biography of diatoms in in Australasia: a preliminary assessment. In: Ebach MC (ed) Handbook of Australasian biogeography. CRC press, Boca Raton, FL, pp 17–46

Zhao Y, Chen X, Yang Y et al (2018) Potential forensic biogeographic application of diatom colony consistency analysis employing pyrosequencing profiles of the 18S rDNA V7 region. Int J Legal Med 132:1611–1620

Zhao J, Liu C, Bardeesi ASA et al (2017) The diagnostic value of quantitative assessment of diatom test for drowning: an analysis of 128 water-related death cases using microwave digestion-vacuum filtrationautomated scanning electron microscopy. J Forensic Sci 62:1638–1642

Zhao J, Liu C, Hu S et al (2013) Microwave digestionvacuum filtration-automated scanning electron microscopy as a sensitive method for forensic diatom test. Int J Legal Med 127:459–463

Zhao J, Ma Y, Liu C et al (2016) A quantitative comparison analysis of diatoms in the lung tissues and the drowning medium as an indicator of drowning. J Forensic Legal Med 42:75–78

Zhou Y, Zhang J, Huang J et al (2019) Digital wholeslide image analysis for automated diatom test in forensic cases of drowning using a convolutional neural network algorithm. Forensic Sci Int 302:109922

Nanotechnology and Its Uses in Forensic Science

36

Prashant Kumar and Paromita Banerjee

Abstract

Nanotechnology has had a significant impact on the field of forensic science. Different nanoparticles, including silver, zinc oxide, silicon dioxide, aluminum oxide, gold (with the silver physical developer), europium, fluorescent carbon, and amphiphilic silica, could be applied to a variety of object surfaces to produce characteristics like distinct ridge details of fingerprints; among all of these, gold is the most frequently used. The fingerprint is regarded as significant evidence at any crime scene, and nanotechnology has enormous potential for fingerprint analysis in the future. This chapter will explore the current and potential applications of nanotechnology in forensics. The chapter will begin by introducing nanotechnology and its potential applications. A discussion of the advantages and limitations of nanotechnology in forensics will follow this. The chapter will then explore the current and emerging nanotechnology techniques used in forensic science. These techniques include nanoscale imaging, nanoscale spectroscopy, nanoscale chemical analysis, and nanoscale DNA sequencing. Finally, the chapter will discuss the implications of nanotechnology for the future of forensics.

Keywords

Nanotechnology · Forensic nanotechnology · Crime detection · Forensic investigation

P. Kumar (✉) · P. Banerjee
Department of Bioinformatics, Kalinga University, Raipur, Chhattisgarh, India
e-mail: kumar.prashant@kalingauniversity.ac.in

A. Puri et al. (eds.), *Fundamentals of Forensic Biology*,
https://doi.org/10.1007/978-981-99-3161-3_36

36.1 Introduction

Forensic nanotechnology is a rapidly emerging field in the forensic sciences. This field focuses on applying nanotechnology to forensic investigations, providing a new level of insight and accuracy to the analysis of evidence. It involves using nanoscale materials and processes to analyze evidence and identify the presence of physical and chemical components that may not be detected using traditional methods. The term ‘nanotechnology’ refers to manipulating matter at the nanometer scale. This scale is smaller than that of a human cell which is around 100 nanometers in size (Pitkethly 2010). Nanotechnology is an interdisciplinary field that combines elements of physics, chemistry, materials science, and engineering. The application of nanotechnology to forensic science can be used to improve the accuracy and speed of investigations. It has the potential to revolutionize the way forensic evidence is collected, analyzed, and interpreted. This chapter will provide an overview of forensic nanotechnology, including its current applications in forensic science, potential future uses, and the associated challenges. Forensic nanotechnology has a wide range of potential applications in forensic science. For example, it can be used to detect and analyze trace amounts of evidence, such as fingerprints, bodily fluids, and drugs (Hussain et al. 2021) It can also identify specific chemicals or substances that may not be visible to the naked eye. Nanotechnology can also be used to analyze evidence at the molecular level, allowing for a more detailed analysis than traditional methods. This can provide more accurate results and enable investigators to make more informed decisions when analyzing evidence. In addition, nanotechnology can detect and analyze DNA, providing a more precise and reliable method of identifying individuals. RNA may also be seen and examined by this technique, which can be used to identify the source of a specific sample (Sinha and Rao 2020) Future Uses of Forensic Nanotechnology The potential applications of nanotechnology in forensic science are far-reaching, and the field is expanding. Nanotechnology can be used to create more accurate and efficient ways to analyze evidence and identify individuals. In the future, nanotechnology could develop devices and sensors that can detect and analyze evidence at the nanoscale. This could allow for more accurate and efficient forensic investigations and provide a new level of insight into cases. Nanotechnology could also create devices capable of detecting and analyzing evidence in real time. This could enable investigators to quickly and accurately identify evidence and speed up the process of investigations (Lodha et al. 2016). There are difficulties with applying nanotechnology to forensic science. One of the biggest challenges is the potential cost of implementing nanotechnology in forensic investigations. Nanotechnology is a complex field, and the materials and equipment needed to use it can be expensive. Another challenge is that nanotechnology is still relatively new, and many unknowns exist about its potential applications. This can make it difficult to accurately predict the benefits and risks of using nanotechnology in forensic investigations. Finally, there are ethical considerations surrounding using nanotechnology in forensic science. For example, the use of nanotechnology could infringe on the privacy of individuals, which is a significant concern.

36.2 Applications of Forensic Nanotechnology

Forensic nanotechnology has a wide range of applications in crime investigation and resolution. These include 1. DNA and other biological material analysis: Nanotechnology can identify DNA, proteins, and other biological material, enabling forensic scientists to quickly and accurately identify individuals from a crime scene. Forensic nanotechnology can also be used to detect and identify nanotoxicity substances (Sinha et al. 2020) 2. Evidence tagging and tracking: Nanotechnology can be used to tag and track evidence from a crime scene. This could help reduce the time it takes to process evidence and ensure it is handled correctly. 3. Finger printing: Nanotechnology can more accurately identify fingerprints from a crime scene than traditional methods. Fingerprinting nanotechnology is used to identify the unique characteristics of nanomaterials and nanosystems. It uses various analytical techniques to characterize the size and shape, elemental composition, surface area, structure, and other physical and chemical properties of nanomaterials and nanosystems. This is done to build a "fingerprint" or a "signature" of the nanomaterial. This process can identify the same nanomaterials in different environments and monitor the changes in nanomaterials over time. This can be important for tracking the safety and environmental impact of nanotechnology. 4. Gunshot residue analysis: Nanotechnology can be used to analyze gunshot residue and link it to a particular weapon. Gunshot residue analysis is an important tool used in forensic science to determine the distance at which a gun is fired. Recently, nanotechnology has been used to improve gunshot residue analysis. Nanostructured materials, such as carbon nanotubes, can be used to detect the presence of gunshot residue particles. These nanostructures can form sensitive and selective sensors that detect particles at low concentrations. They can also be used to detect a variety of different types of particles, such as lead, copper, and zinc. Nanostructured materials can also detect the presence of other compounds that may be associated with gun firings, such as nitrates and chlorides. In addition, nanotechnology can be used to create devices that can analyze gunshot residue more quickly and accurately. For example, nanotechnology can be used to develop devices that can detect the presence of gunshot residue particles in a much shorter amount of time. This can be especially useful in crime scenes, where time is of the essence. Nanotechnology can also be used to create devices that can analyze gunshot residue particles more accurately, leading to more accurate conclusions about the distance and angle of fire. Overall, nanotechnology has the potential to revolutionize the field of gunshot residue analysis. Using nanostructured materials, researchers can develop sensitive and selective sensors that detect gunshot residue particles at low concentrations. They can also be used to detect a variety of different types of particles, such as lead, copper, and zinc. Nanotechnology can also be used to create devices that can analyze gunshot residue more quickly and accurately, leading to more accurate conclusions about the distance and angle of fire. 5. Drug testing: Nanotechnology can be used to analyze and identify illicit drugs and trace amounts of drugs in a person's system. Nanotechnology is currently being used to develop new and improved methods for drug testing. The use of nanotechnology in drug testing provides a more sensitive and accurate way to detect the presence of drugs in a sample. It can detect the presence of a wide range of drugs, including drugs of abuse and illicit drugs. This technology is also being used to develop methods for drug screening that are quicker and more cost effective than traditional methods. For example, nanotechnology can detect a drug's presence in a sample without requiring expensive and time-consuming laboratory analysis. Additionally, nanotechnology can be used to develop rapid drug screens that can be used in the field to detect the presence of drugs quickly (Paikrao et al. 2022)0.6. Trace evidence analysis: Nanotechnology can analyze trace evidence, such as fibers, paint chips, and glass, to identify a perpetrator (Sharma et al. 2022). Drug testing in nanotechnology is still in its early stages. However, it is becoming increasingly important as nanotechnology is used in various consumer products, from cosmetics to medical devices. Drug testing in nanotechnology is designed to ensure that nanoparticles used in these products are safe, effective, and reliable. It involves testing the properties of particles to ensure they meet safety and quality requirements. This includes testing for toxicity, biocompatibility, and performance. Drug testing in nanotechnology also involves assessing nanoparticles' potential environmental and health impacts and their interactions with other materials. Both government and private organizations conduct this type of testing to ensure the safety of nanotechnology products. Trace evidence analysis in nanotechnology is the analysis of particles and other materials that are too small to be seen with the naked eye. This type of evidence is often used to aid in investigating criminal cases and in the study of the properties and behavior of nanomaterials. Trace evidence analysis in nanotechnology involves the use of various analytical techniques to detect and identify particles, such as scanning electron microscopy (SEM), energy-dispersive X-ray spectroscopy (EDS), and transmission electron microscopy (TEM). These techniques allow investigators to measure particles' size, shape, composition, and surface characteristics and determine their origin and distribution. In addition to providing important evidence for criminal investigations, trace evidence analysis in nanotechnology has also been used to study the behavior of nanoparticles in the environment and to develop new materials and products. For example, researchers have used trace evidence analysis to study the interactions between nanoparticles and cells and assess the potential toxicity of nanomaterials. In addition, trace evidence analysis has been used to develop and optimize the performance of nanoscale devices, such as transistors, sensors, and medical implants. 7. Explosive detection: Nanotechnology can detect explosives, reducing the risk of a terrorist attack. The forensic application of explosive detection nanotechnology (EDN) is a growing field used to examine materials and items for traces of explosives. EDN can detect many explosives, including C-4, TNT, and Semtex. EDN also screens items for explosives residue to identify potential threats (Pandya et al. 2012).

36.3 Conclusion

Forensic nanotechnology is a rapidly emerging field in the forensic sciences that has the potential to revolutionize the way evidence is collected, analyzed, and interpreted. It has a wide range of potential applications, including the detection and analysis of trace amounts of evidence, the identification of individuals, and the detection and analysis of DNA and RNA. However, the use of nanotechnology in forensic science has its challenges. These include the potential cost of implementing nanotechnology, the lack of knowledge about its possible applications, and ethical considerations. In conclusion, nanotechnology can significantly improve the accuracy and speed of forensic investigations, but more research is needed to fully understand its potential and challenges.

Acknowledgments None.

Conflicts of Interest The author declares that there are no conflicts of interest.

References

Hussain CM, Rawtani D, Pandey G, Tharmavaram M (2021) Chapter 19–Nanotechnology in forensic science. In: Hussain CM, Rawtani D, Pandey G, Tharmavaram M (eds) Handbook of analytical techniques for forensic samples. Elsevier, Amsterdam, pp 359–379. ISBN 9780128223000. https://doi.org/10.1016/B978-0-12-822300-0.00019-7

Lodha AS, Pandya A, Shukla RK (2016) Nanotechnology: an applied and robust approach for forensic investigation. Forensic Res Criminol Int J 2(1):35–37. https://doi.org/10.15406/frcij.2016.02.00044

Paikrao HM, Tajane DS, Patil AS, Dipale AD (2022) Applications of nanotechnology in forensic science. In: Sarma H, Gupta S, Narayan M, Prasad R, Krishnan A (eds) Engineered nanomaterials for innovative therapies and biomedicine. Nanotechnology in the life sciences. Springer, Cham. https://doi.org/10.1007/978-3-030-82918-6_11

Pandya A, Goswami H, Lodha A, Menon S (2012) A novel nanoaggregation detection technique of TNT using selective and ultrasensitive nanocurcumin as a probe. Analyst 137:1771

Pitkethly M (2010) Nanotechnology and forensics, mat. Today 12(6):6. https://www.materialstoday.com/nanomaterials/articles/s1369702109701671/

Sharma B, Bhat A, Singh AP, Kumar P, Khanna S, Vita D, Kaur E, Singh J (2022) Railway disaster: a track of forensic investigation. Neuro Quantol 20(9):5313–5319. https://doi.org/10.14704/nq.2022.20.9.NQ44617

Sinha M, Rana M, Kushwaha P (2020) Applications of mitochondrial DNA in forensic science. In: Shrivastava P, Dash HR, Lorente JA, Imam J (eds) Forensic DNA typing: principles, applications and advancements. Springer, Singapore. https://doi.org/10.1007/978-981-15-6655-4_16

Sinha M, Rao IA (2020) SNP testing in forensic science. In: Shrivastava P, Dash HR, Lorente JA, Imam J (eds) Forensic DNA typing: principles, applications and advancements. Springer, Singapore. https://doi.org/10.1007/978-981-15-6655-4_18

www.ingramcontent.com/pod-product-compliance
Ingram Content Group UK Ltd.
Pitfield, Milton Keynes, MK11 3LW, UK
UKHW022217051026
15624UKWH00001B/209
9789819931606